Modular Forms and String Duality

FIELDS INSTITUTE COMMUNICATIONS

THE FIELDS INSTITUTE FOR RESEARCH IN MATHEMATICAL SCIENCES

Modular Forms and String Duality

Noriko Yui
Helena Verrill
Charles F. Doran
Editors

American Mathematical Society
Providence, Rhode Island

The Fields Institute for Research in Mathematical Sciences
Toronto, Ontario

FIELDS

2000 *Mathematics Subject Classification.* Primary 11Fxx, 14Gxx, 14J32, 14N35, 33Cxx, 81T30.

Library of Congress Cataloging-in-Publication Data

Modular forms and string duality / Noriko Yui, Helena Verrill, Charles F. Doran, editors.
p. cm. — (Fields Institute Communications, ISSN 1069-5265 ; 54)
Proceedings of a workshop held at the Banff International Research Station, June 3–8, 2006.
Includes bibliographical references.
ISBN 978-0-8218-4484-7 (alk. paper)
1. Forms, Modular—Congresses. 2. Duality (Mathematics)—Congresses. 3. Mirror symmetry—Congresses. 4. Number theory—Congresses. 5. String theory—Congresses. 6. Particles (Nuclear physics)—Congresses. I. Yui, Noriko. II. Verrill, Helena. III. Doran, Charles F., 1971–

QA243.M695 2008
512.7′3—dc22 2008028173

∞ The paper used in this book is acid-free and falls within the guidelines
established to ensure permanence and durability.
This publication was prepared by the Fields Institute.
`http://www.fields.utoronto.ca`
Visit the AMS home page at `http://www.ams.org/`

10 9 8 7 6 5 4 3 2 1 13 12 11 10 09 08

Contents

Aspects of Physics and String Theory

Acknowledgments

The editors wish to express their appreciation to all the contributors for preparing their manuscripts for the Fields Communication Series, which required extra effort presenting not only current developments but also the history of the subjects treated in their articles.

All papers in this volume were refereed very rigorously. We are deeply grateful to all our referees for their time-consuming effort and discipline in evaluating the articles.

All papers were copy-edited by Arthur Greenspoon of Mathematical Reviews. The editors and the Fields Institute are grateful for his help smoothing out, English and mathematical presentations.

The workshop was supported by the Banff International Research Station (BIRS) through the five-day workshops program. We thank BIRS for their financial support. We enjoyed the excellent support of the staff at BIRS, and are grateful for their hospitality. In addition, some young participants from the United States were supported in part by Mathematical Sciences Research Institute (MSRI) Berkeley.

Last but not least, we thank Debbie Iscoe of the Fields Institute for her help reformatting articles and assembling this volume for publication.

Noriko Yui, Helena Verrill and Charles Doran
June 2008

Introduction

Modular forms have long played a key role in the theory of numbers, including most famously the proof of Fermat's Last Theorem. Through its quest to unify the spectacularly successful theories of quantum mechanics and general relativity, string theory has long suggested deep connections between branches of mathematics such as topology, geometry, representation theory, and combinatorics. Less well-known are the emerging connections between string theory and number theory – the subject of the workshop *Modular Forms and String Duality* held at the Banff International Research Station (BIRS), June 3-8, 2006. Mathematicians and physicists alike converged on the Banff Station for a week of introductory lectures, designed to educate one another in relevant aspects of their subjects, and of research talks at the cutting edge of this rapidly growing field.

The workshop was organized by Charles F. Doran, Helena Verrill and Noriko Yui. The workshop was a huge success. Altogether thirty-seven mathematicians and physicists converged at the BIRS for the five day workshop. Twenty-six one hour talks were presented. Some were introductory lectures by mathematicians designed to prepare physicists in modular forms, quasimodular forms, modularity of Galois representations, and toric geometry. At the same time, introductory lectures by physicists were intended to educate mathematicians on some aspects of mirror symmetry and string theory in connection with number theory. These introductory lectures were scheduled in the mornings of the early days of the workshop. Research talks were scheduled in the afternoons and later days. They covered recent advances on various aspects of modular forms, differential equations, conformal field theory, topological strings and Gromov–Witten invariants, holomorphic anomaly equations, motives, mirror symmetry, homological mirror symmetry, construction of Calabi–Yau manifolds, among others.

Summary of scientific and other objectives

Physical duality symmetries relate special limits of the various consistent string theories (Types I, II, Heterotic string and their cousins, including M-theory and F-theory) one to another. The comparison of the mathematical descriptions of these theories often reveals quite deep and unexpected mathematical conjectures. The best known string duality to mathematicians, Type IIA/IIB duality, also called **mirror symmetry**, has inspired many new developments in algebraic and arithmetic geometry, number theory, toric geometry, Riemann surface theory, and infinite dimensional Lie algebras. Other string dualities such as Heterotic/Type II duality and F-Theory/Heterotic string duality have also, more recently, led to a series of mathematical conjectures, many involving elliptic curves, K3 surfaces, and modular forms. Modular forms and quasi-modular forms play a central role in mirror symmetry, in particular as generating functions counting the number of curves on Calabi–Yau manifolds and describing Gromov–Witten invariants. This has led to a realization that the time is ripe to assess the role of number theory, in particular that of modular forms, in mirror symmetry and string dualities in general.

One of the principal goals of this workshop was to look at modular and quasi-modular forms, congruence zeta-functions, Galois representations, and L-series for dual families of Calabi–Yau varieties with the aim of interpreting duality symmetries in terms of arithmetic invariants associated to the varieties in question. Over the last two decades, a great deal of work has been done on these problems. In particular it appears that we need to modify the classical theories of Galois representations (in particular, the question of modularity) and modular forms, among others, for families of Calabi–Yau varieties (threefolds) in order to accommodate "quantum corrections".

As dictated by the research interests of the participating members, the research activities were focused on the following themes:

(A) Arithmetic of Calabi–Yau varieties defined over number fields: Arithmetic of elliptic curves, K3 surfaces, Calabi–Yau threefolds, and higher dimensional Calabi–Yau varieties defined over number fields in connection with string dualities. These included the following topics and problems: Interpretation of string duality phenomena of Calabi–Yau varieties in terms of zeta-functions and L-series of the varieties in question, the modularity conjectures for Calabi–Yau varieties, the conjectures of Birch and Swinnerton-Dyer for elliptic curves and Abelian varieties, the conjectures of Beilinson-Bloch on special values of L-series and algebraic cycles, and intermediate Jacobians of Calabi–Yau threefolds. Calabi-Yau varieties of CM (complex multiplication) type and their possible connections to rational conformal field theories and, in the elliptic fibered case, behavior under F-Theory/Heterotic string duality.

(B) Mirror symmetry for families of Calabi–Yau varieties: Characterization of mirror maps in connection with the mirror moonshine phenomenon and, via the Fourier-Laplace transform, the classification of Q-Fano threefolds. In particular, differential equations associated to modular and quasi-modular forms related to GKZ-hypergeometric systems and, more generally, to Picard-Fuchs differential systems were investigated.

(C) Modular and quasi-modular forms in string duality: Modular forms and quasi-modular forms have appeared frequently in mirror symmetry contexts, e.g., in the generating functions counting the number of simply ramified covers of elliptic curves with marked points, in Gromov–Witten invariants, and also as mirror maps. The appearance of modular and quasi-modular forms in string dualities, e.g., in the Harvey-Moore conjectures of Heterotic-Type II duality, were investigated. Understanding why modular and quasi-modular forms play central roles in string dualities was one of our goals.

Talks presented at the workshop may be classified into not clearly disjoint sets of the following seven subjects. They are:

(a) Modular, quasimodular, bimodular forms, and their applications (D. Zagier, M. Kaneko, J. Stienstra)

(b) Topological string theory and modular forms (E. Scheidegger, A. Clingher, A. Klemm, V. Bouchard)

(c) Modularity, and arithmetic questions (R. Livné, S. Gukov, N. Yui, R. Schimmrigk)

(d) Mirror symmetry: various versions (S. Hosono, J. Walcher, M. Aldi)
(e) Toric geometry (H. Verrill, C. Doran)
(f) Differential equations (G. Almkvist, Masahiko Saito, A. Sebbar)
(g) Miscellaneous topics (N.-H. Lee, J. Stienstra, C. Herzog)

About this Volume

This volume consists of the proceedings of the workshop, presenting articles on recent developments in the interface of number theory and high energy physics, in particular, modular forms and string duality. Some of the articles are written-up versions of the talks presented at the workshop, while others report on subsequent developments on the subject matter of the workshop. In addition, some articles were solicited by the editors from non-participants, i.e., invitees who could not make the workshop due to various constraints.

All thirteen contributions included here have been refereed rigorously, and a number of articles went through extensive revisions to reach the standard imposed by the Fields Communications Series. In particular, all contributors were asked to make extra efforts to include expository components in their articles to reach out to a wider range of readers.

Here is a brief description of the contributions assembled in this volume. The papers are divided roughly into three categories, namely, Arithmetic and Modular Forms, Geometric and Differential Equations, and Physics and String Theory.

• **Arithmetic and Modular Forms**: The paper of Kadir and Yui considers motives arising from certain Calabi–Yau orbifolds of monomial deformations of Fermat hypersurfaces, and establishes the monomial-motive correspondence, at the Fermat point in the moduli space.

The article of Schimmrigk et al. discusses the mirrors of rigid Calabi–Yau varieties, in particular, it considers the cubic sevenfold and the quartic fivefold, showing that their motivic L-functions agree with the L-functions of their rigid mirror Calabi–Yau varieties.

Lee presents an update on the modularity of Calabi–Yau threefolds defined over the field of rational numbers, in particular, he constructs a non-rigid Calabi–Yau threefold as a quotient of the Schoen quintic threefold, and establishes its modularity.

The article of Long reports on aspects of modular forms for noncongruence arithmetic subgroups: the unbounded denominator property, modularity of the Galois representations for noncongruence cuspforms, and Atkin and Swinnerton-Dyer congruences.

• **Geometric and Differential Equations**: Almkvist, van Straten and Zudilin consider Apéry limits of differential equations of order 4 and 5 of Calabi–Yau type, identifying them with some known transcendental numbers.

Stienstra's paper is devoted to surveying some relations between plane models of statistical mechanics and hypergeometric systems in two variables. The study of embeddings of quivers (i.e., oriented graphs) into Riemann surfaces turned out to be crucial in several areas. One direction is Grothendieck's theory of dessins d'enfants which relates the combinatorics of such embeddings with the Galois group of the field of rational numbers. Another direction is dimer models, which are periodic

2-dimensional statistical systems. The paper explains several exciting interrelations of these different directions.

The article of Zagier and Zinger shows that certain hypergeometric series used to formulate mirror symmetry for Calabi–Yau hypersurfaces, in string theory and algebraic geometry, satisfy a number of interesting properties. These properties are used, for instance, to verify the Bershadsky–Cecotti–Ooguri–Vafa (BCOV) prediction for the genus one Gromov–Witten invariants of a quintic threefold.

Zudilin's paper gives a survey on methods of proofs of Ramanujan-type formulae for $1/\pi$. The methods include elliptic curves, modular forms, creative telescoping, and hypergeometric series.

• **Physics and String Theory**: Ballard's article gives a survey on homological mirror symmetry, focusing on the simplest examples of curves of genus zero and one. Bouchard presents a survey on the computation of orbifold Gromov–Witten invariants using mirror symmetry and topological string theory, focusing on the orbifold $\mathbf{C}^3/\mathbf{Z}_3$.

Gannon treats rational conformal field theory and mapping class groups. Rational conformal field theories produce a tower of finite-dimensional representations of surface mapping class groups, acting on the conformal blocks of the theory. Gannon gives reviews on this formalism with explicit examples.

Gukov and Murakami take the reader to the realm of the colored Jones polynomial and the Chern–Simons gauge theory. It is a physics prediction that the asymptotic behavior of the colored Jones polynomial is equal to the perturbative expansion of the Chern–Simons gauge theory with complex gauge group $SL(2, \mathbf{C})$ on the hyperbolic knot complement. The paper makes the first step toward verifying this prediction.

Walcher gives an exposition on recent progress in mirror symmetry, in particular, extending the classical closed string results to the open string sector. The ideas are worked out for the particular example of the quintic hypersurface and the mirror quintic. The paper also explains the relevance of mirror symmetry for open strings to homological mirror symmetry.

List of Participants

We had in total 37 participants for the workshop, of which ten were either graduate students or postdoctoral fellows, and two were official observers. We had five last minute cancellations Terry Gannon (Alberta, Canada), Amer Iqbal (Washington), Bong H. Lian (Brandeis, USA), John McKay (Concordia, Canada), and Andrey Todorov (Santa Cruz, USA and MPIM Bonn, Germany) for various reasons.

1. Marco Aldi (Northwestern University, USA) (Ph.D. student)
2. Gert Almkvist (Lunds University, Sweden)
3. Maiia Bakhova (Louisiana State University, USA) (Ph.D. student)
4. Matthew Ballard (University of Washington, USA) (Ph.D. student)
5. Vincent Bouchard (MSRI, USA) (Postdoctoral fellow)
6. Adrian Clingher (Stanford University, USA)
7. Chuck F. Doran (University of Washington, USA)
8. George Elliott (University of Toronto, Canada) (Observer)
9. Sharon Frechette (College of the Holy Cross, USA)
10. Sergei Gukov (California Institute of Technology, USA)
11. Christopher Herzog (University of Washington, USA)
12. Sinobu Hosono (University of Tokyo, Japan)
13. Simon Judes (Columbia University, USA) (Ph.D. student)
14. Shabnam Kadir (University of Hannover, Germany) (Postdoctoral fellow)
15. Masanobu Kaneko (Kyushu University, Japan)
16. Albrecht Klemm (University of Wisconsin, USA)
17. Edward Lee (UCLA, USA) (Postdoctoral fellow)
18. Nam-Hoon Lee (KIAS, Korea) (Postdoctoral fellow)
19. Ron Livné (Hebrew University of Jerusalem, Israel)
20. Ling Long (Iowa State University, USA)
21. Stephen Lu (Université du Québec à Montrèal, Canada)
22. Richard Ng (Iowa State University, USA) (Observer)
23. Matthew Papanikolas (Texas A & M University, USA)
24. Mike Roth (Queen's University, Canada)
25. Masahiko Saito (Kobe University, Japan)
26. Emanuel Scheidegger (Università del Piemonte Orientale Amadeo Avogadro, Italy)
27. Rolf Schimmrigk (Indiana University South Bend, USA)
28. Abdellah Sebbar (University of Ottawa, Canada)
29. Ahmed Sebbar (University of Bordeaux I, France)
30. Jan Stienstra (University of Utrecht, the Netherlands)
31. Hiroyuki Tsutsumi (Osaka University of Health & Sport Sciences, Japan)
32. Helena Verrill (Louisiana State University, USA)
33. Johannes Walcher (Institute for Advanced Study, USA)

34. Ursula Whitcher (University of Washington, USA) (Ph.D. student)
35. Jeng-Daw Yu (Harvard University, USA) (Postdoctoral fellow)
36. Noriko Yui (Queen's University, Canada)
37. Don Zagier (Max-Planck Institute for Mathematics Bonn, Germany/ Collège de France, France)

Schedule of Workshops

JUNE 4, 2006

9:00am–10:00am **Don Zagier** (MPIM Bonn and College de France)

Introduction: Modular Forms and Differential Equations

10:15am–11:15am **Ron Livné** (Hebrew and IAS Princeton)

Modularity of Galois Representations: Overview

11:30am–12:30pm **Helena Verrill** (Louisiana State)

An Introduction to the Batyrev–Borisov Construction of Toric Calabi–Yau Varieties

3:00pm–4:00pm **Shinobu Hosono** (Tokyo)

Introduction to Differential Equations in Mirror Symmetry

4:15pm–5:15pm **Emanuel Scheidegger** (TU Vienna)

Topological Strings on $K3$ Fibrations and Modular Forms

5:30pm–6:30pm **Jan Stienstra** (Utrecht)

From Multi-grid to Multi-helix; Remarkable Geometries from AdS/CFT

JUNE 5, 2006

9:00am–10:00am **Don Zagier** (MPIM Bonn and College de France)

Quasimodular Forms, Rankin–Cohen Brackets and Related Algebraic Structures

10:15am–11:15am **Masanobu Kaneko** (Kyushu)

Modular and Quasimodular Forms and their Applications

11:30am–12:30pm **Johannes Walcher** (IAS Princeton)

Opening Mirror Symmetry on the Quintic

2:30pm–3:30pm **Gert Almkvist** (Lund)

Apéry-like Limits Connected with Calabi-Yau Differential Equations

4:00pm–5:00pm **Christopher Herzog** (Washington)

How Exceptional Collections Stack Up

5:15pm–6:15pm **Nam-Hoon Lee** (KIAS)

Constructing Calabi–Yau Manifolds

8:00pm–9:00pm **Ling Long** (Iowa State)

Modular Forms for Noncongruence Subgroups

JUNE 6, 2006

9:00am–10:00am **Adrian Clingher** (Stanford)

Geometry Underlying the F-Theory/Heterotic String Duality in Eight Dimensions

10:15am–11:15am **Sergei Gukov** (CalTech)

Strings, Fields, and Arithmetic, Part I

11:30am–12:30pm **Noriko Yui** (Queen's)

Motives, Mirror Symmetry and Modularity

8:00pm–9:00pm **Sergei Gukov** (CalTech)

Strings, Fields, and Arithmetic, Part II

JUNE 7, 2006

9:00am–10:00am **Albrecht Klemm** (Wisconsin)

Modular, Quasimodular Forms and Gromov–Witten Invariants

10:15am–11:15am **Ron Livné** (Hebrew and IAS Princeton)

Explicit Descriptions of Universal $K3$ Families over Shimura Curves

11:30am–12:30pn **Chuck Doran** (Washington)

Algebraic Topology of Calabi–Yau Threefolds in Toric Varieties

2:30pm–3:30pm **Marco Aldi** (Northwestern)

Twisted Homogeneous Crings of Abelian Surfaces via Mirror Symmetry

4:00pm–5:00pm **Masahiko Saito** (Kobe)

Painlevé Property of ODEs and Deformation of Logarithmic Symplectic Varieties

5:15pm–6:15pm **Rolf Schimmrigk** (Indiana, South Bend)

String Modular Ω–Motives and Aspect of Mirror Symmetry

8:00pm–9:00pm **Ahmed Sebbar** (Bordeaux)

Differential Theta Relations and Galois Theory for Riccati Equations

JUNE 8, 2006

9:00am–10:00am **Jan Stienstra** (Utrecht)

Bimodular Forms and Holomorphic Anomaly Equation

10:15am–11:15am **Vincent Bouchard** (MSRI/Perimeter)

Topological Strings, Holomorphic Anomaly, and (Almost) Modular Forms

A number of invitees had to cancel their participation to the workshop at the last minute by various reasons. They were

- Terry Gannon (Alberta): **The Monster, Modular Functions and RCFT**
- Amer Iqbal (Washington)
- Bong H. Lian (Brandeis): **Introduction to Mirror Symmetry**
- John McKay (Concordia)
- Andrey Todorov (Santa Cruz/MPIM Bonn): **Regularized Determinants of CY Metrics and Applications to Mirror Symmetry**

Aspects of Arithmetic and Modular Forms

Fields Institute Communications
Volume **54**, 2008

Motives and Mirror Symmetry for Calabi–Yau Orbifolds

Shabnam Kadir
Institute for Mathematics
University of Hannover
Welfengarten 1, 30167, Hannover, Germany
kadir@math.uni-hannover.de

Noriko Yui
Department of Mathematics
Queen's University
Kingston, ON K7L 3N6 Canada
yui@mast.queensu.ca; yui@fields.utoronto.ca

Abstract. We consider certain families of Calabi–Yau orbifolds and their mirror partners constructed from Fermat hypersurfaces in weighted projective 4-spaces. Our focus is the topological mirror symmetry. There are at least three known ingredients to describe the topological mirror symmetry, namely, integral vertices in reflexive polytopes, monomials in graded polynomial rings (with some group actions), and periods (and Picard–Fuchs differential equations). In this paper, we will introduce Fermat motives associated to these Calabi–Yau orbifolds and then use them to give a motivic interpretation of the topological mirror symmetry phenomenon between mirror pairs of Calabi–Yau orbifolds. We establish, at the Fermat (the Landau–Ginzburg) point in the moduli space, the one-to-one correspondence between the monomial classes and Fermat motives. This is done by computing the number of $\mathbb{F}_q$-rational points on our Calabi–Yau orbifolds over $\mathbb{F}_q$ in two different ways: Weil's algebraic number theoretic method involving Jacobi (Gauss) sums, and Dwork's p-adic analytic method involving Dwork characters and Gauss sums. We will discuss specific examples in detail.

2000 *Mathematics Subject Classification.* Primary 14J32, 14G10, 11G40, 11F80.
Key words and phrases. Calabi–Yau threefolds, orbifoldings, Fermat motives, Mirror symmetry, monomials, Gauss (Jacobi) sums, motive-monomial correspondence, Fermat point.
This work is partially supported by N. Yui's Discovery Grant from the Natural Science Research Council of Canada (NSERC).

CONTENTS

1 Introduction

This is a sequel to the article of Yui [Y05] where Calabi–Yau orbifolds of Fermat hypersurface threefolds in weighted projective 4-spaces were constructed, and their L-series (associated to the ℓ-adic Galois representations) were determined. It was often the case that the Galois representations had very high rank, which made it rather impossible to carry out the required calculations. To remedy this, we introduced Fermat motives, and then decomposed the Calabi–Yau threefolds into Fermat motives. Via cohomological realizations of these motives, we were able to calculate the motivic L-series for each motive. The global L-series was then obtained by gluing the motivic results together.

These Calabi–Yau orbifolds are all non-rigid ($h^{2,1} > 0$), and their mirror Calabi–Yau threefolds exist satisfying the following conjecture.

Conjecture 1.1 (Topological Mirror Symmetry Conjecture) Given a family of Calabi–Yau threefolds X, there is a mirror family of Calabi–Yau threefolds $\hat{X}$ such that

$$h^{2,1}(\hat{X}) = h^{1,1}(X) \quad \text{and} \quad h^{1,1}(\hat{X}) = h^{2,1}(X)$$

so that the Euler characteristics are subject to the relation

$$\chi(\hat{X}) = -\chi(X).$$

The topological mirror symmetry described above can be reformulated in the toric geometric setting à la Batyrev [Ba94]. In Batyrev's theory, mirror symmetry is described in terms of pairs of reflexive polytopes. Integral vertices of reflexive polytopes are the main ingredients in Batyrev's theory, and they correspond to monomials in graded polynomial rings. By Aspinwall, Greene and Morrison [AGM93], there is the monomial–divisor mirror map (for the corresponding cohomology groups, $H^{1,1}_{toric}$ and $H^{2,1}_{poly}$), which yields a one-to-one correspondence between toric divisors of a Calabi–Yau family and monomials in the mirror Calabi–Yau family.

Based on the theory of Dwork, Katz and Griffiths on periods (see Cox and Katz [CK99]), Candelas et al. [CORV00, CORV03] (resp. Kadir [Ka05]) established explicitly a one-to-one correspondence between monomials and periods via Picard–Fuchs differential equations, for the quintic one-parameter (resp. the octic two-parameter) family of Calabi–Yau threefolds. In their calculations, periods decomposed into the product of subperiods. This seems to suggest that there should be a motivic interpretation for such factorizations.

The concept of motives has been emerging in the physics literature, and the purpose of this paper is to give a mathematically rigorous discussion on motives, restricting our attention to specific examples of Calabi–Yau threefolds. We follow the notion of motives due to Grothendieck. Starting with Fermat hypersurfaces in weighted projective 4-spaces, we define and construct explicitly the so-called Fermat motives from algebraic correspondences, as described in Shioda [Sh87]. Our goal is to interpret the topological mirror symmetry phenomenon for the mirror pairs of specific Calabi–Yau orbifolds in terms of Fermat motives and their cohomological realizations. As our main result, we establish a one-to-one correspondence between monomials and Fermat motives. This correspondence determines Fermat motives which are invariant under the mirror map. Since Fermat motives are defined only at the Fermat (the Landau–Ginzburg) point in the moduli space, this correspondence is established only at the Fermat point.

There are more monomials than motives, and we observe that monomials associated to conifold points seem to be associated to (mixed) Tate motives.

Incidentally, at the Fermat point, our Calabi–Yau threefolds capture the structure of CM type varieties (see Yui [Y05]), and hence our motives are also CM type motives. (G. Moore [Mo07] defined "attractive" Calabi–Yau threefolds. Among our Calabi–Yau orbifolds, there is only one such threefold, namely, $m = 6, Q = (1,1,1,1,2)$. In fact, this is the only Calabi–Yau orbifold whose weight motive is rigid, i.e., $h^{2,1}(\mathcal{M}_Q) = 0$ and $B_3(\mathcal{M}_Q) = 2$.)

Now we will describe the contents of this paper.

Section 2 is devoted to the definition of Fermat motives and their cohomological realizations. We use the definition of motives due to Grothendieck, and Manin [Ma70], which is based on algebraic correspondences and projectors. We follow the exposition of Shioda [Sh87] and Gouvêa and Yui [GY95].

In Section 3, we construct Calabi–Yau orbifolds in weighted projective 4-spaces. The starting point is the Fermat hypersurface V of degree $m \geq 5$ and dimension 3, and a finite abelian group (which is a subgroup of the automorphism group of V). This group will determine a weight. We take the quotient of V by such a group. This gives rise to a quotient threefold with singularities. We then resolve singularities by taking the crepant resolution (which is guaranteed to exist uniquely (up to flops) for dimension ≤ 3). The smooth threefold thus obtained is our Calabi–Yau threefold. There are altogether 147 such Calabi–Yau orbifolds.

In Section 4, we describe the Greene–Plesser orbifolding construction of mirror partners of the Calabi–Yau orbifolds constructed in the previous section. We review the mirror construction from the paper of Greene and Plesser [GP90]. The mirror symmetry is interpreted as the duality between the two finite abelian groups associated to the mirror pair of Calabi–Yau threefolds.

In Section 5, we will construct Fermat motives for the Calabi–Yau orbifolds in Section 3, and compute their invariants (e.g., motivic Hodge numbers, motivic Betti numbers) via their cohomological realizations. For each mirror pair of Calabi–Yau threefolds, we also determine Fermat motives which are invariant under the mirror map. In particular, we observe that for each Calabi–Yau orbifold, the motive associated to the weight is always invariant under the mirror map. If $h^{1,1} = 1$, the weight motive is the only motive invariant under the mirror map. However, when $h^{1,1} > 1$, there are other motives apart from the weight motive that remain invariant under the mirror map.

In Section 6, we review the construction of mirror pairs of Calabi–Yau hypersurfaces in toric geometry due to Batyrev [Ba94]. We will confine ourselves to Calabi–Yau threefolds. Reflexive polytopes and their dual polytopes are the main players in Batyrev's toric mirror symmetry. A pair of reflexive polytopes (Δ, Δ^*) gives rise to a mirror pair of Calabi–Yau hypersurfaces. It is noted that the origin is the only integral point contained both in the reflexive polytope and its dual polytope. (This fact plays a pivotal role in proving our main result.) Integral points correspond to monomials in graded polynomial rings. We will discuss, in particular, the monomial–divisor mirror map of Aspinwall, Greene and Morrison [AGM93], which gives the isomorphism between the two spaces $H^{1,1}_{toric}(X)$ and $H^{2,1}_{poly}(\hat{X})$ for a mirror pair $(X, \hat{X})$. This establishes a one-to-one correspondence between integral points in the reflexive polytope of X and monomials in the polynomial ring of $\hat{X}$.

In Section 7, we will discuss a one-to-one correspondence between monomials and periods via Picard–Fuchs differential equations. The method of Dwork–Katz–Griffiths determines the Picard–Fuchs differential equations for Calabi–Yau hypersurfaces in weighted projective spaces (of any dimension) (see Cox and Katz [CK99]). In this section, we will illustrate this correspondence, focusing on the concrete calculations of periods of Candelas et al. [CORV00, CORV03] for the one-parameter deformation of the quintic Calabi–Yau threefold in the ordinary projective 4-space $\mathbb{P}^4$, and of Kadir [Ka05] for the two-parameter deformation of the octic Calabi–Yau threefold in the weighted projective space $\mathbb{P}^4(Q)$ with $Q = (1, 1, 2, 2, 2)$. Here we observe that the Picard–Fuchs differential equation decomposes into the product of lower order Picard–Fuchs differential equations. This suggests that such a decomposition ought to have its origin in the motivic decomposition of the manifold.

Section 8 contains our main examples and the main result on the monomial–motive correspondence (Theorem 8.1). We establish a one-to-one correspondence between the class of monomials and Fermat motives at the Fermat point for the Calabi–Yau threefolds of Section 3. We prove that the motives which are invariant under the mirror map correspond to the class of the constant monomial (and hence to the origin in the polytopes). We illustrate the monomial–motive correspondence for the quintic and the octic Calabi–Yau orbifolds.

Section 9 contains our proof for the monomial–motive correspondence. We compute the number of $\mathbb{F}_q$-rational points (and hence congruence zeta-functions) for our Calabi–Yau threefolds over finite fields $\mathbb{F}_q$ in two different ways. On the one hand, we compute them with Weil's method using Jacobi (Gauss) sums. On the other hand, we compute them with Dwork's p-adic method using Dwork's characters and Gauss sums. We then show that the two approaches yield the same result at the Fermat point.

Away from the Fermat point, Calabi–Yau orbifolds with deformation parameters yield more monomials than motives. A conifold point on these Calabi–Yau orbifolds with deformation parameters is a singularity locally isomorphic to the projective quadric surface $X^2 + Y^2 + Z^2 + T^2 = 0$, and we may associate a motive (e.g., a mixed Tate motive) employing the same line of arguments as in Bloch–Esnault–Kreimer [BEK05].

Finally, Section 10 presents conclusions of this work, and further problems and future projects. The main conclusion is to bring motives into the realm of topological mirror symmetry for our Calabi–Yau orbifolds.

In a recent paper [K07], Kloosterman has extended the motive-monomial correspondence (established at the Fermat point in this article) to any quasismooth member of a one-parameter monomial deformation of a Fermat hypersurface in a weighted projective space, using a p-adic (e.g., Monsky–Washnitzer) cohomology theory. Also he has given explicit solutions to the p-adic Picard–Fuchs differential equation associated with monomial deformations of Fermat hypersurfaces.

Acknowledgments

Shabnam Kadir held a Fields Postdoctoral Fellowship in 2004–2005 supported by the Fields Institute Thematic Program "Geometry of String Theory", and N. Yui's NSERC Discovery Grant.

During the course of this work, Noriko Yui held visiting positions at a number of institutions. These include University of Leiden (Winter 2004), Max-Planck-Institut-für-Mathematik Bonn (Spring 2004, Spring 2006), University of Hannover (Spring 2004, 2006), University of Mainz (Spring 2004), Tsuda College (Spring 2005) and Kavli Institute for Theoretical Physics (KITP) (December 2005) at University of California Santa Barbara. The research at KITP was supported in part by the National Science Foundation under the Grant No. PHY99-07949. She is grateful for the hospitality of these institutions.

Finally, it is our pleasure to thank the referee for useful comments, remarks and suggestions for the improvement of the paper.

2 Fermat motives

We will employ the definition of motives due to Grothendieck and Manin [Ma70]. The most recommended references for the generality of motives might be Deligne–Milne [DM82], and Soulé [So84]. In this paper we will confine ourselves to the so-called "Fermat" motives arising from Fermat hypersurfaces. We will recall the construction of Fermat motives from Shioda [Sh87] and Gouvêa and Yui [GY95]. The construction works for any dimension, not only for dimension 3.

We start with the Fermat hypersurface of degree m and of dimension n in the projective space $\mathbb{P}^{n+1}$:

$$V : Z_1^m + Z_2^m + \cdots + Z_{n+2}^m = 0 \subset \mathbb{P}^{n+1}.$$

Let μ_m be the group of m-th roots of unity and let

$$\mathfrak{G} := (\mu_m)^{n+2}/(\text{diagonal}) = \{\mathbf{g} := (g_1, g_2, \cdots, g_{n+2}) \,|\, g_i \in \mu_m \quad \forall i\}/\{(g, g, \cdots, g)\}$$

be a subgroup of the automorphism group $\mathrm{Aut}(V)$ of V. The group $\mathfrak{G}$ is a finite group of order m^{n+1}, and it acts on V by componentwise multiplication of g_i for each i. The character group of $\mathfrak{G}$ is identified with the set

$$\hat{\mathfrak{G}} := \{\mathbf{a} = (a_1, a_2, \cdots, a_{n+2}) \,|\, a_i \in (\mathbb{Z}/m\mathbb{Z}) \;\forall\, i, \; \sum_{i=1}^{n+2} a_i \equiv 0 \pmod{m}\,\}$$

and the duality between $\mathfrak{G}$ and $\hat{\mathfrak{G}}$ is given by:

$$\hat{\mathfrak{G}} \times \mathfrak{G} \mapsto \mathbb{L} := \mathbb{Q}(\zeta_m) \quad (\mathbf{a}, \mathbf{g}) \mapsto \mathbf{a}(\mathbf{g}) := \prod_{i=1}^{n+2} g_i^{a_i}$$

where $\zeta_m = e^{2\pi i/m}$ is a primitive m-th root of unity in $\mathbb{C}^\times$. For each $\mathbf{a} \in \hat{\mathfrak{G}}$, let $A = [\mathbf{a}]$ be the $(\mathbb{Z}/m\mathbb{Z})^\times$-orbit of $\mathbf{a}$. Put $\mathbb{L}_\mathbf{a} = \mathbb{L}_A := \mathbb{Q}(\zeta_m^d)$ where $d := \gcd(\mathbf{a}, m)$.

Definition 2.1 For each character $\mathbf{a} \in \hat{\mathfrak{G}}$, define

$$p_\mathbf{a} := \frac{1}{\#\hat{\mathfrak{G}}} \sum_{\mathbf{g}\in\mathfrak{G}} \mathbf{a}(\mathbf{g})^{-1}\mathbf{g} = \frac{1}{m^{n+1}} \sum_{\mathbf{a}\in\mathfrak{G}} \mathbf{a}(\mathbf{g})^{-1}\mathbf{g},$$

and for each $(\mathbb{Z}/m\mathbb{Z})^\times$-orbit $A = [\mathbf{a}]$ of $\mathbf{a}$, define

$$p_A := \sum_{\mathbf{a}\in A} p_\mathbf{a} = \frac{1}{m^{n+1}} \sum_{\mathbf{g}\in\mathfrak{G}} (\mathrm{trace}_{\mathbb{L}_\mathbf{a}/\mathbb{L}}(\mathbf{a}(\mathbf{g})^{-1}))\,\mathbf{g}.$$

Then it is easy to see that $p_\mathbf{a}$ and p_A are elements of the group rings $\mathbb{L}[\mathfrak{G}]$ and $\mathbb{Z}[\frac{1}{m}][\mathfrak{G}]$, respectively.

Proposition 2.1 *$p_\mathbf{a}$ and p_A are projectors (idempotents), that is,*

$$p_\mathbf{a} \cdot p_\mathbf{b} = \begin{cases} p_\mathbf{a} & \text{if } \mathbf{a} = \mathbf{b} \\ 0 & \text{if } \mathbf{a} \neq \mathbf{b} \end{cases}$$

and

$$p_A \cdot p_B = \begin{cases} p_A & \text{if } A = B \\ 0 & \text{if } A \neq B \end{cases}$$

Furthermore, we have the decomposition

$$\sum_{\mathbf{a}\in\hat{\mathfrak{G}}} p_\mathbf{a} = 1, \quad \text{and} \quad \sum_{A\in O(\hat{\mathfrak{G}})} p_A = 1$$

where $O(\hat{\mathfrak{G}})$ denotes the set of $(\mathbb{Z}/m\mathbb{Z})^\times$-orbits in $\hat{\mathfrak{G}}$.

Definition 2.2 Identifying each automorphism $\mathbf{g} \in \mathrm{Aut}(V)$ with its graph, the projector p_A may be regarded as an algebraic n-cycle on $V \times V$ with coefficients in $\mathbb{Z}[\frac{1}{m}]$. The pair

$$(V, p_A) =: \mathcal{M}_A$$

may be called the *Fermat motive* corresponding to the $(\mathbb{Z}/m\mathbb{Z})^\times$-orbit A of $\hat{\mathfrak{G}}$. Furthermore, there is the motivic decomposition of the Fermat hypersurface:

$$(V, \mathbf{1}) = \oplus_{A\in O(\hat{\mathfrak{G}})}(V, p_A) = \oplus_{A\in O(\hat{\mathfrak{G}})}\mathcal{M}_A$$

corresponding to $\sum_{A\in O(\hat{\mathfrak{G}})} p_A = 1$. (Here $\mathbf{1} := \Delta_V$.)

Note that Fermat motives are well-defined over $\mathbb{Q}$ or finite fields $\mathbb{F}_p$ for prime p such that $(m, p) = 1$.

Now we will discuss the cohomological realizations of Fermat motives.

Definition 2.3 Let $H^\bullet(V, \star)$ be any Weil cohomology group. For a Fermat motive $\mathcal{M}_A = (V, p_A)$, define

$$H^\bullet(\mathcal{M}_A, \star) := H^\bullet(V, \star)^{p_A}$$

as the image of p_A (or equivalently, the kernel of $p_A - 1$) acting on $H^\bullet(V, \star)$.

The motivic decomposition yields the decomposition of cohomology groups $H^\bullet(V, \star)$ with various coefficients $\star$, e.g., ℓ-adic étale cohomology, de Rham cohomology, crystalline cohomology, etc. However, since crystalline cohomology is defined only for smooth varieties, here we will confine our discussions on de Rham and Betti realizations.

Proposition 2.2 *Let*

$$\mathfrak{A} := \{\mathbf{a} = (a_1, a_2, \cdots, a_{n+2}) \in \hat{\mathfrak{G}} \mid a_i \neq 0 \in (\mathbb{Z}/m\mathbb{Z}) \,\forall\, i\}$$

be a subset of $\hat{\mathfrak{G}}$. *For* $\mathbf{a} = (a_1, a_2, \cdots, a_{n+2}) \in \mathfrak{A}$, *let*

$$\|\mathbf{a}\| := \left(\frac{1}{m}\sum_{\mathbf{a}\in A} a_i\right) - 1.$$

(a) The $(i, n-i)$*-th motivic Hodge number is given by*

$$h^{i,n-i}(\mathcal{M}_A) = \begin{cases} \#\{\mathbf{a} \in A \mid \|\mathbf{a}\| = i\} & \text{for } i,\, 0 \leq i \leq n \text{ and } A \subset \mathfrak{A} \\ 1 & \text{for } i = \frac{n}{2} \text{ and } A = [(0, 0, \cdots, 0)] \\ 0 & \text{otherwise} \end{cases}$$

In particular, the geometric genus of $\mathcal{M}_A$ *is* $p_g(\mathcal{M}_A) := h^{0,n}(\mathcal{M}_A)$. *We have the Hodge decomposition*

$$h^{i,n-1}(V) = \sum_A h^{i,n-i}(\mathcal{M}_A)$$

and in particular,

$$p_g(V) = \sum_A p_g(\mathcal{M}_A).$$

(b) The i-th motivic Betti number is given by

$$B_i(\mathcal{M}_A) = \begin{cases} \#A & \text{for } i = n \text{ and } A \subset \mathfrak{A} \\ 1 & \text{for } i \text{ even and } A = [(0, 0, \ldots, 0)] \\ 0 & \text{otherwise} \end{cases}$$

For $i = n$, *we have*

$$B_n(\mathcal{M}_A) \leq \varphi(m)/\gcd(\mathbf{a}, m)$$

where $\varphi(m)$ *denotes Euler's phi-function. (We may call* $B_n(\mathcal{M}_A)$ *the dimension of the motive* $\mathcal{M}_A$.*)*

The global i-th Betti number $B_i(V)$ *is given by*

$$B_i(V) = \sum_A B_i(\mathcal{M}_A).$$

For $h^{i,n-i}(V)$ *and* $B_i(V)$, *the sum runs over A in the set* $O(\mathfrak{A})$ *of* $(\mathbb{Z}/m\mathbb{Z})^\times$*-orbits in* $\mathfrak{A}$ *if n is odd, and the set* $O(\mathfrak{A} \cup \{0\})$ *if n is even.*

3 Construction of Calabi–Yau orbifolds

From now on, we will confine ourselves to threefolds ($n = 3$). We will construct mirror pairs of Calabi–Yau threefolds by the orbifolding method due to Greene and Plesser [GP90].

Fix a Fermat hypersurface threefold $V = V(m, 3, \mathbf{1})$ of degree m and coefficients $\mathbf{1}$ defined over $\mathbb{Q}$:

$$V : Z_1^m + Z_2^m + Z_3^m + Z_4^m + Z_5^m = 0 \subset \mathbb{P}^4_{\mathbb{Q}}.$$

Let $Q = (q_1, q_2, q_3, q_4, q_5) \in \mathbb{N}^5$ be a vector consisting of 5-tuples of positive rational integers. We call Q a *weight* if it satisfies the following two conditions:

(1) $\gcd(q_1, \cdots, \widehat{q_j}, \cdots, q_5) = 1$ for every j, $1 \leq j \leq 5$, and

(2) $q_j \mid m = \deg(V)$ for every j, $1 \leq j \leq 5$.

(Here ^ means the exclusion of that component.)

For each j, $1 \leq j \leq 5$, let μ_{q_j} denote the finite group scheme of q_j-th roots of unity over $\mathbb{Q}$, namely,

$$\mu_{q_j} := \operatorname{Spec}(\mathbb{Q}[T]/(T^{q_j} - 1)).$$

Take the direct product of the μ_{q_j}'s and define

$$G = G_Q := \mu_{q_1} \times \cdots \times \mu_{q_5}.$$

Then $G = G_Q$ is a finite automorphism group scheme of $V_{\mathbb{Q}}$, whose action over the closure $\bar{\mathbb{Q}}$ is described as follows:

$$(Z_1, \cdots, Z_5) \mapsto (\zeta_{q_1}^{e_1} Z_1, \cdots, \zeta_{q_5}^{e_5} Z_5), \quad e_j \in \mathbb{Z}/q_j\mathbb{Z} \text{ for each } j$$

where ζ_{q_j} denotes a primitive q_j–th root of unity in $\bar{\mathbb{Q}}$ for every j, $1 \leq j \leq 5$.

Take the quotient, $Y := V/G_Q$. Then Y is, in general, a singular variety defined over $\mathbb{Q}$ in the weighted projective 4-space $\mathbb{P}^4(Q)$. Let $\Sigma(Y)$ denote the singular locus of Y. Then $\Sigma(Y)$ is finite, and all singularities are at most abelian (indeed, cyclic) quotient singularities.

Next we wish to construct a smooth Calabi–Yau threefold from Y. For this we first look for the sufficient condition under which this quotient variety Y is a singular Calabi–Yau threefold, that is, the canonical sheaf ω_Y is trivial. The so-called Calabi–Yau condition is imposed on the weight $Q = (q_1, q_2, \cdots, q_5)$ by

(3) $q_1 + q_2 + q_3 + q_4 + q_5 = m$ or equivalently $\sum_{i=1}^{5} \frac{q_i}{m} = 1.$

Indeed, in the weighted projective 4-space $\mathbb{P}^4(Q)$, the canonical sheaf is given by (see Dolgachev [Dol82])

$$\omega_Y \simeq \mathcal{O}(m - q_1 - q_2 - q_3 - q_4 - q_5).$$

We require the triviality of ω_Y, which gives rise to the sufficient condition (3). With the condition (3), Y has trivial canonical bundle away from its singularities.

Definition 3.1 For a fixed m, we call $Q = (q_1, q_2, q_3, q_4, q_5)$ an *admissible* weight if it satisfies the conditions (1), (2) and (3), and we will call $\langle m, Q \rangle$ an *admissible* pair.

Now we look for a smooth resolution of singularities without disturbing the triviality of the canonical bundle. Such a resolution is called a *crepant* resolution. The existence of a crepant resolution for our singular Calabi–Yau threefold Y is proved by Greene, Roan and Yau [GRY91]. In fact, it is unique up to flops. We summarize their results reformulated by Yui [Y05] suitable for the arithmetic discussion and applications.

Theorem 3.1 *Let $V = V(m, 3, \mathbf{1})$ be the Fermat hypersurface threefold of degree $m \geq 5$ with coefficients $\mathbf{1} = (1, 1, \cdots, 1)$ defined over $\mathbb{Q}$. Let $Q = (q_1, q_2, q_3, q_4, q_5)$ be a weight.*

(a) There are 147 admissible pairs $\langle m, Q \rangle$ giving rise to singular Calabi–Yau threefolds Y. The smallest is $\langle 5, (1, 1, 1, 1, 1) \rangle$ and the largest is $\langle 1807, (1, 42, 258, 602, 903) \rangle$.

In the weighted projective 4-space $\mathbb{P}^4_{\mathbb{Q}}(Q)$, Y is defined by the equation

$$Y_1^{m/q_1} + Y_2^{m/q_2} + Y_3^{m/q_3} + Y_4^{m/q_4} + Y_5^{m/q_5} = 0 \subset \mathbb{P}^4_{\mathbb{Q}}(Q)$$

with $\deg(Y_j) = q_j$ for each j, $1 \leq j \leq 5$.

(b) Y has at most abelian (cyclic) quotient singularities. Let $\Sigma(Y)$ be the singular locus of Y. Let ℓ be an integer ranging over the set $\{0, 1, \cdots, m-1\}$. For each ℓ, we define the two sets of indices

$$S_\ell := \{\, j \,|\, 1 \le j \le 5, \ell/m_j \in \mathbb{Z}\,\} \quad \text{and} \quad I_\ell = \{1, 2, 3, 4, 5\} \setminus S_\ell.$$

Further, put $c_{I_\ell} := \gcd\{q_j \,|\, j \notin I_\ell\} = \gcd\{q_j \,|\, j \in S_\ell\}$. Let

$$Y_{I_\ell} := Y \cap \{Y_j = 0,\ j \in I_\ell\}.$$

Then $Y_{I_\ell} \in \Sigma(Y)$, and it is defined over $\mathbb{Q}$.

More concretely, the following assertions hold:

(b1) $\#S_\ell \neq 4$ for any ℓ, and $\#S_\ell = 5$ for $\ell = 0$.

(b2) If $\#S_\ell = 3$ for some ℓ, then there is a set I_ℓ with $\#I_\ell = 2$, $c_{I_\ell} \ge 2$. In this case $\Sigma(Y)$ contains a dimension 1 singularity.

Suppose that C is an irreducible component of $Y_{I_\ell} = Y \cap \{Y_j = 0 \,|\, j \in I_\ell\}$; then C is a smooth weighted diagonal curve defined over $\mathbb{Q}$ of degree $m' := m/c_{I_\ell}$ and reduced weight $Q' := (q'_i, q'_j, q'_k)$, where $i, j, k \notin I_\ell$ and $q'_\bullet = q_\bullet/c_{I_\ell}$. The multiplicity of C in Y_{I_ℓ} is $m_C := c_{I_\ell} - 1$. The genus $g(C)$ of C is given by the coefficient of $t^{m'-(q'_i+q'_j+q'_k)}$ in the formal power series

$$\frac{1 - t^{m'}}{(1 - t^{q'_i})(1 - t^{q'_j})(1 - t^{q'_k})}.$$

(b3) If $\#S_\ell = 2$ for some ℓ, then there are dimension 0 singularities. Let $P \in \Sigma(Y)$ be a singular point. Then P is a cyclic quotient singularity, and $\pi^{-1}(P)$ is copies of smooth rational surfaces birationally equivalent to $\mathbb{P}^2$. It is defined over a finite Galois extension of $\mathbb{Q}$; its Galois orbit is defined over $\mathbb{Q}$.

(b4) If $\#S_\ell = 1$, Y has no singularity.

(c) There exists the unique crepant resolution $\pi : X \to Y$, and X is a smooth Calabi–Yau threefold defined over $\mathbb{Q}$. The third Betti number of X can be computed by

$$B_3(X) = B_3(Y) + \sum_C m_C\, B_1(C)$$

where C runs over all the smooth irreducible curves in $\Sigma(Y)$ with multiplicity $m_C = c_{I_\ell} - 1$. ($\pi^{-1}(C)$ is copies of smooth ruled surfaces birationally equivalent to $C \times \mathbb{P}^1$ and it is defined over $\mathbb{Q}$.)

Furthermore, $B_3(Y)$ can be computed by Vafa's formula (see Roan [Ro90]) as follows. For each integer $\ell \in \{0, 1, 2, \cdots, m-1\}$, let $S_\ell = \{\, j \,|\, 1 \le j \le 5, \ell\frac{q_i}{m} \in \mathbb{Z}\}$ be the set of indices defined in Proposition 3.1. and define the integer β_ℓ by

$$\beta_\ell := \frac{1}{m} \sum_{r=0}^{m-1} \prod_{\ell\frac{q_i}{m} \in \mathbb{Z},\, r\frac{q_i}{m} \in \mathbb{Z}} \left(1 - \frac{m}{q_i}\right).$$

Here we employ the convention that the product $\prod_{\ell\frac{q_i}{m}, r\frac{q_i}{m}} (1 - \frac{m}{q_i}) = 1$ if there is no q_i with $\ell\frac{q_i}{m} \in \mathbb{Z}, r\frac{q_i}{m} \in \mathbb{Z}$. Then

$$B_3(Y) = -\beta_0. \quad \textit{and} \quad B_3(X) = B_3(Y) - \sum_{\#S_\ell = 3} \beta_\ell.$$

where the sum runs over the set S_ℓ consisting of three elements. The Euler characteristic of X is given by

$$\chi(X) = \sum_{\ell=0}^{m-1} \beta_\ell.$$

Consequently,

$$B_2(X) = h^{1,1}(X) = -1 + \frac{1}{2}\sum_{\ell=0}^{m-1} \sum_{\#S_\ell<3} \beta_\ell.$$

In particular, all cycles generating $H^{1,1}(X)$ arise from singular points on Y.

Remark 3.1 Here we used Vafa's method developed for computing orbifold Euler characteristics. (At a glance, it is not transparent why β_ℓ is an integer. Hendrik Lenstra gave an elementary proof to establish the integrality of the β_ℓ's.) It should be noted that Vafa's method is a special case of the calculation of *stringy* Euler characteristics by the toric method developed by Batyrev [Ba94]. In Section 5 we will give another approach for the computation of Betti numbers and Euler characteristics of the 147 Calabi–Yau orbifolds using Fermat motives.

4 Construction of mirror Calabi–Yau orbifolds

The Calabi–Yau orbifolds constructed in Theorem 3.1 are not rigid, i.e., $h^{1,2} \neq 0$. Therefore, their mirror Calabi–Yau threefolds exist in the sense of the Topological Mirror Symmetry Conjecture 1.1. In this section, we will construct mirror partners of these Calabi–Yau orbifolds by applying again the orbifolding construction on perturbed Calabi–Yau hypersurfaces by certain (finite abelian) groups of discrete symmetries. This construction is due to Greene and Plesser [GP90] (see also Greene–Roan and Yau [GRY91], Roan [Ro90], [Ro91], [Ro94]). Now we will describe the mirror construction.

Theorem 4.1 *Let X be a Calabi–Yau orbifold corresponding to $\langle m, Q\rangle$ with $Q = (q_1, q_2, q_3, q_4, q_5)$ in Theorem 3.1 obtained from the weighted Fermat hypersurface:*

$$Y_1^{m_1} + Y_2^{m_2} + Y_3^{m_3} + Y_4^{m_4} + Y_5^{m_5} = 0 \subset \mathbb{P}^4(Q)$$

where we put $m_j = m/q_j$ for $1 \le j \le 5$. Let $\mathcal{W}_Q$ be the generic weighted Fermat hypersurface with deformation parameters, and let $\{\mathcal{W}_Q = 0\}$ be its zero locus in $\mathbb{P}^4(Q)$. Then $\{\mathcal{W}_Q = 0\}$ is a Calabi–Yau threefold. Let Ω denote the unique (up to scalar multiplication) holomorphic 3-form on $\mathcal{W}_Q$. Then

$$\Omega = \mathrm{Res}_{\mathcal{W}_Q=0}\frac{d\mu}{\mathcal{W}_Q}$$

where

$$d\mu = \sum_{j=1}^{5}(-1)^j q_j Y_j dY_1 \cdots \wedge \widehat{dY_j} \wedge \cdots \wedge dY_5.$$

For each j, $1 \le j \le 5$, let g_{m_j} be a primitive m_j-th root of unity in $\mathbb{C}^\times$ and let μ_{m_j} be the cyclic group generated by $g_j = e^{2\pi i/m_j}$. Define the group $\hat{G}$ by

$$\hat{G} = \{\, \mathbf{g} = (g_1, g_2, g_3, g_4, g_5) \,|\, g_i^{m_i} = 1, \prod_{i=1}^{5} g_i = 1\}/\{g, g, \cdots, g)\}.$$

Then $\hat{G}$ is a finite abelian group (scheme) of order $\prod_{j=1}^5 m_j/m^2 = m^3/\prod_{i=1}^5 q_i$, and it leaves Ω invariant.

This gives rise to a $\hat{G}$-invariant hypersurface

$$\widehat{\mathcal{W}}_Q := \sum_{I=(i_1,\cdots,i_5)} c_I Y_1^{i_1}\cdots Y_5^{i_5}$$

in $\mathbb{P}^4(Q)$ quotiented out by $\hat{G}$, where the sum runs over

$$I = (i_1,\cdots,i_5) \in (\mathbb{Z}/m_1\mathbb{Z})\times\cdots\times(\mathbb{Z}/m_5\mathbb{Z})$$

subject to the relation $q_1 i_1 + \cdots + q_5 i_5 = m$. The zero locus $\{\widehat{\mathcal{W}}_Q = 0\}$ defines a Calabi–Yau threefold.

Then there exists a resolution $\hat{X} := \{\widehat{\mathcal{W}}_Q = 0\} \subset \mathbb{P}^4(Q)/\hat{G}$ which is a mirror partner of the family of Calabi–Yau orbifolds $X := \{\mathcal{W}_Q = 0\}$ satisfying the Topological Mirror Symmetry Conjecture 1.1:

$$h^{2,1}(\hat{X}) = h^{1,1}(X),\ h^{1,1}(\hat{X}) = h^{2,1}(X),\ \chi(\hat{X}) = -\chi(X).$$

The orbifolding construction was described by Greene and Plesser [GP90]. The calculations of Hodge numbers and Euler characteristics were carried out in Candelas-Lynker-Schimmrigk [CLS90]. Roan [R091] gave a mathematical proof of these results.

Remark 4.1 In Theorem 4.1, defining equations for mirror Calabi–Yau threefolds are given by deformations of weighted Fermat hypersurface threefolds. Generically, these defining equations may have as many independent deformation parameters as $h^{2,1}_{poly} < h^{2,1}$. (The Hodge number $h^{2,1}_{poly}$ is defined in terms of toric geometry. For the definition, see Sections 6 and 7 below.) The Hodge numbers are invariant under the Kähler deformations, i.e., the deformations preserve the Kähler structure of the manifold in question (see Nakamura [Na75]), so $h^{1,1}$ remains unchanged under deformations. In particular, by taking a special point of the generic deformation, we can have a defining equation for our Calabi–Yau family with a small number of deformation parameters.

Example 4.1 We consider one-parameter deformations of Calabi–Yau threefolds obtained from weighted Fermat hypersurfaces corresponding to a pair $\langle m, Q\rangle$ in Theorem 4.1:

$$Y_1^{m/q_1} + Y_2^{m/q_2} + Y_3^{m/q_3} + Y_4^{m/q_4} + Y_5^{m/q_5} - m\lambda Y_1Y_2Y_3Y_4Y_5 = 0 \subset \mathbb{P}^4(Q)$$

where λ is a parameter (subject to the relation $\lambda^m = 1$). Then its zero locus $\mathcal{W}_Q = 0$ gives rise to a family of Calabi–Yau threefolds.

Table 1 lists Calabi–Yau threefolds $\{\mathcal{W}_Q = 0\}$ corresponding to the pairs $\langle m, Q\rangle$ with $h^{1,1}(X) = 1$, and their mirror families of Calabi–Yau threefolds $\widehat{\mathcal{W}}_Q$. Here a generator $\mathbf{g} = (g_1,\cdots,g_5)$ for $\hat{G}$ stands for $(e^{2\pi i g_1/m},\cdots,e^{2\pi i g_5/m})$. The action of $\hat{G}$ on X is

$$(Y_1,Y_2,Y_3,Y_4,Y_5) \mapsto (g_1Y_1, g_2Y_2, g_3Y_3, g_4Y_4, g_5Y_5).$$

For instance, for $m = 5$ and the generator $(1,0,0,4,0)$, the action reads as

$$(Y_1,Y_2,Y_3,Y_4,Y_5) \mapsto (e^{2\pi i/5}Y_1, Y_2, Y_3, e^{2\pi i 4/5}Y_4, Y_5).$$

The above construction of mirror pairs of Calabi–Yau orbifolds may be characterized in terms of the "duality" between the two finite abelian groups. See Greene–Plesser [GP90], and for a nice exposition, Morrison [Mor97].

Table 1 Mirror Calabi–Yau orbifolds I

m	Q	G_Q	$\hat{G}$	generators
5	$(1,1,1,1,1)$	$\{1\}$	$(\mathbb{Z}/5\mathbb{Z})^3$	$(1,0,0,4,0),(1,0,4,0,0),(1,4,0,0,0)$
6	$(1,1,1,1,2)$	$(\mathbb{Z}/2\mathbb{Z})$	$(\mathbb{Z}/3\mathbb{Z})\times(\mathbb{Z}/6\mathbb{Z})^2$	$(0,2,2,2,0),(5,0,0,1,0),(0,5,0,1,0)$
8	$(1,1,1,1,4)$	$(\mathbb{Z}/4\mathbb{Z})$	$(\mathbb{Z}/2\mathbb{Z})\times(\mathbb{Z}/8\mathbb{Z})^2$	$(7,0,0,1,0),(0,7,0,1,0),(0,0,7,1,0)$
10	$(1,1,1,2,5)$	$(\mathbb{Z}/2\mathbb{Z})\times(\mathbb{Z}/5\mathbb{Z})$	$(\mathbb{Z}/10\mathbb{Z})^2$	$(9,0,1,0,0,),(0,9,1,0,0)$

Table 2 Mirror Calabi–Yau threefolds II

m	Q	$\hat{G}$	generator
8	$(1,1,2,2,2)$	$(\mathbb{Z}/8\mathbb{Z})^2$	$(1,7,0,0,0),(7,1,0,0,0)$
12	$(1,1,2,2,6)$	$(\mathbb{Z}/2\mathbb{Z})\times(\mathbb{Z}/6\mathbb{Z})^2$	$(6,6,0,0,0),(4,0,2,2,0),(0,4,2,2,0)$
12	$(1,2,2,3,4)$	$(\mathbb{Z}/6\mathbb{Z})^2$	$(4,2,2,0,0),(0,2,2,0,4)$
14	$(1,2,2,2,7)$	$(\mathbb{Z}/7\mathbb{Z})^2$	$(2,2,2,2,4),(4,4,4,4,2)$
18	$(1,1,1,6,9)$	$(\mathbb{Z}/3\mathbb{Z})\times(\mathbb{Z}/6\mathbb{Z})^2$	$(6,6,6,0,0),(3,3,3,0,9),(3,3,3,3,6)$

Theorem 4.2 *For the above mirror pairs of Calabi–Yau orbifolds $(X,\hat{X})$, the mirror symmetry can be described as the "duality" between the two finite abelian groups, that is, there is the group $G=G_Q$ associated to the original Calabi–Yau orbifold X:*

$$G=G_Q=\prod_{j=1}^{5}\mu_{q_j}\quad \textit{of order}\quad \textstyle\prod_{j=1}^{5}q_j.$$

On the mirror side, there is the group $\hat{G}$ associated to the mirror Calabi–Yau orbifold $\hat{X}$:

$$\hat{G}=\{\mathbf{g}=(g_1,\cdots,g_5)\,|\,g_j^{m_j}=1,\ \prod_{j=1}^{5}g_j=1\}/\mu_m\quad \textit{of order}\quad m^3/\textstyle\prod_{j=1}^{5}q_j.$$

These two abelian groups are subject to the "duality" relation:

$$\#G_Q\times\#\hat{G}=m^3.$$

Example 4.2 Now we consider two-parameter deformations of Calabi–Yau orbifolds with $h^{1,1}(X)=2$. Note that for these examples listed in Table 2, the groups G_Q are more complicated than the cases $h^{1,1}(X)=1$ discussed above, but in any event, we have the "duality" relation: $\#G_Q\times\#\hat{G}=m^3$.

For any of these Calabi–Yau orbifolds X, its mirror Calabi–Yau threefold is defined by a hypersurface $\mathcal{W}_Q=0$ in the weighted projective space $\mathbb{P}^4(Q)/G$ where $\mathcal{W}_Q$ is of the form:

$$\mathcal{W}_Q=Y_1^{m_1}+\cdots+Y_5^{m_5}-m\lambda Y_1Y_2\cdots Y_5-k\phi\begin{cases}Y_1^4Y_2^4 & \text{for } m=8\\ Y_1^6Y_2^6 & \text{for } m=12\\ Y_1^6Y_4^2 & \text{for } m=12\\ Y_1^7Y_5 & \text{for } m=14\\ Y_1^6Y_2^6Y_3^6 & \text{for } m=18\end{cases}$$

where λ and ϕ are parameters and k is a positive integer such that $k|m$. The mirror of the Calabi–Yau threefolds with $m=8$ and $m=12$, $Q=(1,1,2,2,6)$ with two parameters was studied in Candelas et al. [COFKM93].

Remark 4.2 Theorem 4.2 is valid even when one passes to a pair of finite groups $(H, \hat{H})$ where H is a subgroup of $(\mathbb{Z}/m\mathbb{Z})^3$ and $\hat{H}$ is a dual group of H in the sense that $\#H \times \#\hat{H} = m^3$. See Klemm and Theisen [KT93] for examples of one-modulus Calabi–Yau threefolds of this type. See Problem 10.3.

Remark 4.3 One-(or multi)-parameter deformations of Calabi–Yau orbifolds constructed in Theorem 3.1 are no longer dominated by product varieties. (Cf. Schoen [Sch96].) Consequently, these deformations are no longer CM type varieties.

5 Fermat motives and mirror maps

In this section, we first construct Fermat motives arising from the Calabi–Yau threefolds constructed in Theorem 3.1 at special points of deformations, i.e., at the Fermat points (where all the deformation parameters are set to zero). We will interpret the mirror symmetry phenomenon for our Calabi–Yau orbifolds and their mirror partners in terms of Fermat motives. In particular, we will determine Fermat motives which are invariant under the mirror map for each mirror pair of Calabi–Yau orbifolds.

Now we will construct Fermat motives for the Calabi–Yau orbifolds X constructed from Fermat hypersurfaces of degree m and weight $Q = (q_1, q_2, q_3, q_4, q_5)$. Recall from Section 2 that associated to the Fermat hypersurface threefold

$$V \,:\, Z_1^m + Z_2^m + Z_3^m + Z_4^m + Z_5^m = 0 \subset \mathbb{P}^4,$$

we have a group

$$\mathfrak{G} := \{\mathbf{g} = (g_1, g_2, g_3, g_4, g_5) \,|\, g_i \in \mu_m\}/\{(g, g, \cdots, g)\} \subset \mathrm{Aut}(V)$$

and its dual group

$$\hat{\mathfrak{G}} = \{\mathbf{a} = (a_1, a_2, a_3, a_4, a_5) \,|\, a_i \in (\mathbb{Z}/m\mathbb{Z}),\ \sum_{i=1}^{5} a_i \equiv 0 \pmod{m}\}.$$

We consider the subset of the dual group $\hat{\mathfrak{G}}$ defined by

$$\mathfrak{A} = \{\mathbf{a} \in \hat{\mathfrak{G}} \,|\, a_i \neq 0 \in (\mathbb{Z}/m\mathbb{Z})\,\forall\, i\}.$$

There is a pairing between $\mathfrak{G}$ and $\mathfrak{A}$ given by

$$(\mathbf{g}, \mathbf{a}) \mapsto \prod_{i=1}^{5} g_i^{a_i} \in \mathbb{L} = \mathbb{Q}(\zeta_m).$$

Now we pass to weighted Fermat Calabi–Yau threefolds. We ought to bring in weights to our discussion.

Theorem 5.1 (Yui [Y05]) *Let Y be the singular Calabi–Yau orbifold corresponding to an admissible pair $\langle m, Q\rangle$ with $Q = (q_1, q_2, q_3, q_4, q_5)$ in Theorem 3.1. Define*

$$\mathfrak{A}(Q) := \{\mathbf{a} = (a_1, a_2, a_3, a_4, a_5) \in \mathfrak{A} \,|\, a_i \in (q_i\mathbb{Z}/m\mathbb{Z})\,\forall\, i\}.$$

For each character $\mathbf{a} \in \mathfrak{A}(Q)$, let $A = [\mathbf{a}]$ be the $(\mathbb{Z}/m\mathbb{Z})^\times$-orbit of $\mathbf{a}$, i.e., $A = \{t\mathbf{a} \,|\, t \in (\mathbb{Z}/m\mathbb{Z})^\times\}$. Let $p_{\mathbf{a}}$ and p_A be the projectors defined as in Definition 2.1. Then $(Y, p_A) =: \mathcal{M}_A$ is the Fermat motive corresponding to A. Furthermore, we have the motivic decomposition

$$(Y, \mathbf{1}) = \oplus_{A \in O(\mathfrak{A}(Q))} \mathcal{M}_A$$

where A runs over the set of $(\mathbb{Z}/m\mathbb{Z})^\times$-orbits in $\mathfrak{A}(Q)$.

Table 3 Fermat motives for the quintic Calabi–Yau threefold

A	dim	mult	$h^{3,0}$	$h^{2,1}$	$\sum B_3(\mathcal{M}_A)$
$[1,1,1,1,1]$	4	1	1	1	4
$[1,1,1,3,4]$	4	20	0	2	80
$[1,1,2,2,4]$	4	30	0	2	120

Then the following assertions hold.
(a) We have

$$h^{3,0}(Y) = 1 = h^{3,0}(\mathcal{M}_{[Q]})$$

$$h^{2,1}(Y) = \sum_{A \in O(\mathfrak{A}(Q))} h^{2,1}(\mathcal{M}_A)$$

and

$$B_3(Y) = \sum_{A \in O(\mathfrak{A}(Q))} B_3(\mathcal{M}_A).$$

(b) Let $\pi : X \to Y$ be the unique crepant resolution of Y, and let $\hat{\Sigma}(Y) := \pi^{-1}(\Sigma(Y))$ be the pullback of the singular locus $\Sigma(Y)$. Let $C \in \Sigma(Y)$ be an irreducible curve of genus $g(C)$. Then

$$h^{2,1}(\hat{\Sigma}(Y)) = \sum_C m_C g(C)$$

where the sum runs over all distinct irreducible curves in $\Sigma(Y)$ with multiplicity m_C.

(c) For the crepant resolution X of Y, we have

$$h^{3,0}(X) = 1 = h^{3,0}(\mathcal{M}_{[Q]})$$

$$h^{2,1}(X) = h^{2,1}(Y) + h^{2,1}(\hat{\Sigma}(Y))$$

and

$$B_3(X) = B_3(Y) + 2\sum_C m_C g(C).$$

Furthermore,

$$B_2(X) = h^{1,1}(X) = 1 + \sum_P m_P$$

where P is a singular point in $\Sigma(Y)$ with multiplicity m_P.

Example 5.1 Let $m = 5$ and $Q = (1,1,1,1,1)$. This is the quintic Calabi–Yau threefold X, which is smooth. The Fermat motives are tabulated in Table 3.

Here each motive has dimension $\varphi(5) = 4$ and the multiplicity is given by the number of permutations of the components. Furthermore, Y is smooth, so $Y = X$. Summing over the motivic Hodge and Betti numbers, we get the global Hodge and Betti numbers:

$$h^{3,0}(X) = 1,\ h^{2,1}(X) = 1 + 2 \times 20 + 2 \times 30 = 101,\ h^{1,1}(X) = 1.$$

The third Betti number and the Euler characteristic are given by

$$B_3(X) = 4 + 80 + 120 = 204,\ E(X) = 2(h^{1,1}(X) - h^{2,1}(X)) = 2(1 - 101) = -200.$$

Table 4 Fermat motives for the octic Calabi–Yau threefold

A	dim	mult	$h^{3,0}$	$h^{2,1}$	$\sum B_3(\mathcal{M}_A)$
$[1,1,2,2,2]$	4	1	1	1	4
$[7,3,2,2,2]$	4	1	0	2	4
$[6,4,2,2,2]$	2	2	0	2	4
$[7,1,2,2,4]$	4	6	0	12	24
$[6,2,2,2,4]$	2	6	0	6	12
$[4,4,2,2,4]$	2	3	0	3	6
$[5,1,2,2,6]$	4	3	0	6	12
$[4,2,2,2,6]$	2	6	0	6	12
$[3,3,2,2,6]$	4	3	0	6	12
$[5,1,2,4,4]$	4	3	0	6	12
$[4,2,2,4,4]$	2	6	0	6	12
$[3,3,2,4,4]$	4	3	0	6	12
$[3,1,2,4,6]$	4	6	0	12	24
$[2,2,2,4,6]$	2	6	0	6	12
$[3,1,4,4,4]$	4	1	0	2	4
$[2,2,4,4,4]$	2	1	0	1	2

Example 5.2 Let $m = 8$ and $Q = (1,1,2,2,2)$. This is an octic Calabi–Yau threefold X. The Fermat motives for the singular Calabi–Yau threefold Y are tabulated in Table 4.

Note that the dimension of each motive is either 4 or 2 and the latter occurs when $\gcd(\mathbf{a}, 8) \neq 1$.

Summing up the motivic Hodge and Betti numbers, we obtain

$$h^{3,0}(Y) = 1,\ h^{2,1}(Y) = 83 \quad \text{and} \quad B_3(Y) = \sum_A B_3(\mathcal{M}_A) = 168.$$

There is one curve singularity $C : Y_3^4 + Y_4^4 + Y_5^4 = 0$ which has genus $g(C) = 3$ with multiplicity $m_C = 1$. Passing to the crepant resolution X of Y, we then get

$$h^{3,0}(X) = 1, \quad h^{2,1}(X) = 3 + 83 = 86, \quad h^{1,1}(X) = 2$$

and

$$B_3(X) = 168 + 2 \times 3 = 174.$$

Finally the Euler characteristic is

$$E(X) = 2(2 - 86) = 2(-84) = -168.$$

Remark 5.1 This gives an alternative method for computing the global Betti numbers and Euler characteristic for the Calabi–Yau orbifolds X corresponding to admissible pairs $\langle m, Q \rangle$, which is different from the method of Vafa described in Theorem 3.1 (c).

Now we pass to the mirror partners of Calabi–Yau orbifolds. We need to bring weights into the duality.

Definition 5.1 Let X be a Calabi–Yau orbifold of degree m and weight $Q = (q_1, q_2, q_3, q_4, q_5)$. Let ζ_m denote a primitive m-th root of unity. Let $m_i = m/q_i$ and let μ_{m_i} denote the group (scheme) of m_i-th roots of unity, for each i, $0 \le i \le 5$. Let

$$\mathfrak{A}(Q) = \{\, \mathbf{a} = (a_1, a_2, a_3, a_4, a_5) \in \mathfrak{A} \mid a_i \in (q_i\mathbb{Z}/m\mathbb{Z}) \,\}.$$

Define

$$\hat{G} = \{\, \mathbf{g} = (g_1, g_2, g_3, g_4, g_5) \mid g_i \in \mu_{m_i}, \prod_{i=1}^{5} g_i = 1 \,\}/\{(g, g, \cdots, g)\,\}.$$

Then $\hat{G}$ is the group of discrete symmetry on X leaving the holomorphic 3-form Ω invariant, and $\hat{G}$ acts on the set $\mathfrak{A}(Q)$ by

$$\hat{G} \times \mathfrak{A}(Q) \to \mathbb{Q}(\zeta_m) \,:\, (\mathbf{g}, \mathbf{a}) \mapsto \prod_{i=1}^{5} g_i^{a_i} = g_1^{a_1} g_2^{a_2} \cdots g_5^{a_5}.$$

(1) Let $\mathbf{a} \in \mathfrak{A}(Q)$, and $\mathbf{g} \in \hat{G}$ be a generator of $\hat{G}$. We say that $\mathbf{a}$ is *invariant under* $\mathbf{g}$ if

$$\prod_{i=1}^{5} g_i^{a_i} = g_1^{a_1} g_2^{a_2} g_3^{a_3} g_4^{a_4} g_5^{a_5} = 1.$$

(2) Let $A = [\mathbf{a}]$ be the $(\mathbb{Z}/m\mathbb{Z})^\times$-orbit of $\mathbf{a}$ and let $\mathcal{M}_A$ be the corresponding Fermat motive. We say that $\mathcal{M}_A$ is *invariant under* $\hat{G}$ if any $\mathbf{a} \in A$ is invariant under every generator $\mathbf{g}$ of $\hat{G}$.

Example 5.3 We consider Calabi–Yau orbifolds X and their mirror partners with Kähler modulus 1, i.e., $h^{1,1}(X) = 1$. There are altogether four such Calabi–Yau orbifolds among the 147 admissible pairs. All these four Calabi–Yau orbifolds have no singularities of dimension 1.

(a) Let $m = 5$ and $Q = (1, 1, 1, 1, 1)$. We know that $h^{2,1} = 101$, $B_3 = 204$, and that $\hat{G} = (\mathbb{Z}/5\mathbb{Z})^3$ is generated by

$$\mathbf{g} = (1, 0, 0, 4, 0), (1, 0, 4, 0, 0), \quad \text{and} \quad (1, 4, 0, 0, 0).$$

Let $\mathbf{a} = (1, 1, 1, 1, 1)$. Then for $\mathbf{g} = (1, 0, 0, 4, 0)$, we have

$$\prod_{i=1}^{5} g_i^{a_i} = (\zeta_5)^1 \cdot 1^1 \cdot 1^1 \cdot (\zeta_5^4)^1 \cdot 1^1 = \zeta_5^5 = 1.$$

and similarly for the other two generators. Now consider the Fermat motive corresponding to

$$[Q] = [(1, 1, 1, 1, 1)] = \{(1, 1, 1, 1, 1), (2, 2, 2, 2, 2), (4, 4, 4, 4, 4), (3, 3, 3, 3, 3)\}.$$

We can compute that all elements in $[Q]$ are invariant under $\hat{G}$. Hence the motive $\mathcal{M}_Q$ is invariant under $\hat{G}$.

On the other hand, let $\mathbf{a} = (1, 1, 1, 3, 4)$. Then for $\mathbf{g} = (1, 0, 0, 4, 0)$, we have

$$\prod_{i=1}^{5} g_i^{a_i} = (\zeta_5)^1 \cdot 1^1 \cdot 1^1 \cdot (\zeta_5^4)^3 \cdot 1^4 = (\zeta_5)^{13} \neq 1,$$

and similarly for the other two generators. Also for $\mathbf{a} = (1,1,2,2,4)$ and $\mathbf{g} = (1,0,0,4,0)$, we have

$$\prod_{i=1}^{5} g_i^{a_i} = (\zeta_5)^1 \cdot 1^1 \cdot 1^2 \cdot (\zeta_5^4)^2 \cdot 1^4 = (\zeta_5)^9 \neq 1,$$

and similarly for the other two generators. Consequently, the Fermat motives $\mathcal{M}_A$ for $A = [(1,1,1,3,4)], [(1,1,2,2,4)]$ are not invariant under $\hat{G}$.

(b) Let $m = 6$ and $Q = (1,1,1,1,2)$. We know that $h^{2,1} = 103$, $B_3 = 208$ and that $\hat{G} = (\mathbb{Z}/3\mathbb{Z}) \times (\mathbb{Z}/6\mathbb{Z})^2$ with generators

$$\mathbf{g} = (0,2,2,2,0), (5,0,0,1,0), (0,5,0,1,0).$$

The Fermat motives for this Calabi–Yau orbifold are defined by characters

$$(1,1,1,1,2), (1,1,1,5,4), (1,1,2,4,4), (1,1,3,3,4), (1,1,3,5,2),$$

$$(1,1,4,4,2), (1,2,2,3,4), (1,2,2,5,2), (1,2,3,4,2), (1,3,3,3,2),$$

$$(2,2,2,2,4), (2,2,2,4,2), (2,2,3,3,2)$$

and they are all of dimension 2.

Among these, the Fermat motives corresponding to $(1,1,1,1,2)$ and $(2,2,2,2,4)$ are invariant under $\hat{G}$. For instance, for $\mathbf{a} = (2,2,2,2,4)$ and $\mathbf{g} = (5,0,0,1,0)$, we have

$$\prod_{i=1}^{5} g_i^{a_i} = (\zeta_6^5)^2 \cdot 1^2 \cdot 1^2 \cdot (\zeta_6)^2 \cdot 1^4 = (\zeta_6)^{12} = 1,$$

and similarly for the other generators. On the other hand, for $\mathbf{a} = (1,1,3,3,4)$ and $\mathbf{g} = (5,0,0,1,0)$, we compute

$$\prod_{i=1}^{5} g_i^{a_i} = (\zeta_6)^8 \neq 1,$$

and also for the other two generators. Therefore, the two Fermat motives, $\mathcal{M}_A$ with $A = [Q] = [(1,1,1,1,2)]$ and $A = [(2,2,2,2,4)]$ are invariant under $\hat{G}$.

(c) Let $m = 8$ and $Q = (1,1,1,1,4)$. Then $h^{2,1} = 149$, $B_3 = 300$, and that $\hat{G} = (\mathbb{Z}/2\mathbb{Z}) \times (\mathbb{Z}/8\mathbb{Z})^2$ with generators $\mathbf{g} = (7,0,0,1,0), (0,7,0,1,0), (0,0,7,1,0)$. After calculating the action of $\hat{G}$ on each Fermat motive, we see that only the motive corresponding to weight Q of dimension 4 is invariant.

(d) Let $m = 10$ and $Q = (1,1,1,2,5)$. We know that $h^{1,1} = 1 = B_2$, $h^{2,1} = 145$ and $B_3 = 284$, and that $\hat{G} = (\mathbb{Z}/10\mathbb{Z})^2$ with generators $\mathbf{g} = (9,0,1,0,0), (0,9,1,0,0)$. We compute the products $\prod_{i=1}^{5} g_i^{a_i}$ for all motives and generators and the outcome is that only the motive $\mathcal{M}_{[Q]}$ corresponding to the weight Q of dimension 4 is invariant under $\hat{G}$.

In each case of (a),(c) and (d), the Fermat motive $\mathcal{M}_{[Q]}$ corresponding to the weight Q is the only motive left invariant under the mirror map. It is of dimension 4, and we have $B_3(\hat{X}) = 4 = 2(1 + h^{1,1}(X))$, while, in the case (b), there are two Fermat motives $\mathcal{M}_{[Q]}$ and $\mathcal{M}_{[2,2,2,2,4]}$ which are invariant under the mirror map, and we have $B_3(\hat{X}) = 2 + 2 = 2(1 + h^{1,1}(X))$.

Example 5.4 We consider examples of Calabi–Yau orbifolds X and their mirror partners with Kähler modulus two, i.e., $h^{1,1}(X) = 2$. There are six such examples among our 147 Calabi–Yau orbifolds. Some of these six Calabi–Yau orbifolds have curve singularities.

(a) Let $m = 8$ and $Q = (1, 1, 2, 2, 2)$. We know that $h^{2,1} = 86$ and $B_3 = 168 + 6$ where 6 comes from the singular locus, and that $G_Q = (\mathbb{Z}/2\mathbb{Z})^3$ and $\hat{G} = (\mathbb{Z}/8\mathbb{Z})^2$ with generators $\mathbf{g} = (1, 7, 0, 0, 0), (7, 1, 0, 0, 0)$. The Fermat motives constituting the singular part of this Calabi–Yau orbifold correspond to

$$(1,1,2,2,2),(1,1,2,6,6),(1,1,4,4,6),(1,3,2,4,6),(1,3,2,2,6),(1,3,4,4,4),$$
$$(1,5,2,2,6),(1,5,2,4,4),(1,5,6,6,6),(1,7,2,2,4),$$

of dimension 4, and

$$(2,2,2,4,6),(2,2,4,4,4),(2,4,2,2,6),(2,4,2,4,4),(2,6,2,2,4),(4,4,2,2,4)$$

of dimension 2.

We compute the product $\prod_{i=1}^5 g_i^{a_i}$ for all these motives and all generators $\mathbf{g}$. For the weight $Q = (1, 1, 2, 2, 2)$ and $\mathbf{g} = (1, 7, 0, 0, 0)$, we have

$$\prod_{i=1}^{5} g_i^{a_i} = (\zeta_8)^1 \cdot (\zeta_8^7)^1 \cdot 1^2 \cdot 1^2 \cdot 1^2 = 1$$

and similarly for $\mathbf{g} = (7, 1, 0, 0, 0)$. Therefore the motive $\mathcal{M}_{[Q]}$ is invariant under $\hat{G}$. The motive $\mathcal{M}_{[Q]}$ has dimension 4. Also for $(2, 2, 4, 4, 4)$ and $\mathbf{g} = (1, 7, 0, 0, 0)$, we have

$$\prod_{i=1}^{5} g_i^{a_i} = (\zeta_8)^2 \cdot (\zeta_8^7)^2 \cdot 1^4 \cdot 1^4 \cdot 1^4 = 1$$

and similarly for $\mathbf{g} = (7, 1, 0, 0, 0)$. The motive $\mathcal{M}_{[2,2,4,4,4]}$ has dimension 2. All the remaining Fermat motives are not invariant under $\hat{G}$.

(b) For $m = 12$ and $Q = (1, 1, 2, 2, 6)$, we know that $h^{2,1} = 128$ and $B_3 = 258$. The weight motive $\mathcal{M}_{[Q]}$ has dimension 4. There is another motive of dimension 2 which is invariant under the mirror operation.

(c) For $m = 12$, $Q = (1, 2, 2, 3, 4)$, we know that $h^{2,1} = 74$, $B_3 = 150$. The weight motive $\mathcal{M}_{[Q]}$ has dimension 4 and it is the only motive which is invariant under the mirror operation.

(d) Let $m = 14$ and $Q = (1, 2, 2, 2, 7)$. Then $h^{2,1} = 122$, $B_3 = 216 + 30$ where 30 comes from the singularity, and $\hat{G} = (\mathbb{Z}/7\mathbb{Z})^2$ with generators $\mathbf{g} = (2, 2, 2, 2, 4)$ and $(4, 4, 4, 4, 2)$. We calculate that only the motive $\mathcal{M}_{[Q]}$ corresponding to the weight Q is invariant under $\hat{G}$, and it has dimension 6.

(e) Let $m = 18$ and $Q = (1, 1, 1, 6, 9)$. Then $h^{2,1} = 272$ and $B_3 = 546$, and $\hat{G} = (\mathbb{Z}/3\mathbb{Z}) \times (\mathbb{Z}/6\mathbb{Z})^2$ with generators $\mathbf{g} = (6, 6, 6, 0, 0), (3, 3, 3, 0, 9), (3, 3, 3, 3, 6)$. We calculate that there is only one motive that is left invariant under $\hat{G}$, namely, the Fermat motive $\mathcal{M}_{[Q]}$ corresponding to the weight Q, and its dimension is 6.

Theorem 5.2 *Let $(X, \hat{X})$ be a mirror pair of Calabi–Yau orbifold corresponding to an admissible pair $\langle m, Q \rangle$ with $Q = (q_1, q_2, q_3, q_4, q_5)$. Let $\mathbf{g} = (g_1, g_2, g_3, g_4, g_5)$ be any generator of $\hat{G}$. Let $\mathcal{M}_A$ be the Fermat motive corresponding to $A = [a_1, a_2, a_3, a_4, a_5]$.*

Suppose that

$$\prod_{i=1}^{5} g_i^{a_i} = 1 \;\forall\; \mathbf{a} \in A$$

then $\mathcal{M}_A$ is invariant under the mirror map.

In particular, if $h^{1,1}(X) = 1$, then the Fermat motive $\mathcal{M}_{[Q]}$ corresponding to the weight Q is the only Fermat motive left invariant under $\hat{G}$.

If $h^{1,1}(X) > 1$, there is an algorithm to determine the other Fermat motives which are invariant under the mirror map.

This theorem will be proved in Section 9 below. We would like to understand possible relations among motives, monomials, and integral points in reflexive polytopes in the toric geometric setting à la Batyrev [Ba94].

Remark 5.2 When $h^{1,1}(X) = 1$, Goto [G06] has recently showed that the formal group arising from X is invariant under the mirror map. In particular, the height h of the formal group is bounded above by 2.

6 Batyrev's mirror symmetry

Batyrev [Ba94] gives a combinatorial construction of mirror pairs of Calabi–Yau hypersurfaces in the toric geometric setting. Here is a brief summary of Batyrev's construction and the main ingredients in his theory. A nice reference on this topic might be Cox and Katz [CK99].

Let $\Delta \subset \mathbb{R}^n$ be an n-dimensional polytope, and let $\Delta^* = \mathrm{Hom}(\Delta, \mathbb{Z})$ be the dual polytope. Denote by $\langle *, * \rangle$ the nondegenerate pairing between the n-dimensional $\mathbb{R}$-vector spaces $\Delta_{\mathbb{R}}$ and $\Delta^*_{\mathbb{R}}$. Then

$$\Delta^* := \{\mathbf{y} = (y_1, y_2, \cdots, y_n) \,|\, \langle \mathbf{y}, \mathbf{x} \rangle := \sum_{i=1}^{n} y_i x_i \geq -1 \,\forall \mathbf{x} = (x_1, x_2, \cdots, x_n) \in \Delta\}.$$

The polytope Δ is said to be *reflexive* if it has the following properties:

(1) Δ is a convex integral polytope, i.e., all vertices of Δ are integral,
(2) Δ contains the origin $v_0 = (0, 0, \cdots, 0)$ as an interior point, and
(3) each codimension one face is of the form

$$\{\mathbf{x} \in \Delta_{\mathbb{R}} \,|\, \langle \mathbf{y}, \mathbf{x} \rangle = -1 \text{ for some } \mathbf{y} \in \Delta^*_{\mathbb{R}}\}.$$

If Δ is reflexive, then its dual Δ^* is again reflexive, and $(\Delta^*)^* = \Delta$. (Indeed, this fact is the basis for Batyrev's mirror symmetry.)

To a reflexive polytope Δ, one associates a complete rational fan $\Sigma(\Delta)$ as follows: For every l-dimensional face $\Theta_l \subset \Delta$, define an n-dimensional cone $\sigma(\Theta_l)$ by

$$\sigma(\Theta_l) := \{\lambda(p' - p) \,|\, \lambda \in \mathbb{R}_{\geq 0}, p \in \Delta, \, p' \in \Theta_\ell\}.$$

That is, $\sigma(\Theta_l)$ consists of all vectors $\lambda(p - p')$ where $\lambda \in \mathbb{R}_{\geq 0}$, $p \in \Delta$, $p' \in \Theta_l$. Then the fan $\Sigma(\Delta)$ is given as the collection of all $(n-l)$-dimensional dual cones $\sigma^*(\Theta_l)$ for $l = 0, 1, \cdots, n$ for all faces of Δ, and the complete fan defines the toric variety $\mathbb{P}_\Delta$. (That is, $\mathbb{P}_{\Sigma(\Delta)}$ is defined by a compactification of the algebraic torus $(\mathbb{C}^*)^n$ using combinatorial data encoded in the fan $\Sigma(\Delta)$.)

Lemma 6.1 (Batyrev [Ba94]) *Denote by v_i $(i = 0, 1, \cdots, s)$ the integral points in Δ. Consider the affine space $\mathbb{C}^{s+1}$ with smooth coordinates $\mathbf{c} = (c_0, c_1, \cdots, c_s)$. Let Z_f denote the zero locus of the Laurent polynomial*

$$f_\Delta(\mathbf{c}, \mathbf{X}) = \sum_{i=0}^{s} c_i \mathbf{X}^\mu \in \mathbb{C}[X_1^{\pm 1}, X_2^{\pm 1}, \cdots, X_n^{\pm 1}]$$

where $\mathbf{X}^\mu = X_1^{\mu_1} X_2^{\mu_2} \cdots X_n^{\mu_n}$ in the algebraic torus $(\mathbb{C}^*)^n \subset \mathbb{P}_\Delta$. Let $\bar{Z}_f$ be the closure of Z_f in $\mathbb{P}_\Delta$. The Δ-regularity conditions for hypersurfaces imply that the singularities of hypersurfaces are induced only by singularities of the ambient toric variety $\mathbb{P}_\Delta$. As a consequence, one obtains a simultaneous resolution of all members of the family of Calabi–Yau hypersurfaces, and hence one has a different family of smooth Calabi–Yau hypersurfaces X_Δ associated to Δ.

Similarly, starting with the dual reflexive polytope Δ^* and carrying out the same construction as above, one obtains a family of smooth Calabi–Yau hypersurfaces X_{Δ^*} associated to Δ^*.

We thus obtain a pair of different families of smooth Calabi–Yau hypersurfaces. The remarkable theorem of Batyrev is formulated as follows.

Theorem 6.2 (Batyrev) *If $n \leq 4$, there is the unique crepant resolution X_Δ (resp. X_{Δ^*}) of singularities for the hypersurface $\bar{Z}_{f_\Delta}$ (resp. $\bar{Z}_{f_{\Delta^*}}$).*

When $n = 4$, the resolutions X_Δ and X_{Δ^*} are mirror symmetric in the sense of Topological Mirror Symmetry Conjecture 1.1, that is,

$$h^{1,1}(X_\Delta) = h^{2,1}(X_{\Delta^*}), \quad h^{2,1}(X_\Delta) = h^{1,1}(X_{\Delta^*}).$$

The Hodge numbers are given combinatorially by

$$h^{1,1}(X_\Delta) = h^{2,1}(X_{\Delta^*}) = l(\Delta^*) - (4+1) - \sum_{\operatorname{codim} \Theta^* = 1} l'(\Theta^*) + \sum_{\operatorname{codim} \Theta^* = 2} l'(\Theta^*) l'(\Theta),$$

$$h^{2,1}(X_\Delta) = h^{1,1}(X_{\Delta^*}) = l(\Delta) - (4+1) - \sum_{\operatorname{codim} \Theta = 1} l'(\Theta) + \sum_{\operatorname{codim} \Theta = 2} l'(\Theta) l'(\Theta^*).$$

Here Θ, Θ^* are faces of Δ and Δ^*, and (Θ, Θ^*) is a dual pair. $l(\Delta)$ denotes the number of integral points in Δ, and $l(\Theta)$ (resp. $l'(\Theta)$) the number of integral points (resp. the number of integral points in the interior) of the face Θ.

Remark 6.1 The formulae for Hodge numbers give yet another method of computing Hodge numbers (and hence Betti numbers) and the Euler characteristic of our Calabi–Yau orbifolds. Via Batyrev's combinational approach, there are altogether 473800776 families of Calabi–Yau threefolds corresponding to 4-dimensional reflexive polytopes. See Kreuzer and Skarke [KS].

Our mirror pairs of Calabi–Yau hypersurfaces obtained in Section 4 by the orbifolding construction can be reformulated in the framework of Batyrev's mirror construction; Cf. Batyrev [Ba94].

Proposition 6.3 *Let $\langle m, Q \rangle$ be an admissible pair where $Q = (q_1, q_2, q_3, q_4, q_5)$ satisfies the Calabi–Yau condition: $m = q_1 + q_2 + \cdots + q_5$, or equivalently $\sum_{i=1}^{5} \frac{q_i}{m} = 1$ given in Theorem 3.1. Let*

$$\Delta(Q) = \{(x_1, x_2, x_3, x_4, x_5) \in \mathbb{Q}^5 \mid \sum_{i=1}^{5} q_i x_i = 0, \ x_i \geq -1 \ (1 \leq i \leq 5)\}.$$

Then $\Delta(Q)$ is a reflexive polytope and the toric variety $\mathbb{P}_{\Delta^*(Q)}$ is isomorphic to the weighted projective 4-space $\mathbb{P}^4(Q)$, and the Calabi–Yau threefold $X_{\Delta(Q)}$ is isomorphic to some Calabi–Yau orbifold in Theorem 4.1.

More precisely, we have the following assertions.

(a) Put $m_i = m/q_i$ for each i, $1 \le i \le 5$. Then the quasi-homogeneous equation

$$Y_1^{m_1} + Y_2^{m_2} + Y_3^{m_3} + Y_4^{m_4} + Y_5^{m_5} = 0$$

defines a $\Delta(Q)$-regular Calabi–Yau hypersurface of Fermat type in the weighted projective 4-space $\mathbb{P}^4(Q)$. Then the family of Calabi–Yau hypersurfaces consists of quotients of deformations of this hypersurface by the fundamental group $\pi_1(\Delta(Q))$. Here $\pi_1(\Delta(Q))$ is isomorphic to the kernel of the surjective homomorphism

$$(\mu_{m_1} \times \mu_{m_2} \times \mu_{m_3} \times \mu_{m_4} \times \mu_{m_5})/\mu_m \to \mu_m$$

and it has order $m^3/q_1q_2\cdots q_5$.

(b) For the dual reflexive polytope $\Delta^*(Q)$, the fundamental group $\pi_1(\Delta^*(Q))$ is isomorphic to the kernel of the surjective homomorphism

$$(\mu_{m_1} \times \mu_{m_2} \times \cdots \times \mu_{m_5})/\mu_m \to \mu_m \quad : \quad \bar{\gamma}_Q(g_1^{a_1}\cdots g_5^{a_5}) = g^{q_1a_1+\cdots q_5a_5}$$

and it has order $q_1q_2q_3q_4q_5 = \#G_Q$.

(c) In the toric geometric setting, mirror pairs of Calabi–Yau hypersurfaces are described by a pair of reflexive polytopes, that is, $\pi_1(\Delta(Q))$ and $\pi_1(\Delta^*(Q))$ are dual finite abelian groups in the sense that

$$\#\pi_1(\Delta(Q)) \times \#\pi_1(\Delta^*(Q)) = m^3$$

coinciding with the mirror duality in the orbifolding construction of Theorem 4.1.

Reflexive polytopes for Calabi–Yau hypersurfaces of Fermat type are discussed in a number of papers in the literature listing integral vertices, e.g., Batyrev [Ba94], Hosono-Lian and Yui [HLY95]. Here are some examples.

Example 6.1 In our examples of one-parameter Calabi–Yau orbifolds with $h^{1,1} = 1$ in Example 4.1, we have $q_1 = 1$ and $q_2 \le q_3 \le q_4 \le q_5$. The convex polytope $\Delta(Q)$ is the convex hull of the integral vertices given as follows

$$(0,0,0,0), (-1,-1,-1,-1), ((\frac{m}{q_2}) - 1, -1, -1, -1), (-1, (\frac{m}{q_3} - 1), -1, -1)$$

$$(-1,-1,(\frac{m}{q_4} - 1), -1), (-1,-1,-1,(\frac{m}{q_5} - 1)).$$

The dual polytope $\Delta^*(Q)$ is the convex hull of the following vectors

$$(0,0,0,0), (1,0,0,0), (0,1,0,0), (0,0,1,0), (0,0,0,1), (-q_2,-q_3,-q_4,-q_5).$$

Example 6.2 For the two-parameter families of Calabi–Yau orbifolds with $h^{1,1} = 2$ in Example 4.2, we again have $q_1 = 1$ and $q_2 \le q_3 \le q_4 \le q_5$. We list the dual polytopes for these examples in Table 5.

Proposition 6.4 *For our mirror paris of Calabi–Yau hypersufaces constructed in Theorem 4.1 and Proposition 6.3, the origin is the only integral point present in both the reflexive polytope $\Delta(Q)$ and its dual reflexive polytope $\Delta^*(Q)$.*

Table 5 Vertices for dual polytopes

m	Q	$\Delta^*(Q)$
8	$(1,1,2,2,2)$	$(0,0,0,0),(1,0,0,0),(0,1,0,0),(0,0,1,0),$ $(0,0,0,1),(-1,-2,-2,-2),(-1,-1,-1,0)$
12	$(1,1,2,2,6)$	$(0,0,0,0),(1,0,0,0),(0,1,0,0),(0,0,1,0),$ $(0,0,0,1),(-1,-2,-2,-6),(-3,-1,-1,0)$
12	$(1,2,2,3,4)$	$(0,0,0,1),(1,0,0,0),(0,1,0,0),(0,0,1,0),$ $(0,0,0,1),(-2,-2,-3,-4),(-1,-1,-1,-2)$
14	$(1,2,2,27)$	$(0,0,0,0),(1,0,0,0),(0,1,0,0),(0,0,1,0),$ $(0,0,0,1),(-2,-2,-2,-7),(-1,-1,-1,-3)$
18	$(1,1,1,6,9)$	$(0,0,0,0),(1,0,0,0),(0,1,0,0),(0,0,1,0),$ $(0,0,0,1),(-1,-1,-6,-9),(0,0,-2,-3)$

In Batyrev's mirror symmetry, the essential information is encoded in the *Newton polyhedron* of the defining equation of the Calabi–Yau hypersurface, which is the convex hull of the monomials appearing in the hypersurface. Suppose that a family of Calabi–Yau hypersurfaces in $\mathbb{P}^4(Q)$ is the zero locus of the generic equation

$$f(\mathbf{c},\mathbf{X})=0\subset\mathbb{P}^4(Q)$$

of degree m and weight $Q=(q_1,q_2,q_3,q_4,q_5)$ with deformation parameters. The Newton polyhedron of f always contains the monomial $x_1\cdots x_5$. Corresponding to this monomial, we have the integral point $(1,1,\cdots,1)$. Translate it to the origin. Then the Newton polyhedron of f is given by

$$\Delta(Q)=\text{convex hull of}\,\{(x_1,x_2,x_3,x_4,x_5)\in\mathbb{Q}^5\mid\sum_{i=1}^{5}q_ix_i=0,\ x_i\geq -1\,(1\leq i\leq 5)\,\}.$$

The shifted by $(1,1,\cdots,1)$ convex polytope is the intersection of $\mathbb{Q}^5$ with the affine hyperplane

$$q_1x_1+q_2x_2+\cdots+q_5x_5=q_1+q_2+\cdots+q_5=m.$$

Since $X_{\Delta(Q)}$ is isomorphic to some of our Calabi–Yau orbifolds in Theorem 4.1, we obtain the following result.

Theorem 6.5 *There is a one-to-one correspondence between integral points in $\Delta(Q)$ and monomials of degree m in the graded polynomial ring*

$$\mathcal{R}=\mathbb{C}[X_1,X_2,X_3,X_4,X_5]$$

of a Calabi–Yau orbifold X associated to an admissible pair $\langle\, m,Q\,\rangle$.

Remark 6.2 In fact, this is a concrete realization of the more general high-power monomial–divisor mirror map constructed by Batyrev [Ba94] and Aspinwall–Greene and Morrison [AGM93] giving the isomorphism between the subspaces $H^{1,1}_{toric}(X)$ and $H^{2,1}_{poly}(\hat{X})$. The space $H^{2,1}_{poly}(\hat{X})\subset H^{2,1}(\hat{X})$ is isomorphic to the first-order polynomial deformations, and can be generated by monomials. Its dimension $h^{2,1}_{poly}(\hat{X})$ is equal to the number of generators of the Mori cone (which is dual to the Kähler cone of the mirror X). The generators of the Mori cone correspond to divisor classes which generate the subspace $H^{1,1}_{toric}(X)$. This gives rise to a natural isomorphism

$$H^{1,1}_{toric}(X)\simeq H^{2,1}_{poly}(\hat{X})$$

and such an isomorphism can be interpreted as the differential of the expected mirror map between the moduli spaces.

7 Monomials and periods

We will briefly describe the Dwork–Katz–Griffiths reduction method [CK99] for determining the Picard–Fuchs equations for hypersurfaces in a weighted projective space $\mathbb{P}^4(Q)$ with weight $Q = (q_1, q_2, q_3, q_4, q_5)$. The Picard–Fuchs differential equation is the device which establishes the one-to-one correspondence between monomials and periods.

Put $\mathbf{X} := (X_1, X_2, X_3, X_4, X_5)$. Suppose that a hypersurface

$$P(\mathbf{X}) := P(X_1, \cdots, X_5) = 0$$

defines a Calabi–Yau threefold X of degree m and weight $Q = (q_1, q_2, \cdots, q_5)$, and let $P_{\mathbf{c}}(\mathbf{X})$, or in short, $P_{\mathbf{c}}$, be a Calabi–Yau threefold $X_{\mathbf{c}}$ with a deformation parameter $\mathbf{c}$ (in practice, these parameters are complex structure moduli.) Let $\Omega_{\mathbf{c}}$ denote the unique (up to scalar multiplication) holomorphic 3-form on $X_{\mathbf{c}}$. Then the period $\Pi(\mathbf{c})$ can be written as

$$\Pi(\mathbf{c}) = \int_{\Gamma_i} \Omega_{\mathbf{c}} = \int_{\gamma} \int_{\Gamma} \frac{\omega}{P_{\mathbf{c}}} \tag{1}$$

where

$$\omega = \sum_{i=1}^{5} (-1)^i q_i X_i dX_1 \wedge \ldots \wedge d\hat{X}_i \wedge \ldots \wedge dX_5$$

is the unique (up to scalars) holomorphic 3-form on $X_{\mathbf{c}}$. Here $\Gamma_i \in H_3(X_{\mathbf{c}}, \mathbb{Z})$ is a topological 3-cycle, and γ is a small curve around the hypersurface $P = 0$ in the 4-dimensional ambient space. Since $H_3(X, \mathbb{Z})$ has rank $B_3(X)$, there are in total $B_3(X)$ periods.

Definition 7.1 Let $\mathbb{C}[X_1, X_2, X_3, X_4, X_5]$ be the weighted polynomial ring with weight $Q = (q_1, q_2, q_3, q_4, q_5)$. Let $\mathbf{X}^{\mathbf{v}} = X_1^{v_1} X_2^{v_2} \cdots X_5^{v_5}$ be a monomial in $\mathbb{C}[X_1, X_2, \cdots, X_5]$. The *degree* of $\mathbf{X}^{\mathbf{v}}$, denoted by $w(\mathbf{v})$, is defined by

$$w(\mathbf{v}) = \sum_{i=1}^{5} q_i v_i = q_1 v_1 + q_2 v_2 + \cdots + q_5 v_5.$$

We see that there are in total $B_3(X)$ monomials, which are divided into monomials of degree $w(\mathbf{v}) = 0, m, 2m, 3m$ with cardinalities $1, h^{2,1}, h^{1,2}, 1$, respectively.

Proposition 7.1 *Let $\mathbf{X}^{\mathbf{v}} = X_1^{v_1} X_2^{v_2} \cdots X_5^{v_5}$ be a monomial of degree $w(\mathbf{v}) = \sum_{i=1}^{5} q_i v_i$ in the weighted polynomial ring $\mathbb{C}[X_1, X_2, X_3, X_4, X_5]$ with weight $(q_1, q_2, \cdots, q_5)$. Then there is a one-to-one correspondence between monomials and periods. That is, (using the notation of [CORV00],[CORV03]) to every monomial $\mathbf{X}^{\mathbf{v}}$, there corresponds a period*

$$\Pi_{\mathbf{v}} := \int_{\Gamma_{\mathbf{v}}} \frac{\mathbf{X}^{\mathbf{v}}}{P_{\mathbf{c}}^{w(\mathbf{v})+1}}$$

where $\Gamma_{\mathbf{v}}$ is a topological cycle corresponding to $\mathbf{v}$ in $H_3(\mathbf{X}_{\mathbf{c}}, \mathbb{Z})$.

Our examples of mirror pairs of Calabi–Yau threefolds in weighted projective 4-spaces are equipped with large automorphism groups. If $(X, \hat{X})$ is a mirror pair of families of Calabi–Yau threefolds corresponding to $\langle m, Q \rangle$, then the automorphism group consists of $\hat{G}$ and permutations of the weighted projective coordinates. We may choose a set of basis elements for the set of monomials under this action.

The action of the automorphisms on monomials can be transferred by Proposition 7.1 to the actions of partial differential operators on periods, and under these actions modulo the exact ones, periods decompose into the disjoint union of orbits.

Observe that $\frac{\partial}{\partial X_i}\left(\frac{g(\mathbf{X})}{P_{\mathbf{c}}^r}\right)\omega$ is exact if $g(\mathbf{X})$ is homogeneous with degree such that the whole expression has degree zero. This leads to the partial integration rule

$$\frac{h\,\partial_i P}{P_{\mathbf{c}}^r} = \frac{1}{r-1}\frac{\partial_i h}{P_{\mathbf{c}}^{r-1}}$$

with $\partial_i = \frac{\partial}{\partial X_i}$. One then takes derivatives of the expressions

$$\Pi_{\mathbf{v}} = \int_\Gamma \frac{\mathbf{X}^{\mathbf{v}}}{P_{\mathbf{c}}^{w(\mathbf{v})+1}}$$

with respect to the moduli $\mathbf{c}$ of the defining equation.

If one produces an expression such that the numerator in the integrand is not one of the basis elements, one relates it, using the equations $\partial_i P = \ldots$, to the basis and uses the relation given above. This leads to a system of first order differential equations (known as 'Gauss–Manin equations') for the $\Pi_{\mathbf{v}}$ which can be rewritten as a system of partial differential equations for the period, which is the Picard–Fuchs equations:

$$\partial_{c_k}\Pi(\mathbf{c}) = M^{(k)}(a)\Pi(\mathbf{c}),\ k = 1,\ldots,h^{2,1}.$$

Furthermore, the Picard–Fuchs differential equations (or equivalently, the periods) are equipped with the following operators:

$$\mathcal{D}_i\left(\frac{\mathbf{X}^{\mathbf{v}}}{P_{\mathbf{c}}^{w(\mathbf{v})+1}}\right) := \partial_i\left(X_i\frac{\mathbf{X}^{\mathbf{v}}}{P_{\mathbf{c}}^{w(\mathbf{v})+1}}\right)$$

for $i = 1, 2, \cdots, 5$. Since

$$\int_\Gamma d^5\mathbf{X}\mathcal{D}_i\left(\frac{\mathbf{X}^{\mathbf{v}}}{P_{\mathbf{c}}^{w(\mathbf{v})+1}}\right) = 0,$$

this gives rise to relations (identities) between the differential form associated to the monomial $\mathbf{X}^{\mathbf{v}}$ and those associated to $\mathbf{X}^{\mathbf{v}}$ multiplied by $\hat{G}$-invariant monomials. Hence under the action of $\mathcal{D}_i$ $(i = 1, 2, \cdots, 5)$, the Picard–Fuchs differential equation (and hence the period) decomposes into the product of differential equations.

Summing up, the Picard–Fuchs differential equation is determined by those complex structure parameters for which there is a monomial perturbation in the defining equation, and there are $h^{2,1}_{poly}$ of these. The number of deformation parameters is equal to the number of generators of the Mori cone (which is dual to the Kähler cone) in the mirror. The generators of the Mori cone come from relations between the divisors. In this way we get the secondary fan which describes the complex structure moduli space in terms of "large structure coordinates".

Based on the theory of Dwork–Katz–Griffiths, these operators are explicitly defined for the one-parameter family of quintic Calabi–Yau threefolds by Candelas et al. [CORV00, CORV03], and for the two-parameter octic family of Calabi–Yau threefolds by Kadir [Ka05]. Also, they have the so-called Picard–Fuchs diagrams to illustrate the decomposition of the Picard–Fuchs differential equation and the

Table 6 Monomials for the quintic Calabi–Yau threefold

#	Monomial $\mathbf{v}$	deg $w(\mathbf{v})$	mult	length of orbit
1	$(0,0,0,0,0)$	0	1	5
2	$(4,1,0,0,0)$	5	20	5
3	$(3,2,0,0,0)$	5	20	5
4	$(3,1,1,0,0)$	5	30	5
5	$(2,2,1,0,0)$	5	30	5
6	$(4,0,3,2,1)$	10	24	1

periods. Clearly, such decompositions ought to have their origin in the motivic decomposition of the manifolds. We obtain the following result:

Theorem 7.2 *(a) Periods are divided into disjoint set of equivalence classes under the action of the partial differential operators $\mathcal{D}_i$ $(i = 1, 2, \cdots, 5)$.*

(b) Equivalently, the Picard–Fuchs differential equation decomposes into the product of the Picard–Fuchs differential equations corresponding to equivalence classes.

(c) There is a one-to-one correspondence between the set of monomial classes under the action of $\hat{G}$-invariant monomials and the equivalence classes of periods under the action of the $\mathcal{D}_i$ operators.

8 The monomial–motive correspondence: Examples

In this section, we will formulate our main findings, namely, a one-to-one correspondence between motives and monomials for mirror pairs of Calabi–Yau hypersurfaces in weighted projective 4-spaces. This correspondence is established at the Fermat points in the moduli spaces.

First we illustrate the correspondence by two examples. A monomial $X_1^{v_1}X_2^{v_2}X_3^{v_3}X_4^{v_4}X_5^{v_5}$ is represented by the exponent $\mathbf{v} = (v_1, v_2, v_3, v_4, v_5)$ and $(0,0,0,0,0)$ represents the constant.

Example 8.1 Consider the quintic hypersurfaces :

$$X_1^5 + X_2^5 + X_3^5 + X_4^5 + X_5^5 - 5\psi X_1X_2X_3X_4X_5 = 0 \subset \mathbb{P}^4$$

with a deformation parameter ψ. So it has the weight $Q = (1,1,1,1,1)$. There are six classes of monomials (cf. Candelas et al. [CORV00, CORV03]). Here the multiplicity is the number of permutations.

In the set of monomials, the $\hat{G}$-invariant monomials are

$$X_i^5 \ (1 \le i \le 5) \quad \text{and} \quad X_1X_2X_3X_4X_5.$$

The (repeated) multiplication by the $\hat{G}$-invariant monomial $X_1X_2\cdots X_5 = (1,1,1,1,1)$ to each monomial and reducing each component modulo 5 establishes the monomial–motive correspondence. The correspondences are:

$$[\mathbf{a}] = [a_1, a_2, a_3, a_4, a_5] \mapsto \text{class of} \quad X_1^{a_1/5}X_2^{a_2/5}X_3^{a_3/5}X_4^{a_4/5}X_5^{a_5/5}$$

and

$$\mathbf{v} = (v_1, v_2, v_3, v_4, v_5) \mapsto [\mathbf{v} + t(1,1,1,1,1)] \in \mathfrak{A}(Q) \quad \text{for } t \in (\mathbb{Z}/5\mathbb{Z})^\times.$$

For instance, we have

Table 7 Monomial–motive correspondence for the quintic Calabi–Yau threefold

#	monomial $\mathbf{v}$	deg $R([\mathbf{v}],t)$	$\lambda_{\mathbf{v}}$	motive	dim	mult
1	(0,0,0,0,0)	4	1	[1,1,1,1,1]	4	1
2	(4,1,0,0,0)	4	20	[1,1,1,3,4]	4	20
3	(3,2,0,0,0)					
4	(3,1,1,0,0)	4	30	[1,1,2,2,4]	4	30
5	(2,2,1,0,0)					
6	(4,0,3,2,1)	0	24	?	?	

Table 8 Monomials for the octic Calabi–Yau threefold

#	monomial $\mathbf{v}$	deg $w(\mathbf{v})$	mult	#	monomial	deg	mult
1	(0,0,0,0,0)	0	1	9	(0,4,0,3,3)	16	3
2	(0,2,1,1,1)	8	2	10	(4,0,1,1,0)	8	3
3	(6,2,0,0,0)	8	1	11	(2,0,3,0,0)	8	6
4	(0,0,0,2,2)	8	3	12	(6,0,1,0,0)	8	6
5	(2,0,1,3,3)	16	6	13	(0,0,3,1,0)	8	6
6	(4,0,2,0,0)	8	3	14	(2,0,2,1,0)	8	6
7	(0,0,2,1,1)	8	3	15	(4,0,3,2,1)	16	6
8	(2,2,1,1,0)	8	3				

$$(0,0,0,0,0) \mapsto (1,1,1,1,1) \leftrightarrow [1,1,1,1,1]$$
$$(4,1,0,0,0) \mapsto (1,3,2,2,2) \mapsto (3,4,1,1,1) \leftrightarrow [1,1,1,3,4]$$
$$(3,2,0,0,0) \mapsto (4,3,1,1,1) \leftrightarrow [1,1,1,3,4]$$
$$(3,1,1,0,0) \mapsto (4,2,2,1,1) \leftrightarrow [1,1,2,2,4]$$
$$(2,2,1,0,0) \mapsto (3,3,2,1,1) \mapsto (1,1,4,2,2) \leftrightarrow [1,1,2,2,4]$$

The length of the orbit of each monomial of these actions is listed in the last column of the table.

The monomial $(4,0,3,2,1)$ only contributes at a conifold point, and hence it is not realizable at the Fermat point.

The new notation introduced in Table 7 and Table 8 ought to be explained: For a monomial $\mathbf{v}$, $R([\mathbf{v}],t)$ denotes the polynomial factor corresponding to the orbit $[\mathbf{v}]$ and $\lambda_{\mathbf{v}}$ its degree.

In Section 4, we observed that the motive corresponding to the weight $[1,1,1,1,1]$ is the only motive that is invariant under the mirror symmetry operation by the group $\hat{G}$. Also, in Section 6, we showed in the toric geometric setting, that the only the origin is left invariant under the mirror map. The monomial–motive correspondence is consistent with these two facts.

Example 8.2 Consider the octic Calabi–Yau hypersurfaces in the weighted projective 4-space $\mathbb{P}^4(Q)$ with $Q = (1,1,2,2,2)$ with two deformation parameters:

$$X_1^8 + X_2^8 + X_3^4 + X_4^4 + X_5^4 - 2\phi X_1^4 X_2^4 - 8\lambda X_1 X_2 X_3 X_4 X_5 = 0$$

in $\mathbb{P}^4(Q)$. In this case, there are 15 classes of monomials (see Kadir [Ka05]).

Here again the monomial $(4,0,3,2,1)$ only contributes at a conifold point, and hence is not realizable at the Fermat point.

In this case, the $\hat{G}$-invariant monomials are

$$X_1^8, X_2^8, X_3^4, X_4^4, X_5^4$$

and

$$X_1^4 X_2^4 \quad \text{and} \quad X_1 X_2 X_3 X_4 X_5.$$

Given a monomial $\mathbf{v} = (v_1, v_2, v_3, v_4, v_5)$, adding $(1,1,1,1,1)$ or $(4,4,0,0,0)$ repeatedly and then reducing the first two components modulo 8 and the last three components modulo 4 yields the monomial–motive correspondence. The correspondences are given as follows with $Q = (q_1, q_2, q_3, q_4, q_5) = (1,1,2,2,2)$.

$$[\mathbf{a}] = [a_1, a_2, a_3, a_4, a_5] \mapsto \text{class of } X_1^{a_1/q_1} X_2^{a_2/q_2} X_3^{a_3/q_3} X_4^{a_4/q_4} X_5^{a_5/q_5}$$

and

$$\mathbf{v} = (v_1, v_2, v_3, v_4, v_5) \mapsto [\mathbf{v} + \prod (1,1,2,2,2)^t (4,4,0,0,0)^s] \in \mathfrak{A}(Q)$$

for some $t, s \in \mathbb{N}$. Here are some examples. First we describe the motives to monomials correspondence.

$$[1,1,2,2,2] \mapsto X_1 X_2 X_3^{2/2} X_4^{2/2} X_5^{2/2} = \prod_{i=1}^{5} X_i \mapsto (0,0,0,0,0)$$

$$[2,2,4,4,4] \mapsto X_1^2 X_2^2 X_3^{4/2} X_4^{4/2} X_5^{4/2} = (\prod_{i=1}^{5} X_i)^2 \mapsto (0,0,0,0,0)$$

$$[5,1,2,4,4] \mapsto X_1^5 X_2 X_3^{2/2} X_4^{4/2} X_5^{4/2} = \prod_{i=1}^{5} X_i (X_1^4 X_4 X_5) \mapsto (4,0,0,1,1)$$

$$[3,1,2,4,6] \mapsto X_1^3 X_2 X_3 X_4^2 X_5^3 = (\prod_{i=1}^{5} X_i)(X_1^2 X_4 X_5^2) \mapsto (2,0,0,1,2)$$

Conversely, here are examples of the monomials to motives correspondence.

$$(6,2,0,0,0) \mapsto [7,3,2,2,2]$$

$$(3,5,2,1,1) \mapsto (7,9,2,1,1) \mapsto (7,1,2,1,1) \mapsto [7,1,4,2,2]$$

In the above examples, the monomial–motive correspondences are obtained by combinatorial matching of monomials and motives. In the rest of this section we will establish this correspondence mathematically.

First we re-access the situation. Let $\mathcal{R} = \mathbb{C}[X_1, X_2, X_3, X_4, X_5]$ be the weighted polynomial ring of the family of Calabi–Yau hypersurfaces in weighted projective 4-space. Let $\mathcal{S}$ denote the set of all monomials $\mathbf{X}^{\mathbf{v}} = X_1^{v_1} X_2^{v_2} \cdots X_5^{v_5}$. Then $\mathcal{S}$ has group of automorphisms consisting of $\hat{G}$ and permutations of weighted coordinates. Under the action of automorphisms, $\mathcal{S}$ can be decomposed into the disjoint sets of orbits of monomials, which was introduced earlier in Theorem 7.2 and called a monomial class. A monomial class consists of monomials $\mathbf{X}^{\mathbf{v}}$ and its multiples by $\hat{G}$-invariant monomials. We may employ the notation $([v_1], [v_2], [v_3], [v_4], [v_5])$ for the monomial class.

Since periods are in one-to-one correspondence with monomials, the decomposition of monomials into the set of orbits under the action of automorphisms carries over to periods. On the set of periods, there are the differential operators

Table 9 Monomial–motive correspondence for the octic Calabi–Yau threefold

#	Monomial $\mathbf{v}$	deg $R([\mathbf{v}],t)$	$\lambda_{\mathbf{v}}$	motive	dim	mult
1	$(0,0,0,0,0)$	6	1	$[1,1,2,2,2]$	4	1
				$[2,2,4,4,4]$	2	1
2	$(0,2,1,1,1)$	4	2	$[3,1,4,4,4]$	4	1
				$[6,4,2,2,2]$	2	2
3	$(6,2,0,0,0)$	4	1	$[7,3,2,2,2]$	4	1
4	$(0,0,0,2,2)$	4	3	$[3,3,2,2,6]$	4	3
5	$(2,0,1,3,3)$	2	6	$[4,2,2,2,6]$	2	6
6	$(4,0,2,0,0)$	4	3	$[5,1,2,2,6]$	4	3
7	$(0,0,2,1,1)$	3	3	$[4,4,2,2,4]$	2	3
8	$(2,2,1,1,0)$	3	3	$[3,3,2,2,4]$	4	3
9	$(0,4,0,3,3)$	4	3	$[6,2,2,2,4]$	2	6
10	$(4,0,1,1,0)$	4	3	$[5,1,2,2,4]$	4	3
11	$(2,0,3,0,0)$	3	6	$[4,2,2,4,4]$	2	6
12	$(6,0,1,0,0)$	3	6	$[7,1,2,2,4]$	4	6
13	$(0,0,3,1,0)$	2	6	$[2,2,2,4,6]$	2	6
14	$(2,0,2,1,0)$	4	6	$[3,1,2,4,6]$	2	12
15	$(4,0,3,2,1)$	0	6	?	?	?

$\mathcal{D}_i$ described in Section 7. Under these differential operators, periods decompose into the product of subperiods, which are in one-to-one correspondence with the disjoint sets of orbits of monomials.

Theorem 8.1 *For a Calabi–Yau threefold X associated to an admissible pair $\langle m, Q\rangle$ with $Q=(q_1,q_2,q_3,q_4,q_5)$, there is a one-to-one correspondence at the Fermat point between monomial classes and Fermat motives. The correspondences are given as follows:*

A motive $\mathcal{M}_A$ with $A=[\mathbf{a}]=[a_1,a_2,a_3,a_4,a_5]$ and $\mathbf{a}\in\mathfrak{A}(Q)$ corresponds to the monomial class

$$X_1^{a_1/q_1}X_2^{a_2/q_2}X_3^{a_3/q_3}X_4^{a_4/q_4}X_5^{a_5/q_5}.$$

Conversely, a monomial class $X_1^{v_1}X_2^{v_2}X_3^{v_3}X_4^{v_4}X_5^{v_5}$ with $([v_1],[v_2],[v_3],[v_4],[v_5])$ representing the class of monomials, corresponds to the motive $\mathcal{M}_A$ with

$$A=[[v_1]q_1,[v_2]q_2,[v_3]q_3,[v_4]q_4,[v_5]q_5]\in\mathfrak{A}(Q).$$

(Note that the constant monomial class contains the monomial $(1,1,1,1,1)$ which corresponds to the weight motive $\mathcal{M}_{[Q]}$ with $Q=(q_1,q_2,q_3,q_4,q_5)$.)

9 Proof of the monomial–motive correspondence

We will prove Theorem 8.1 by comparing the two expressions for the congruence zeta-function of X over $\mathbb{F}_q$ obtained in two different ways. One is based on Weil's method and involves Fermat motives. The other is based on Dwork's method and involves monomials. We will show that the two approaches agree at the Fermat point giving the same congruence zeta-function.

Let X be a Calabi–Yau threefold defined over $\mathbb{Q}$. We wish to compute the number of $\mathbb{F}_q$-rational points on the reduction $X_{\mathbb{F}_q}$ of X to the finite field $\mathbb{F}_q$ of characteristic p. (We need to choose a "good prime" p; for our examples, good primes are those primes $p\nmid m$.) Denote by $\#X(\mathbb{F}_q)$ the number of $\mathbb{F}_q$-rational points on $X_{\mathbb{F}_q}$. The congruence zeta-function of $X_{\mathbb{F}_q}$ is defined by organizing the numbers $\#X_{\mathbb{F}_{q^r}}$ for all integers $r \geq 1$:

$$Z(X_{\mathbb{F}_q}, t) = \exp\left(\sum_{r=1}^{\infty} \#X_{\mathbb{F}_{q^r}} \frac{t^r}{r}\right).$$

Let $\bar{\mathbb{F}}_q$ denote the algebraic closure of $\mathbb{F}_q$ and write $\bar{X} = X \otimes_{\mathbb{F}_q} \bar{\mathbb{F}}_q$. Choose a prime $\ell \neq p$. Then by the work developed by Grothendieck and taken to the summit by Deligne [D74], $Z(X_{\mathbb{F}_q}, t)$ has an interpretation in terms of ℓ-adic étale cohomology groups $H^i_{et}(\bar{X}, \mathbb{Q}_\ell)$ as

$$Z(X_{\mathbb{F}_q}, t) = \prod_{i=0}^{6} P_i(X_{\mathbb{F}_q}, t)^{(-1)^{i-1}}$$

where

$$P_i(X_{\mathbb{F}_q}, t) := \det\left(1 - t\,\mathrm{Frob}_q \mid H^i(\bar{X}, \mathbb{Q}_\ell)\right) \in 1 + \mathbb{Z}[t]$$

is an integral polynomial of degree equal to the i-th Betti number $B_i(X)$, and it satisfies an analogue of the Riemann Hypothesis, namely, its reciprocal roots are algebraic integers with absolute value $q^{i/2}$.

For diagonal hypersurfaces, the congruence zeta-functions were explicitly described in Weil [We49]. The numbers $\#X(\mathbb{F}_{q^r})$ were computed using Jacobi sums and Gauss sums. For our Calabi–Yau orbifolds associated to admissible pairs $\langle m, Q \rangle$, the congruence zeta-functions are of the form

$$Z(X_{\mathbb{F}_q}, t) = \frac{P_3(X_{\mathbb{F}_q}, t)}{(1-t)(1-qt)^{h^{1,1}(X)}(1-q^2t)^{h^{2,2}(X)}(1-q^3t)}.$$

Only the middle cohomology group H^3 (and hence P_3) contains essential information.

Proposition 9.1 (Yui [Y05]) *Let X be a Calabi–Yau orbifold corresponding to an admissible pair $\langle m, Q \rangle$ with $Q = (q_1, q_2, q_3, q_4, q_5)$. Let q be the (smallest) power of prime p such that $q \equiv 1 \pmod{m}$. Then*

$$P_3(X_{\mathbb{F}_q}, t) = \prod_{\mathbf{a} \in \mathfrak{A}(Q)} (1 - J(\mathbf{a})\, t)$$

with

$$\mathfrak{A}(Q) = \{\mathbf{a} \in (\mathbb{Z}/m\mathbb{Z})^5 \mid a_i \neq 0, \sum_{i=1}^{5} a_i \equiv 0 \pmod{m}\}.$$

Here $J(\mathbf{a})$ is the Jacobi sum defined as

$$J(\mathbf{a}) = (-1)^3 \sum \chi(x_1)^{a_1}\chi(x_2)^{a_2}\chi(x_3)^{a_3}\chi(x_4)^{a_4}\chi(x_5)^{a_5}$$

where $\chi : \mathbb{F}_q \to \mu_m$ is a multiplicative character of order m, and the sum runs over $(x_1, x_2, \cdots, x_5) \in (\mathbb{F}_q^*)^5$ such that $\sum_{i=1}^5 x_i = 0 \in \mathbb{F}_q$.

Furthermore, the Jacobi sum is expressed as the product of Gauss sums as follows:

$$J(\mathbf{a}) = (-1)^3 \frac{1}{q} G(\chi^{a_1}) G(\chi^{a_2}) \cdots G(\chi^{a_5})$$

where $G(\chi)$ is the Gauss sum defined by

$$G(\chi) = G(\chi, \psi) := \sum_{x \in \mathbb{F}_q^*} \chi(x)\psi(x)$$

with ψ an additive character of $\mathbb{F}_q$.

(If we use a different normalization by putting a minus sign in front of the sum in the definition of Gauss sums, then $G(\chi_1, \phi) = 1$ for the trivial character χ_1. In this case, the minus sign in front of $J(\mathbf{a})$ can be dropped. There are two types of normalizations in the literature.)

Furthermore, $P_3(X, t)$ factors into the product:

$$P_3(X_{\mathbb{F}_q}, t) = \prod_{A \in O(\mathfrak{A}(Q))} P_3(\mathcal{M}_A, t)$$

where

$$P_3(\mathcal{M}_A, t) = \prod_{\mathbf{a} \in A} (1 - J(\mathbf{a})\, t) \in 1 + \mathbb{Z}[t] \quad \textit{with } \deg P_3(\mathcal{M}_A, t) = B_3(\mathcal{M}_A).$$

By Weil [We49], we know that Gauss sums are algebraic integers with absolute value $q^{1/2}$ in the field $\mathbb{Q}(\zeta_m, \zeta_q)$ over $\mathbb{Q}$, where ζ_m (resp. ζ_q) is a primitive m-th (resp. q-th) root of unity. Since Jacobi sums are expressed as the product of conjugate Gauss sums over $\mathbb{Q}(\zeta_m)$, Jacobi sums $J(\mathbf{a})$ are algebraic integers in $\mathbb{L} = \mathbb{Q}(\zeta_m)$ with absolute value $q^{3/2}$. Weil's formula for $\#X(\mathbb{F}_q)$ given in terms of Gauss and Jacobi sums.

Lemma 9.2 (Weil's formula) *Let X be a Calabi–Yau threefold corresponding to an admissible pair $\langle m, Q \rangle$. Then the number of $\mathbb{F}_q$-rational points on X is given by the formula:*

$$\#X(\mathbb{F}_q) = 1 + q + q^2 + q^3 - \sum_{\mathbf{a} \in \mathfrak{A}(Q)} J(\mathbf{a}) = 1 + q + q^2 + q^3 + \frac{1}{q} \sum_{\mathbf{a} \in \mathfrak{A}(Q)} \prod_{i=1}^{5} G(\chi^{a_i}).$$

We may further factor the last term into sums involving Fermat motives as

$$N_{motive} := \sum_{\mathbf{a} \in \mathfrak{A}(Q)} J(\mathbf{a}) = \sum_{A \in O(\mathfrak{A}(Q))} \sum_{\mathbf{a} \in A} J(\mathbf{a}).$$

Now we consider a mirror partner $\hat{X}$ of a Calabi–Yau orbifold X corresponding to $\langle m, Q \rangle$, and its congruence zeta-function at the Fermat point.

Theorem 9.3 *Let $(X, \hat{X})$ be a mirror pair of Calabi–Yau orbifolds over $\mathbb{Q}$ corresponding to an admissible pair $\langle m, Q \rangle$ constructed in Section 4. Let $X_{\mathbb{F}_q}$ (resp. $\hat{X}_{\mathbb{F}_q}$) be the reduction of X (resp. $\hat{X}$) to the finite field $\mathbb{F}_q$ of characteristic p (where p is a good prime). Then the congruence zeta-function of $\hat{X}$ at the Fermat point is given as follows:*

$$Z(\hat{X}_{\mathbb{F}_q}, t) = \frac{P_3(\hat{X}_{\mathbb{F}_q}, t)}{(1-t)(1-qt)^{h^{2,1}}(1-q^2t)^{h^{1,2}}(1-q^3t)}$$

with

$$P_3(\hat{X}_{\mathbb{F}_q}, t) = \prod_{A} P_3(\mathcal{M}_A, t)$$

where the product runs over all A such that corresponding Fermat motives $\mathcal{M}_A$ are invariant under the mirror map (i.e., the action of $\hat{G}$).

There is another method of calculating the number of $\mathbb{F}_q$-rational points on the family of Calabi–Yau hypersurfaces. This method was developed by Dwork [D60]. The number of $\mathbb{F}_q$-rational points is computed using the p-adic analytic method, namely, the p-adic Gamma function, the Dwork character, and the Gross–Koblitz formula. This method works for more general hypersurfaces, e.g., deformations of diagonal hypersurfaces.

For this, we review the Dwork character, Gauss sums, and the Gross–Koblitz formula from Candelas et al. [CORV00] (tailored especially for physicists).

Let $\mathbb{F}_p$ be the prime field of p elements and let $\mathbb{Z}_p^* \simeq \mu_{p-1}$ be the group of $(p-1)$-st roots of unity. The Teichmüller character $\omega_p : \mathbb{F}_p^* \to \mathbb{Z}_p^*$ is a multiplicative character of $\mathbb{F}_p^*$ of order $p-1$, so that $\omega_p(x) \equiv x \pmod{p}$. Let π be an element in the algebraic closure $\bar{\mathbb{Q}}_p$ such that $\pi^{p-1} = -p$, and let F be the function defined by $F(x) = \exp(X + X^p/p)$. The Dwork character Θ is an additive character of $\mathbb{F}_p$ defined by

$$\Theta(x) = \Theta_{p^0}(x) = F(\pi\omega_p(x)).$$

Then we can define a Gauss sum

$$G_n = G(\Theta, \omega_p^n) = \sum_{x \in \mathbb{F}_p^*} \Theta(x)\omega_p^n(x).$$

We have the inversion formula

$$\Theta(x) = \frac{1}{p-1} \sum_{k=0}^{p-2} G_{-k}\omega_p^k(x).$$

For a finite extension $\mathbb{F}_q$ $(q = p^r))$ of $\mathbb{F}_p$, the Teichmüller character ω_q is defined as a multiplicative character of $\mathbb{F}_q^* \to \mu_{q-1}$ of the maximal order $q-1$, and the Dwork character Θ_r is an additive character of $\mathbb{F}_q$ defined by composing Θ_{p^0} with the trace $\mathrm{tr} : \mathbb{F}_q \to \mathbb{F}_p$, $\mathrm{tr}(x) = x + x^p + x^{p^2} + \cdots + x^{p^{r-1}}$; that is,

$$\Theta_r(x) = \prod_{i=0}^{r-1} \Theta_{p^i}(x).$$

A Gauss sum is defined by

$$G_{r,n} = G(\Theta_r, \omega_q^n) = \sum_{x \in \mathbb{F}_q^*} \Theta_r(x)\omega_q^n(x).$$

Lemma 9.4 (The Gross–Koblitz formula) *Let ω_p be the Teichmüller character of $\mathbb{F}_p^*$. Let Γ_p be the p-adic Gamma function defined by*

$$\Gamma_p(n) := (-1)^n \prod_{1 \le i < n, p \nmid i} i$$

taking values in the p-adic integer ring $\mathbb{Z}_p$. Then the Gauss sum $G(\Theta, \omega_p)$ is expressed by the Gross–Koblitz formula for the case $p - 1 \nmid n$:

$$G_n = p(p)^{-\langle \frac{n}{p-i} \rangle}\Gamma_p\left(1 - \left\langle \frac{n}{p-1} \right\rangle\right).$$

Here $\langle x \rangle$ denotes the fractional part for $x \in \mathbb{R}$.

For a finite extension $\mathbb{F}_q$ $(q = p^r)$, *the Gross–Koblitz formula for the case* $q-1 \nmid n$ *is given by*

$$G_{r,n} = G(\Theta_r, \omega_q^n) = (-1)^{r+1} q \pi^{-S(n)} \prod_{i=0}^{r-1} \Gamma_p \left(1 - \left\langle \frac{p^i n}{q-1} \right\rangle\right)$$

in $\mathbb{Q}_p(\pi)$ *where* $S(n)$ *denotes the sum of the p-adic digits of* n.

Lemma 9.5 (Dwork's formula) *Let* X *be a variety defined as the zero set of a polynomial* $P(\mathbf{X}) \in \mathbb{F}_q[X_1, X_2, \cdots, X_5]$. *Then*

$$\sum_{y\in\mathbb{F}_q} \Theta(yP(X)) = \begin{cases} 0 & \text{if } P(X) \neq 0 \\ q & \text{if } P(X) = 0 \end{cases}$$

Hence we have

$$q\#X(\mathbb{F}_q) = \sum_{X_i\in\mathbb{F}_q} \sum_{y\in\mathbb{F}_q} \Theta(yP(\mathbf{X})).$$

The congruence zeta-function $Z(X_{\mathbb{F}_q}, t)$ of $X_{\mathbb{F}_q}$ is defined same as above by counting the number $\#X_{\mathbb{F}_{q^r}}$ for each $r \geq 1$. The Dwork theory has established the rationality of $Z(X_{\mathbb{F}_q}, t)$ ([Dw60].) We can compute the congruence zeta-function for our mirror pairs of Calabi–Yau threefolds.

Let $\langle m, Q \rangle$ be an admissible pair, and let $\Delta(Q)$ be a reflexive polytope. Let $f(\mathbf{c}, \mathbf{X}) = 0 \in \mathbb{P}^4(Q)$ be a defining equation of degree m over $\mathbb{Q}$ with deformation parameter $\mathbf{c}$. Let $X_{\Delta(Q)}$ denote a family of Calabi–Yau threefolds. Let $\mathbf{v} = (v_1, v_2, v_3, v_4, v_5)$ denote a monomial $X_1^{v_1} X_2^{v_2} \cdots X_5^{v_5}$ in the weighted polynomial coordinate ring $\mathbb{C}[X_1, X_2, \cdots, X_5]$ of $X_{\Delta(Q)}$, and let $\lambda_{\mathbf{v}}$ stands for its multiplicity, i.e., the number of weighted permutations. Let $\mathcal{S}$ be the set of all monomials. Then we know that $\mathcal{S}$ decomposes into the disjoint sets $[\mathbf{v}]$ of orbits of monomials $\mathbf{v}$ under the action of the automorphism group. Denote by $O(\mathcal{S})$ the set of orbits.

Theorem 9.6 *Let* $\langle m, Q \rangle$ *be an admissible pair, and let* X_Δ *be a family of Calabi–Yau threefolds. For a good prime p, let* $X_{\Delta/\mathbb{F}_q}$ *denote the reduction of* X_Δ *over the finite field* $\mathbb{F}_q$. *Then the congruence zeta-function of* $X_{\Delta/\mathbb{F}_q}$ *is a rational function over* $\mathbb{Z}$ *and is given by*

$$Z(X_{\Delta/\mathbb{F}_q}, t) = \frac{R(X_{\Delta/\mathbb{F}_q}, t)}{(1-t)(1-qt)^{h^{1,1}(X_\Delta)}(1-q^2t)^{h^{2,2}(X_\Delta)}(1-q^3t)}.$$

Here $R(X_{\Delta/\mathbb{F}_q}, t)$ *is an integral polynomial of degree* $B_3(X_\Delta)$ *and has the form:*

$$R(X_{\Delta(Q)/\mathbb{F}_q}, t) = \prod_{[\mathbf{v}]\in O(\mathcal{S})} R([\mathbf{v}], t) = R(\mathbf{0}, t) \prod_{[\mathbf{v}]\neq[\mathbf{0}]} R([\mathbf{v}], t).$$

Furthermore, $R([\mathbf{v}], t)$ *is an integral polynomial (not necessarily irreducible), and*

$$\sum_{[\mathbf{v}]\in O(\mathcal{S})} \lambda_{\mathbf{v}} \deg(R) = B_3(X_{\Delta(Q)})$$

where $\lambda_{\mathbf{v}}$ *denotes the multiplicity of* $\mathbf{v}$.

Now consider a mirror partner $X_{\Delta^*(Q)}$ and its congruence zeta-function.

Theorem 9.7 *Let* $(X_{\Delta(Q)}, X_{\Delta^*(Q)})$ *be a mirror pair of Calabi–Yau threefolds defined over* $\mathbb{Q}$ *corresponding to a pair of reflexive polytopes* $(\Delta(Q), \Delta^*(Q))$ *constructed in Section 6. Let* $X_{\Delta(Q)/\mathbb{F}_q}$ *(resp.* $X_{\Delta^*(Q)/\mathbb{F}_q}$*) be the reduction of* $X_{\Delta(Q)}$ *(resp.* $X_{\Delta^*(Q)}$*) to the finite field* $\mathbb{F}_q$ *of characteristic p (where p is a good prime).*

Then the congruence zeta-function of $X_{\Delta^*(Q)/\mathbb{F}_q}$ at the Fermat point is given as follows:

$$Z(X_{\Delta^*(Q)/\mathbb{F}_q},t)=\frac{R([\mathbf{0}],t)}{(1-t)(1-qt)^{h^{2,1}(X_{\Delta(Q)})}(1-q^2t)^{h^{1,2}(X_{\Delta(Q)})}(1-q^3t)}$$

where $R([\mathbf{0}],t)$ is defined as in Theorem 9.6.

We compare the expression for the congruence zeta-function for X over $\mathbb{F}_q$ obtained from Weil's method and that from Dwork's method at the Fermat point in the moduli space (i.e., at the point where all parameters are set to zero).

Theorem 9.8 *Let $\langle m,Q\rangle$ be an admissible pair. Let $(X,\hat{X})$ be a mirror pair of Calabi–Yau threefolds. Then for a good prime p, the two expressions for the denominator of the congruence zeta-function for X coincide at the Fermat point. That is,*

$$P_3(X_{\mathbb{F}_q},t)=\prod_{A\in O(\mathfrak{A}(Q))}P_3(\mathcal{M}_A,t)^{\dim(A)}=R(X_{\Delta(Q)/\mathbb{F}_q},t)=\prod_{[\mathbf{v}]\in O(S)}R(\mathbf{v},t)^{\lambda(\mathbf{v})}.$$

In particular, the factor $R([\mathbf{0}],t)$ coincides with the factor $\prod_A P_3(\mathcal{M}_A,t)$, where A runs over the $\hat{G}$-invariant motives. Moreover, this is the only factor present in the denominator of the congruence zeta-function of X and $\hat{X}$.

For the calculation of the polynomials $P_3(\mathcal{M}_A,t)$, we have the following results for the cases when the polynomials become powers of linear polynomials over $\mathbb{Q}$, and also when the polynomials are irreducible over $\mathbb{Q}$.

Proposition 9.9 (Yui [05]) *Fix an admissible pair $\langle m,Q\rangle$, and let $\mathcal{M}_A$ be a Fermat motive of $\mathbf{a}\in\mathfrak{A}(Q)$. Let p be a good prime and let f be the order of p mod m, i.e., f is the smallest positive integer such that $p^f\equiv 1 \pmod m$.*

(a) If f is even and $p^{f/2}+1\equiv 0\pmod m$, then putting $q=p^f$, we have

$$P_3(\mathcal{M}_A,t)=(1-q^{3/2}t)^{\varphi(m)}.$$

(b) Let $\mathcal{M}_Q$ be the motive corresponding to the weight $Q=(q_1,q_2,q_3,q_4,q_5)$. Then for $p\equiv 1\pmod m$, $P_{\mathcal{M}_Q}(t)$ is always irreducible over $\mathbb{Q}$.

Example 9.1 For $m=5$ and $Q=(1,1,1,1,1)$, the polynomials $P_3(\mathcal{M}_A,t)$ are given as follows. By the (Newton) slopes, we mean the (normalized) p-adic orders of the reciprocal roots of $P_3(\mathcal{M}_A,t)$, so that they take rational values in $[0,3]$ with multiplicity $(*)$.

(a) When $p\equiv 1\pmod 5$ (e.g., $p=11$), we have

$$\begin{aligned}P_3([1,1,1,1,1],t)&=1+89t+3^3\cdot 11\cdot 13t^2+11^3\cdot 89t^3+11^6t^4\\&\qquad\text{with slopes } 0,1,2,3\\P_3([1,1,1,3,4],t)&=1-11t-3^2 11^2t^2-11^4t^3+11^6t^4=P_3([1,1,2,2,4],t)\\&\qquad\text{with slopes } 1(2),2(2).\end{aligned}$$

Both polynomials are irreducible over $\mathbb{Q}$.

(b) When $p\not\equiv 1\pmod 5$, then $p\equiv 2,3$ or $4\pmod 5$. If $p\equiv 2,3\pmod 5$, then $p^4\equiv 1\pmod 5$ and $p^2\equiv -1\pmod 5$. In this case, for every motive $\mathcal{M}_A$, we have

$$P_3(\mathcal{M}_A,t)=(1-p^{4\cdot\frac{3}{2}}t)^4 \text{ with slopes } 3/2(4).$$

If $p \equiv 4 \pmod 5$, then $p^2 \equiv 1 \pmod 5$ and $p \equiv -1 \pmod 5$. In this case, for every motive $\mathcal{M}_A$, we have

$$P_3(\mathcal{M}_A, t) = (1 - p^{2\cdot\frac{3}{2}}t)^4 \text{ with slopes } 3/2(4).$$

On the other hand, using Dwork's method, the polynomials $R([\mathbf{0}], t)$, $R([4,1,0,0,0], t)$ and $R([3,2,0,0,0], t)$ are computed by Candelas et al. [CORV00, CORV03] as follows. For $p = 11$,

$$R([\mathbf{0}], t) = 1 + 89t + 3^3 \cdot 13 \cdot 11t^2 + 89 \cdot 11^3t^3 + 11^6t^4 \text{ with slopes } 0, 1, 2, 3$$

and

$$R([4,1,0,0,0], t) = R([3,2,0,0,0], t) = 1 - 11t - 3^2 11^2 t^2 - 11^4t^3 + 11^6t^4$$

with slopes $1(2), 2(2)$.

Example 9.2 For $m = 8$ and $Q = (1,1,2,2,2)$, the polynomials $P_3(\mathcal{M}_A, t)$ are given as follows.

(a) When $p \equiv 1 \pmod 8$ (e.g., $p = 17$), we have

$$P_3([1,1,2,2,2], t) = 1 - 2^2 3^2 \cdot 5t + 2 \cdot 17 \cdot 467t^2 - 2^2 3^2 \cdot 5 \cdot 17^3t^3 + 17^6t^4 \text{ with slopes } 0, 1, 2, 3$$

$$P_3([1,1,2,6,6], t) = (1 + 2 \cdot 3 \cdot 17t + 17^3t^2)^2 = P_3([1,1,4,4,6], t)$$

$$P_3([1,3,2,4,6], t) = (1 - 2 \cdot 3 \cdot 17t + 17^3t^2)^2 = P_3([1,3,4,4,4], t)$$

$$P_3([1,5,2,2,6], t) = (1 - 2 \cdot 17t + 17^3t^2)^2 = P_3([1,5,2,4,4], t) = P_3([1,5,6,6,6], t)$$

all with slopes $1(2), 2(2)$. For all the remaining motives (see Table 4),

$$P_3(\mathcal{M}_A, t) = (1 + 2 \cdot 17t + 17^3t^2)^2 \text{ with slopes } 1(2), 2(2).$$

In particular,

$$P_3([2,2,4,4,4], t) = (1 + 2 \cdot 17t + 17^3t^2)^2.$$

(b) When $p \not\equiv 1 \pmod 8$, $p \equiv 3, 5$ or $7 \pmod 8$, and in all cases $p^2 \equiv 1 \pmod 8$.

If $p \equiv 7 \pmod 8$, we have, for every motive $\mathcal{M}_A$,

$$P_3(\mathcal{M}_A, t) = (1 - p^3t)^4 \text{ with slopes } 3/2(4).$$

If $p \equiv 3 \pmod 8$ (e.g., for $p = 11$), we have

$$P_3([1,1,2,2,2], t) = (1 - 2 \cdot 7 \cdot 11^2t + 11^6t^2)^2 = P_3([1,1,2,6,6], t) = P_3([1,1,4,4,6], t)$$

$$P_3([1,3,2,4,6], t) = (1 + 2 \cdot 7 \cdot 11^2t + 11^6t^2)^2 = P_3([1,3,4,4,4], t)$$

all with slopes $1(2), 2(2)$. For the remaining motives $\mathcal{M}_A$,

$$P_3(\mathcal{M}_A, t) = (1 \pm 11^3t)^4 \text{ with slopes } 3/2(4).$$

In particular,

$$P_3([2,2,4,4,4], t) = (1 - 11^3t^2)^4.$$

If $p \equiv 5 \pmod 8$ (e.g., for $p = 13$), we have

$$P_3([1,1,2,2,2], t) = (1 + 2 \cdot 7 \cdot 13 \cdot 17t + 13^6t^2)^2 \text{ with slopes } 1(2), 2(2)$$

$$\begin{aligned} P_3([1,1,2,6,6], t) = (1 - 13^3t)^4 &= P_3([1,1,4,4,6], t) = P_3([1,3,2,4,6], t) \\ &= P_3([1,3,4,4,4], t) \end{aligned}$$

with slopes $3/2(4)$. For the remaining motives $\mathcal{M}_A$,

$$P_3(\mathcal{M}_A, t) = (1 \pm 2 \cdot 5 \cdot 13^2 + 13^6t^2)^2 \text{ with slopes } 1(2), 2(2).$$

Table 10 Polynomials for $p = 11, 13, 17$

Monomial $\mathbf{v}$	$p = 11$	$p = 13$	$p = 17$
$(0,0,0,0,0)$	$(0)_2(0, 2.7.11^2)_4$	$(6.13)_2(0, -238.13)_4$	$(-2.17)_2(180, 934.17)_4$
$(0,2,1,1,1)$	$(0)_2[(-6.11)_2(6.11)_2]^{1/2}$	$(0)_2(-6.13)_2$	$(-2.17)_2(6.17)_2$
$(6,2,0,0,0)$	$(0)_2^2$	$(-4.13)_2(4.13)_2$	$(2.17)_2^2$
$(0,0,0,2,2)$	$(0, 2.7.11^2)_4$	$(0)_2$	$(-6.17)_2^2$
$(2,0,1,3,3)$	$(0)_2$	$(-6.13)_2$	$(-2.17)_2$
$(4,0,2,0,0)$	$(0)_2^2$	$(-4.13)_2(4.13)_2$	$(2.17)_2^2$
$(0,0,2,1,1)$	$(0)_2^{1/2}(0, 2.7.11^2)_4^{1/2}$	$(0)_2(6.13)_2$	$(-6.17)_2^2(-2.17)_2$
$(2,2,1,1,0)$	$(0)_2^{1/2}(0, 2.7.11^2)_4^{1/2}$		
$(6,0,1,0,0)$	$(0)_2^{1/2}(11\sqrt{11})_1(-11\sqrt{11})_1$	$(-4.13)_2(4.13)_2(6.13)_2$	$(-2.17)_2^3$
$(2,0,3,0,0)$	$(0)_2^{1/2}(11\sqrt{11})_1(-11\sqrt{11})_1$		
$(0,4,0,3,3)$	$(0)_2^2$	$(-4.13)_2(4.13)_2\times$	$(-2.17)_2^2(2.17)_2^2$
$(4,0,1,1,0)$	$(0)_2^2$	$(-6.13)_2(6.13)_2$	
$(0,0,3,1,0)$	$(0)_2$	$(-6.13)_2$	$(-2.17)_2$
$(2,0,2,1,0)$	$[(-6.11)_2(6.11)_2]^{1/2}$	$(0)_2$	$(6.17)_2$
$(4,0,2,3,1)$	1	1	1

In particular,

$$P_3([2,2,4,4,4], t) = (1 + 2 \cdot 5 \cdot 13^2 t + 13^6 t^2)^2.$$

We compute the polynomials $R([\mathbf{v}], t)$ using Dwork's method; the results are shown in Table 10. The following notation is used:

Notation	Polynomial, for prime p
$(a)_1$	$(1 + at)$
$(a)_2$	$(1 + at + p^3 t^2)$
$(a, b)_4$	$(1 + at + bt^2 + ap^3 t^3 + p^6 t^4)$

Alternatively, since the basic constituents of the congruence zeta-function are the numbers of $\mathbb{F}_{q^r}$-rational points on X, we compare the expression for the numbers of $\mathbb{F}_q$-rational points, N_{motive}, computed by Weil's method with the number of $\mathbb{F}_q$-rational points, $N_{monomial}$, computed by Dwork's method.

Proof of Theorem 9.8. We now prove Theorem 9.8 by showing that

$$N_{motive} = N_{monomial}.$$

We count the number of points on the variety excluding the contribution coming from any exceptional divisor required to smooth the ambient weighted projective space. The piece coming from the exceptional divisor needs to be added by hand using Weil's method, but emerges naturally from toric geometry using Dwork's method (namely toric geometry forbids the vanishing of coordinates associated to points lying on the same cone; see [CORV00, CORV03] and [Ka05] for examples). Using Dwork's method, as outlined above, we obtain the following expression for the case when we have an admissible pair $\langle m, Q \rangle$ with $Q = (q_1, q_2, q_3, q_4, q_5)$:

$$q\nu = \sum_{y \in \mathbb{F}_q} \sum_{\mathbf{x} \in (\mathbb{F}_q)^5} \Theta\left(y\left(\sum_{i=1}^{5} x_i^{\frac{m}{q_i}}\right)\right) = q^5 + \sum_{y \in \mathbb{F}_q^*} \sum_{\mathbf{x} \in (\mathbb{F}_q)^5} \Theta\left(y\left(\sum_{i=1}^{5} x_i^{\frac{m}{q_i}}\right)\right)$$
$$+ \nu_{exceptional}.$$

Dividing both sides by q and bringing in Gauss sums, we get:

$$\begin{aligned}
\nu &= q^4 + \frac{1}{q(q-1)^5} \sum_{y\in\mathbb{F}_q^*} \sum_{\mathbf{x}\in(\mathbb{F}_q^*)^5} \prod_{i=1}^{5} G_{-s_i}\omega_q^{\frac{ms_i}{q_i}}(x_i)\omega_q^{\sum s_i}(y) \\
&-G_0 \frac{1}{q(q-1)^4} \sum_{j=1}^{5} \sum_{y\in\mathbb{F}_q^*} \sum_{\substack{x_i\in\mathbb{F}_q^* \\ i\neq j}} \prod_{i\neq j} G_{-s_i}\omega_q^{\frac{ms_i}{q_i}}(x_i)\omega_q^{\sum s_i}(y) \\
&+G_0^2 \frac{1}{q(q-1)^3} \sum_{j,k} \sum_{y\in\mathbb{F}_q^*} \sum_{\substack{x_i\in\mathbb{F}_q^* \\ i\neq j}} \prod_{i\notin\{j,k\}} G_{-s_i}\omega_q^{\frac{ms_i}{q_i}}(x_i)\omega_q^{\sum s_i}(y) \\
&-G_0^3 \frac{1}{q(q-1)^2} \sum_{j,k,l} \sum_{y\in\mathbb{F}_q^*} \sum_{\substack{x_i\in\mathbb{F}_q^* \\ i\neq j}} \prod_{i\notin\{j,k,l\}} G_{-s_i}\omega_q^{\frac{ms_i}{q_i}}(x_i)\omega_q^{\sum s_i}(y) \\
&+G_0^4 \frac{1}{q(q-1)} \sum_{j,k,l,n} \sum_{y\in\mathbb{F}_q^*} \sum_{\substack{x_i\in\mathbb{F}_q^* \\ i\neq j}} \prod_{i\notin\{j,k,l,n\}} G_{-s_i}\omega_q^{\frac{ms_i}{q_i}}(x_i)\omega_q^{\sum s_i}(y) \\
&-G_0^5 \frac{q-1}{q} + \nu_{exceptional}.
\end{aligned}$$

Further, we pass to the following expression for ν:

$$\begin{aligned}
\nu &= q^4 + \frac{q-1}{q} \sum_{s_i=1}^{q-2} \prod_{i=1}^{5} G_{-s_i} - G_0 \frac{q-1}{q} \sum_{j=1}^{5} \sum_{\substack{s_i=1 \\ i\neq j}}^{q-2} \prod_{i\neq j} G_{-s_i} \\
&+G_0^2 \frac{q-1}{q} \sum_{j,k} \sum_{\substack{s_i=1 \\ i\notin\{j,k\}}}^{q-2} \prod_{i\notin\{j,k\}} G_{-s_i} - G_0^3 \frac{q-1}{q} \sum_{j,k,l} \sum_{\substack{s_i=1 \\ i\notin\{j,k,l\}}}^{q-2} \prod_{i\notin\{j,k,l\}} G_{-s_i} \\
&+G_0^4 \frac{q-1}{q} \sum_{j,k,l,n} \sum_{\substack{s_i=1 \\ i\notin\{j,k,l,n\}}}^{q-2} \prod_{i\notin\{j,k,l,n\}} G_{-s_i} - G_0^5 \frac{q-1}{q} + \nu_{exceptional},
\end{aligned}$$

where passing from the former to the latter expression we had to impose the following conditions:

$$\begin{aligned}
q-1 | & \quad \frac{ms_i}{q_i} \qquad \forall i = 1, \cdots, 5 \\
q-1 | & \quad \textstyle\sum_{i=1}^{5} s_i
\end{aligned}$$

Hence introduce a vector of integers, $\mathbf{v^t} = (v_1, v_2, v_3, v_4, v_5)$, such that:

$$\begin{aligned}
v_i(q-1) &= \frac{s_i m}{q_i} \qquad \forall i = 1, \cdots, 5 \\
v(q-1) &= \textstyle\sum_{i=1}^{5} s_i
\end{aligned}$$

and of course $0 \leq s_i \leq q-2, \quad i = 1, \cdots, 5$. It is easy to see that we need to consider vectors $\mathbf{v}$ such that:

$$\sum_{i=1}^{5} q_i v_i = mv, \quad 0 \leq q_i v_i \leq \left\lfloor \frac{m(q-2)}{q-1} \right\rfloor \quad (\forall i = 1, \ldots, 5)$$

$$0 \leq v \leq \left\lfloor \frac{5(q-2)}{q-1} \right\rfloor \quad (= 4 \quad \text{for } q \geq 7).$$

Hence,

$$\nu = q^4 + \frac{q-1}{q} \sum_{\prod^{\mathbf{v}} v_i \neq 0} \prod_{i=1}^{5} G_{-\left(\frac{q-1}{m} v_i q_i\right)} + \nu_{exceptional}.$$

The reason that all monomials $\mathbf{v}$ with zero entries are excluded is as follows: Consider a monomial $\mathbf{v}$ with exactly r zero entries. In the second equation for ν, it will appear exactly $\binom{r}{0}$ times in the first sum, and $(-1)^u \binom{r}{u}$ times in the $(u+1)$-st sum. Hence in total it appears $\sum_{u=0}^{r} (-1)^u \binom{r}{u} = 0$ times. Clearly this argument works in any dimension.

Finally we obtain:

$$N_{monomial} = \frac{1}{q} \sum_{\prod^{\mathbf{v}} v_i \neq 0} \prod_{i=1}^{5} G_{-\left(\frac{q-1}{m} v_i q_i\right)} = \frac{1}{q} \sum_{\prod^{\mathbf{v}} v_i \neq 0} \prod_{i=1}^{5} G_{\left(\frac{q-1}{m}\right)(m - v_i q_i)},$$

where the last equality comes from the fact that $G_{r,n} = G_{r,q-1+n}$. If we define the following:

$$q_i v_i = a_i', \quad a_i = m - a_i', \quad \forall i = 1, \cdots 5$$

we see that

$$vm = \sum_{i=1}^{5} a_i = \sum_{i=1}^{5} (m - a_i') = m(5 - v').$$

It is clear that the vectors with components a_i are precisely those in $\mathfrak{A}(Q)$; hence

$$N_{monomial} = \frac{1}{q} \sum_{\prod^{\mathbf{v}} v_i \neq 0} \prod_{i=1}^{5} G_{\left(\frac{q-1}{m} a_i\right)}.$$

On the other hand, we know that the expression for the sum $\sum_{\mathbf{a} \in \mathfrak{A}(Q)} J(\mathbf{a})$ in terms of Weil's method is

$$\frac{1}{q} \sum_{\mathbf{a} \in \mathfrak{A}(Q)} \prod_{i=1}^{5} G(\chi^{a_i}).$$

Therefore, we obtain:

$$N_{motive} = N_{monomial}.$$

This completes the proof of Theorem 9.8.

As a consequence of the proof of Theorem 9.8, we can now establish the motive–monomial correspondence at the Fermat point.

For a character $\mathbf{a} = (a_1, a_2, a_3, a_4, a_5) \in \mathfrak{A}(Q)$, there corresponds the monomial

$$X_1^{a_1/q_1} X_2^{a_2/q_2} X_3^{a_3/q_3} X_4^{a_4/q_4} X_5^{a_5/q_5}.$$

Now the $(\mathbb{Z}/m/Z)^{\times}$-orbit of $\mathbf{a}$ gives rise to the Fermat motive $\mathcal{M}_A$, while, the multiplication by $\hat{G}$-invariant monomials yields the corresponding monomial class. These two actions are compatible.

Conversely, starting with a monomial $X_1^{v_1}X_2^{v_2}X_3^{v_3}X_4^{v_4}X_5^{v_5}$, let $([v_1],[v_2],[v_3],[v_4],[v_5])$ be the equivalence class of monomials under multiplication of the $\hat{G}$-invariant monomials. Then there corresponds the motive

$$[[v_1]q_1,[v_2]q_2,[v_3]q_3,[v_4]q_4,[v_5]q_5]\in\mathfrak{A}(Q).$$

Note that the constant monomial class corresponds to the weight motive $\mathcal{M}_{[Q]}=\mathcal{M}_{[q_1,q_2,q_3,q_4,q_5]}$.

This establishes the motive-monomial correspondence at the Fermat point.

A cohomological proof of the motive-monomial correspondence has been given by Kloosterman [K07].

Example 9.3 For instance, for the quintic threefolds the monomial classes $(4,1,0,0,0)$ and $(3,2,0,0,0)$ combine to give only one motive $[1,1,1,3,4]$:

Monomials	Motives	Monomials	Motives
$(4,1,0,0,0)$		$(2,3,0,0,0)$	
$(0,2,1,1,1)$		$(3,4,1,1,1)$	$\in[1,1,1,3,4]$
$(1,3,2,2,2)$	$\in[1,1,1,3,4]$	$(4,0,2,2,2)$	
$(2,4,3,3,3)$	$\in[1,1,1,3,4]$	$(0,1,3,3,3)$	
$(3,0,4,4,4)$		$(1,2,4,4,4)$	$\in[1,1,1,3,4]$

Example 9.4 For the octic threefolds the monomial classes $(2,0,3,0,0)$ and $(6,0,1,0,0)$ together correspond to two different motives, $[4,2,2,4,4]$ and $[7,1,2,2,4]$:

Monomials	Motives	Monomials	Motives
$(2,0,3,0,0)$		$(6,0,1,0,0)$	
$(3,1,0,1,1)$		$(7,1,2,1,1)$	$\in[7,1,4,2,2]$
$(4,2,1,2,2)$	$\in[4,2,2,2,4]$	$(0,2,3,2,2)$	
$(5,3,2,3,3)$	$\in[7,1,2,2,4]$	$(1,3,0,3,3)$	
$(6,4,3,0,0)$		$(2,4,1,0,0)$	
$(7,5,0,1,1)$		$(3,5,2,1,1)$	$\in[7,1,4,2,2]$
$(0,6,1,2,2)$		$(4,6,3,2,2)$	$\in[4,2,2,2,4]$
$(1,7,2,3,3)$	$\in[7,1,4,2,2]$	$(5,7,0,3,3)$	

Example 9.5 It should be noted that whenever a monomial class only makes a contributions at special points in the moduli space, such as conifold points (i.e. not at the Fermat point), every monomial in the calss have at least one zero entry (see the example below), and hence, as expected, no correspondence with Fermat motives can be made.

$Q=(1,1,2,2,2)$
$(4,0,3,2,1)$
$(5,1,0,3,2)$
$(6,2,1,0,3)$
$(7,3,2,1,0)$
$(0,4,3,2,1)$
$(1,5,0,3,2)$
$(2,6,1,0,3)$
$(3,7,2,1,0)$

$Q=(1,1,1,1,1)$
$(4,0,3,2,1)$
$(0,1,4,3,2)$
$(1,2,0,4,3)$
$(2,3,1,0,4)$
$(3,4,2,1,0)$

The above correspondence can be easily seen in the existing expressions derived for the number of points for the following cases: the one-parameter family of quintic Calabi–Yau threefolds by Candelas et al. [CORV00, CORV03] and for two-parameter family of the octic Calabi–Yau threefolds by Kadir [Ka05].

Example 9.6 Let $m = 5$ and $Q = (1, 1, 1, 1, 1)$. Then the $N_{monomial}$ computed in terms of Gauss sums formed from the Dwork character as follows.

(a) When $5|q-1$

$$N_{monomial} = \frac{1}{q}\sum_{a=0}^{3}\sum_{\substack{\mathbf{v} \\ v_i+a\neq 0}} \lambda_{\mathbf{v}} \prod_{i=1}^{5} G_{-(v_i+a)k}$$

with $k = \frac{q-1}{5}$

(b) When $5 \nmid q-1$

$$N_{monomial} = \frac{1}{q}G_0^5.$$

The corresponding expression in terms of Jacobi sums from Weil's method is as follows:

$$\sum_{\mathbf{a}\in\mathfrak{A}(Q)} J(\mathbf{a}) = \frac{1}{q}\sum_{\mathbf{a}\in\mathfrak{A}(Q)}\prod_{i=1}^{5} G(\chi^{a_i}).$$

Therefore,

$$N_{motive} = N_{monomial}.$$

Example 9.7 For the two-parameter model in with $Q = (1, 1, 2, 2, 2)$ the expression for N_{mon} is more complicated. At the Fermat point, they are given as follows.

(a) When $8|q-1$

$$N_{monomial} = \frac{1}{q}\sum_{b=0}^{1}\sum_{a=0}^{3}\sum_{\mathbf{v}} \lambda_{\mathbf{v}} \prod_{i=1,2} G_{-(v_i+a+4b)k} \prod_{i=3,4,5} G_{-2(v_i+a)k}.$$

There is a condition on the sum over the monomials that $v_i + a + 4b \neq 0$ for $i = 1, 2$, and $v_i + a \neq 0$ for $i = 3, 4, 5$.

(b) When $4|q-1$, $8 \nmid q-1$

$$N_{monomial} = \frac{1}{q}\sum_{a=0}^{3}\sum_{\mathbf{v}} \lambda_{\mathbf{v}} \prod_{i=1,2} G_{-\frac{(v_i+2a)}{2}l} \prod_{i=3,4,5} G_{-(v_i+2a)l}.$$

There is a condition on the sum over the monomials that $v_i + 2a \neq 0$ for $i = 1, \ldots 5$.

(c) When $4 \nmid q-1$

$$N_{monomial} = \frac{1}{q}\sum_{b=0}^{1}\sum_{\substack{\mathbf{v} \\ v_i+4b\neq 0}} \lambda_{\mathbf{v}} \prod_{i=1,2} G_{-\frac{(v_i+4b)}{4}m} \prod_{i=3,4,5} G_{-\frac{v_i}{2}m}.$$

There is a condition on the sum over the monomials that $v_i + 4b \neq 0$ for $i = 1, 2$, and $v_i \neq 0$ for $i = 3, 4, 5$.

10 Conclusions and further problems

In the above we had four objects to describe the Topological Mirror Symmetry for Calabi–Yau hypersurfaces of dimension 3. They are:

(a) Reflexive polytopes and their integral vertices (Toric geometry);

(b) Monomials in the graded polynomial ring, and the action of the group of automorphisms of the coordinate ring, which gives rise to monomial classes (Toric geometry via coordinate rings);

(c) Picard–Fuchs differential equations (and hence periods), and the action of the differential operators $\mathcal{D}_i$, which gives rise to Picard–Fuchs differential systems, and

(d) Fermat motives (i.e., characters in $\mathfrak{A}(Q)$ and the action of the Galois group $(\mathbb{Z}/m\mathbb{Z})^\times$ of the m-th cyclotomic field $L = \mathbb{Q}(\zeta_m)$ over $\mathbb{Q}$) (Algebraic number theory).

The bijection between (a) and (b) was established by Batyrev [Ba94], and Aspinwall–Greene and Morrison [AGM93], the bijection between (b) and (c) follows from the work of Dwork, Katz, Griffiths (see Cox and Katz [CK99]); the explicit bijections were constructed for the quintic Calabi–Yau threefolds by Candelas et al. [CORV00, CORV03], and for the octic Calabi–Yau threefolds with two deformation parameters by Kadir [Ka05].

The point of this article is to establish a one-to-one correspondence between Fermat motives and any of the above three objects in (a), (b) and (c) in the equivariant fashion compatible with the various actions. This correspondence was established at the Fermat (the Landau–Ginzburg) point in the moduli space of the family of Calabi–Yau threefolds.

Though Fermat motives are defined only at the Fermat point (and also motives are algebraic in nature), through their correspondences to monomials, they appear to contain some information about Calabi–Yau orbifolds with deformation parameters.

This raises the following questions.

Problem 10.1 We now introduce parameters into the defining hypersurface equations of our Calabi–Yau orbifolds. Batyrev's method produces monomials for these Calabi–Yau hypersurfaces. Our motives are defined at the Fermat point on putting deformation parameters equal to zero. When the deformation parameters are non-zero, we still have monomials which correspond to Fermat motives, but acquire more monomials. Some monomials do arise from conifold singularities. How can one interpret these extra monomials from the motivic point of view?

Naïvely, one way to start would be to group the monomials multiplied by weight under their transformation properties under $(\mathbb{Z}/m\mathbb{Z})^\times$. For instance, considering Example 9.4 once more and multiplying the monomials by the weight $(1,1,2,2,2)$ we get:

Weighted Monomials	Motives	Weighted Monomials	Motives
(2, 0, 6, 0, 0) (6, 0, 2, 0, 0)	[2, 0, 6, 0, 0]	(5, 3, 4, 6, 6) (3, 5, 4, 2, 2) (7, 1, 4, 2, 2) (1, 7, 4, 6, 6)	[7, 1, 2, 2, 4]
(3, 1, 0, 2, 2) (1, 3, 0, 6, 6) (7, 5, 0, 2, 2) (5, 7, 0, 6, 6)	[3, 1, 0, 2, 2]	(6, 4, 6, 0, 0) (2, 4, 2, 0, 0)	[6, 4, 6, 0, 0]
(4, 2, 2, 4, 4) (4, 6, 6, 4, 4)	[4, 2, 2, 2, 4]	(0, 6, 2, 4, 4) (0, 2, 6, 4, 4)	[0, 6, 2, 4, 4]

Hence we obtain 4 new "fictitious motives" containing zero entries, [2, 0, 6, 0, 0], [3, 1, 0, 2, 2], [6, 4, 6, 0, 0] and [0, 6, 2, 4, 4], in addition to the two found at the Fermat point, [4, 2, 2, 2, 4] and [7, 1, 2, 2, 4].

The "fictitious motives" containing zeroes may be related to lower dimensional genuine motives of Fermat varieties of the same degree (there may be a twisting by a character). For instance, we drop the component 0 in both [3, 1, 0, 2, 2] and [0, 6, 2, 4, 4] and we obtain [3, 1, 2, 2] and [6, 2, 4, 4], which may be considered as motives arising from the Fermat surface $Z_1^8 + Z_2^8 + Z_3^8 + Z_4^8 = 0 \subset \mathbb{P}^4$ of geometric genus $p_g = 35$. The motive [3, 1, 2, 2] has dimension 4 and multiplicity 6. The Hodge numbers are $h^{2,0} = 2, h^{1,1} = 4, h^{2,0} = 2$ so that the 2nd Betti number is $B_2 = 8$. The motive [6, 2, 4, 4] has dimension 2 and multiplicity 1. The Hodge numbers are $h^{2,0} = 0, h^{1,1} = 2, h^{0,2} = 0$. However, it is not clear how these Fermat motives of lower dimensions come into the picture of the octic Calabi–Yau threefold.

When there is a conifold singularity, it is locally isomorphic to the quadric $X^2 + Y^2 + Z^2 + T^2 = 0$, and there should correspond a twisted Tate motive (an extension of a Tate motive), which in turn should come from some monimial class.

Exploring the role played by the deformation parameters will be a project in the future. It is well known that Fermat hypersurfaces are dominated by the product of Fermat hypersurfaces of lower dimensions of the same degree (see, for instance Hunt and Schimmrigk [HS99]). However, the situation is totally different if one passes to deformations of Fermat hypersurfaces. Chad Schoen [Sch96] showed that the one-parameter deformation of the quintic Calabi–Yau threefold

$$Z_1^5 + Z_2^5 + Z_3^5 + Z_4^5 + Z_5^5 - 5\lambda Z_1 Z_2 Z_3 Z_4 Z_5 = 0 \subset \mathbb{P}^4 \times \mathbb{P}^1$$

is not dominated by product varieties. This would imply that extra monomials (4, 0, 3, 2, 1) of degree 10 and multiplicity 24, which corresponds to the conifold singular point $\lambda = 1$, are not arising from lower dimensional Fermat surfaces of degree 5. At $\psi = 1$, they acquire 125 nodes, and resolving them with $\mathbb{P}^1$ yields the rigid Calabi–Yau threefold considered by Chad Schoen [Sch86]. Candelas et al. [CORV03] explained the role of these extra monomials in the mirror construction. From their calculations of the local zeta-functions, clearly we see the appearance of the Tate motives corresponding to these extra monomials.

In fact, Kloosterman [K07] has extended the motive-monomial correspondence beyond the Fermat point for a Fermat hypsurface with one-parameter deformation paramter in a weighted projective space.

Problem 10.2 The study of the zeta-functions and L-series of mirror pairs of Calabi–Yau orbifolds with deformation parameters is an ongoing project with Y. Goto and R. Kloosterman ([GKY]). One of the challenging tasks in this project is the resolution of singularities. We use toric resolutions to get smooth Calabi–Yau

Table 11 $m = 5$ and $Q = (1, 1, 1, 1, 1)$: Mirror pairs for subgroups

	H	generators	$h^{1,2}$	$h^{1,1}$	χ
1	$\{1\}$		101	1	-200
2	$(\mathbb{Z}/5\mathbb{Z})$	$(1,0,0,4,0)$	49	5	-88
3	$(\mathbb{Z}/5\mathbb{Z})$	$(1,2,3,4,0)$	21	1	-40
4	$(\mathbb{Z}/5\mathbb{Z})^2$	$(1,0,0,4,0),(1,2,3,4,0)$	21	17	-8
$\hat{4}$	$(\mathbb{Z}/5\mathbb{Z})$	$(1,2,2,0,0)$	17	21	8
$\hat{3}$	$(\mathbb{Z}/5\mathbb{Z})^2$	$(1,2,3,4,0),(1,0,2,2,0)$	1	21	40
$\hat{2}$	$(\mathbb{Z}/5\mathbb{Z})^2$	$(1,0,0,4,0),(1,0,4,0,0)$	5	49	88
$\hat{1}$	$(\mathbb{Z}/5\mathbb{Z})^3$	$(1,0,0,4,0),(1,0,4,0,0),(1,4,0,0,0)$	1	101	200

threefolds. In several examples, we have already carried out resolutions, and have computed the zeta-functions and L-series. For these examples, we have established the motivic modularity.

Problem 10.3 The duality described between the two finite abelian groups $G = G_Q$ and $\hat{G}$ in Section 4 may be extended further to pairs of quotients $(X/H, \hat{X}/\hat{H})$ where $H \subset \hat{G}$ and $\hat{H}$ is the complement of H in $\hat{G}$ such that $\#H \times \#\hat{H} = m^3$. H and $\hat{H}$ act on X and $\hat{X}$ respectively. Indeed, mirror symmetry can be extended to the mirror pairs of Calabi–Yau orbifolds corresponding to $(H, \hat{H})$ (Klemm and Theisen [KT93]), where H is normalized in some cases. In this setting, mirror symmetry may be viewed as a Galois correspondence. (Cf. Batyrev [Ba94]).

Establish motive-monomial correspondence for these Calabi–Yau threefolds. For this, one should calculate which monomials (or equivalently, motives) are preserved under the orbifolding operation by various different groups.

Example 10.3.1: Consider the case

$$m = 5,\ Q = (1,1,1,1,1)\ :\ Y_1^5 + Y_2^5 + Y_3^5 + Y_4^5 + Y_5^5 = 0$$

with the groups $(G_Q, \hat{G}) = (\{1\}, (\mathbb{Z}/5\mathbb{Z})^3)$, and the relevant dual pairs of subgroups. They are tabulated below, where the number $\hat{\#}$ indicates a mirror partner of $\#$.

Example 10.3.2: Next, we consider Calabi–Yau orbifolds corresponding to

$$m = 10,\ Q = (1,1,1,2,5)\ :\ Y_1^{10} + Y_2^{10} + Y_3^{10} + Y_4^5 + Y_5^2 = 0$$

with the groups $(G_Q, \hat{G}) = ((\mathbb{Z}/10\mathbb{Z}), (\mathbb{Z}/10\mathbb{Z})^2)$. The relevant subgroups H are normalized by factoring out the group $G_Q = (\mathbb{Z}/10\mathbb{Z})$, that is, $(\mathbb{Z}/2\mathbb{Z})$ actually means $(\mathbb{Z}/2\mathbb{Z}) \times (\mathbb{Z}/10\mathbb{Z})$.

Table 12 $m = 10$ and $Q = (1, 1, 1, 2, 5)$: Mirror pairs for subgroups

	H	generators	$h^{2,1}$	$h^{1,1}$	χ
1	$\{1\}$		145	1	-288
2	$(\mathbb{Z}/2\mathbb{Z})$	$(0,5,5,0,0)$	99	3	-192
3	$(\mathbb{Z}/2\mathbb{Z})^2$	$(0,5,5,0,0),(5,5,0,0,0)$	67	7	-120
4	$(\mathbb{Z}/5\mathbb{Z})$	$(0,4,4,1,0)$	47	11	-72
5	$(\mathbb{Z}/5\mathbb{Z})$	$(0,8,2,0,0)$	37	13	-48
6	$(\mathbb{Z}/10\mathbb{Z})$	$(9,0,1,0,0)$	39	15	-48
7	$(\mathbb{Z}/10\mathbb{Z})$	$(0,5,3,1,0)$	29	17	-24
$\hat{7}$	$(\mathbb{Z}/10\mathbb{Z})$	$(0,7,1,1,0)$	17	29	24
$\hat{6}$	$(\mathbb{Z}/10\mathbb{Z})$	$(0,3,3,2,0)$	15	39	48
$\hat{5}$	$(\mathbb{Z}/2\mathbb{Z}) \times (\mathbb{Z}/10\mathbb{Z})$	$(5,5,0,0,0),(0,5,3,1,0)$	13	37	48
$\hat{4}$	$(\mathbb{Z}/10\mathbb{Z}) \times (\mathbb{Z}/2\mathbb{Z})$	$(9,0,1,0,0),(0,5,5,0,0)$	11	47	72
$\hat{3}$	$(\mathbb{Z}/\mathbb{Z})^2$	$(0,8,2,0,0),(8,0,2,0,0)$	7	67	120
$\hat{2}$	$(\mathbb{Z}/10\mathbb{Z}) \times (\mathbb{Z}/5\mathbb{Z})$	$(9,0,1,0,0),(0,8,2,0,0)$	3	99	192
$\hat{1}$	$(\mathbb{Z}/10\mathbb{Z})^2$	$(9,0,1,0,0),(0,9,1,0,0)$	1	145	288

References

[AGM93] Aspinwall, P., Greene, B., and Morrison, D., *The monomial-divisor mirror map*, Internat. Math. Res. Notices (1993), pp. 319–337.

[Ba94] Batyrev, V., *Dual polyhedra and mirror symmetry for Calabi–Yau hypersurfaces in toric varieties*, J. Alg. Geom. **3** (1994), pp. 493–535.

[CLS90] Candelas, P., Lynker, M., and Schimmrigk, R., *Calabi–Yau manifolds in weighted* $\mathbb{P}^4$, Nucl. Phys. **B341** (1990), pp. 383–402.

[COFKM93] Candelas, P., de la Ossa, X., Font, A., Katz, S., and Morrison, D., *Mirror symmetry for two-parameter models - I*, Nuclear Phys. B **416** (1994), pp. 481–538. (Also in [GYau94] pp. 483–543.)

[CORV00] Candelas P., de la Ossa X., and Rodriguez Villegas F, *Calabi–Yau manifolds over finite fields I*, arXiv:hep-th/0012233.

[CORV03] Candelas, P., de la Ossa, X., and Rodriguez–Villegas, F., *Calabi–Yau manifolds over finite fields, II*, Fields Institute Communications **38** (2003), pp. 121–157.

[CK99] David A. Cox and Sheldon Katz, *Mirror Symmetry and Algebraic Geometry* Mathematical Surveys & Monographs **68**, American Mathematical Society, 1999.

[D74] Deligne, P., *La Conjecture de Weil I*, Pub. Math. IHES **43** (1974), pp. 273–307.

[DM82] Deligne, P., and Milne, J.S., *Tannakian categories, Hodge cycles, motives and Shimura varieties*, LNM **900**, Springer–Verlag, 1982.

[Dol82] Dolgachev, I., *Weighted projective spaces* in "Group Actions and Vector Fields" LNM **956** (1982), pp. 24–71. Springer–Verlag.

[Dw60] Dwork, B., *On the rationality of the zeta-function of an algebraic variety*, Amer. J. Math. **82** (1960), pp. 631–648.

[G06] Goto, Y., *On the height of Calabi–Yau threefolds*, in preparation.

[GKY] Goto, Y., Kloosterman, R., and Yui, N., *Zeta-functions and L-series of certain K3-fibered Calabi–Yau threefolds*, in preparation.

[GY95] Gouvêa, F., and Yui, N., *Arithmetic of Diagonal Hypersurfaces over Finite Fields*, London Math. Soc. Lecture Notes Series **209**, Cambridge University Press 1995.

[GP90] Greene, B., and Plesser, M.R., *Duality in Calabi–Yau moduli spaces*, Nuclear Physics B **338** (1990), no. 1, pp. 15–37.

[GRY91] Greene, B. R., Roan, S.-S. and Yau, S.-T., *Geometric singularities and spectra of Landau–Ginzburg models*, Commun. Math. Phys. **142** (1991), pp. 245–259.

[GYau96] Greene, R., and Yau, S.-T., *Mirror Symmetry II*, AMS/IP Studies in Advanced Mathematics, **1**, American Mathematical Society and International Press 1996.

[HLY95] Hosono, S., Lian, B.-H., and Yau, S.-T., *Picard–Fuchs differential operators for Calabi–Yau hypersurfaces with* $h^{1,1} \leq 3$–Optional appendix to *GKZ–generalized hypergeometic system in mirror symmetry for Calabi–Yau hypersurfaces*, Commun. Math. Phys. **182** (1996), pp. 535–577. (The appendix does not appear in the text of the article, though available in the source file of the article in the arXiv:alg-geom/9511001.)

[HS99] Hunt, B., and Schimmrigk, R., *K3-fibered Calabi–Yau threefolds. I. The twist map*, Internat. J. Math. **10** (1999), no.7, pp. 833–869.

[Ka05] Kadir, S., *Arithmetic mirror symmetry for a two-parameter family of Calabi–Yau manifolds*, in "Mirror Symmetry V", AMS/IP Studies in Advanced Mathematics **38** (2006), pp. 35–86.

[K07] Kloosterman, R., *The zeta-function of monomial deformations of Fermat hypersurfaces*, Algebra & Number Theory, **1** (2007), no.4, pp. 421–450.

[KS] Kreuzer, M., and Skarke, H., *Calabi–Yau data*, http://hep.itp.tuwien.ac.at/kreuzer/CY/CYcy.html.

[KT93] Klemm, A., and Theisen, S., *Considerations of one-modulus Calabi–Yau compactifications: Picard-Fuchs equations, Kähler potentials and mirror maps*, Nucl. Physics **B 389** (1993), pp. 153–180.

[Ma70] Manin, Yu. I., *Correspondences, motives and monoidal transformations*, Math. USSR-Sb. **77** (1970), pp. 475–507.

[Mor97] Morrison, D. M., *Mathematical aspects of mirror symmetry*, in "Complex Algebraic Geometry (Park City, Utah, 1993), AMS/Park City Math. Ser. **3** (1997), pp. 265–327.

[Mo07] Moore, G., *Strings and Arithmetic, Lecture II: Arithmetic and Attractors*, in *Frontiers in Number Theory, Physics and Geometry II*, Springer–Verlag, 2007, pp. 303–359. [hep-th/9807087].

[Na75] Nakamura, I., *Complex parallelisable manifolds and their small deformations*, J. Diff. Geom. **10** (1975), pp. 85–112.

[Ro90] Roan, S.-S., *On Calabi–Yau orbifolds in weighted projective spaces*, International J. Math., **1**, No.2 (1990), pp. 211–232.

[Ro91] Roan, S.-S., *The mirror of Calabi–Yau orbifold*, Internat. J. Math., **2**, No.4 (1991), 439–455.

[Ro94] Roan, S.-S., *On* $c_1 = 0$ *resolution of quotient singularities*, Internat. J. Math., **4** (1994), pp. 523–536.

[Sch86] Schoen, C., *On the geometry of a special determinantal hypersurface associated to the Mumford-Horrocks vector bundle*, J. Reine Angew. Math. **364** (1986), pp. 85–111.

[Sch96] Schoen, C., *Varieties dominated by product varieties*, International J. Math. **7**, No. 4 (1996), pp. 541–571.

[Sh87] Shioda, T., *Some observations on Jacobi sums*, Adv. Stud. Pure Math. **12** (1987), pp. 119–135.

[So84] Soulé, C., *Groupes de Chow et K-théorie de varietés sur un corps fini*, Math. Ann. **268** (1984), pp. 317–345.

[We49] Weil, A., *Number of solutions of equations over finite fields*, Bull. Amer. Math. Soc. **55** (1949), pp. 497–508.

[Y05] Yui, N., *The L-series of Calabi–Yau orbifolds of CM type* with Appendix by Goto, Y., *The L-series of cubic hypersurface fourholds*, in *Mirror Symmetry V*, AMS/IP Studies in Advanced Mathematics **38** (2006), pp. 185–252.

Fields Institute Communications
Volume 54, 2008

String Modular Motives of Mirrors of Rigid Calabi-Yau Varieties

Savan Kharel
Department of Physics and Astronomy
Indiana University South Bend
South Bend, IN 46634 USA
skharel@iusb.edu

Monika Lynker
Department of Physics and Astronomy
Indiana University South Bend
South Bend, IN 46634 USA
mlynker@iusb.edu

Rolf Schimmrigk
Department of Physics and Astronomy
Indiana University South Bend
South Bend, IN 46634 USA
netahu@yahoo.com, rschimmr@iusb.edu

Abstract. The modular properties of some higher dimensional varieties of special Fano type are analyzed by computing the L-function of their $\Omega-$motives. It is shown that the emerging modular forms are string theoretic in origin, derived from the characters of the underlying rational conformal field theory. The definition of the class of Fano varieties of special type is motivated by the goal to find candidates for a geometric realization of the mirrors of rigid Calabi-Yau varieties. We consider explicitly the cubic sevenfold and the quartic fivefold, and show that their motivic L-functions agree with the L-functions of their rigid mirror Calabi-Yau varieties. We also show that the cubic fourfold is string theoretic, with a modular form that is determined by the L-series of an exactly solvable K3 surface.

2000 *Mathematics Subject Classification.* Primary 11G40; Secondary 11F23, 14G10, 14J32.

M.L. and R.S. were supported in part by the NSF under Grant No. PHY99-07949. They are grateful to the Kavli Institute for Theoretical Physics, Santa Barbara, for hospitality and support through KITP Scholarships during the course of part of this work. R.S. was also supported through IUSB faculty research grants.

CONTENTS

1 Introduction

In this paper we continue the program of applying methods from arithmetic geometry to the problem of understanding the question of how spacetime emerges in string theory, more precisely, the relation between the physics on the worldsheet and the nontrivial geometric component of spacetime. The focus here will be on a class of varieties that have been discussed in the context of mirror symmetry for rigid Calabi-Yau spaces [26, 6, 27]. These manifolds are higher dimensional spaces, characterized by the existence of a nontrivial cohomology group $H^{n-(Q-1),(Q-1)}(X)$, where $n = \dim_{\mathbb{C}} X$ and $Q \geq 1$ is an integer. In particular $H^{n-i,i}(X) = 0$ for $0 \leq i < Q-1$ if $Q > 1$, and for $Q = 1$ this class of varieties reduces to manifolds of Calabi-Yau type. For $Q > 1$ spaces of this type were called Fano manifolds of special type in [26, 27], and generalized Calabi-Yau manifolds in [6]. The latter name has in the meantime been adopted for Hitchin's generalization of Calabi-Yau varieties. We will call these spaces special Fano manifolds of charge Q. In the context of exactly solvable models manifolds of this type are distinguished by the fact that the number of tensor factors of $N = 2$ supersymmetric minimal models exceeds the number of variables normally associated with the standard relation between the central charge and the dimension of the variety.

Our goal is to show that, in complete analogy to the case of Calabi-Yau varieties considered in [30, 31] and references therein, the arithmetic geometry of Fano varieties of special type encodes features of the underlying conformal field theories. In particular the L-function of the Ω−motive, considered in [30, 32] in the context of Calabi-Yau varieties, generalizes to the class of Fano varieties of special type. We show that for those members in this class that are expected to be mirrors of rigid Calabi-Yau manifolds the inverse Mellin transform of the L-function of the Ω−motive with a Tate twist of charge $(Q-1)$ is a modular form of weight four which decomposes into factors derived from the underlying conformal field theory.

In the context of finding mirrors of rigid Calabi-Yau manifolds two varieties are of particular interest. The first is the cubic sevenfold

$$X_7^3 = \left\{ (z_0 : \cdots : z_8) \in \mathbb{P}_8 \;\middle|\; \sum_{i=0}^{8} z_i^3 = 0 \right\} \tag{1.1}$$

whose Hodge decomposition of the intermediate cohomology is given by

$$\begin{aligned} h^{7,0}(X_7^3) = h^{6,1}(X_7^3) &= 0 \\ h^{5,2}(X_7^3) &= 1 \\ h^{4,3}(X_7^3) &= 84 \end{aligned} \tag{1.2}$$

The second variety is the quartic fivefold

$$X_5^4 = \left\{ (z_0 : \cdots : z_6) \in \mathbb{P}_{(1,1,1,1,1,1,2)} \;\middle|\; \sum_{i=0}^{5} z_i^4 + z_6^2 = 0 \right\}, \tag{1.3}$$

with a Hodge decomposition of the intermediate cohomology that is given by

$$\begin{aligned} h^{5,0}(X_5^4) &= 0 \\ h^{4,1}(X_5^4) &= 1 \\ h^{3,2}(X_5^4) &= 90. \end{aligned} \tag{1.4}$$

The motivation for considering these hypersurfaces is that their cohomology suggests that they are related to conformal field theories at central charge $c = 9$. The geometric cohomology projection of [26, 27] and the period computation of [6] furthermore suggest that they are of relevance in the context of rigid mirrors of Calabi-Yau threefolds.

We first show that the Tate twisted Ω–motives of these varieties are modular in terms of Hecke eigenforms of weight four which are induced by string theoretic modular forms of elliptic type. The latter factor into Hecke indefinite modular forms of weight one associated to the affine Lie algebra $A_1^{(1)}$. Let X_n^d be a hypersurface of degree d and dimension n, and denote by $L_\Omega^{Q-1}(X_n^d, s)$ the L-function of the Ω–motive $M_\Omega(X_n^d)$ of the variety X_n^d with the necessary Tate twist of charge $(Q-1)$. Denote by $\Theta_{\ell,m}^k(\tau) = \eta^3(\tau) c_{\ell,m}^k(\tau)$ the theta functions defined in terms of the Dedekind eta function $\eta(\tau)$ and the Kac-Peterson string functions $c_{\ell,m}^k(\tau)$. The Hecke indefinite modular forms $\Theta_{\ell,m}^k(\tau)$ are associated to the affine Lie algebra $A_1^{(1)}$ of the underlying rational conformal field theory. Finally, we denote by χ_2 the Legendre character, and set $q = e^{2\pi i \tau}$.

Theorem 1.1 *Let $(n,d) \in \{(7,3),(5,4)\}$. The inverse Mellin transforms $f_\Omega^{Q-1}(X_n^d, q)$ of the L-functions $L_\Omega^{Q-1}(X_n^d, s)$ of the Ω–motive with a Tate twist of charge $(Q-1)$ are modular forms in $S_4(\Gamma_0(N))$ with $N \in \{9, 64\}$, described by the algebraic Hecke characters ψ_{d^3}*

$$f_\Omega^{Q-1}(X_n^d, q) = f(\psi_{d^3}^3, q). \tag{1.5}$$

The modular forms $f(\psi_{d^3}, q) \in S_2(\Gamma_0(d^3))$ are of elliptic type, and factor into string theoretic Hecke indefinite modular forms as

$$\begin{aligned} f(\psi_{27}, q) &= \Theta_{1,1}^1(q^3)\Theta_{1,1}^1(q^9) \\ f(\psi_{64}, q) &= \Theta_{1,1}^2(q^4)^2 \otimes \chi_2. \end{aligned} \tag{1.6}$$

This result provides an interpretation of the L-function of the twisted Ω–motive of these varieties in terms of modular forms on the string worldsheet theory of exactly solvable Gepner models at central charge $c = 9$, with nine and six minimal $N = 2$ supersymmetric factors, respectively. The twist character χ_2 is the quadratic character of the field of quantum dimensions of the underlying affine Lie algebra $A_1^{(1)}$ [28, 23, 29].

The characters ψ_{27} and ψ_{64} are the complex multiplication Hecke characters associated to the elliptic Brieskorn-Pham curves $E^3 \subset \mathbb{P}_2$ and $E^4 \subset \mathbb{P}_{(1,1,2)}$ of conductor 27 and 64, respectively. These tori are exactly solvable and their string theoretic modular forms have been determined in [33] and [23], respectively, to factor as

$$\begin{aligned} f(E^3, q) &= \Theta^1_{1,1}(q^3)\Theta^1_{1,1}(q^9) \\ f(E^4, q) &= \Theta^2_{1,1}(q^4)^2 \otimes \chi_2. \end{aligned} \tag{1.7}$$

Their appearance in the context of the two Fano varieties X_7^3 and X_5^4 suggests to consider resolutions X_3^d of quotients of triple products of elliptic curves $(E^d)^3$ for $d = 3, 4$, and to compare their L-functions with those of the special Fano varieties determined in Theorem 1.1. Define $\mathbb{Z}_d := \mathbb{Z}/d\mathbb{Z}$. It turns out that there are actions of $\mathbb{Z}_d \times \mathbb{Z}_d$ on $(E^d)^3$ such that the resolved manifolds X_3^d of the quotient threefolds $(E^d)^3/\mathbb{Z}_d \times \mathbb{Z}_d$ have the appropriate mirror cohomology of the special Fano varieties X_n^d for $d = 3, 4$:

$$\begin{aligned} h^{1,1}(X_3^3) &= 84 \\ h^{1,1}(X_3^4) &= 90. \end{aligned} \tag{1.8}$$

Combining Theorem 1.1 with results by Cynk and Hulek [7] leads to the following corollary.

Theorem 1.2 *The varieties X_3^d, $d = 3, 4$, are rigid Calabi-Yau manifolds whose L-functions of the intermediate cohomology are modular and agree with the L-functions of the special Fano varieties X_n^d with $(d, n) \in \{(3, 7), (4, 5)\}$, respectively.*

We interpret the string theoretic modularity of the special Fano varieties X_7^3 and X_5^4, together with the agreement of their motivic L-functions with those of their corresponding rigid Calabi-Yau spaces, as further support for the interpretation of certain higher dimensional Fano varieties of special type as mirrors of rigid Calabi-Yau manifolds. It would be of interest to extend our arithmetic analysis of the mirror pairs (X_n^d, X_3^d) to families by using the methods developed in [3, 4], and further considered in [19, 20, 21, 5].

The modular analysis described here generalizes to other dimensions and central charges. To illustrate this we consider a third Fano variety of special type, defined by the cubic fourfold

$$X_4^3 = \left\{ (z_0 : \cdots : z_5) \in \mathbb{P}_5 \;\middle|\; \sum_{i=0}^{5} z_i^3 = 0 \right\}, \tag{1.9}$$

whose Hodge decomposition of the intermediate cohomology is given by

$$\begin{aligned} h^{4,0}(X_4^3) &= 0 \\ h^{3,1}(X_4^3) &= 1 \\ h^{2,2}(X_4^3) &= 21. \end{aligned} \tag{1.10}$$

For this variety we obtain the following string theoretic interpretation. Let $\vartheta(q)$ denote a modular form of weight one to be defined later in the paper.

Theorem 1.3 *The inverse Mellin transform $f^1_\Omega(X_4^3, q)$ of the L-series of the Tate twisted Ω-motive of the cubic fourfold X_4^3 is a modular form of weight three and level 27, which factors into Hecke indefinite modular forms as*

$$f^1_\Omega(X_4^3, q) = \vartheta(q^3)\Theta^1_{1,1}(q^3)\Theta^1_{1,1}(q^9) \in S_3(\Gamma_0(27)). \tag{1.11}$$

This form is identical to the modular form of the Ω*–motive of the K3 surface*

$$S^6 = \left\{(z_0 : \cdots : z_3) \in \mathbb{P}_{(1,1,1,3)} \;\middle|\; z_0^6 + z_1^6 + z_2^6 + z_3^2 = 0\right\}, \tag{1.12}$$

i.e. $f_\Omega^1(X_4^3, q) = f_\Omega(S_6, q)$. *It has complex multiplication and is determined by the Hecke character* ψ_{27} *of weight 2 associated to the Eisenstein field* $\mathbb{Q}(\mu_3)$

$$f_\Omega^1(X_4^3, q) = f(\psi_{27}^2, q). \tag{1.13}$$

The modularity of the cubic fourfold has independently been considered by Goto [9] and also by Hulek and Kloosterman [12]. The string theoretic interpretation based on the affine algebra $A_1^{(1)}$ follows from the results of [30].

2 Fano varieties of special type

In this section we briefly review the characterization of Fano varieties of special type introduced in [26, 6, 27], adopting the notation of [26, 27]. Let $Q, D_{\mathrm{crit}} \in \mathbb{N}$ and set $s = D_{\mathrm{crit}} + 2Q - 1$. We consider the class of hypersurfaces in weighted projective space $\mathbb{P}_{(k_0,\ldots,k_s)}$ with weights k_i chosen such that

$$d = \frac{1}{Q}\sum_{i=0}^{s} k_i \tag{2.1}$$

is an integer, defining the degree of the polynomial that determines the variety

$$X_{s-1}^d = \left\{(z_0 : \cdots : z_s) \in \mathbb{P}_{(k_0,\ldots,k_s)} \;\middle|\; p(z_0, \ldots, z_s) = 0,\ \deg p = d\right\}. \tag{2.2}$$

The integer Q has the physical interpretation of a charge of the underlying physical theory, and D_{crit} can be viewed as the critical complex dimension of the corresponding Calabi-Yau manifold in those cases where it can be constructed. It is determined by the central charge c of the conformal field theory as $c = 3D_{\mathrm{crit}}$.

Alternatively, varieties X of this type can be characterized by their first Chern class

$$c_1(X) = (Q-1)c_1(\mathcal{N}), \tag{2.3}$$

where $c_1(\mathcal{N})$ is the first Chern class of the normal bundle $\mathcal{N}$ of $X \subset \mathbb{P}_{(k_0,\ldots,k_s)}$. For $Q = 1$ this construction therefore recovers the class of Calabi-Yau hypersurfaces for arbitrary dimensions, while for $Q > 1$ these hypersurfaces have positive first Chern class and hence do not a priori describe consistent string vacua, but instead define a sub-class of Fano varieties. It was shown in [26, 27] that these varieties nevertheless are closely related to string vacua in the critical dimension $D_{\mathrm{crit}} = s + 1 - 2Q$ in the sense that it is possible to derive the massless spectrum of critical vacua from the cohomology of the special Fano $(s-1)$–folds via a geometric projection. For particular classes of these special Fano varieties it is also possible to construct the critical Calabi-Yau manifold explicitly via this projection, in which case the charge Q has the meaning of the codimension of the critical manifold. More intrinsically, we can therefore write

$$D_{\mathrm{crit}}(X) = \dim_{\mathbb{C}} X + 2(1-Q). \tag{2.4}$$

The essential ingredient of our arithmetic analysis below is the existence of a nontrivial cohomology group $H^{n-(Q-1),(Q-1)}(X)$. We therefore use this feature as the characterizing property.

Definition 2.1 A Fano variety is called special and of charge Q if it has a nonvanishing cohomology group $H^{n-(Q-1),(Q-1)}(X)$.

Interest in this class of manifolds arose originally because they provide a geometric framework for $N = 2$ exactly solvable Gepner models which do not have Kähler deformations, hence they provide candidates for the mirrors of rigid Calabi-Yau varieties [26, 6, 27]. A toric description of these special Fano spaces was subsequently given by Batyrev and Borisov [1], and Batyrev and Dais [2]. A discussion of Lagrangian subvarieties in the context of the Strominger-Yau-Zaslow mirror conjecture [36] was given by Leung [22].

3 String theoretic modular forms

3.1 The simplest class of $N = 2$ supersymmetric exactly solvable theories are built in terms of the affine Lie algebras. A construction of such non-twisted affine Kac-Moody algebras is provided by the extension [16]

$$\hat{\underline{G}} = L\underline{G} \oplus \mathbb{C}k \oplus \mathbb{C}d \tag{3.1}$$

of the loop algebra

$$L\underline{G} = \underline{G} \otimes \mathbb{C}[t, t^{-1}] \tag{3.2}$$

by the central extension k and $d = t\frac{d}{dt}$. In terms of the generators $J^a \otimes t^m$ the algebra takes the form

$$[J^a \otimes t^m, J^b \otimes t^n] = if_c^{ab} J^c \otimes t^{m+n} + km\delta^{ab}\delta_{m+n,0}. \tag{3.3}$$

The representations of this algebra can be parametrized by affine roots $\hat{\lambda} = (\lambda, k, n)$ of the Cartan-Weyl subalgebra $\{H_0^i, E_0^\alpha, L_0\}$, with $i = 1, ..., r$, where r denotes the rank of the underlying Lie algebra $\underline{G}$. For fixed affine level k of the theory the characters $\chi_{\hat{\lambda}}$ are essentially parametrized by the weight λ of the representation. For the reduced characters of an affine Lie algebra $\hat{\underline{G}}$ at level k the characters transform as

$$\chi_{\hat{\lambda}}(-1/\tau) = \sum_{\hat{\mu} \in P_+^k} S_{\hat{\lambda},\hat{\mu}} \chi_{\hat{\mu}}(\tau), \tag{3.4}$$

where the modular S-matrix takes the form

$$S_{\hat{\lambda},\hat{\mu}} = \frac{i^{|\Delta_+|}}{\sqrt{|P/Q^\vee|(k+g)^r}} \sum_{w \in W} \epsilon(w) e^{-2\pi i \frac{<w(\lambda+\rho).\mu+\rho>}{k+g}}. \tag{3.5}$$

Here $P = \sum_i \mathbb{Z}\omega_i$ denotes the lattice with fundamental weights ω_i defined by $< \omega_i, \alpha_j^\vee >= \delta_{ij}$ via co-roots $\alpha_j^\vee = \frac{2\alpha}{<\alpha,\alpha>}$. $Q^\vee = \sum_i \mathbb{Z}\alpha_i^\vee$ is the co-root lattice and $P/Q^\vee$ denotes the lattice points of P lying in an elementary cell of $Q^\vee$, while $|P/Q^\vee|$ describes the number of points in this set. P_+^k is the set of all dominant weights at level k, and $\epsilon(w) = (-1)^{\ell(w)}$ is the signature of the Weyl group element w, where $\ell(w)$ is the minimum number of simple Weyl reflections that w decomposes into. Δ_+ is the number of positive roots in the Lie algebra $\underline{G}$.

3.2 Supersymmetric string models can be constructed in terms of conformal field theories with $N = 2$ supersymmetry. For the present discussion the important structure is determined by the affine Lie algebra $A_1^{(1)}$, which provides the essential building block of the supersymmetric theory. It turns out that of particular importance are the Hecke indefinite modular forms which can be defined as

$$\Theta_{\ell,m}^k(\tau) = \sum_{\substack{-|x| < y \leq |x| \\ (x,y) \text{ or } (\frac{1}{2}-x, \frac{1}{2}+y) \\ \in \mathbb{Z}^2 + \left(\frac{\ell+1}{2(k+2)}, \frac{m}{2k}\right)}} \text{sign}(x) e^{2\pi i \tau((k+2)x^2 - ky^2)} \tag{3.6}$$

Related to these forms are the string functions $c^k_{\ell,m}(\tau) = \Theta^k_{\ell,m}(\tau)/\eta^3(\tau)$ of Kac-Peterson [17, 18], which are of immediate physical relevance because they appear in the $N = 2$ superconformal characters $\chi^k_{\ell,q,s}(\tau)$,

$$\chi^k_{\ell,q,s}(\tau, z, u) = \sum c^k_{\ell,q+4j-s}(\tau)\theta_{2q+(4j-s)(k+2),2k(k+2)}(\tau, z, u), \tag{3.7}$$

where the theta functions θ_m are defined as

$$\theta_{n,m}(\tau, z, u) = e^{-2\pi imu} \sum_{\ell\in\mathbb{Z}+\frac{n}{2m}} e^{2\pi im\ell^2\tau+2\pi i\ell z}. \tag{3.8}$$

These characters in turn define the partition function of the conformal field theory.

The theta functions $\Theta^k_{\ell,m}(\tau)$ are associated to quadratic number fields determined by the level of the affine theory. These are modular forms of weight one and cannot, therefore, be identified with geometric modular forms. It turns out, however, that appropriate products do lead to interesting forms [33, 23, 29, 30, 31].

4 L-functions of $\Omega-$motives

For a general smooth algebraic variety X reduced mod p the congruence zeta function of $X/\mathbb{F}_p$ is defined by

$$Z(X/\mathbb{F}_p, t) = \exp\left(\sum_{r\in\mathbb{N}} \#\left(X/\mathbb{F}_{p^r}\right)\frac{t^r}{r}\right). \tag{4.1}$$

Here the sum is over all finite extensions $\mathbb{F}_{p^r}$ of $\mathbb{F}_p$ of degree r. Per definition $Z(X/\mathbb{F}_p, t) \in 1 + \mathbb{Q}[[t]]$, but the expansion can be shown to be integer valued by writing it as an Euler product.

The proof by Grothendieck [10] of part of the Weil conjectures [40] asserts that the zeta function is a rational function determined by the cohomology of the variety

$$Z(X/\mathbb{F}_p, t) = \frac{\prod_{j=1}^n \mathcal{P}_p^{2j-1}(t)}{\prod_{j=0}^n \mathcal{P}_p^{2j}(t)}, \tag{4.2}$$

where $\dim_{\mathbb{C}}X = n$, and $\mathcal{P}_p^i(t)$ is a polynomial $\mathcal{P}_p^i(t) = \sum_{j=0}^{b_i}\beta_j^i(p)t^j$ associated to the ith cohomology group with degrees b_i given by the ith Betti number $b_i = \dim H^i(X)$. This result motivates the introduction of L−functions associated to the polynomials $\mathcal{P}_p^i(t)$, thereby reducing the complexity of the congruent zeta function.

In the context of the string theoretic modularity problem the L-function associated to the full cohomology group of a variety is a physically useful object only in the case of elliptic curves. In higher dimensions these functions are in general not modular. This motivates the consideration of a factorization of L−functions, and to ask whether modular forms emerge from the emerging pieces, and if so, whether these modular forms admit a Kac-Moody theoretic interpretation.

It was shown in refs. [30, 32], in the context of establishing string theoretic modularity of certain Calabi-Yau varieties, that a useful way to factorize cohomological L-functions of Calabi-Yau varieties is to consider subspaces of the cohomology defined by Galois orbits of the holomorphic $n-$forms Ω, $n = \dim_{\mathbb{C}}X$, where the Galois groups are defined in an inherent way by the arithmetic of the variety. The resulting L-functions have rational coefficients, and therefore can in principle admit a string theoretic interpretation along the lines discussed in [33, 23, 29, 30, 31, 32]. In this paper we generalize this strategy to the case of Fano varieties of special type by considering the orbit motive of a Galois representation determined by the generator of the cohomology group $H^{n-(Q-1),(Q-1)}(X)$ with $Q > 1$. We continue to denote the generator of this group by Ω, and denote the motive by $M_\Omega(X)$. The goal is to compute the L-series associated to this motive, and to find a string theoretic interpretation of this function, following the strategy used in [30, 32]. More precisely, we need to consider motives with Tate twists, introduced by Tate in the context of the Tate conjectures, which assert that certain subgroups of the ith twists $H^{2i}(X)(i)$ of the cohomology groups $H^{2i}(X)$ are generated by classes of algebraic cycles [37, 38] (see also [34]). In the present context we consider for any special Fano variety of charge Q the twisted omega motive $M_\Omega(X)(Q-1)$ and its L-function, which we will denote by $L_\Omega^{Q-1}(X,s) = L(M_\Omega(X)(Q-1),s)$. We will call $M_\Omega(X)(Q-1)$ the $\Omega-$motive with a Tate twist of charge $(Q-1)$, or simply the omega motive of charge $(Q-1)$.

In analogy to the general considerations in [30, 32] for the class of Calabi-Yau varieties in arbitrary dimensions, we can ask the following for the class of Fano varieties of special type.

Question 4.1 Is the L-function $L_\Omega^{Q-1}(X,s)$ of the $\Omega-$motive $M_\Omega(X)(Q-1)$ of charge $(Q-1)$ of a special Fano variety of charge Q always modular?

Question 4.2 Can modular inverse Mellin transforms of $L_\Omega^{Q-1}(X,s)$ be expressed in terms of Hecke indefinite modular forms associated to Kac-Moody algebras?

It will become clear below that the answer to this question is affirmative, at least for certain examples. In such cases the L-function can be viewed as a map that takes motives and turns them into conformal field theoretic objects. More challenging, in part because of the lack of definition of some of the terms, therefore would be the following picture.

Conjecture. The L-function is a functor from the additive category of motives of special Fano type to the multiplicative category of $N = 2$ supersymmetric conformal field theories.

The basic outline described below is a generalization of the method based on Jacobi sums of weighted projective spaces described in [30] in the context of K3 surface modularity, and in [32] for higher dimensional Calabi-Yau varieties.

5 Modularity of Hecke L-series

The modularity of the L-series determined in this paper follows from the fact that they can be interpreted in terms of Hecke L-series associated to algebraic Hecke character, or Größencharaktere, defined by Jacobi sums. A Größencharakter can be associated to any number field. Hecke's modularity discussion [11] has been extended by Shimura [35] and Ribet [24].

Let K be a number field and $\sigma: K \longrightarrow \mathbb{C}$ denote an embedding.

Definition 5.1 A Größencharakter is a homomorphism $\psi : I_\mathfrak{m}(K) \rightarrow \mathbb{C}^\times$ of the fractional ideals of K prime to the congruence integral ideal $\mathfrak{m}$ such that $\psi((z)) = \sigma(z)^{w-1}$, $\forall K^\times \ni z \equiv 1(\text{mod}^\times \mathfrak{m})$.

In the present case the cyclotomic Jacobi sums determined by the finite field Jacobi sums computed above arise from imaginary quadratic fields, in which case the structure of these characters simplifies. Consider an imaginary quadratic field

K with discriminant $-D$ and denote by φ the Dirichlet character associated to K, viewed as a mod D character. Define a second Dirichlet character λ defined mod $\mathrm{N}\mathfrak{m}$ by setting

$$\lambda(a) = \psi((a))/\sigma(a)^{w-1}, \qquad a \in \mathbb{Z}. \tag{5.1}$$

The product of φ and λ then determines the nebentypus character of the modular form determined by the Hecke character ψ.

Denote by $\mathrm{N}\mathfrak{p}$ the norm of the prime ideal $\mathfrak{p}$, and by

$$L(\psi, s) = \prod_{\mathfrak{p}\in \mathrm{Spec}\ \mathcal{O}_K} \left(1 - \psi(\mathfrak{p})\mathrm{N}\mathfrak{p}^{-s}\right)^{-1} \tag{5.2}$$

the Hecke L-series associated to the character ψ. The modularity of the corresponding $q-$series $f(\psi, q) = \sum_n a_n q^n$ associated to the L-series via the Mellin transform is characterized by the following result of Hecke.

Theorem 5.2 *The power series $f(\psi,q)$ is a cusp form of weight w and character $\epsilon = \varphi\lambda$ on $\Gamma_0(D\mathrm{N}\mathfrak{m})$. If $p \not| D\mathrm{N}\mathfrak{m}$ then $f|_{T_p} = a_p f$, where T_p are the Hecke operators.*

6 Special Fano varieties with critical dimension three

The focus in this paper will be on Fano varieties of Brieskorn-Pham type, hence the computation of their L-functions can proceed via their Jacobi sums [40, 41].

Theorem 6.1 *For a smooth weighted projective hypersurface with degree vector $\underline{n} = (n_0, ..., n_s)$*

$$X^{\underline{n}} = \{b_0 z_0^{n_0} + b_1 z_1^{n_1} + \cdots + b_s z_s^{n_s} = 0\} \subset \mathbb{P}_{(k_0,k_1,\ldots,k_s)}, \tag{6.1}$$

defined over the finite field $\mathbb{F}_q$, for $q = p^r, r \in \mathbb{N}$, set $d_i = (n_i, q-1)$ and denote by $\mathcal{A}_s^{q,\underline{n}}$ the set of rational vectors $\alpha = (\alpha_0, \alpha_1, \ldots, \alpha_s)$ given by

$$\mathcal{A}_s^{q,\underline{n}} = \left\{\alpha \in \mathbb{Q}^{s+1} \;\Big|\; 0 < \alpha_i < 1,\ d_i\alpha_i = 0 \bmod 1,\ \sum_{i=0}^{s} \alpha_i \equiv 0 \pmod 1\right\}. \tag{6.2}$$

For each $(s+1)-$tuple α define the Jacobi sum

$$j_q(\alpha) = \frac{1}{q-1} \sum_{\substack{u_i\in\mathbb{F}_q \\ u_0+u_1+\cdots+u_s=0}} \chi_{\alpha_0}(u_0)\chi_{\alpha_1}(u_1)\cdots\chi_{\alpha_s}(u_s), \tag{6.3}$$

where $\chi_{\alpha_i}(u_i) = e^{2\pi i \alpha_i m_i}$ with integers m_i determined via $u_i = g^{m_i}$, where $g \in \mathbb{F}_q$ is a generator. Then the cardinality of $X^{\underline{n}}/\mathbb{F}_q$ is given by

$$N_q(X^{\underline{n}}) = 1 + q + \cdots + q^{s-1} + \sum_{\alpha\in\mathcal{A}_s^{q,\underline{n}}} j_q(\alpha). \tag{6.4}$$

With these Jacobi sums one defines for primes p such that $d \mid (p^f - 1)$ for some $f \in \mathbb{N}$ the polynomials

$$\mathcal{P}_p^{s-1}(t) = \prod_{\alpha\in\mathcal{A}_s^n} \left(1 - (-1)^{s-1} j_{p^f}(\alpha) \prod_i \bar{\chi}_{\alpha_i}(b_i) t\right)^{1/f} \tag{6.5}$$

and the associated L-function is given as a product over all good primes p,

$$L^{(j)}(X, s) = \prod_p (\mathcal{P}_p^j(p^{-s}))^{-1}. \tag{6.6}$$

The completion of the L-function at the bad primes will not be of relevance in the following.

The two varieties considered here in the context of $D_{\rm crit}$, i.e. Gepner models with central charge $c = 9$, are odd-dimensional, hence the rational form of the congruence zeta function for a projective hypersurface of odd dimension s takes the form

$$Z(X_{s-1}/\mathbb{F}_p, t) = \frac{\mathcal{P}_p^{s-1}(t)}{(1-t)(1-pt)\cdots(1-p^{s-1}t)}. \tag{6.7}$$

We also consider a Fano variety corresponding to a K3 surface, i.e. a Gepner model with central charge $c = 6$. In this case the smooth Fano variety is of complex dimension four and the zeta function is of the form

$$Z(X_4^3/\mathbb{F}_p, t) = \frac{1}{(1-t)(1-pt)\mathcal{P}_p^4(t)(1-p^3t)(1-p^4t)}. \tag{6.8}$$

In all these examples there is therefore only one interesting cohomological L-function. It turns out that these L-functions are not themselves of interest and that it is more useful to consider the L-function of the Ω–motive. In all three examples the order of the Galois group is two, hence the motive is of dimension two. Denoting the Jacobi sum associated to the Ω–form by $j_p(\alpha_\Omega)$ the polynomial of the Ω–motive takes the form

$$\mathcal{P}_p^\Omega(t) = \left(1 - j_{p^f}(\alpha_\Omega)t^f\right)^{1/f}\left(1 - j_{p^f}(\bar{\alpha}_\Omega)t^f\right)^{1/f}, \tag{6.9}$$

where $\bar{\alpha}_\Omega := \mathbf{1} - \alpha_\Omega$ is the dual vector and $\mathbf{1}$ is the unit vector. For primes p such that $d \mid (p-1)$ this simplifies and the local L-function factors of the twisted omega motive of charge $(Q-1)$ take the form $L_{\Omega,p}^{Q-1}(X,s) = (\mathcal{P}_p^{\Omega,Q-1}(p^{-s}))^{-1}$ with

$$\mathcal{P}_p^{\Omega,Q-1}(t) = 1 + \beta(p)p^{-(Q-1)}t + p^{n-2(Q-1)}t^2,$$

where $\beta(p) = j_p(\alpha_\Omega) + j_p(\bar{\alpha}_\Omega)$.

6.1 The cubic Fermat sevenfold X_7^3.

6.1.1 *The motivic L-function.* For the cubic Fermat sevenfold the Hodge decomposition takes the form $h^{i,i}(X_7^3) = 1$, $i = 0,1,2,3$ with all other Hodge numbers $h^{i,j}(X_7^3) = 0$ for $i,j < 4, i \neq j$. The intermediate cohomology decomposes as in (1.2), i.e. we have $Q = 3$ and $D_{\rm crit} = 3$. The zeta function therefore has the form $Z(X_7^3/\mathbb{F}_p, t) = \mathcal{P}_p^7(t)/\prod_{i=0}^7(1-p^it)$, leading to the L-function

$$L(X_7^3, s) := \prod_p 1/\mathcal{P}_p^7(p^{-s}). \tag{6.10}$$

The cyclotomic field defined by this manifold is $\mathbb{Q}(\mu_3)$, hence we have

$$\mathrm{Gal}(\mathbb{Q}(\mu_3)/\mathbb{Q}) = (\mathbb{Z}/3\mathbb{Z})^\times,$$

and the L-series of the Ω–motive of X_7^3 can therefore be computed from the orbit of the Jacobi sums $j_p(\alpha_\Omega)$ of length two, with $\alpha_\Omega = \frac{1}{3}(1,1,1,1,1,1,1,1,1)$. The L-series of the twisted Ω–motive of charge two of X_7^3 is defined as $L_\Omega^2(X_7^3, s) = \prod_p 1/\mathcal{P}_p^{\Omega,2}(p^{-s})$ with $\mathcal{P}_p^{\Omega,2}(t) = 1 + a_pt + p^3t^2$ for $n \mid (p-1)$, $\beta(p) = j_p(\alpha_\Omega) + \bar{j}_p(\alpha_\Omega)$ and $a_p = \beta(p)/p^2$ because the Tate twist is of charge two. The computation of these sums for the first few nontrivial primes is sufficient to determine the L-function of the Tate twisted Ω–motive

$$L_\Omega^2(X_7^3, s) \doteq 1 + \frac{20}{7^s} - \frac{70}{13^s} + \frac{56}{19^s} + \frac{308}{31^s} + \cdots \tag{6.11}$$

and the associated $q-$series

$$f_{\Omega}^{2}(X_{7}^{3},q) \doteq q + 20q^{7} - 70q^{13} + 56q^{19} + 308q^{31} + \cdots \tag{6.12}$$

completely.

It will become clear below that this $q-$series describes a modular cusp form of weight four and modular level nine with respect to the Hecke congruence group.

It was suggested in [26, 6, 27] that the cubic Fermat sevenfold can be interpreted as a geometric model associated to the tensor model $1^{\otimes 9}$ of nine copies of $N = 2$ superconformal models [8]. If this is indeed the case, then we would expect, in analogy to the results obtained in [33, 23, 29, 30, 31, 32], that we can express the $q-$series of the $\Omega-$motive in terms of the Hecke indefinite modular forms $\Theta_{\ell,m}^{k}(\tau)$ associated to the underlying conformal field theory model. At the affine level $k = 1$ there is a single such form, hence this is a make or break situation which is very constrained, much like the first example established in this framework [33]. It turns out that the geometric modular form of the $\Omega-$motive of the cubic sevenfold does admit a string theoretic interpretation. There are several ways this can be described. The most direct is that the modular form $f_{\Omega}^{2}(X_{7}^{3},q)$ can be written as

$$f_{\Omega}^{2}(X_{7}^{3},q) = \Theta_{1,1}^{1}(q^{3})^{4} \in S_{4}(\Gamma_{0}(9)). \tag{6.13}$$

This modular form has the expected weight of a Calabi-Yau threefold, consistent with the central charge of the model. The absence of a twist character in this result is consistent with the fact that the field of quantum dimensions of the underlying field theory is trivial [28].

6.1.2 *CM interpretation of the L-series and modularity.* The motivic modular form $f_{\Omega}^{2}(X_{7}^{3},q)$ of the cubic sevenfold is a form with complex multiplication by the Eisenstein field $\mathbb{Q}(\sqrt{-3})$, and therefore can be described by a Hecke character associated to this field. Define a character ψ_{27} by the following congruence condition

$$\psi_{27}(\mathfrak{p}) = \alpha_{\mathfrak{p}}, \qquad \alpha_{\mathfrak{p}} \equiv 1 (\mathrm{mod}\ 3), \tag{6.14}$$

where the generator of the prime ideal $\mathfrak{p} = (\alpha_{\mathfrak{p}})$ is determined uniquely by the congruence constraint. As described in the previous section, associated to any Hecke character ψ is a Hecke series $L(\psi,s)$, and we find

$$L_{\Omega}^{2}(X_{7}^{3},s) = L(\psi_{27}^{3},s). \tag{6.15}$$

It follows from Hecke's theorem that the $q-$series of this L-function is modular, hence the data provided above is sufficient to prove that the Mellin transform of the L-function $L_{\Omega}^{2}(X_{7}^{3},s)$ is a modular form, and therefore the Tate twisted $\Omega-$motive of the cubic sevenfold is modular.

This result shows that the basic arithmetic building block of the cubic sevenfold is the elliptic cubic Fermat curve

$$E^{3} = \left\{(z_{0}:z_{1}:z_{2}) \in \mathbb{P}_{2} \mid z_{0}^{3} + z_{1}^{3} + z_{2}^{3} = 0\right\}. \tag{6.16}$$

The Hasse-Weil L-series of E^{3} agrees with the L-series of the algebraic Hecke character ψ_{27} :

$$L_{\mathrm{HW}}(E^{3},s) = L(\psi_{27},s). \tag{6.17}$$

The Mellin transform of $L_{\mathrm{HW}}(E^{3},s)$ has weight 2 and level 27, and it factors into string theoretically induced modular forms as [33]

$$f(E^{3},q) = \Theta_{1,1}^{1}(q^{3})\Theta_{1,1}^{1}(q^{9}). \tag{6.18}$$

This proves the part of Theorem 1.1 concerned with the cubic Fermat sevenfold.

The relation between the twisted Ω–motivic L-series of X_7^3 and the L-series of the elliptic curve E^3 that follows from these considerations can be formulated in a more direct way. Consider the expansions $L_\Omega^2(X_7^3, s) = \sum_n a_n n^{-s}$ for the sevenfold, and $L_{\rm HW}(E^3, s) = \sum_n b_n n^{-s}$ for the elliptic curve. The Hecke character interpretation means that $a_p = \alpha_{\mathfrak{p}}^3 + \bar{\alpha}_{\mathfrak{p}}^3$ and $b_p = \alpha_{\mathfrak{p}} + \bar{\alpha}_{\mathfrak{p}}$. Therefore for split primes p

$$a_p = b_p^3 - 3pb_p. \tag{6.19}$$

6.1.3 *X_7^3 as the mirror of a rigid Calabi-Yau variety.* The fact that the CM character of the elliptic curve $E^3 \subset \mathbb{P}_2$ is the building block of the L-function of the cubic sevenfold suggests that its Ω–motivic L-function should be identical to that of the resolution of a quotient of the triple product $(E^3)^3$ which produces the correct mirror cohomology of the critical part of the spectrum of the sevenfold. Such a quotient variety can be obtained by considering the $\mathbb{Z}_3 \times \mathbb{Z}_3$ action defined on $(E^3)^3$ as

$$\mathbb{Z}_3 \times \mathbb{Z}_3 \ni g_1 \times g_2 : \begin{bmatrix} 1 & 0 & 0 & 0 & 0 & 0 & 2 & 0 & 0 \\ 0 & 0 & 0 & 1 & 0 & 0 & 2 & 0 & 0 \end{bmatrix} \tag{6.20}$$

with obvious notation for the generators g_1, g_2. The resolution X_3^3 of the quotient variety $(E^3)^3/\mathbb{Z}_3 \times \mathbb{Z}_3$ is rigid and has $h^{1,1}(X_3^3) = 84$, as needed for the mirror cohomology of the relevant part of the cohomology of the sevenfold. Higher dimensional varieties defined by the resolution of quotients of the type E^n/G, for elliptic curves and certain finite groups G were considered by Cynk and Hulek [7].

Theorem 6.2 *Let $n \in \mathbb{N}$ be an odd integer, E be the elliptic curve with an automorphism of order 3, and denote by $\bar{X}_n$ the quotient of E^n by the action of the group*

$$\left\{ (\xi^{a_1} \times \cdots \times \xi^{a_n}) \in \mathrm{End}(E^n) \mid \sum_{i=1}^n a_i \equiv 0 (\mathrm{mod}\ 3) \right\}.$$

Then $\bar{X}_n$ has a smooth model X_n, which is a Calabi-Yau manifold such that $\dim H^n(X_n) = 2$ *and $L(X_n, s) \doteq L(g_{n+1}, s)$, where g_{n+1} is the weight $n+1$ cusp form with complex multiplication by $\mathbb{Q}(\sqrt{-3})$, associated to the $n^{\rm th}$ power of the Größencharakter of E.*

It follows by combining this result with those of Theorem 1.1 that the L-function of the Ω–motive of the cubic sevenfold, with a Tate twist of charge two, is identical to the L-function of the intermediate cohomology of the variety X_3^3. This completes the proof of the part of Theorem 1.2 concerned with the cubic Fermat sevenfold.

6.1.4 *Relatives and correspondences.* The agreement between the motivic L-function of X_7^3 and the L-function of the intermediate cohomology of X_3^3 identifies these two varieties as "relatives" in the sense of ref. [13]. A variety X_1 with a 2-dimensional Galois representation ρ_1 is defined to be a relative of a variety X_2 if ρ_1 occurs in the cohomology of X_2. This notion identifies X_3^3 as a relative of X_7^3. Other relatives of Calabi-Yau type for X_3^3 have been considered by van Geemen in [42], Saito-Yui [25], Verrill [39], and Yui [43].

A stronger relation is predicted by the Tate conjecture, which in the present context can be formulated as a statement that an isomorphism between two 2-dimensional Galois representations which occur in the étale cohomology of two varieties defined over $\mathbb{Q}$ should be induced by a correspondence between the two varieties defined over $\mathbb{Q}$, i.e. there should be an algebraic cycle on the product of the two varieties [43]. The Tate conjecture therefore implies that there should exist a correspondence between X_3^3 and X_7^3. An explicit construction of such a correspondence should provide better insight into the precise relation between the pair of manifolds (X_7^3, X_3^3).

The fact that physically the L-function is determined by the modular form $\Theta_{1,1}^1(\tau)$ of the underlying conformal field theory on the worldsheet leads to the conclusion that from a four-dimensional point of view these two varieties contain motivic sectors that are indistinguishable.

6.2 The quartic Brieskorn-Pham fivefold X_5^4.

6.2.1 *The L-series of the $\Omega-$motive.* The quartic fivefold X_5^4 defined in (1.3) has a Hodge decomposition of the form $h^{i,i}(X_5^4) = 1, i = 0,1,2$, with all other $H^{i,j}(X_5^4) = 0$ for $i,j < 5$ and $i \neq j$. The decomposition of the intermediate cohomology group is given by (1.4), i.e. $Q = 2$ and $D_{\mathrm{crit}} = 3$.

The congruence zeta function therefore takes the form

$$Z(X_5^4/\mathbb{F}_p, t) = \mathcal{P}_p^5(t)/\prod_{i=0}^{5}(1-p^i t),$$

leading to the L-function $L(X_5^4, s) = \prod_p(\mathcal{P}_p^5(p^{-s}))^{-1}$. We are interested in the factor describing the twisted $\Omega-$motive of charge one, which is determined by the Jacobi sums $j_p(\alpha_\Omega)$ with $\alpha_\Omega = \frac{1}{4}(1,1,1,1,1,1,2)$. The computation of the first few Jacobi sums for low primes leads to the $q-$expansion of the L-function of the twisted $\Omega-$motive of the quartic fivefold given by

$$f_\Omega^1(X_5^4, q) \doteq q - 22q^5 + 18q^{13} - 94q^{17} + 130q^{29} + \cdots \tag{6.21}$$

This is a modular cusp form of weight four and level 64, $f_\Omega^1(X_5^4, q) \in S_4(\Gamma_0(64))$.

6.2.2 *CM interpretation of the L-series and modularity.* The motivic modular form $f_\Omega^1(X_5^4, q)$ has CM by the Gauss field $\mathbb{Q}(\sqrt{-1})$, and therefore can be described by a Hecke character associated to this field as well. Consider the twisted character $\psi_{64} = \psi_{32}\chi_2$ where ψ_{32} is defined by

$$\psi_{32}(\mathfrak{p}) = \alpha_\mathfrak{p}, \qquad \alpha_\mathfrak{p} \equiv 1(\mathrm{mod}\ (2+2i)), \tag{6.22}$$

where the generator of the prime ideal $\mathfrak{p} = (\alpha_\mathfrak{p})$ is determined uniquely by the constraint.

The Hecke series associated to ψ_{64} then is again the building block of the geometric L-series

$$L_\Omega^1(X_5^4, s) = L(\psi_{64}^3, s) \tag{6.23}$$

hence modularity of this L-function follows. Similar to the cubic sevenfold this result shows that the quartic fivefold has an elliptic building block, in this case the Brieskorn-Pham type curve

$$E^4 = \left\{(z_0 : z_1 : z_2) \in \mathbb{P}_{(1,1,2)} \;\middle|\; z_0^4 + z_1^4 + z_2^2 = 0\right\}. \tag{6.24}$$

It can be checked that the Hasse-Weil L-series of E^4 is determined by the character ψ_{64}

$$L_{\mathrm{HW}}(E^4, s) = L(\psi_{64}, s) \tag{6.25}$$

and it was shown in [23] that the Mellin transform of $L_{\mathrm{HW}}(E^4, s)$ factors into string theoretic Hecke indefinite modular forms as

$$f(E^4, q) = \Theta_{1,1}^2(q^4)^2 \otimes \chi_2. \tag{6.26}$$

This completes the proof of Theorem 1.1.

6.2.3 *A rigid Calabi-Yau mirror.* The appearance of the quartic elliptic curve E^4 again suggests to consider a resolved quotient of the triple product $(E^4)^3$, defined by some discrete symmetry group action such that the resolved manifold has a Hodge decomposition that is mirror to the critical part of the spectrum of the quartic fivefold. It turns out that the group that produces such a resolution is given by $\mathbb{Z}_4 \times \mathbb{Z}_4$, with the action

$$\mathbb{Z}_4 \times \mathbb{Z}_4 \ni g_1 \times g_2 : \begin{bmatrix} 1 & 0 & 0 & 0 & 0 & 0 & 3 & 0 & 0 \\ 0 & 0 & 0 & 1 & 0 & 0 & 3 & 0 & 0 \end{bmatrix}. \tag{6.27}$$

Denote by X_3^4 the resolution of the quotient variety $(E^4)^3/\mathbb{Z}_4 \times \mathbb{Z}_4$. This is a rigid Calabi-Yau variety which has $h^{1,1}(X_3^4) = 90$, as needed for the mirror cohomology of the relevant part of the cohomology of the fivefold. Again we can apply a result of Cynk and Hulek [7] to the mirror just constructed.

Theorem 6.3 *Let $n \in \mathbb{N}$ be an odd integer, E be an elliptic curve with an automorphism of order 4, and denote by $\bar{X}_n$ the quotient of E^n by the action of the group*

$$\left\{ (\xi^{a_1} \times \cdots \times \xi^{a_n}) \in \mathrm{End}(E^n) \mid \sum_{i=1}^{n} a_i \equiv 0 (\mathrm{mod}\ 4) \right\}.$$

Then $\bar{X}_n$ has a smooth model X_n, which is a Calabi-Yau manifold such that $\dim H^n(X_n) = 2$ *and $L(X_n, s) \doteq L(g_{n+1}, s)$, where g_{n+1} is the weight $n+1$ cusp form with complex multiplication by $\mathbb{Q}(\sqrt{-1})$, associated to the nth power of the Größencharakter of E.*

It follows from Theorems 1.1 and 6.3 that the L-functions of the quartic fivefold and the variety X_3^4 are identical. This completes the proof of Theorem 1.2.

6.2.4 *Relatives and correspondences.* As in the case of the mirror pair (X_7^3, X_3^3), the pair of varieties (X_5^4, X_3^4) form a pair of relatives in that the Galois representation of X_3^4 is contained in the cohomology of the quartic fivefold. Again, the Tate conjecture implies that there is a correspondence between X_5^4 and X_3^4. The fact that the L-function has a physical interpretation in terms of the worldsheet modular form $\Theta_{1,1}^2(\tau)$ leads again to the conclusion that from a low energy perspective the corresponding motives of these two varieties are indistinguishable.

7 A special Fano variety of critical dimension two

7.1 Arithmetic of the cubic Fermat fourfold X_4^3. The relation between special Fano varieties and critical string model manifolds exists in all dimensions. An interesting lower-dimensional case is that of K3 surfaces. Among the Gepner models there exists an example that is the direct analog of the cubic 7−fold, defined by the cubic fourfold (1.9). The Hodge diamond structure of this variety is $h^{i,i}(X_4^3) = 1,\ i = 0, 1$, with all other $H^{i,j}(X_4^3) = 0$ for $i, j < 4$ and $i \neq j$, while the intermediate cohomology group decomposes as described above in eq. (1.10), i.e. $Q = 2$ and $D_{\mathrm{crit}} = 2$.

The middle dimensional cohomology therefore is that of the K3 surface, modulo the sector inherited from the ambient space $\mathbb{P}_5$. The zeta function of this cubic fourfold therefore takes the form

$$Z(X_4^3, t) = \frac{1}{(1-t)(1-pt)\mathcal{P}_4^{(p)}(t)(1-p^3t)(1-p^4t)}, \tag{7.1}$$

and it is natural to ask whether the $\Omega-$motive of this variety leads to the L-function of an $\Omega-$motive of K3-type.

The Galois group associated to this variety is again $(\mathbb{Z}/3\mathbb{Z})^\times$ and the computation of the Tate twisted $\Omega-$motive obtained from the Jacobi sums $j_p(\alpha_\Omega)$ with $\alpha_\Omega = \frac{1}{3}(1,1,1,1)$ leads to the $q-$expansion

$$f_\Omega^1(X_4^3, q) \doteq q - 13q^7 - q^{13} + 11q^{19} - 46q^{31} + \cdots \tag{7.2}$$

This series describes a weight 3 modular form of level 27 which can be described in closed form as

$$f_\Omega^1(X_4^3, q) = \vartheta(q^3)\Theta_{1,1}^1(q^3)\Theta_{1,1}^1(q^9), \tag{7.3}$$

where $\vartheta(q)$ is the Eisenstein series

$$\vartheta(q) = \sum_{z\in\mathcal{O}_K} q^{\mathrm{N}z} \tag{7.4}$$

associated to the field $K = \mathbb{Q}(\sqrt{-3})$.

It was shown in [30] that (7.3) is precisely the form which describes the Mellin transform of the L-series of the $\Omega-$motive of the degree six K3 surface

$$S^6 = \left\{(z_0 : \cdots : z_3) \in \mathbb{P}_{(1,1,1,3)} \;\middle|\; z_0^6 + z_1^6 + z_2^6 + z_3^2 = 0\right\}. \tag{7.5}$$

Hence we have

$$f_\Omega^1(X_4^3, q) = f_\Omega(S^6, q). \tag{7.6}$$

It is therefore natural to ask whether these two varieties can be related in some explicit way. This can be analyzed via the twist map.

7.2 Twist map construction. In the notation of [15, 14] consider the map

$$\Phi:\ \mathbb{P}_{(w_0,\ldots,w_m)} \times \mathbb{P}_{(v_0,\ldots,v_n)} \longrightarrow \mathbb{P}_{(v_0w_1,\ldots,v_0w_m,w_0v_1,\ldots,w_0v_n)} \tag{7.7}$$

defined as

$$\begin{aligned}((x_0,\ldots,x_m),(y_0,\ldots,y_n)) \;&\mapsto\; (y_0^{w_1/w_0}x_1,\ldots,y_0^{w_m/w_0}x_m, x_0^{v_1/v_0}y_1,\ldots,x_0^{v_n/v_0}y_n)\\ &=: (z_1,\ldots,z_m,t_1,\ldots,t_n).\end{aligned} \tag{7.8}$$

This map restricts on the subvarieties

$$\begin{aligned} X_1 &= \{x_0^\ell + p(x_i) = 0\} \subset \mathbb{P}_{(w_0,\ldots,w_m)}\\ X_2 &= \{y_0^\ell + q(y_j) = 0\} \subset \mathbb{P}_{(v_0,\ldots,v_n)}\end{aligned} \tag{7.9}$$

to the hypersurface

$$X = \{p(z_i) - q(t_j) = 0\} \subset \mathbb{P}_{(v_0w_1,\ldots,v_0w_m,w_0v_1,\ldots,w_0v_n)} \tag{7.10}$$

as a finite map. The degrees of the hypersurfaces X_i are given by

$$\deg X_1 = w_0\ell,\ \deg X_2 = v_0\ell, \tag{7.11}$$

leading to the degree $\deg X = v_0w_0\ell$.

Applying this construction to the hypersurfaces

$$S_\pm^6 = \left\{(z_0 : \cdots : z_3) \in \mathbb{P}_{(1,1,1,3)} \;\middle|\; z_0^6 + z_1^6 + z_2^6 \pm z_3^2 = 0\right\} \tag{7.12}$$

leads to the cubic fourfold embedded in $\mathbb{P}_5$, explaining geometrically the L-function result.

7.3 CM interpretation of the L-series and modularity. The modular form $f_\Omega^1(X_4^3, q)$ of the twisted omega motive of the cubic fourfold has complex multiplication by the Eisenstein field $\mathbb{Q}(\sqrt{-3})$ with the same underlying basic character ψ_{27} defined in (6.14) as the cubic sevenfold, except that in the present case we have to consider the square of the character

$$L_\Omega^1(X_4^3, s) = L(\psi_{27}^2, s). \tag{7.13}$$

Hence the $\Omega-$motive of the cubic fourfold is modular. The basic arithmetic building block is again the elliptic curve E^3, described in string theoretic terms in §6.

References

[1] V.V. Batyrev and L.A. Borisov, *Dual cones and mirror symmetry for generalized Calabi-Yau manifolds*, in MIRROR SYMMETRY, II, 71 – 86, AMS/IP Stud. Adv. Math., 1, 1997, [arXiv: alg-geom/9402002]

[2] V.V. Batyrev and D.I. Dais, *Strong McKay correspondence, string-theoretic Hodge numbers and mirror symmetry*, Topology **35** (1996) 901 – 929, [arXiv: alg-geom/9410001]

[3] P. Candelas, X. de la Ossa and F. Rodriguez-Villegas, *Calabi-Yau manifolds over finite fields I*, [arXiv: hep-th/0012233]

[4] P. Candelas, X. de la Ossa and F. Rodriguez-Villegas, *Calabi-Yau manifolds over finite fields II*, Fields Inst. Commun. **38** (2003) 121 – 157, [arXiv: hep-th/0402133]

[5] P. Candelas and X. de la Ossa, *The zeta-function of a p-adic manifold, Dwork theory for physicists*, [arXiv: 0705.2056 (hep-th)]

[6] P. Candelas, E. Derrick and L. Parkes, *Generalized Calabi-Yau manifolds and the mirror of a rigid manifold*, Nucl. Phys. **B407** (1993) 115 – 154, [arXiv: hep-th/9304045]

[7] S. Cynk and K. Hulek, *Construction and examples of higher-dimensional modular Calabi-Yau manifolds*, [arXiv: math.AG0509424], to appear in Canad. Math. Bull.

[8] D. Gepner, *Spacetime supersymmetry in compactified string theory and superconformal models*, Nucl. Phys. **B296** (1988) 757 – 778

[9] Y. Goto, *The L-series of cubic hypersurface fourfolds*, Appendix to [44]

[10] A. Grothendieck, *Formulé de Lefschetz et rationalité de fonction de L*, Séminaire Bourbaki **279**, 1964/1965, 1 – 15

[11] E. Hecke, *Über Dirichlet-Reihen mit Funktionalgleichung und ihre Nullstellen auf der Mittelgeraden*, Sitzungsberichte d. Bayerischen Akademie der Wissenschaften, Mathematisch-naturwissenschaftliche Abteilung, 1937, 73 – 95

[12] K. Hulek and R. Kloosterman, *The L-series of a cubic fourfold*, [arXiv: math.AG/ 0601207], to appear in Manuscripta Math.

[13] K. Hulek, J. Spandaw, B. van Geemen and D. van Straten, *The modularity of the Barth-Nieto quintic and its relatives*, Adv. Geometry **1** (2001) 263 – 289

[14] B. Hunt and R. Schimmrigk, *Heterotic gauge structure of type* II *K3 Fibrations*, Phys. Lett. **B381** (1996) 427 – 436, [arXiv: hep-th/9512138]

[15] B. Hunt and R. Schimmrigk, *K3 fibered Calabi-Yau threefolds I: The twist map*, Int. J. Math. **10** (1999) 833 – 866, [arXiv: math.AG/9904059]

[16] V.G. Kac, *Infinite-dimensional Lie algebras*, Cambridge University Press, 1990

[17] V.G. Kac and D.H. Peterson, *Affine Lie algebras and Hecke modular forms*, Bull. Amer. Math. Soc. **3** (1980) 1057 – 1061

[18] V.G. Kac and D.H. Peterson, *Infinite-dimensional Lie algebras, theta functions and modular forms*, Adv. Math. **53** (1984) 125 – 264

[19] S. Kadir, *The arithmetic of Calabi-Yau manifolds and mirror symmetry*, Ph.D. thesis, 2004, [arXiv: hep-th/0409202]

[20] S. Kadir, *Arithmetic mirror symmetry for a two-parameter family of Calabi-Yau manifolds*, in MIRROR SYMMETRY V, eds. N. Yui, S.-T. Yau and J. Lewis, AMS/IP, 2006

[21] S. Kadir and N. Yui, *Motives and mirror symmetry for Calabi-Yau orbifolds*, preprint, [arXiv: alg-geom/0606707]

[22] N.C. Leung, *Geometric aspects of mirror symmetry (with SYZ for rigid CY manifolds)*, in the proceedings of ICCM 2001, [arXiv: math.DG/0204168]

[23] M. Lynker and R. Schimmrigk, *Geometric Kac–Moody modularity*, J. Geom. Phys. **56** (2006) 843 – 863, [arXiv: hep-th/0410189]

[24] K.A. Ribet, *Galois representations attached to eigenforms with nebentypus*, in MODULAR FUNCTIONS OF ONE VARIABLE V, eds. J.-P. Serre and D. Zagier, LNM 601, Springer, 1977

[25] M. Saito and N. Yui, *The modularity conjecture for rigid Calabi-Yau threefolds over* $\mathbb{Q}$, J. of Math. Kyoto Univ. **41** (2001) 403 – 419

[26] R. Schimmrigk, *Critical superstring vacua from noncritical manifolds*, Phys. Rev. Lett. **70** (1993) 3688 – 3691, [arXiv: hep-th/9210062]

[27] R. Schimmrigk, *Mirror symmetry and string vacua from a special class of Fano varieties*, Int. J. Mod. Phys. **A11** (1996) 3049 – 3096, [arXiv: hep-th/9405086]

[28] R. Schimmrigk, *Calabi-Yau arithmetic and rational conformal field theories*, J. Geom. Phys. **44** (2003) 555 – 569, [arXiv: hep-th/0111226]

[29] R. Schimmrigk, *Arithmetic spacetimes from string theory*, Int. J. Mod. Phys. **A21** (2006) 6323 – 6350, [arXiv: hep-th/0510091]

[30] R. Schimmrigk, *The Langlands program and string modular K3 surfaces*, Nucl. Phys. **B771** (2007) 143 – 166, [arXiv: hep-th/0603234]

[31] R. Schimmrigk, *A modularity test for elliptic mirror symmetry*, Phys. Lett. **B655** (2007) 84 – 89, [arXiv: 0705.2427 (hep-th)]

[32] R. Schimmrigk, *Emergent spacetime from modular motives*, in preparation.

[33] R. Schimmrigk and S. Underwood, *The Shimura-Taniyama conjecture and conformal field theories*, J. Geom. Phys. **48** (2003) 169 – 189, [arXiv: hep-th/0211284]

[34] J.-P. Serre, *Sur les représentations modulaires de degré 2 de* Gal($\overline{\mathbb{Q}}/\mathbb{Q}$), Duke Math. J. **54** (1987)179 – 230

[35] G. Shimura, *On elliptic curves with complex multiplication as factors of the Jacobians of modular function fields*, Nagoya Math. J. **43** (1971) 199 – 208

[36] A. Strominger, S.-T. Yau and E. Zaslow, *Mirror symmetry is T–duality*, Nucl. Phys. **B479** (1996) 243 – 259, [arXiv: hep-th/9606040]

[37] J. Tate, *Algebraic cycles and poles of zeta functions*, in: ARITHETICAL ALGEBRAIC GEOMETRY, ed. O.F.G. Schilling, 1965

[38] J. Tate, *Conjectures on algebraic cycles in* ℓ*-adic cohomology*, in MOTIVES, Proceedings Symp. Pure Math. **55**, Part 1, Amer. Math. Soc., 1994

[39] H. Verrill, *The L-series of rigid Calabi-Yau threefolds from fibre products of elliptic curves*, Appendix to [43]

[40] A.Weil, *Number of solutions of equations in finite fields*, Bull. Am. Math. Soc. **55** (1949) 497 – 508

[41] A. Weil, *Jacobi sums as "Grössencharaktere"*, Trans. Amer. Math. Soc. **73** (1952) 487 – 495

[42] J. Werner and B. van Geemen, *New examples of threefolds with* $c_1 = 0$, Math. Z. **203** (1990) 211 – 225

[43] N. Yui, *Update on the modularity of Calabi-Yau varieties*, with an appendix by H. Verrill, in CALABI-YAU VARIETIES AND MIRROR SYMMETRY, eds. N. Yui and J.D. Lewis, Fields Institute Communications, 38, 2003

[44] N. Yui, *The L-series of Calabi-Yau orbifolds of CM type*, with an appendix by Y. Goto, in MIRROR SYMMETRY V, 185 – 252, AMS/IP Studies in Advanced Mathematics, Vol. 38, 2006

Fields Institute Communications
Volume **54**, 2008

Update on Modular Non-Rigid Calabi–Yau Threefolds

Edward Lee
Department of Mathematics
UCLA
Los Angeles, CA 90095 USA
lee@math.ucla.edu

Abstract. We review some recent results on the modularity of non-rigid Calabi–Yau threefolds.

1 Introduction

1.1 Calabi–Yau varieties. This note can be considered a sequel to a previous survey article by Yui [38]. The main objects of discussion are Calabi–Yau varieties and their cohomological L-series. We will review recent constructions of Calabi–Yau varieties which are proven to be modular in the sense that their cohomological L-series are associated to modular forms.

We recall first some basic definitions.

Definition 1.1 Let X be a complex projective variety of dimension d. X is a *Calabi–Yau variety* if it satisfies the conditions

1. $H^i(X, \mathcal{O}_X) = 0$ for $0 < i < d$;
2. The canonical sheaf K_X of X is trivial.

The Hodge cohomology groups and Hodge numbers are defined by

$$H^{i,j}(X) = H^j(X, \Omega^i_X), \qquad h^{i,j}(X) = \dim H^{i,j}(X).$$

If X is smooth, the cohomology groups of X admit the Hodge decomposition

$$H^k(X, \mathbb{C}) = \oplus_{i+j=k} H^{i,j}(X).$$

Furthermore, by complex conjugation we have $\overline{H^{i,j}(X)} \cong H^{j,i}(X)$, and by Serre duality and the triviality of the canonical sheaf we have $H^{d-i,d-j}(X) \cong H^{i,j}(X)$.

2000 *Mathematics Subject Classification.* Primary 11F23; Secondary 11G40, 14J32, 14G10, 11F80.

Key words and phrases. Non-rigid Calabi–Yau threefolds, modularity, Galois representation, modular forms, automorphic forms.

The Betti numbers $b^i(X)$ and the Euler characteristic $\chi(X)$ are defined by $b^i(X) = \dim H^i(X, \mathbb{C})$ and

$$\chi(X) = \sum_{i=0}^{2d} (-1)^i b^i(X).$$

Smooth one-dimensional Calabi–Yau varieties are simply elliptic curves, with Hodge diamond

$$\begin{matrix} & 1 & \\ 1 & & 1 \\ & 1 & \end{matrix}$$

Smooth two-dimensional Calabi–Yau varieties are $K3$ surfaces, with Hodge diamond

$$\begin{matrix} & & 1 & & \\ & 0 & & 0 & \\ 1 & & 20 & & 1 \\ & 0 & & 0 & \\ & & 1 & & \end{matrix}$$

In this article, we will be most interested in three-dimensional Calabi–Yau varieties. Smooth Calabi–Yau threefolds have Hodge diamond

$$\begin{matrix} & & & 1 & & & \\ & & 0 & & 0 & & \\ & 0 & & h^{1,1} & & 0 & \\ 1 & & h^{1,2} & & h^{1,2} & & 1 \\ & 0 & & h^{1,1} & & 0 & \\ & & 0 & & 0 & & \\ & & & 1 & & & \end{matrix}$$

though we shall have occasion to discuss results in other dimensions.

Calabi–Yau threefolds can be classified as *rigid* or *non-rigid* according to whether $h^{2,1}$ is zero or nonzero respectively. By Kodaira–Spencer theory, the infinitesimal deformations of X are classified by $H^1(X, T_X)$, which is isomorphic to $H^1(X, \Omega_X^2)$ by the triviality of K_X. A rigid Calabi–Yau threefold has $h^{2,1} = 0$ and thus admits no infinitesimal deformations, whence the name. Since the infinitesimal deformations of any Calabi–Yau variety X are unobstructed [31] [32], X moves in an $h^{2,1}$-dimensional family.

1.2 ***L*-series of a Calabi–Yau variety.** We now introduce the cohomological L-series of a Calabi–Yau variety X, which encodes a vast amount of arithmetic information about X.

Suppose that the Calabi–Yau variety X is defined over a number field K, and let $\mathcal{O}_K$ be the ring of integers of K. Then X always has a model defined over $\mathcal{O}_K$. Fix some model over $\mathcal{O}_K$, and let $\mathfrak{p}$ be a prime ideal of $\mathcal{O}_K$. We say that the prime $\mathfrak{p}$ is a *good* prime of X if the reduction $X_\mathfrak{p}$ is smooth; otherwise we say that it is a *bad* prime of X. Let q be the norm of the prime ideal $\mathfrak{p}$ and let $\mathbb{F}_q$ be the field of q elements.

If $\mathfrak{p}$ is a good prime of X, let $\mathrm{Frob}_\mathfrak{p}$ denote the Frobenius morphism on $X_\mathfrak{p}$ induced by the coordinate map $x \to x^q$. Then $\mathrm{Frob}_\mathfrak{p}$ acts on the étale cohomology groups $H^i_{et}(\overline{X}_\mathfrak{p}, \mathbb{Q}_l)$, where $\overline{X}_\mathfrak{p}$ denotes base change of $X_\mathfrak{p}$ to $\overline{\mathbb{F}}_q$ and l is a rational prime which is coprime to q. By abuse of notation we will often continue to write $X_{\mathfrak{p}}$ for its base change to $\overline{\mathbb{F}}_{\mathfrak{p}}$.

The Frobenius morphism induces an endomorphism of the étale cohomology groups, which we denote by $\mathrm{Frob}_{\mathfrak{p}}^{*}$. Let

$$P_i^{\mathfrak{p}}(T) = \det(1 - \mathrm{Frob}_{\mathfrak{p}}^{*}T | H_{et}^i(\overline{X}, \mathbb{Q}_l))$$

be the characteristic polynomial of the Frobenius endomorphism on $H_{et}^i(\overline{X}, \mathbb{Q}_l)$. We will be most interested in the middle-dimensional cohomology group of X.

If $\mathfrak{p}$ is a prime of bad reduction for X, then we can still define the polynomial

$$P_i^{\mathfrak{p}}(T) = \det(1 - \mathrm{Frob}_{\mathfrak{p}}^{*}T | H_{et}^i(\overline{X}, \mathbb{Q}_l)^{I_{\mathfrak{p}}})$$

where $I_{\mathfrak{p}}$ is the inertia group at $\mathfrak{p}$. However, unlike the elliptic curve case, our current understanding of bad reduction of higher dimensional Calabi–Yau varieties is very thin. In most of the examples below we will determine L-series modulo the Euler factors at bad primes.

Étale cohomology is a Weil cohomology theory, and thus satisfies a Lefschetz fixed-point theorem:

Theorem 1.1 *If f is an endomorphism of the smooth variety X, then the number of fixed points of f satisfies the formula*

$$\mathrm{Fix}\,(f, X) = \sum_{i=0}^{2d} (-1)^i \mathrm{tr}\, f^*(H_{et}^i(X, \mathbb{Q}_l)). \tag{1.1}$$

In the case $f = \mathrm{Frob}_{\mathfrak{p}}$, the number of fixed points will simply be the number of $\mathbb{F}_q$-rational points of X, and this equals the alternating sum of the traces of the Frobenius action on the cohomology groups. In the Calabi–Yau threefold case, the traces of $\mathrm{Frob}_{\mathfrak{p}}$ on H^0 and H^6 are equal to 1 and q^3 respectively, and $H^1 = H^5 = 0$. If one understands the Frobenius action on the groups H^2 and H^4, then the trace of Frobenius on H^3 can be determined by counting points (often using a computer). For example, if H^2 is spanned by algebraic cycles defined over $\mathcal{O}_K/\mathfrak{p}$, then Frobenius acts by multiplication by q on the whole space of algebraic cycles, and thus the trace of Frobenius on H^2 is just $qb^2(X)$.

Furthermore, Deligne's proof of the Weil conjectures tells us that

Theorem 1.2 *$P_i^{\mathfrak{p}}(T)$ is a polynomial with integer coefficients, whose reciprocal roots are all algebraic integers with absolute value $p^{\frac{i}{2}}$.*

The absolute Galois group $G(\overline{\mathbb{K}}/\mathbb{K})$ acts on X. If l is a prime, we have the l-adic Galois representation

$$\rho_{X,l}^i : \mathrm{Gal}(\overline{\mathbb{K}}/\mathbb{K}) \to GL(H_{et}^i(X, \mathbb{Q}_l))$$

and we define its L-series $L(\rho_{X,l}^i, s) = L(H_{et}^i(X, \mathbb{Q}_l))$:

Definition 1.2 Let X be a Calabi–Yau variety of dimension d defined over a number field $\mathbb{K}$. For each i, the ith cohomological L-series of X is defined by the Euler product

$$L(H_{et}^i(X, \mathbb{Q}_l), s) = \prod_{\mathfrak{p}} P_i^{\mathfrak{p}}(q^{-s})^{-1}, q = \mathrm{Norm}\,\mathfrak{p}.$$

However, in the sequel the factors corresponding to primes of bad reduction $\mathfrak{p}$ for X will often be omitted.

Definition 1.3 A Calabi–Yau variety X of dimension d defined over $\mathbb{Q}$ is *modular* if there is a modular form (or more generally an automorphic form) such that its associated Galois representation is equivalent to $\rho^d_{X,l}$.

In particular, the L-series of X coincides with the L-series of the modular or automorphic form, possibly up to a finite number of Euler factors.

2 Elliptic curves

The dimension 1 case is of course the case of elliptic curves, and the question of modularity of elliptic curves over $\mathbb{Q}$ was famously settled by Wiles and his collaborators [1] [30]. Here the only nontrivial cohomology group is $H^1_{et}(E, \mathbb{Q}_l)$, and the Lefschetz formula gives us

$$a_p = \mathrm{tr}\,(\mathrm{Frob}_p^*|H^1_{et}(E_p, \mathbb{Q}_l)) = p + 1 - \#E_p.$$

The conductor N of E is divisible exactly by the bad primes of E and measures the extent to which E_p is singular:

$$N = \prod_p p^{f_p}$$

where

$$f_p = \begin{cases} 0 & \text{if E has good reduction at } p, \\ 1 & \text{if } E_p \text{ has a node}, \\ 2 & \text{if } E_p \text{ has a cusp and } p \neq 2, 3, \\ 2 + \delta_p & \text{if } E_p \text{ has a cusp and } p = 2 \text{ or } p = 3 \end{cases}$$

and δ_p measures the wild ramification in the action of the inertia subgroup I_p of $\mathrm{Gal}(\overline{\mathbb{Q}}/\mathbb{Q})$ on $H^1_{et}(E, \mathbb{Q}_l)$.

The modularity theorem then states that there exists a weight 2 cusp form f of level N such that

$$L(E, s) = L(f, s).$$

3 K3 surfaces

In dimension 2, the Calabi–Yau varieties are exactly the $K3$ surfaces. The only nontrivial cohomology group is H^2, which has dimension 22.

Let $NS(X)$ be the Néron-Severi group of X, i.e. the group of divisors on X modulo algebraic equivalence; its rank is called the *Picard number* of X. Since the divisors of X live in $H^{1,1}(X) \cap H^2(X, \mathbb{Z})$, the maximum Picard number of X is 20. $K3$ surfaces achieving this maximum are called *singular* $K3$ surfaces.

If X is a singular $K3$ surface, the lattice $H^2(X, \mathbb{Z})$ under cup product decomposes as $NS(X) \oplus T(X)$ (up to finite index), where T is called the *transcendental lattice* of $H^2(X, \mathbb{Z})$. This induces a decomposition of the L-series of X as

$$L(X, s) = L(NS(X) \otimes \mathbb{Q}_l, s)L(T(X) \otimes \mathbb{Q}_l, s).$$

Theorem 3.1 *Let X be a singular $K3$ surface over $\mathbb{Q}$. Up to Euler factors at the bad primes, the L-series of $T(X)$ is given by*

$$L(T(X) \otimes \mathbb{Q}_l, s) = L(f, s)$$

where f is a cusp form of weight 3 on a congruence subgroup of $PSL_2(\mathbb{Z})$. The level N can be computed explicitly. The algebraic part $L(NS(X), s)$ depends on the fields of definition of the divisors of X.

The proof of this result depends on the work of Inose and Shioda on the classification of singular $K3$ surfaces [17] and on the work of Livné on the modularity of motivic orthogonal two-dimensional Galois representations [20].

If the Picard number of X is less than 20, much less is known about the modularity of X. In this case the transcendental part $T(X)$ of H^2 has dimension higher than 2.

4 Modularity of rigid Calabi–Yau threefolds

Perhaps the first example of a rigid Calabi–Yau threefold which was shown to be modular is a resolution $\tilde{Q}$ of the Schoen quintic threefold

$$Q = \{\mathbf{x} \in \mathbb{P}^4 : x_0^5 + x_1^5 + x_2^5 + x_3^5 + x_4^5 - 5x_0x_1x_2x_3x_4 = 0\};$$

in [25] it was shown that the middle cohomology of $\tilde{Q}$ is associated to the weight 4 form

$$\eta(z)^4\eta(5z)^4 + 5\eta(z)^3\eta(5z)^4\eta(25z) + 20\eta(z)^2\eta(5z)^4\eta(25z)^2 + \eta(z)\eta(5z)^4\eta(25z)^3 + \eta(5z)^4\eta(25z)^4$$

of level 25, where η is the Dedekind eta-function. Since then, many clever constructions of other rigid modular Calabi–Yau threefolds have been found; for reasons of space we must refer the reader to [24].

In the rigid case, it is expected that H^3 will always correspond to a weight 4 modular form. Under mild conditions, Dieulefait and Manoharmayum have proven this to be true:

Theorem 4.1 [9] [10] *Suppose that X is a rigid Calabi–Yau threefold over $\mathbb{Q}$. Assume that one of the following statements is true:*

1. *X has good reduction at 3 or 7 or*
2. *X has good reduction at 5 or*
3. *X has good reduction at 3 and the trace of* Frob_3 *on U is not divisible by 3.*

Then X is modular.

We might hope that the conditions of good reduction at the specific primes 3,5,7 can eventually be removed, as in the gradual proof of modularity for elliptic curves.

5 Modular threefolds containing elliptic ruled surfaces

We now turn our attention to modular non-rigid Calabi–Yau threefolds; the first such example discovered can be found in [33].

The simplest cases in which the modularity of a nonrigid Calabi–Yau threefold X can be proven occur when the semisimplification of $H^3(X)$ decomposes as $U \oplus V$. Here U is two-dimensional and V is of the form $\oplus_{i=1}^{h^{1,2}} H^3(S_i)$, where the S_i are ruled surfaces over elliptic curves. (The surfaces S_i may be said to *span* $H^{2,1}(X) \oplus H^{1,2}(X)$ in this case). If the surfaces S_i are defined over $\mathbb{Q}$ (and even if only the scheme-theoretic union $\cup S_i$ is defined over $\mathbb{Q}$), then it essentially follows from the theorem of Dieulefait–Manoharmayum that U is modular:

Theorem 5.1 [16] *Let X be a smooth projective threefold defined over $\mathbb{Q}$ such that $h^{3,0}(X) = h^{0,3}(X) = 1$ and assume that X contains birational ruled surfaces $S_i, i = 1, \ldots, h^{1,2}(X)$ which are defined over $\mathbb{Q}$ and which span $H^{2,1}(X) \oplus H^{1,2}(X)$.*

Let ρ be the 2-dimensional Galois representation given by U, and suppose we have an exact sequence

$$0 \to U \to H^3_{et}(X, \mathbb{Q}_l) \to \oplus_{i=1}^{h^{1,2}} H^3_{et}(S_i, \mathbb{Q}_l) \to 0.$$

Assume that one of the following conditions holds:

1. *X has good reduction at 3 or 7 or*
2. *X has good reduction at 5 or*
3. *X has good reduction at 3 and the trace of* Frob_3 *on U is not divisible by 3.*

Then X is modular, and the L-function of X is given by

$$L(X,s) \doteq L(f_4, s) \prod_{i=1}^{h^{1,2}} L(g_2^i, s-1) \tag{5.1}$$

where f_4 is a weight 4 modular form, the g_2^i are the weight 2 forms associated to the base curves E_i, and $\doteq$ denotes equality up to finitely many Euler factors.

Examples of modular threefolds such that $H^3(X)$ decomposed in this manner were constructed by Hulek–Verrill, Cynk–Meyer, Schütt and Lee. In each of these cases U is a two-dimensional Galois representation whose modular form can be determined by counting points on X over finite fields.

5.1 Hypersurfaces in toric varieties. In [15] Hulek and Verrill consider a family $\{X_{\mathbf{a}}, \mathbf{a} \in \mathbb{P}^5\}$ of Calabi–Yau threefolds which are obtained from anticanonical hypersurfaces in toric varieties; they consider a reflexive polytope obtained from the A_4 root lattice.

Let M_{A_4} be the sublattice of points $(x_1, x_2, x_3, x_4, x_5) \in \mathbb{Z}^5$ such that $\sum x_i = 0$, and define the polytope Δ_{A_4} to be the convex hull in $M \otimes \mathbf{R}$ of the points $(e_i - e_j), 1 \le i \ne j \le 5$. They then take the dual polytope Δ^*, consider a subdivision $\tilde{\Delta}^*$ and consider hypersurfaces in the resulting toric variety $\tilde{P}$. The toric variety $\tilde{P}$ contains an open 4-torus T, which we describe as $\{\mathbf{X} \in \mathbb{P}^4 : \prod X_i \ne 0\}$.

Let $X_{\mathbf{a}}$ be the anticanonical hypersurface of $\tilde{P}$ whose equation, when restricted to the open torus T, is

$$(X_1 + \cdots + X_5)(\frac{a_1}{X_1} + \cdots + \frac{a_5}{X_5}) = t \tag{5.2}$$

for $\mathbf{a} = (a_1 : a_2 : a_3 : a_4 : a_5 : t) \in \mathbb{P}^5$. In general, $X_{\mathbf{a}}$ has 30 ordinary double point singularities, and special values of $\mathbf{a}$ yield threefolds with extra nodes, so one takes a big resolution $\tilde{X}_{\mathbf{a}}$. (Big resolutions of ordinary double point singularities are non-crepant resolutions, so $\tilde{X}_{\mathbf{a}}$ is only birationally Calabi–Yau.)

For $\mathbf{a} = (1:1:1:1:1:25)$,$(1:1:1:9:9:9)$ and $(1:1:4:4:4:16)$, $\tilde{X}_{\mathbf{a}}$ is shown to contain ruled surfaces S_i over elliptic curves E_i as subvarieties; these elliptic curves E_i are defined over $\mathbb{Q}$ and their modular forms can be determined explicitly.

The inclusion map $i : \cup S_i \hookrightarrow \tilde{X}_{\mathbf{a}}$ is proven to induce an embedding $i_* : \oplus_i H_3(S_i) \hookrightarrow H_3(\tilde{X}_{\mathbf{a}})$. By using Poincaré duality and using comparison theorems to pass to étale cohomology, we get the desired splitting of $H^3(\overline{X_{\mathbf{a}}})$ (up to semisimplification)

$$H^3_{et}(X, \mathbb{Q}_l) \cong \oplus H^3_{et}(S_i, \mathbb{Q}_l) \oplus U.$$

It remains to prove that U is modular. Now the trace of Frob_p on U will simply be the trace of Frob_p on $H^3_{et}(X)$ minus the trace on the $H^3_{et}(S_i)$. The traces on

$H^3_{et}(S_i)$ will just be p times the coefficients of the modular forms of the elliptic curves E_i, and the traces of Frob_p on $\tilde{X}$ are obtained by counting points and using the Lefschetz theorem as described earlier. Hence the traces of Frobenius on U can be effectively computed. By inspection, they find a candidate modular form of weight 4 whose coefficients agree with the traces of Frobenius for many values of p. They then use a well-known theorem of Livné (following ideas of Faltings and Serre) which essentially states that two two-dimensional Galois representations U_1 and U_2 with even trace can be proven to be isomorphic if their traces are the same at some finite set of primes T, where T depends on the bad primes of U_1 and U_2:

Theorem 5.2 [20] *Let ρ_1 and ρ_2 be two 2-adic 2-dimensional Galois representations, unramified outside a finite set of primes S. Let K_S be the compositum of all quadratic extensions of $\mathbb{Q}$ ramified at primes in S, and let T be a set of primes disjoint from S. Then if*

1. $\operatorname{tr}\rho_1 \equiv \operatorname{tr}\rho_2 \equiv 0$ *and* $\det\rho_1 \equiv \det\rho_2 \pmod 2$*;*
2. $\{\mathrm{Frob}_p | K_S : p \in T\}$ *is equal to the set* $\mathrm{Gal}(K_S/K) - \mathrm{Id}$*;*
3. *for all* $p \in T$, $\operatorname{tr}\rho_1(\mathrm{Frob}_p) = \operatorname{tr}\rho_2(\mathrm{Frob}_p)$*;*

   then ρ_1 and ρ_2 have isomorphic semisimplifications.

The upshot of this theorem is that checking that U and the Galois representation associated to a modular form g are isomorphic reduces to the problem of checking that their coefficients are even and that the traces of U are equal to the coefficients of g at some *finite* set of primes T.

Thus we have

Theorem 5.3 *For the special values of* **a** *listed above, the threefold $\tilde{X}_{\mathbf{a}}$ is modular, with L-series equal to*

$$L(H^3_{et}(\tilde{X}_{\mathbf{a}}), s) \mathrel{\dot{=}} L(g, s) \prod_{i=1}^{h^{1,2}} L(f_i, s-1) \tag{5.3}$$

where the f_i are modular forms of elliptic curves over $\mathbb{Q}$ and g is some modular form of weight 4.

5.2 Products of elliptic fibrations. In [16] and [27] Hulek–Verrill and Schütt consider fiber products of relatively minimal, regular, rational elliptic surfaces with section over $\mathbb{P}^1$ (in fact, the previous example is birational to a threefold of this type). Given two such surfaces $\pi_1 : Y_1 \to \mathbb{P}^1$ and $\pi_2 : Y_2 \to \mathbb{P}^1$, it was first shown in [26] that if the fibers of both Y_1 and Y_2 over the set of common cusps are all of type I_n, the fiber product $W = Y_1 \times_{\mathbb{P}^1} Y_2$ of two such elliptic fibrations is a nodal Calabi–Yau threefold.

Assume now that all degenerate fibers of π_1 and π_2 are of type I_n, and consider a point $p \in \mathbb{P}^1$ which is a cusp for exactly one of the fibrations, say π_1. Then the fiber W_p is a product of an elliptic curve E and a degenerate fiber of type I_n, i.e. a union of elliptic ruled surfaces. It can be shown by intersection theory that the inclusion induces an embedding

$$\imath_* : \oplus_{i=1}^{h^{1,2}} H_3(E_i \times \mathbb{P}^1) \to H_3(X).$$

After using duality and comparison theorems, we obtain a surjection

$$\imath^* : H^3_{et}(X, \mathbb{Q}_l) \to \oplus_{i=1}^{h^{1,2}} H^3_{et}(E_i \times \mathbb{P}^1, \mathbb{Q}_l)$$

whose kernel is 2-dimensional, so we are in the situation of Theorem 5.1.

In the examples of [16] the representation U has even trace, so once a candidate modular form is found, the theorem of Faltings-Serre-Livné can be used to establish modularity.

In the examples of [27], the representations U do not have even trace, so the argument to prove the equivalence of U with the Galois representation of a modular forms f is more subtle. Here Schütt works out in detail a method first suggested by Serre in [25] to prove that two two-dimensional 2-adic representations ρ_1 and ρ_2 are isomorphic. The idea is to first prove that the mod 2 reductions are isomorphic and absolutely irreducible. Then under the assumption (to be contradicted) that ρ_1 and ρ_2 are isomorphic mod 2^α but not mod $2^{\alpha+1}$, one constructs an auxiliary Galois representation $\tilde{\rho} : \mathrm{Gal}(\overline{\mathbb{Q}}/\mathbb{Q}) \to M_2(\mathbb{F}_2) \rtimes GL_2(\mathbb{F}_2)$ which measures the degree to which ρ_1 and ρ_2 would not be isomorphic. Carrying out this method requires one to know the Galois extensions K of $\mathbb{Q}$ with Galois group S_3 or C_3 unramified outside S (one way is to check the tables in [18]), then to know the Galois extensions L/K with $\mathrm{Gal}(L/\mathbb{Q})$ isomorphic to $S_3 \times C_2$ or S_4 and unramified outside S.

A particularly interesting example in [27] is of a Calabi–Yau threefold $\tilde{W}_2$ which has bad reduction at 2,5,11 and 359. Numerical evidence suggests that the non-elliptic part of its third cohomology should be associated with a newform f_{55} of weight 4 and level 55, but unfortunately this cannot be proven yet due to the difficulty of computing all cubic and quartic Galois extensions of $\mathbb{Q}$ unramified outside $\{2, 5, 11, 359\}$. However, assuming that modularity can be proven, this would be an example where the level of the modular form is not divisible by all the primes of bad reduction, unlike the case of elliptic curves.

5.3 A Horrocks–Mumford quintic. The Horrocks–Mumford vector bundle HM is essentially the only known rank 2 indecomposable vector bundle on $\mathbb{P}^4$, introduced in [11] and studied extensively since then (see [12] for an introduction). It was shown in [11] that the zero locus of an arbitrary section of HM is an abelian surface of degree 10. Now $\wedge^2 HM \cong \mathcal{O}(5)$, so a determinantal section of the bundle $\wedge^2 HM$ is a (singular) quintic threefold which is the total space of a pencil of abelian surfaces.

In [19] the Horrocks–Mumford quintic X with equation

$$\sum_{j \in \mathbb{Z}/5} (z_j^3 z_{j+1} z_{j-1} - z_j^3 z_{j+2} z_{j-3} - z_j z_{j+1}^2 z_{j-1}^2 + z_j z_{j+2}^2 z_{j-2}^2) = 0$$

was studied (where the indices are read modulo 5). It was shown that the singular locus of X contains the elliptic normal curves E_1 and E_2, where E_1 is cut out by the five equations $iz_j^2 + z_{j+1}z_{j-1} + z_{j+2}z_{j-2} = 0$ and E_2 is the complex conjugate of E_1, and there exists a big resolution $\tilde{X}$ of X where the inverse images of E_1 and E_2 are ruled surfaces over each elliptic curve. The following modularity result was shown:

Theorem 5.4 [19] *There exists a resolution $\tilde{X}$ of X such that, up to semisimplification,*

$$H^3_{et}(\tilde{X}, \mathbb{Q}_l) \cong U \oplus \mathrm{Ind}\,^{\mathrm{Gal}(\overline{\mathbb{Q}}/\mathbb{Q})}_{\mathrm{Gal}(\overline{\mathbb{Q}}/\mathbb{Q}(i))} H^1((E_1), \mathbb{Q}_l)(-1)$$

where U is a two-dimensional Galois representation unramified outside the bad primes $\{2, 5\}$ of X. The representation U corresponds to the unique normalized cusp form f_5 of weight 4 and level 5.

The present example is notable because of the presence of elliptic surfaces which are not defined over $\mathbb{Q}$ (only their scheme-theoretic union is) and also because the level of the modular form does not divide all of the bad primes, as was conjecturally the case in the previous example of Schütt.

6 Kummer-type constructions

Recently several examples of modular non-rigid Calabi–Yau threefolds have been obtained via Kummer-type constructions.

6.1 Quotients of type $(K3 \times E)/\langle -1, -1 \rangle$. In [22] Livné and Yui construct four classes of non-rigid Kummer-fibred Calabi–Yau threefolds which achieve the strong Arakelov–Yau bound. They start by considering an elliptic modular surface Y associated to a torsion-free congruence subgroup Γ of $PSL_2(\mathbb{Z})$ of genus 0 and index 24 and lifting it to a subgroup $\tilde{\Gamma}$ of $SL_2(\mathbb{Z})$. Let $Y(\tilde{\Gamma})$ be the elliptic modular surface associated to $\tilde{\Gamma}$; it is an elliptic fibration over the genus-zero modular curve $X(\tilde{\Gamma})$. Now let E be an elliptic curve defined over $\mathbb{Q}$, and consider the threefold $Y(\tilde{\Gamma}) \times E$ as a fibration of abelian surfaces over $X(\tilde{\Gamma})$. Let $X(\tilde{\Gamma}) \to X(\tilde{\Gamma})$ be the associated Kummer family obtained by quotienting out by the involution -1 on each fiber and blowing up the fixed locus. Livné and Yui prove that

Theorem 6.1 *In the cases $\Gamma = \Gamma(4), \Gamma_0(3) \cap \Gamma(2), \Gamma_0(8) \cap \Gamma(2), \Gamma_1(8; 4, 1, 2)$, the above construction yields a non-rigid Calabi–Yau threefold $X(\tilde{\Gamma})$ whose generic fiber is a Kummer surface and which achieves the strong Arakelov–Yau bound*

$$\deg f_* \omega_{X/\mathbb{P}^1} \leq \frac{1}{2} \deg \Omega^1_{\mathbb{P}^1}(\log S),$$

where S is the set of cusps of Γ. The L-series of X is

$$\begin{aligned} L(H^3(X), s) &= L(T(Y) \times H^1(E), s) L(N_- \times H^1(E), s) \\ &= L(g_3 \otimes g_2, s) L(g_E, s-1)^k \prod_\delta L(g_E \otimes \chi_\delta, s-1)^{k_\delta}. \end{aligned}$$

Here N_- is the motive of algebraic cycles on Y which are odd under the action of the involution -1, g_3 is the weight 3 modular form associated to the rank 2 motive T, g_E is the weight 2 modular form of the elliptic curve $E/\mathbb{Q}$, k is the number of cycles in N_- defined over $\mathbb{Q}$, k_δ is the number of cycles in N_- defined over $\mathbb{Q}(\sqrt{\delta})$ and χ_δ is the quadratic character cut out by $\mathbb{Q}(\sqrt{\delta})$.

The proof depends on the fact that for the given groups Γ, $Y(\tilde{\Gamma})$ is a singular $K3$ surface: it has Picard number 20, as discussed in Section 3. Define the rank 2 motive of transcendental cycles on $Y(\tilde{\Gamma})$ to be $T = H^2(Y(\tilde{\Gamma}))/NS(Y(\tilde{\Gamma}))$. Then the modular form g_3 of weight 3 is the modular form associated to the rank 2 motive T. The intuition here is that $H^3(X)$ is the part of $H^3(Y \times E)$ which is even under the involution $(-1, -1)$, and this is equal to (odd part of $H^2(Y) \otimes$ odd part of $H^1(E) \oplus$ even part of $H^2(Y) \otimes$ even part of $H^1(E)$), since $H^1(Y) = 0$. Since the holomorphic 1-form on E is odd, $H^1(E)$ has no even part, and we are left with (odd part of $H^2(Y) \otimes$ odd part of $H^1(E)$). The odd part of $H^2(Y)$ consists of the transcendental part of Y (the holomorphic 2-form) plus the odd algebraic cycles on Y.

Borcea and Voisin [4], [35] independently gave a construction of mirror families of Calabi–Yau threefolds of type $K3 \times E/\langle -1, -1 \rangle$. It would be interesting to compute and compare the L-series and zeta functions of modular threefolds in mirror families and compare them with recent results of Wan [36].

6.2 A quotient of the Schoen quintic. In this section we consider a threefold X obtained by resolving a $\mathbb{Z}/2$ quotient of the Schoen quintic threefold Q in $\mathbb{P}^4$ given by the equation $x_0^5 + x_1^5 + x_2^5 + x_3^5 + x_4^5 - 5x_0x_1x_2x_3x_4 = 0$; it does not seem to have been studied in the current literature. We will discuss its modularity here. In [25] it was shown that Q has the following properties:

1. Q is the total space of a $\mathbb{P}^1$ family of abelian surfaces arising from the Horrocks–Mumford vector bundle on $\mathbb{P}^4$;
2. Q is modular in the sense that the middle-dimensional cohomology group of a resolution of Q has L-series equal to the Mellin transform of a weight 4 modular form of level 25, up to an Euler factor at the bad prime 5.

We show that X has the following properties:

1. X is a Calabi–Yau variety;
2. X is the total space of a $\mathbb{P}^1$ family of Kummer surfaces;
3. A big resolution $\tilde{X}$ of X is modular, and the middle-dimensional cohomology group of $\tilde{X}$ is, up to semisimplification, the direct sum of $H^3(Q)$ and $H^3(S)$ where S is a $\mathbb{P}^1$-bundle over an elliptic curve E defined over $\mathbb{Q}$.

Our starting point is the observation in [11] that the zero set of a generic section s of HM is an abelian surface A_s of degree 10, and that the involution ι on $\mathbb{P}^4$ defined by $\iota[x_0 : x_1 : x_2 : x_3 : x_4] = [x_0 : x_4 : x_3 : x_2 : x_1]$ induces the negation morphism on A_s if the origin of A_s is chosen to be one of the fixed points of ι.

The involution ι has fixed loci $\mathbb{P}^1_- = \{x_0 = x_1 + x_4 = x_2 + x_3 = 0\}$ and $\mathbb{P}^2_+ = \{x_1 - x_4 = x_2 - x_3\} = 0$. It was proven by Horrocks–Mumford that the restriction map $\Gamma(HM) \to \Gamma(HM_{\mathbb{P}^1_-})$ is an injection, that $\Gamma(HM_{\mathbb{P}^1_-}) \cong \mathcal{O}(-1) \oplus \mathcal{O}(6)$ and that the image of this restriction map is spanned by the sections

$$\lambda(\lambda^5 + 2\mu^5), \mu(\mu^5 - 2\lambda^5), 5\lambda^4\mu^2, 5\lambda^2\mu^4.$$

Hence the generic A_s intersects $\mathbb{P}^1_-$ in 6 points. A_s intersects $\mathbb{P}^2_+$ in 10 points, and if one of these 16 points is chosen to be the origin for A_s then these 16 points are exactly the points of order 2. Taking the quotient of A_s by ι and blowing up the 16 nodes yields a Kummer surface.

A Horrocks–Mumford quintic is a threefold $Q_{s_1 \wedge s_2}$ given by a determinantal section $s_1 \wedge s_2$ of the bundle $\wedge^2 HM \cong \mathcal{O}(5)$ on $\mathbb{P}^4$. One can easily check that in the generic case, the singularities of this threefold are double points at the 100 points of intersection of A_{s_1} and A_{s_2}. A small resolution $\hat{Q}_{s_1 \wedge s_2}$ can be obtained by considering the section of the bundle $HM \boxtimes \mathcal{O}(1) \to \mathbb{P}^4 \times \mathbb{P}^1$ given by $c_1 s_1 + c_2 s_2$, where $[c_1 : c_2]$ are coordinates on $\mathbb{P}^1$. This small resolution of $Q_{s_1 \wedge s_2}$ is a fibration over $\mathbb{P}^1$ whose generic fiber is an abelian surface.

In [25] it was shown that the quintic Q defined by the equation $x_0^5 + x_1^5 + x_2^5 + x_3^5 + x_4^5 - 5x_0x_1x_2x_3x_4$ is a Horrocks–Mumford quintic. It is a special member of the family of HM-quintics, having 125 nodes at the points $[1 : \zeta^a : \zeta^b : \zeta^c : \zeta^{-c-b-a}]$ where ζ is a primitive 5th root of unity, instead of just the usual 100.

The fixed locus of the involution ι on Q consists of the line $\mathbb{P}^1_-$ and the curve $E = \mathbb{P}^2_+ \cap Q = \{[x_0 : x_1 : x_2] : x_0^5 + 2x_1^5 + 2x_2^5 - 5x_0x_1^2x_2^2 = 0\}$. The images of the five nodes $[1 : \zeta^a : \zeta^{-a} : \zeta^{-a} : \zeta^a]$ lie on E. Since E is a curve of arithmetic genus 6 with 5 nodes, its normalization is an elliptic curve. Of course, Q/ι still has $\frac{125-5}{2} = 60$ nodes away from $\mathbb{P}^1_-$ and E.

For computational purposes, it is convenient to consider a linear change of coordinates $[y_0 : y_1 : y_2 : y_3 : y_4] = [x_0 : x_1 + x_4 : x_2 + x_3 : x_2 - x_3 : x_1 - x_4]$ on $\mathbb{P}^4$. The Schoen quintic then has the equation

$$16y_0^5 + (y_1^5 + 10y_1^3y_4^2 + 5y_1y_4^4) + (y_2^5 + 10y_2^3y_3^2 + 5y_2y_3^4) - 5y_0(y_1^2 - y_4^2)(y_2^2 - y_3^2) = 0.$$

In y-coordinates, the action of ι is given by the matrix

$$\begin{pmatrix} 1 & & & & \\ & 1 & & & \\ & & 1 & & \\ & & & -1 & \\ & & & & -1 \end{pmatrix}.$$

The homogeneous ring of invariants of $\mathbb{P}^4$ is simply $k[y_0, y_1, y_2, y_3^2, y_4^2, y_3y_4] = k[Y_0, Y_1, Y_2, Y_3, Y_4, Y_5]/Y_3Y_4 - Y_5^2$. Hence $\mathbb{P}^4/\iota$ can be exhibited as the hypersurface $Y_3Y_4 - Y_5^2 = 0$ in the toric variety $\mathbb{P}(1,1,1,2,2,2)$, and Q/ι is a complete intersection in $\mathbb{P}(1,1,1,2,2,2)$ given by the two equations

$$\begin{aligned} 16Y_0^5 + (Y_1^5 + 10Y_1^3Y_4 + 5Y_1Y_4^2) + (Y_2^5 + 10Y_2^3Y_3 + 5Y_2Y_3^2)& \\ - 5Y_0(Y_1^2 - Y_4)(Y_2^2 - Y)3) = 0&, \\ Y_3Y_4 - Y_5^2 = 0&. \end{aligned}$$

Simple but messy calculations show that the singularities of Q/ι along $\mathbb{P}^1_-$ are transverse A_1 singularities, and that the singularities along E are also transverse A_1 singularities except at the 5 special points of E, which are A_3 singularities (blowing up E yields a $\mathbb{P}^1$-bundle over E which still has 10 double points lying over the original 5 singularities on E). We blow up the $\mathbb{P}^1_-$ and blow up the ambient $\mathbb{P}^2_+$ containing E; the resulting threefold X has the original 60 nodes from Q/ι and 10 more nodes lying in the exceptional divisor over E. After making an appropriate choice of small resolution of the 60+10 nodes, we obtain a smooth threefold X which is a fibration over $\mathbb{P}^1$ with generic fiber a Kummer surface.

Working over $\mathbb{C}$, the holomorphic 3-form on Q (away from the nodes) is given by

$$\Omega = \operatorname{Res} \frac{\sum_j (-1)^j x_j dx_0 \wedge \cdots \wedge \widehat{dx_j} \wedge \cdots \wedge dx_n}{x_0^5 + \cdots + x_4^5 - 5x_0x_1x_2x_3x_4}.$$

It is simple to check that Ω is invariant under ι and thus descends to a nonvanishing holomorphic 3-form on Q/ι. Since nonvanishing holomorphic 3-forms extend across transverse A_1-singularities and small resolutions of nodes, we obtain a nonvanishing 3-form on X. Hence X is Calabi–Yau. The bad primes of X are checked to be 2 and 5.

Now let $\tilde{X}$ be a big resolution of X, where we replace the exceptional $\mathbb{P}^1$s over nodes with $\mathbb{P}^1 \times \mathbb{P}^1$. We now compute the Euler characteristic of $\tilde{X}$. A smooth quintic threefold has Euler number -200. Q is obtained by shrinking 125 S^3 cycles to points, so the Euler number of Q is -75.

Q is a branched cover of Q/ι which is generically 2:1 and branched over $\mathbb{P}^1_- \cup E$. Now E is a curve of arithmetic genus 6 with 5 S^1-cycles shrunk to points, so its Euler number is -5, while the Euler number of $\mathbb{P}^1_-$ is 2. By the Riemann-Hurwitz formula, $2\chi(Q/\iota) - (2-5) = -75$, so the Euler characteristic of Q/ι is -39.

$\tilde{X}$ is obtained from Q/ι by replacing $\mathbb{P}^1_-$ with a $\mathbb{P}^1$-bundle S over $\mathbb{P}^1_-$, replacing E with a $\mathbb{P}^1$-bundle over E, and taking a big resolution of a total of $60 + 5 \cdot 2 = 70$ nodes. The resulting Euler characteristic of $\tilde{X}$ is 168.

To compute the Hodge numbers of $\tilde{X}$, we use a trick due to van Geemen and Werner [34] (which has in fact been implicitly used in some of the examples discussed above). Applying the Lefschetz theorem to the Frobenius morphism, we have

$$\#X_p = \sum_{i=0}^{6} (-1)^i \mathrm{tr}\,\mathrm{Frob}_p(H^i(\tilde{X})).$$

Here $\#\tilde{X}_p$ is the number of fixed points of X under the Frobenius morphism, i.e. the number of $\mathbb{F}_p$-rational points of X.

For $p \equiv 1 (\mathrm{mod}\ 20)$, $H^2(\tilde{X})$ is spanned by algebraic cycles defined over $\mathbb{F}_p$: the only thing to check is that the rulings on the exceptional $\mathbb{P}^1 \times \mathbb{P}^1$s over the nodes are defined over $\mathbb{F}_p$. The coordinates of the nodes are defined whenever i is in $\mathbb{F}_p$, and local calculations show that the rulings are defined whenever -5 is a square in $\mathbb{F}_p$. Hence $\mathrm{tr}\,\mathrm{Frob}_p(H^2(X)) = ph^2$, where h^2 is the second Betti number of X. By Poincaré duality, $\mathrm{tr}\,\mathrm{Frob}_p(H^4(X)) = p^2h^2$. Now we know that

$$|\mathrm{tr}\,\mathrm{Frob}_p(H^3(X))| \leq p^{\frac{3}{2}} h^3 = p^{\frac{3}{2}}(168 - 2 - 2h^2).$$

By counting points (using a computer) on $\mathbb{F}_p$, we can hope to determine h^2. In fact, choosing $p = 421$ yields $h^2(X) = 85$ and $h^3(X) = 4$.

Consider again the curve $E = \{x_0^5 + 2x_1^5 + 2x_2^5 - 5x_0x_1^2x_2^2 = 0\}$ in $\mathbb{P}^2_+$. It has nodes at $[1 : \zeta^a : \zeta^{-a}]$. One can count points on the normalization $\hat{E}$ of E over $\mathbb{F}_p$: one simply needs to keep track of whether the nodes and their branches are defined over $\mathbb{F}_p$. As before, counting points on E enables us to determine the values of $a_p = \mathrm{tr}\,\mathrm{Frob}_p(H^1(\hat{E}))$; the bad primes of $\hat{E}$ are 2 and 5.

We want to find the modular form associated to $\hat{E}$. For good primes $p < 100$, the values of a_p are the same as the values of the newform 50B1 of level 50 and weight 2. We would like to say that the Galois representations associated to E and to the newform 50B1 are isomorphic, but the traces are not even. However, by the method of Serre explained above and worked out in [25], a two-dimensional Galois representation ramified at $\{2, 5\}$ is uniquely determined by its determinant and its traces at $p \in T = \{3, 7, 11, 13, 17, 29, 31\}$. The traces do indeed agree at this set of primes T, so we have found the correct modular form.

We have a rational map from the $\mathbb{P}^1$-bundle S over $\hat{E}$ into $\hat{X}$ which can be resolved by blowups. Hence it induces a nontrivial map on H^3, so we have a decomposition (up to semisimplification)

$$\begin{aligned} H^3(X) &\cong H^3(S) \oplus L \\ &\cong H^1(\hat{E})(-1) \oplus L. \end{aligned}$$

Now the quotient map defines a correspondence between Q and $\hat{X}$. Since the holomorphic three-form on Q is preserved by the involution, the quotient operation preserves the motive $H^3(Q)$. Hence the same level 25 newform of Q must occur in the decomposition of $H^3(\tilde{X})$. We therefore have

Theorem 6.2 *The threefold $\tilde{X}$ is modular, and up to Euler factors at the bad primes 2 and 5, the L-series of X is given by*

$$L(H^3(\hat{X}, s)) = L(f_{25}, s)L(g_{50}, s-1) \tag{6.1}$$

where f_{25} is a weight 4, level 25 newform and g_{50} is a weight 2, level 50 form.

7 Double octics

In [7] Cynk and Meyer considered examples of Calabi–Yau threefolds constructed as double covers of $\mathbb{P}^3$ branched over a degenerate octic (an arrangement of eight planes) D and proved their modularity by several of the methods used above; further computation was done in [24]. Using a computer search, they found degenerate arrangements of D with $h^{1,2} = 1$ and found candidate modular forms for the resulting threefolds. Some were proven to be modular by finding an elliptic surface S in X and thus splitting $H^3(X)$ into an elliptic piece and a piece U corresponding to a weight 4 modular form, using the technique of Hulek–Verrill. Others were proven to be modular by exhibiting birational maps between them and known modular rigid Calabi–Yaus or the Kummer-type non-rigid Calabi–Yaus of Livné–Yui.

For some examples Cynk and Meyer found an involution of $\mathbb{P}^3$ preserving the arrangement D, leading to an involution on the resulting Calabi–Yau threefold X. They then use a trick of van Geemen and Nygaard [33] to split H^3 into two two-dimensional pieces: if both 1 and -1 are eigenvalues of the action of the involution on H^3, then H^3 splits. Since the splitting commutes with the Frobenius morphism, it splits the Galois representation H^3 into two two-dimensional subrepresentations.

If Φ is a $\mathbb{Q}$-linear involution on $\mathbb{P}^3$ which preserves the arrangement of planes D, then it induces an involution on the resulting Calabi–Yau X. One can apply the Lefschetz theorem to $\mathrm{Frob}_p \circ \Phi$. The fixed points of $\mathrm{Frob}_p \circ \Phi$ are simply the $\mathbb{F}_{p^2}$-rational points of X, which can be counted via computer. The map $\mathrm{Frob}_p \circ \Phi$ acts by multiplication by $\pm p$ on H^2 and by $\pm p^2$ on H^4. The trace of $\mathrm{Frob}_p \circ \Phi$ will be ± 1 times the trace of Frob_p on the ± 1 eigenspaces of H^3. Thus computing the traces of $\mathrm{Frob}_p \circ \Phi$ by counting points on $\mathbb{F}_{p^2}$ and computing the traces of Frob_p will yield the traces of Frob_p on the ± 1 eigenspaces of H^3. Cynk and Meyer show that for some of their examples, the group H^3 does indeed split via the action of some involution Φ, and the two-dimensional pieces are proven to be modular using the Livné theorem.

7.1 Higher-dimensional Kummer constructions. Higher-dimensional modular Calabi–Yau varieties have been considered by Cynk–Hulek and and also by Cynk–Schütt.

In [6] Cynk–Hulek consider products of elliptic curves with automorphisms of finite order:

Theorem 7.1 *Let $E_1, E_2, \ldots, E_n$ be elliptic curves defined over $\mathbb{Q}$ with involutions on each given as $x \to -x$. Consider the subgroup $\mathbb{Z}_2^{n-1} \subset \mathbb{Z}_2^n$ of vectors $(c_1, c_2, \ldots, c_n)$ with $\sum c_i = 0$, and put $X^n = (E_1 \times \cdots \times E_n)/\mathbb{Z}_2^{n-1}$. Then X^n admits a crepant resolution $\hat{X}^n$ which is a smooth Calabi–Yau variety.*

If n is odd, then $H^n(\hat{X}^n) \cong H^n(E_1 \times \cdots \times E_n)^{\mathbb{Z}_2^{n-1}} \cong H^1(E_1) \otimes \cdots \otimes H^1(E_n)$. Then $\hat{X}^n$ is modular, with $L(H^n(\hat{X}^n), s) = L(g_{E_1} \otimes \cdots \otimes g_{E_n}, s)$, where the g_{E_i} are the modular forms associated to the E_i.

If n is even, then the invariant submotive $H^n(E_1 \times \cdots \times E_n)^{\mathbb{Z}_2^{n-1}}$ is isomorphic to the direct sum of a submotive generated by cycles of products of $n/2$ fibers and a submotive $I(\hat{X}^n) \cong H^1(E_1) \otimes \cdots \otimes H^1(E_n)$, and $I(\hat{X}^n)$ contains the submotive of transcendental cycles. In this case $L(I(\hat{X}^n), s) = L(g_{E_1} \otimes \cdots \otimes g_{E_n}, s)$.

If one wants a rigid Calabi–Yau variety, one must divide out by automorphisms of higher order:

Theorem 7.2 *Let E be the elliptic curve with automorphism η of order 3, and let X^n be the quotient of E^n by the group $\{(\eta^{a_1}, \ldots, \eta^{a_n}) : \sum a_i = 0\}$. Then X^n has a crepant resolution $\hat{X}^n$ which is Calabi–Yau, and* $\dim H^n(\hat{X}^n) = 2$ *if n is odd, resp.* $\dim T(\hat{X}^n) = 2$ *if n is even, where $T(\hat{X}^n)$ is the transcendental part of the middle cohomology.*

Moreover, $\hat{X}^n$ is defined over $\mathbb{Q}$ and $L(H^n(\hat{X}^n), s) = L(g_{n+1}, s)$ resp. $L(T(\hat{X}^n), s) = L(g_{n+1}, s)$, where g_{n+1} is the weight $n+1$ cusp form with complex multiplication in $\mathbb{Q}(\sqrt{-3})$ associated to the nth power of the Grössencharakter of E.

Theorem 7.3 *Let E be the elliptic curve with Weierstrass equation $y^2 = x^3 - Dx$, where D is a square-free integer, E has complex multiplication in $\mathbb{Q}(i)$ and the map $\rho : (x, y) \to (-x, iy)$ is an automorphism of order 4.*

Let X^n be the quotient of E^n by the group $\{(\eta^{a_1} \times \cdots \times \eta^{a_n}) : \sum a_i = 0\}$. Then X^n has a crepant resolution $\hat{X}^n$ defined over $\mathbb{Q}$ which is Calabi–Yau. We have $\dim H^n(\hat{X}^n) = 2$ *if n is odd, resp.* $\dim T(\hat{X}^n) = 2$ *if n is even, where $T(\hat{X}^n)$ is the transcendental part of the middle cohomology.*

Then $L(H^n(\hat{X}^n), s) = L(g_{n+1}, s)$ resp. $L(T(\hat{X}^n), s) = L(g_{n+1}, s)$, where g_{n+1} is a weight $n+1$ cusp form with complex multiplication in $\mathbb{Q}(i)$.

In [8] Cynk and Schütt generalize this higher-dimensional Kummer construction. Let L be a number field, $L(\alpha)/L$ an extension of order n, E_1 an elliptic curve over $L(\alpha)$, and $E_2, \ldots, E_n$ the conjugates of E_1. Construct the product $A = E_1 \times \cdots \times E_n$ and the Kummer quotient $X = A/\mathbb{Z}_2^{n-1}$ as before. They then show that A and X descend to the base field L (Weil restriction). In the case $n = 3$, they construct a crepant resolution $\tilde{X}$ of X defined over the base field L.

Now suppose K is a quadratic imaginary field of class number 3, $d = -p_0$ its discriminant, E_1 an elliptic curve with complex multiplication in $\mathcal{O}_K$, F the field of definition of E. Construct the Calabi–Yau threefold $\tilde{X}$ as above. We then have

Theorem 7.4 *The Calabi–Yau threefold $\tilde{X}$ is modular, and*

$$L(\tilde{X}/\mathbb{Q}, s) = L(g, s)L(E/F, s-1)$$

where g is the unique CM-newform of level d^2 and weight 4 with rational coefficients, or equivalently the corresponding Hecke character ϕ of K with conductor $(\sqrt{d})$ and ∞-type 3.

8 A Hilbert-modular Calabi–Yau threefold

In [5] Consani and Scholten study a quintic threefold X given by the homogenization of the equation $P_5(x, y) - P_5(z, w)$ where

$$P_5(x, y) = (x^5 + y^5) - 5xy(x^2 + y^2) + 5xy(x + y) + 5(x^2 + y^2) - 5(x + y).$$

This Calabi–Yau has 120 ordinary double points as singularities. Let $\tilde{X}$ be the big resolution of X. It is shown that $h^3(\tilde{X}) = 4$ and $\tilde{X}$ has bad reduction at $\{2, 3, 5\}$. They prove that the Galois representation ρ associated to $H^3_{et}(\tilde{X}, \mathbb{Q}_l(\sqrt{5}))$ splits into two two-dimensional pieces. They also construct a Hilbert modular newform f of weight $(2, 4)$ and conductor 30 on a finite extension E_λ of $\mathbb{Q}_l$ with ring of integers $\mathcal{O}_\lambda$ and an associated 2-dimensional λ-adic Galois representation $\sigma_{f,\lambda} \to GL_2(\mathcal{O}_\lambda)$ and give numerical evidence that $H^3_{et}(\tilde{X}, \mathbb{Q}_l(\sqrt{5}))$ is associated to the forms f and f^σ. In 2005 Yi announced a proof of this conjecture [37]. However, his proof has not yet been published.

9 Further questions

We conclude this paper with some open problems.

Question 9.1 For which weight 4 newforms f does there exist a Calabi–Yau threefold X such that $L(f, s)$ occurs in the L-series of X?

This question was posed independently by Mazur and van Straten. When f is a newform of weight 2 with coefficents in the field F, a construction of Shimura [28] associates to f an abelian variety $A/\mathbb{Q}$ of dimension $[F:\mathbb{Q}]$. No such construction yet exists for higher weight. The known examples typically arise as highly symmetric members in families of Calabi–Yau varieties.

Question 9.2 If X is a modular Calabi–Yau threefold, can one predict the level of the associated modular form f?

In the elliptic curve case, the level of the newform is given by the conductor N, which is described in terms of the bad reduction of N. However, we have seen examples of Calabi–Yau threefolds where the bad primes do not appear in the level of the modular form. Meyer has formulated the following rule of thumb:

Conjecture 9.1 Let X be a modular Calabi–Yau threefold and let $p \geq 5$ be a prime. Let f be the twist of minimal level N of the weight four newform associated with X. If the exponent of p in N is 2 then X develops non-isolated singularities over $\mathbb{F}_p$. If the exponent is 1 then X is singular modulo p and there exists a birational model of X with only isolated singularities modulo p. If the exponent is 0 then there is a birational model of X with good reduction at p.

Hulek, Kloosterman and Schütt also pose the following question

Question 9.3 [13] Let X be a modular Calabi–Yau variety over $\mathbb{Q}$ with corresponding newform f. Under what conditions do the bad Euler factors of the L-series of X and f coincide?

Question 9.4 More generally, what other kinds of automorphic forms can occur in the L-series of a Calabi–Yau threefold over $\mathbb{Q}$? What about over other number fields?

References

[1] C. Breuil, B. Conrad, F. Diamond and R. Taylor, On the Modularity of Elliptic Curves over $\mathbb{Q}$: Wild 3-adic Exercises, *J. Amer. Math. Soc.* **14**, 2001, 843-939.

[2] W. Barth, K. Hulek and R. Moore, Degenerations of Horrocks-Mumford Surfaces, *Math. Ann.* **277**, 1987, 735-755.

[3] W. Barth and R. Moore, Geometry in the Space of Horrocks-Mumford Surfaces, *Topology* **28**, 1989, 231-245.

[4] C. Borcea, K3 surfaces with involution and mirror pairs of Calabi–Yau manifolds, *Mirror Symmetry II*, AMS/IP Stud. Adv. Math. **1**, 1997, 717-743.

[5] K. Consani and J. Scholten, Arithmetic on a Quintic Threefold, *Internat. J. Math.* **12**, no. 3, 2001, 943-972.

[6] S. Cynk and K. Hulek, Construction and Examples of Higher-Dimensional Modular Calabi–Yau Manifolds, *Canad. Math. Bull.* **50** no. 4, 2007, 486-503.

[7] S. Cynk and C. Meyer, Modularity of Some Non-Rigid Double Octic Calabi–Yau Threefolds, to appear in *Rocky Mountain J. Math.*

[8] S. Cynk and M. Schütt, Generalised Kummer Constructions and Weil Restrictions, `math.AG/0710.4565`.

[9] L. Dieulefait, Improvements on Dieulefait-Manoharmayum and applications, `math.AG/0508163`.

[10] L. Dieulefait and J. Manoharmayum, Modularity of Rigid Calabi–Yau Threefolds over $\mathbb{Q}$, *Calabi–Yau Varieties and Mirror Symmetry.* Fields Inst. Commun. **38**, 2003, 159-166.

[11] G. Horrocks and D. Mumford, A Rank 2 Vector Bundle on $\mathbb{P}^4$ with $15,000$ Symmetries, *Topology* **12**, 1973, 63-81.

[12] K. Hulek, The Horrocks-Mumford Bundle, *Vector Bundles in Algebraic Geometry*, London Math. Soc. Lecture Note Ser. **208**, 1995, 139-177.

[13] K. Hulek, R. Kloosterman and M. Schütt, Modularity of Calabi–Yau Varieties, *Global Aspects of Complex Geometry* (2006), 271-309.

[14] K. Hulek and H. Lange, The Hilbert Modular Surface for the Ideal $(\sqrt{5})$ and the Horrocks-Mumford Bundle, *Math. Z.* **198**, 1988, 95-116.

[15] K. Hulek and H. Verrill, On Modularity of Rigid and Non-Rigid Calabi–Yau Varieties Associated to the Root Lattice A_4, *Nagoya Math. J.* **179** (2005), 103-146.

[16] K. Hulek and H. Verrill, On the Modularity of Calabi–Yau Threefolds Containing Elliptic Ruled Surfaces, *Mirror Symmetry V*, AMS/IP Stud. Adv. Math. **38** (2006), 19-34.

[17] H. Inose and T. Shoida, On Singular $K3$ Surfaces, *Complex Analysis and Algebraic Geometry* (1977), 119-136.

[18] J. Jones, Table of Number Fields with Prescribed Ramification, `http://math.la.asu.edu/ jj/numberfields`.

[19] E. Lee, A Modular Non-Rigid Calabi–Yau Threefold, *Mirror Symmetry V*, AMS/IP Stud. Adv. Math. **38** (2006), 89-122.

[20] R. Livné, Cubic Exponential Sums and Galois Representations, *Contemporary Mathematics* **67**, 1987, 247–261.

[21] R. Livné, Motivic Orthogonal Two-Dimensional Representations of $\mathrm{Gal}(\overline{\mathbb{Q}}/\mathbb{Q})$, *Israel J. Math.* **92**, no. 1-3 (1995), 148-156.

[22] R. Livné and N. Yui, The Modularity of Certain Non-Rigid Calabi–Yau Threefolds, *J. Math. Kyoto Univ.* **45**, no. 4, 645-665.

[23] N. Manolache, The Equations of the Abelian Surfaces Embedded in $\mathbb{P}^4(\mathbb{C})$, *J. Reine. Angew. Math.* **394**, 1989, 196-202.

[24] C. Meyer, *Modular Calabi–Yau Threefolds*, Fields Inst. Monographs **22** (2005).

[25] C. Schoen, On the Geometry of a Special Determinantal Hypersurface Associated to the Mumford-Horrocks Vector Bundle, *J. Reine. Angew. Math.* **364**, 1986, 85-111.

[26] C. Schoen, On Fiber Products of Rational Elliptic Surfaces with Section, *Math. Z.* **197**, no. 2 (1988), 177–199.

[27] M. Schütt, On the Modularity of Three Calabi–Yau Threefolds with Bad Reduction at 11, *Canad. Math. Bull.* **49**, no. 2 (2006), 296–312.

[28] G. Shimura, *Introduction to the Arithmetic Theory of Automorphic Forms*, Princeton Univ. Press, Princeton, 1971.

[29] W. Stein, The Modular Forms Database, `http://modular.fas.harvard.edu/Tables/`

[30] R. Taylor and A. Wiles, Ring-theoretic Properties of Certain Hecke Algebras, *Ann. Math. (2)* **141**, 1995, 553-572.

[31] G. Tian, Smoothness of the Universal Deformation Space of Compact Calabi–Yau Manifolds and its Petersson-Weil Metric, *Mathematical Aspects of String Theory*, 629-646, ed. S.-T. Yau, World Scientific, Singapore, 1987.

[32] A. Todorov, The Weil-Petersson Geometry of the Moduli Space of SU(n ≥ 3) (Calabi–Yau) Manifolds I, *Commun. Math. Phys* 126 (1989) 325-346.

[33] N. Nygaard and B. van Geemen, On the Geometry and Arithmetic of some Siegel Modular Threefolds, *J. Number Theory* **53** no. 1 (1995), 45-87.
[34] B. van Geemen and J. Werner, New Examples of Threefolds with $c_1 = 0$, *Math. Z.* **203**, 1990, 211-225.
[35] C. Voisin, Miroirs et involutions sur les surfaces K3, *Astérisque* **218** (1993), 273-323.
[36] D. Wan, Mirror Symmetry for Zeta Functions, *Mirror Symmetry V*, AMS/IP Stud. Adv. Math. **38** (2006), 159-184
[37] Y.-C. Yi, On the Hilbert Modularity of a Certain Quintic Threefold, preprint.
[38] N. Yui, Update on the Modularity of Calabi–Yau Varieties, *Calabi–Yau Varieties and Mirror Symmetry*. Fields Inst. Commun. **38**, 2003, 307-362.

Fields Institute Communications
Volume **54**, 2008

Finite Index Subgroups of the Modular Group and their Modular Forms

Ling Long
Department of Mathematics
Iowa State University
Ames, IA 50011 USA
linglong@iastate.edu

Abstract. Classically, congruence subgroups of the modular group, which can be described by congruence relations, play important roles in group theory and the study of modular forms. In reality, the majority of finite index subgroups of the modular group are noncongruence. These groups as well as their modular forms are central players of this survey article. Differences between congruence and noncongruence subgroups and modular forms will be discussed. We will mainly focus on three interesting aspects of modular forms for noncongruence subgroups: the unbounded denominator property, modularity of the Galois representation arising from noncongruence cuspforms, and Atkin and Swinnerton-Dyer congruences.

1 Introduction

The special linear group $SL_2(\mathbb{Z})$ consists of all 2-by-2 integral matrices with determinant 1. It is one of the most well-known and important discrete groups. A finite index subgroup of the modular group is said to be *congruence* if it contains the kernel of the natural modulo N homomorphism from $SL_2(\mathbb{Z})$ to $SL_2(\mathbb{Z}/N\mathbb{Z})$ for some positive integer N; otherwise, it is called a *noncongruence* subgroup. The existence of noncongruence subgroups of $SL_2(\mathbb{Z})$ was confirmed by Fricke [Fri86] and Pick [Pic86]. More noncongruence subgroups were constructed by Reiner [Rei58], Newman [New65], Rankin [Ran67], etc. In contrast, any finite index subgroup of $SL_n(\mathbb{Z})$ with $n \geq 3$ is congruence [BLS64]. A famous theorem of Belyĭ implies that any compact complex smooth irreducible curve defined over $\overline{\mathbb{Q}}$, the algebraic closure of $\mathbb{Q}$, can be realized (in many ways) as a modular curve (defined in 2.3) for a finite index subgroup of $SL_2(\mathbb{Z})$ [Bel79]. Consequently, finite index subgroups arise

2000 *Mathematics Subject Classification.* Primary 11F11; Secondary 11F06.

The author was supported in part by an NSF-AWM mentoring travel grant for women. She would like to thank the Pennsylvania State University for hosting her visit during May-June 2006.

naturally in many fields such as the theory of dessins d'enfants which includes Galois coverings of the projective line [Bir94], triangular groups [Jon79, JS94], as well as the theory of translation surfaces [HL05]. As a matter of fact, noncongruence subgroups of $SL_2(\mathbb{Z})$ outnumber congruence subgroups (Section 2.5 below).

Modular forms are generalizations of elliptic functions (doubly periodic functions) and enjoy many wonderful properties. Many important one variable formal power series arising in combinatorics or physics, such as the partition function, turn out to be closely related to modular forms.

The theory of congruence (elliptic) modular forms is now well-developed due to the efforts of many mathematicians. It has a very broad impact including some important applications to information technology [Sar90, Li96]. In the early days, computational observations and conjectures were significant driving forces for the development of this theory. For example, Ramanujan observed that the coefficients of $q\prod_{n\geq 1}(1-q^n)^{24} = \sum \tau(n)q^n$ satisfy special three-term recursions for small primes p,

$$\tau(np) - \tau(p)\tau(n) + p^{11}\tau(n/p) = 0, \forall n \in \mathbb{N}, \tag{1.1}$$

where $\tau(n/p) = 0$ if $p \nmid n$. In [Mor17] Mordell showed that these recursions hold for all primes p by using essentially Hecke operators. In [Hec37] Hecke defined Hecke operators and proved that they satisfy $(2d+1)$-term relations on each d-dimensional space of congruence modular forms. Shortly after, Petersson introduced the Petersson inner product via which he showed that the Hecke operators can be diagonalized simultaneously [Pet39]. Readers are referred to [LLY05a, Section 2] for a summary of the theory of newforms (congruence modular forms with particular properties) developed by Atkin and Lehner [AL70], Miyake [Miy71], and Li [Li75]. The current considerable interest in modular forms is largely due to the vital role they played in the celebrated proof of Fermat's Last Theorem (FLT) by Wiles [Wil95]. In his proof, Wiles used a result of Frey and Ribet which says that the Taniyama-Shimura-Weil (TSW) conjecture implies FLT. The TSW conjecture asserts that the l-adic Galois representations attached to an elliptic curve (a genus one smooth complex curve) defined over $\mathbb{Q}$ are isomorphic to the l-adic Deligne representations attached to a certain newform. The final proof of TSW ([Wil95, TW95, BCDT01]) is a major breakthrough in the very influential Langlands program, which is a system of conjectures connecting the l-adic cohomology theory of algebraic varieties with the representation theory of cuspidal automorphic forms [Gel84, Mur93]. In the sequel, we say an l-adic Galois representation is "modular" when it arises from a congruence automorphic form.

Modular forms for noncongruence subgroups have already been considered by Fricke and Klein. However, there was no serious investigation of these functions till the late 1960's as it was commonly believed that the Hecke theory was missing. Indeed, Atkin conjectured that Hecke operators defined classically have trivial actions on noncongruence modular forms and his conjecture was proven in part by Serre [Tho89] and Scholl [Sch97] and in general by Berger [Ber94].

However, investigations in the past 40 years have revealed that the mathematical structure of noncongruence modular forms is still wonderfully rich. In [ASD71], Atkin and Swinnerton-Dyer used computers together with algebra and combinatorics to work with noncongruence subgroups and noncongruence modular forms. They put forward many interesting and fundamental ideas and original observations. In particular, they studied the arithmetic of the Fourier expansions

$\sum a(n)w^n$ of noncongruence modular forms and observed that for some forms with algebraic coefficients the sequences $\{a(n)\}$ have unbounded denominators. On the other hand, it is known that if f is a congruence holomorphic modular form, the sequence $\{a(n)\}$ must admit bounded denominators [Shi71]. This observation gives a simple and clear criterion for distinguishing noncongruence modular forms. Atkin and Swinnerton also found a p-adic analogue of the Hecke three-term recursion by demonstrating that the coefficients $\{a(n)\}$ of some noncongruence modular forms satisfy

$$a(np) - A(p)a(n) + B(p)a(n/p) \equiv 0, \forall n \geq 1 \tag{1.2}$$

modulo a power of p depending on $\mathrm{ord}_p(n)$ for some constants $A(p)$ and $B(p)$. In [ASD71] a proof of the weight 2 dimension 1 case was provided. The modularity of this case would have followed from the Taniyama-Shimura-Weil conjecture. For a few examples of dimension 2 or more, Atkin and Swinnerton-Dyer found that such three-term congruences still exist with the forms diagonalized p-adically for each separate p and the $A(p)$ being over algebraic number fields, in general different for each p. Several people ([Dit76, Sti85, etc]) studied Atkin and Swinnerton-Dyer (ASD) congruences. The general situation for the weight 2 cases has been confirmed by several papers [Hon70, Car71, Dit76, Kat81].

In [Sch85], using deep algebraic geometry and representation theory, Scholl associated each d-dimensional space of noncongruence cuspforms with $2d$-dimensional l-adic Scholl representations and showed the existence of $(2d+1)$-term p-adic recursions under general conditions. When $d = 1$ Scholl fully established the 3-term ASD congruences (1.2). However, when $d > 1$, the existence of a fixed basis each of which satisfies 3-term ASD congruences for all p (referred to as simultaneous diagonalization) is rare.

Around the same time interesting connections between modular forms for congruence and noncongruence subgroups began to emerge. In [Sch88, Sch93] Scholl gave examples of 1-dimensional spaces of cuspforms with the $A(p)$ being eigenvalues of congruence cuspforms by showing the l-adic Scholl representations associated to those cases are "modular". In [Ber00], Berger found a quadratic relation satisfied by 2 noncongruence cuspforms and 2 congruence cuspforms.

In [ASD71, Sch88] noncongruence subgroups with small indices were given considerable attention. Recently, investigation has been focused on the so-called noncongruence character groups, which are "almost" congruence. These noncongruence subgroups are constructed by assigning characters to the generators of congruence subgroups. The lattice groups considered by Rankin [Ran67] are examples of them. Klein and Fricke used "Wurzelmodul" to describe a Hauptmodul of a genus zero noncongruence character group defined by a cyclic character. Here, we will associate the Fermat curve $x^n + y^n = 1$ with the modular curve for the intersection of two particular character groups.

In [LLY05b], Li et al. gave a 2-dimensional space of cuspforms for a noncongruence character group which satisfies simultaneous diagonalization. Moreover, it was shown that the $A(p)$ are eigenvalues of explicit newforms. Later, Fang et al. found a few more examples of this kind [F$^+$05]. As a matter of fact, simultaneous diagonalization can rarely be achieved. More recently, Atkin et al. [ALL08] and Long [Lon08] fully exhibited two cases which resemble the case in [LLY05b] but are slightly more general. In each case, a basis depending on the congruence class of p modulo a fixed integer is found, for each of the basis functions the coefficients satisfy a 3-term ASD congruence relation. Meanwhile, modularity of each case was verified individually. These modularity results rely heavily on the decompositions of the associated l-adic Scholl representations or their restrictions to suitable Galois subgroups. In contrast, Scholl exhibited other cases where the l-adic Scholl representations are not reducible [Sch06].

This expository paper is organized in the following order. Some background topics are introduced in Section 2. Section 3 is devoted to some differences between congruence and noncongruence subgroups. Modular forms and some related topics will be addressed briefly in Section 4. In Section 5, we will show that the Fourier coefficients of some noncongruence modular forms have unbounded denominators. In Section 6 we will address l-adic Scholl representations attached to noncongruence cuspforms and related modularity results briefly. Section 7 is about the three-term Atkin-Swinnerton-Dyer congruence relations. Throughout this paper, several examples will be discussed and all proofs presented are elementary. The author has not made an attempt to give a comprehensive review of the vast literature. This survey mostly reflects the author's perspectives on these topics based on her recent research activities.

This paper has grown out of the author's talk delivered at the Banff conference. The author is indebted to A.O.L. Atkin. The introduction of this article is written partly based on the valuable historic recounting and analysis of the development of noncongruence modular forms by A.O.L. Atkin (communicated through private correspondence). His wisdom and original ideas have sparked much of the discussion here. It is the author's great pleasure to thank Wenching Winnie Li for her constant discussions and many insightful comments. Some results below are cited from some unpublished notes of Li and the author. The author would also like to thank the conference organizers and in particular Noriko Yui for inviting her to attend this wonderful conference and giving her an opportunity to present this talk. She is also very grateful to many people such as Ron Livné, Siu-Hung Ng, Jan Stienstra, Yifan Yang, and Don Zagier for their valuable discussions and comments. The author specially thanks Chris Kurth for his remarks on an earlier version of this paper, Tonghai Yang for pointing out several useful references, and the referee for many suggestions on improving the clarity of this paper.

2 Backgrounds

In this article, we use p and l for prime numbers and $\mathbb{Q}_l$ for the field of l-adic numbers. We adapt mostly common notations. We use w for all local parameters at infinity instead of q^μ, where μ varies in different cases.

2.1 Finite index subgroups of $PSL_2(\mathbb{Z})$. It is well-known that $SL_2(\mathbb{Z})$ is generated by two elements $E = \begin{pmatrix} 0 & 1 \\ -1 & 0 \end{pmatrix}$ and $V = \begin{pmatrix} 1 & 1 \\ -1 & 0 \end{pmatrix}$ [Ran77]. Identifying elements in $SL_2(\mathbb{Z})$ which differ by a sign, one obtains the modular group $PSL_2(\mathbb{Z})$. It is well-known that $PSL_2(\mathbb{Z})$ is the free product of two elements with orders 2 and 3 respectively, namely the images of E and V in $PSL_2(\mathbb{Z})$ [Ran77]. In this article, we use Γ to denote finite index subgroups of $SL_2(\mathbb{Z})$. We also use Γ to denote the image of $\Gamma/\{\pm I_2\}$ in $PSL_2(\mathbb{Z})$ if there is no confusion.

Congruence subgroups are a class of subgroups which are easy to describe. For example, given a positive integer n, the following are some well-known congruence subgroups.

$$\begin{aligned} \Gamma^0(n) &= \left\{\gamma \in SL_2(\mathbb{Z}), \gamma = \begin{pmatrix} * & 0 \\ * & * \end{pmatrix} \mod n\right\}. \\ \Gamma^1(n) &= \left\{\gamma \in SL_2(\mathbb{Z}), \gamma = \begin{pmatrix} 1 & 0 \\ * & 1 \end{pmatrix} \mod n\right\}. \\ \Gamma(n) &= \left\{\gamma \in SL_2(\mathbb{Z}), \gamma = \begin{pmatrix} 1 & 0 \\ 0 & 1 \end{pmatrix} \mod n\right\}. \end{aligned}$$

In [Mil69a, Mil69b], Millington showed that up to isomorphism there is a one-to-one correspondence between finite index subgroups of $PSL_2(\mathbb{Z})$ and a pair of permutations of orders 2 and 3 respectively which generate a finite transitive permutation group.

Example 1 The group $\Gamma^1(3)$ is an index 4 subgroup of $SL_2(\mathbb{Z})$. It has 4 right coset representatives $\gamma_1 = \begin{pmatrix} 1 & 1 \\ 0 & 1 \end{pmatrix}, \gamma_2 = \begin{pmatrix} 1 & 2 \\ 0 & 1 \end{pmatrix}, \gamma_3 = \begin{pmatrix} 1 & 3 \\ 0 & 1 \end{pmatrix}, \gamma_4 = \begin{pmatrix} 0 & -1 \\ 1 & 0 \end{pmatrix}$. For simplicity, we will use 1,2,3,4 to denote the corresponding cosets. Multiplying by E on the right will permute the right cosets. More explicitly, $E = (12)(34)$ is an order 2 permutation. Similarly, V corresponds to an order 3 permutation $V = (134)(2)$. Millington's theorem says $\Gamma^1(3)$ corresponds uniquely (up to isomorphism) to the pair of permutations $(12)(34)$ and $(134)(2)$. Note that in this case $\begin{pmatrix} 1 & -1 \\ 0 & 1 \end{pmatrix} = VE = (123)(4)$. Since the group of stabilizers of the cusp ∞ is generated by $\begin{pmatrix} 1 & -1 \\ 0 & 1 \end{pmatrix}$, the corresponding cycle (123) is called the marked cycle in the disjoint cycle representation of VE with respect to ∞.

2.2 KFarey. The permutation representation of these groups is of great importance for the computational aspects of finite index groups of the modular group. For example, in [Hsu96], Hsu gave an algorithm for identifying congruence subgroups based on the permutation representations. Recently, Kurth has developed a SAGE package called "KFarey" which has implemented many useful functions for finite index subgroups such as Hsu's algorithm mentioned above, and Farey symbols by Kulkarni [Kul85], determining whether a given element is in a group specified by a given permutation representation and a "marked" cycle in the permutation representation of VE corresponding to infinity, etc. Please refer to a recent expository article on the basic algorithms used in "KFarey" [KL07]. The "KFarey" package can be downloaded from the following website
`http://www.public.iastate.edu/~kurthc/research/index.html`

2.3 Modular curves. The elements $\begin{pmatrix} a & b \\ c & d \end{pmatrix} \in \Gamma$ act on the upper half plane $\mathfrak{H} = \{z \in \mathbb{C} \mid \operatorname{Im} z > 0\}$ via linear fractional transformations:

$$\begin{pmatrix} a & b \\ c & d \end{pmatrix} \cdot z = \frac{az+b}{cz+d}, \quad \forall z \in \mathfrak{H}.$$

An element $\gamma = \begin{pmatrix} a & b \\ c & d \end{pmatrix} \in SL_2(\mathbb{Z})$ is said to be elliptic (resp. parabolic or hyperbolic) if the value of $|a+d|$ is < 2 (resp. $= 2$ or > 2). Each elliptic (resp. parabolic) element has a unique fixed point on $\mathfrak{H}$ (resp. $\mathbb{Q} \cup \{\infty\}$). Hyperbolic elements have no fixed points on $\overline{\mathfrak{H}} = \mathfrak{H} \cup \mathbb{Q} \cup \{i\infty\}$. (Whether a group has elliptic elements or not can be easily read off from its permutation representation. For example, $\Gamma^1(3)$ has an order 3 elliptic point as the left coset labeled by 2 is fixed by V.) The orbit space $\mathfrak{H}/\Gamma$ can be compactified canonically by adding a few isolated points called the cusps of Γ (equivalence classes in $(\{\infty\} \cup \mathbb{Q})/\Gamma$). The compactified curve, denoted by X_Γ, is called the *modular curve* associated with Γ. By convention, the genus of X_Γ is also called the genus of Γ. For convenience, we will assume that all modular curves here are defined over $\mathbb{Q}$ unless specified otherwise. When the curve X_Γ has genus 0, the field of meromorphic functions on X_Γ is generated by a single element, which is called a Hauptmodul of Γ. For example, the classical modular j-function is a Hauptmodul of $SL_2(\mathbb{Z})$.

The importance of finite index subgroups of $SL_2(\mathbb{Z})$ is seen in the following theorem.

Theorem 1 (Belyĭ [Bel79]) *Any smooth compact complex projective curve defined over $\overline{\mathbb{Q}}$ is isomorphic to the modular curve of some finite index subgroup of $SL_2(\mathbb{Z})$.*

On the other hand, one should know that the above correspondence is not unique. Please refer to Section 5.5 for a connection between Fermat curves and modular curves which demonstrates Belyĭ 's theorem above.

Up to conjugation, the stabilizer of each cusp c of Γ is generated by a matrix of the form $\pm\begin{pmatrix} 1 & m \\ 0 & 1 \end{pmatrix}$, where $m \geq 1$ is called the cusp width of c. The *level* of the group Γ, denoted by lev(Γ), is the least common multiple of all cusp widths of Γ. It was defined by Wohlfahrt [Woh64] which generalizes Klein's level definition for congruence subgroups.

2.4 Elliptic modular surface. An elliptic surface

$$\pi : S \to C$$

is a two-dimensional complex variety S together with a morphism π to the base curve C such that for almost all $t \in C$, $\pi^{-1}(t)$ is an elliptic curve. When $\pi^{-1}(c)$ is not an elliptic curve, it is called a special fiber. Special fibers of the Néron models of elliptic surfaces have been classified by Kodaira [Kod63] and Néron [Nér64]. An elliptic surface $\pi : S \to C$ is called semistable if all of its special fibers are of I_n type (in Kodaira's notation) and the list of the (un-ordered) numbers $[n_1, n_2, \cdots, n_r]$ in the subscripts of the special fibers is called the configuration of the semistable elliptic surface.

Let $\mathbb{C}$ be the ground field. Given an elliptic fibration $\pi : S \to C$, let Σ be the finite collection of points on C corresponding to special fibers. Via a standard procedure (cf. [Kod63]) one obtains the monodromy representation of the fundamental group $\pi_1(C \setminus \Sigma, t_0)$:

$$\rho_\pi : \pi_1(C \setminus \Sigma, t_0) \to \mathrm{GL}_2(\mathbb{C}),$$

where t_0 is a fixed reference point on $C \setminus \Sigma$. The image of ρ_π, well-defined up to conjugation corresponding to the different choices of t_0, is called the monodromy group of π. Nori gave criteria for when the monodromy group of an elliptic surface $\pi : S \to C$ gives rise to a finite index subgroup of the modular group [Nor85]. By Nori's theorem, the monodromy group of a semistable elliptic surface with maximal possible Picard number is a finite index subgroup of $SL_2(\mathbb{Z})$. Here, for a surface S, the $\mathbb{Z}$-rank of its group of divisors modulo linear equivalence is called the Picard number of S.

In [Shi72], for every finite index subgroup Γ of $SL_2(\mathbb{Z})$ not containing $-I_2$, Shioda constructed an elliptic modular surface $\pi : \mathcal{E}_\Gamma \to X_\Gamma$ associated with Γ. In particular, for an index 24 genus zero subgroup Γ, $\mathcal{E}_\Gamma$ is an elliptic K3 surface. K3 surfaces are simply connected compact complex surfaces with trivial canonical bundles. They are two-dimensional Calabi-Yau manifolds and occupy an important role in the classification of compact complex surfaces (cf. [BPVdV84]). Miranda and Persson have classified all configurations of semistable elliptic K3 surfaces with Picard number 20 [MP89]. Each K3 surface of this kind corresponds to an index 24 subgroup with 6 cusps and its cusp widths are given by the configuration of the surface. (Montanus and Beukers computed algebraic models of those semistable elliptic K3 surfaces defined over $\mathbb{Q}$.) Among the 112 possible extremal semistable K3 configurations, 9 cases correspond to congruence subgroups. In [TY07], Top and Yui have mentioned that among the remaining 103 configurations corresponding to noncongruence subgroups, 10 can be obtained via double covers of the semistable extremal rational elliptic surfaces classified by Beauville [Bea82] ramifying at only two points. For example, a semistable K3 with a configuration $[1, 1, 1, 1, 10, 10]$ is a double cover of a semistable rational surface with a configuration $[1, 1, 5, 5]$ (cf. Example 2 below).

2.5 Most finite index subgroups are noncongruence. Rademacher conjectured that there are only finitely many genus zero congruence subgroups (c.f [KN65]). In [Den75], Dennin showed that there are only finitely many congruence subgroups of a given genus. In [Seb01], Sebbar classified all elliptic element free genus zero congruence subgroups. There are 33 such groups in total. In [CLY04], Chua, Lang, and Yang gave explicit descriptions for all the genus zero congruence subgroups. In [CP03], Cummins and Pauli classified all congruence subgroups with genus less than 25.

On the other hand, Newman showed that there are infinitely many genus zero noncongruence subgroups [New65]. In [ASD71], Atkin and Swinnerton-Dyer pointed out that noncongruence subgroups outnumber congruence subgroups. In [Jon79], Jones proved that there are infinitely many noncongruence subgroups of a given genus (cf. also [Jon86]). Let $N_c(n)$ (resp. $N_{nc}(n)$) be the number of congruence (resp. noncongruence) subgroups of index n in $PSL_2(\mathbb{Z})$. Stothers gave an upper bound for $N_c(n)$ and showed that $N_c(n)/N_{nc}(n) \to 0$ when $n \to \infty$ [Sto84].

2.6 Noncongruence character groups. It is well-known that for any positive integer $n > 2$, the finite index subgroup $\Gamma^1(n)$ of $SL_2(\mathbb{Z})$ is a normal subgroup of $\Gamma^0(n)$ with the quotient group $\Gamma^0(n)/\Gamma^1(n)$ isomorphic to $(\mathbb{Z}/n\mathbb{Z})^\times$. Such a relation provides a canonical decomposition for the spaces of modular forms (cf. Section 4) for $\Gamma^1(n)$ in terms of the characters of $(\mathbb{Z}/n\mathbb{Z})^\times$ (cf. [Kob93]). In the same spirit, character groups are defined as follows:

Definition 1 A group Γ is called a *character group* of a finite index subgroup Γ^0 if Γ is normal in Γ^0 with abelian quotient. Equivalently, a character group is the kernel of an onto homomorphism

$$\varphi : \Gamma^0 \to G \tag{2.1}$$

where G is a finite abelian group written multiplicatively.

By the Fundamental Theorem of Finite Abelian Groups, it suffices to consider the case when Γ^0/Γ is cyclic.

When working with character groups, we distinguish the following two types of character groups.

Definition 2 A character group $\Gamma = \ker\varphi$ is said to be of type I if $\varphi(\gamma) \neq 1$ for some parabolic $\gamma \in \Gamma^0$; otherwise, it is said to be of type II. Γ is said to be of type II(A) if the images of all elliptic and parabolic elements of Γ^0 are 1 under φ.

Note that if Γ is a type II character group of Γ^0 then the levels of Γ and Γ^0 are the same; this is not true for type I character groups in general. It has been shown that almost all type II character groups are noncongruence [KL08]. If Γ is a type II(A) character group of Γ^0, then the modular curve for Γ is a strictly $[\Gamma^0 : \Gamma]$ to 1 unramified covering map of the modular curve for Γ^0. By definition, the image of any elliptic element γ in Γ^0 under φ has to divide the order of γ, which is a finite positive integer.

In [LLY05b, ALL08, Lon08], the following family of type I character groups is considered.

Example 2 Let

$$\Gamma^1(5) = \left\{\gamma \in SL_2(\mathbb{Z}) | \gamma \equiv \begin{pmatrix} 1 & 0 \\ * & 1 \end{pmatrix} \mod 5\right\}.$$

This is an index 12 elliptic element free congruence subgroup of $SL_2(\mathbb{Z})$ with 4 cusps: $0, -5/2, -2, \infty$. The surface $\mathcal{E}_{\Gamma^1(5)}$ is semistable with configuration $[1,1,5,5]$. It is one of the 6 semistable extremal rational surfaces classified by Beauville [Bea82]. This group is generated by 4 elements, denoted by $\gamma_\infty, \gamma_{-2}, \gamma_0, \gamma_{-5/2}$ (each stabilizes one cusp as indicated by the subscripts) subject to one relation

$$\gamma_\infty \gamma_{-2} \gamma_0 \gamma_{-5/2} = I_2.$$

Let φ_n be the homomorphism from $\Gamma^1(5)$ to $\mathbb{C}^\times$ sending γ_∞ to ω_n, γ_{-2} to ω_n^{-1}, and $\gamma_0, \gamma_{-5/2}$ both to 1, where $\omega_n = e^{2\pi i/n}$. The kernel of φ_n, denoted by Γ_n, is a type I character group of $\Gamma^1(5)$. We list a few properties of Γ_n here:

1. Γ_n is noncongruence when $n \neq 1, 5$ (cf. Section 5).
2. Γ_n is elliptic element free with index $12n$ in $SL_2(\mathbb{Z})$ and has $2+2n$ cusps.
3. Γ_5 is the last group in Sebbar's list [Seb01].
4. The elliptic modular surface for Γ_2 is a K3 surface with configuration $[1,1,1,1,10,10]$.

Example 3 The group $\Gamma^0(11)$ is genus 1, elliptic element free, and index 12 in $SL_2(\mathbb{Z})$. It has two cusps: ∞ and 0. It is generated by 4 generators, g_∞ and g_0, which stabilize ∞ and 0 respectively, and two parabolic generators A and B, subject to one condition

$$g_\infty g_0 A B A^{-1} B^{-1} = I_2.$$

Then the kernel of any homomorphism which sends g_∞ and g_0 to the identity and A and B to roots of unity is a type II character of $\Gamma^0(11)$. Among all such groups, there is only one, namely $\Gamma^1(11)$, which is congruence. By the Hurwitz formula, the genus of such a group remains 1 and as we have remarked before the level of this group is 11. For any type II character group Γ of $\Gamma^0(11)$, its modular curve X_Γ is an elliptic curve which is isogenous to $X_{\Gamma^0(11)}$.

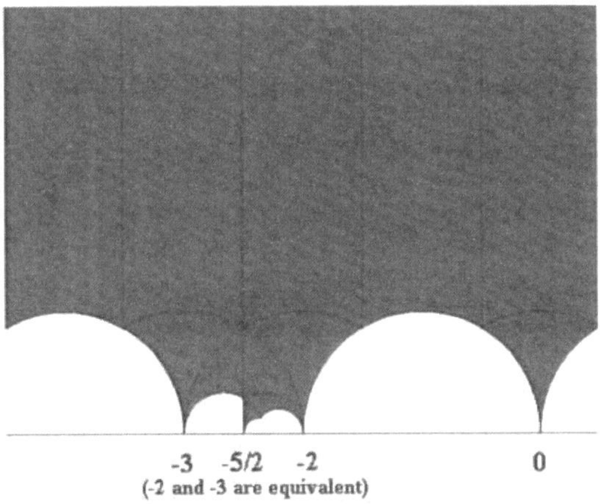

Figure 1 Fundamental domain for $\Gamma^1(5)$, drawn by the Fundamental Domain Drawer of Verrill

3 Some distinctions between congruence and noncongruence subgroups

3.1 One fundamental difference between congruence and noncongruence subgroups is that elements of congruence subgroups can be described by congruence relations while the elements of noncongruence subgroups cannot.

3.2 Conjecturally, Fourier coefficients of modular forms for these two types of groups also behave differently. Please refer to Section 5 for a more detailed discussion.

3.3 The following is a special case of the finiteness theorem of Thompson [Tho80]. Let Γ be a finite index subgroup of the modular group and $\mathcal{K}$ the set of congruence subgroups of Γ; then

$$\lim_{K\in\mathcal{K}} \frac{g(K)}{|\Gamma : K|} = \frac{1}{2}\chi(\Gamma), \tag{3.1}$$

in the sense that for any $\varepsilon > 0$, there are only finitely many elements $K \in \mathcal{K}$ such that

$$\left| \frac{g(K)}{|\Gamma : K|} - \frac{1}{2}\chi(\Gamma) \right| > \varepsilon.$$

Here $g(\Gamma)$ is the genus of Γ,

$$\chi(\Gamma) = 2(g(\Gamma) - 1) + t + \sum_{i=1}^{m} \left(1 - \frac{1}{e_i}\right), \tag{3.2}$$

is the Euler characteristic of Γ, t is the number of cusps of Γ, m is the number of elliptic points of Γ, and the e_i's are the orders of the elliptic points in $PSL_2(\mathbb{Z})$.

By comparison, there are infinitely many noncongruence subgroups of a given genus [Jon79]. Thus, the finiteness theorem of Thompson [Tho80] does not hold if the congruence condition on the set of elements in $\mathcal{K}$ is removed.

3.4 Let $A(\mathfrak{H}/\Gamma)$ denote the hyperbolic area of a fundamental domain of Γ. Let

$$\Delta = -4(\operatorname{Im} z)^2 \frac{\partial^2}{\partial z \partial \bar{z}}$$

be the Laplace operator of the Poincaré metric on $\mathfrak{H}$. It admits a self-adjoint non-negative extension in the space of square integrable functions on $\mathfrak{H}/\Gamma$, which is denoted by Δ_Γ. Let λ_1 be the smallest positive discrete eigenvalue of Δ_Γ (if it exists).

Theorem 2 (Selberg [Sel65]) *If Γ is a congruence subgroup, then $\lambda_1 \geq 3/16$.*

Theorem 3 (Zograf [Zog91]) *Let Γ be a discrete subgroup of $PSL_2(\mathbb{R})$ with hyperbolic area $A(\mathfrak{H}/\Gamma) < \infty$, and assume $A(\mathfrak{H}/\Gamma) \geq 32\pi \cdot (g(\Gamma)+1)$. Then λ_1 exists and*

$$\lambda_1 < \frac{8\pi \cdot (g(\Gamma)+1)}{A(\mathfrak{H}/\Gamma)}. \tag{3.3}$$

Theorem 4 *For any $\varepsilon > 0$, there exists a noncongruence subgroup with $0 < \lambda_1 < \varepsilon$.*

Proof As there are genus 0 noncongruence subgroups with $A(\mathfrak{H}/\Gamma)$ arbitrarily large (cf. Example 2), the claim follows from (3.3). □

4 Modular forms

4.1 A holomorphic (or meromorphic) *modular form* f for Γ is a holomorphic (or meromorphic) function defined on $\bar{\mathfrak{H}}$ satisfying

$$f(\frac{az+b}{cz+d}) = (cz+d)^k f(z), \quad \forall z \in \mathfrak{H}, \begin{pmatrix} a & b \\ c & d \end{pmatrix} \in \Gamma. \tag{4.1}$$

The number k is called the weight of the modular form f. A weight 0 meromorphic modular form is called a *modular function* for Γ. All modular functions for Γ together form a field, denoted by $\mathfrak{M}_\Gamma$, which is transcendental over $\mathbb{C}$. When Γ has genus zero, $\mathfrak{M}_\Gamma$ is generated by any Hauptmodul for Γ.

A holomorphic modular form is called a cuspform if it vanishes at every cusp of Γ. Let $S_k(\Gamma)$ denote the space of holomorphic weight k cuspforms for Γ. It is a finite-dimensional vector space over $\mathbb{C}$. Explicit formulae for the dimension of $S_k(\Gamma)$ are available in [Shi71]. Cuspforms are important invariants of Γ and X_Γ. For instance, weight 2 cuspforms for Γ are invariant differential 1-forms of X_Γ. A cuspform of integral weight larger than 2 is a differential form on X_Γ with coefficients in a line bundle.

For any $\gamma = \begin{pmatrix} a & b \\ c & d \end{pmatrix} \in SL_2(\mathbb{Z})$, define the stroke operator $|_\gamma$ action on a weight k modular form f to be

$$f|_\gamma = (cz+d)^{-k} f(\frac{az+b}{cz+d}).$$

Theta functions and Eisenstein series constitute important classes of modular forms. They are described in most books on modular forms ([Ogg69, Shi71, Kob93, DS05, etc]). In this paper, we use w to denote $e^{2\pi i z/\mu}$, where μ is the width of the cusp at infinity. Accordingly, any modular form has an expansion of the form $\sum_{n \geq n_0} a(n) w^n$.

Example 4 There are two particular holomorphic weight 3 Eisenstein series for $\Gamma^1(5)$: E_1 and E_2, which vanish at all the cusps of $\Gamma^1(5)$ except ∞ and -2 respectively. Both E_1 and E_2 have integral Fourier expansions at ∞. Moreover, $E_1 \cdot E_2 = \eta(5z)^3 \cdot \eta(z)^9$; here $\eta(z) = q^{1/24}\prod_{n\geq 1}(1-q^n)$, with $q = e^{2\pi i z}$ is the Dedekind eta function. One interesting expression for E_1 is

$$E_1 = 1 - \sum_{n\geq 1} b_n \frac{n^2 w^n}{1-w^n},$$

where $b_n = 2, 1, -1, -2, 0, 2, 1, -1, -2, 0\cdots$ [Sti06].

A Hauptmodul of $\Gamma^1(5)$ is

$$t = \frac{E_1}{E_2} = w^{-1} + 5 + 10w + 5w^2 - 15w^3 - 24w^4 + 15w^5 + \cdots, \quad w = e^{2\pi i z/5}. \tag{4.2}$$

Another formula for t is (cf. [Seb02])

$$t = \frac{1}{w}\prod_{n\geq 1}(1-w^n)^{-5\left(\frac{n}{5}\right)},$$

where $\left(\frac{\cdot}{5}\right)$ is the Legendre symbol.

A Hauptmodul for Γ_n is

$$t_n = \sqrt[n]{t}. \tag{4.3}$$

In particular, it is also known that t_5 is a Hauptmodul for $\Gamma(5)$ (when $w = e^{2\pi i z}$).

4.2 Cuspforms for noncongruence character groups. Due to the definition of noncongruence character groups, modular forms for such a group Γ are closely related to modular forms for Γ^0. The following example illustrates such a relation. When $k > 3$ we have the following result.

Lemma 1 *Let Γ_n be as in Example 2. Let $k > 3$ and $f_{n,k} \in S_k(\Gamma^1(5))$ with simple zeros at the cusps $-2, 0, -5/2$ and an order $k-3$ zero at infinity (and no other zeros). Then*

$$f_{n,k} \cdot t_n^j, j = 1-n, \cdots, (2k-3)n-1.$$

form a basis of $S_k(\Gamma_n)$.

Proof Using the dimension formula in [Shi71], we have $\dim S_k(\Gamma_n) = [2n(k-1)] - 1$. By the Riemann-Roch theorem, such a cuspform $f_{n,k}$ exists and for any $j \in \{1-n, \cdots, (2k-3)n-1\}$, $f_{n,k}\cdot t_n^j \in S_{2k}(\Gamma_n)$. As these functions are linearly independent, they form a basis. □

4.3 Hecke operators. Many good properties of congruence modular forms are derived from Hecke operators. The major difficulty in understanding modular forms for noncongruence subgroups is the lack of satisfactory Hecke operators. For a finite index subgroup Γ, define the pth Hecke operator $T_{p,\Gamma}$ associated to Γ as follows: Let $\alpha_p = \begin{pmatrix} p & 0 \\ 0 & 1 \end{pmatrix}$. We decompose the set $\Gamma\alpha_p\Gamma = \{\gamma\alpha_p\gamma' | \gamma, \gamma' \in \Gamma\}$ into a finite disjoint union $\bigcup_i \Gamma\alpha_i$. For any modular form f for Γ, define

$$T_{p,\Gamma} f = \sum f|_{\alpha_i}.$$

It was conjectured by Atkin and has been shown by Serre (for Γ normal in $SL_2(\mathbb{Z})$ [Tho89]), Scholl (for certain cases [Sch97]), and Berger (in general [Ber94]) that

Theorem 5 *For any prime $p \nmid lev(\Gamma)$ and all $f \in S_k(\Gamma)$*

$$T_{p,\Gamma} f = T_{p,\Gamma^0} \circ \mathrm{Tr}_{\Gamma}^{\Gamma^0} f,$$

where $\mathrm{Tr}_{\Gamma}^{\Gamma^0} : S_k(\Gamma) \to S_k(\Gamma^0)$ is the trace map.

Consequently, for all $f \in S_k(\Gamma) \setminus S_k(\Gamma^0)$, $T_{p,\Gamma} f = 0$.

5 Unbounded denominators

5.1 In what follows, we only consider modular forms $\sum_{n \ge n_0} a(n) w^n$ with coefficients $a(n)$ in a fixed number field. Due to the Hecke operators, each space of holomorphic congruence modular forms has a basis consisting of forms with integral coefficients [Shi71]. Consequently, for every congruence holomorphic modular form with algebraic coefficients, the sequence $\{a(n)\}$ has bounded denominators in the sense that there exists an algebraic number M such that $M \cdot a(n)$ is an algebraic integer for all n. In [ASD71], Atkin and Swinnerton observed that this is no longer the case for noncongruence modular forms. Therefore, the sequence $\{a(n)\}$ having unbounded denominators clearly implies $f = \sum_{n \ge n_0} a(n) w^n$ is noncongruence. At present, it remains unknown whether $\{a(n)\}$ having bounded denominators will imply $\sum a(n) w^n$ to be congruence. We will formulate a more precise conjecture as follows:

Let (UBD) refer to the following property:

> *Let f be an arbitrary holomorphic integral weight modular form for Γ but not for any congruence subgroup containing Γ with algebraic Fourier coefficients at infinity. Then the Fourier coefficients of f have unbounded denominators.*

An open question is:

> *Does every noncongruence subgroup satisfy the condition (UBD)?*

Remark 1 In [ASD71] and [Sch85], the coefficients $a(n)$ are normalized so that they are contained in smaller fields. In the following discussion, we have avoided the normalization on purpose for the following two reasons. First, such a procedure is unnecessary for all cases below. Second, it is more appropriate to address the bounded denominator property prior to the normalization, as it was pointed out by D. Zagier that all forms with algebraic coefficients can be normalized to have integral coefficients.

5.2 For any $n \ge 1$ and with a principal branch fixed, we formally write

$$(1+x)^{1/n} = \sqrt[n]{1+x} = \sum_{m \ge 0} \frac{\left(\frac{1}{n}\right)_m}{m!} x^m, \tag{5.1}$$

where $(\frac{1}{n})_m = \frac{1}{n}(\frac{1}{n} - 1) \cdots (\frac{1}{n} - m + 1)$. The next lemma provides a simple but useful criterion for detecting the unbounded denominator property in simple cases.

Lemma 2 *Let n be any natural number and $f = 1 + \sum_{m \ge 1} a(m) w^m, a(m) \in \mathbb{Z}$ for all m. In terms of (5.1), we expand $\sqrt[n]{f} = \sum_{m \ge 0} b(m) w^m, b(m) \in \mathbb{Z}[1/n]$. Let p be a prime factor of n. If there exists one $b(m)$ which is not p-integral, then*

$$\limsup_{m \to \infty} -\mathrm{ord}_p(b(m)) \to \infty.$$

In other words, $\{b(m)\}$ has unbounded denominators.

Proof It is a special case of [KL08, Lemma 11]. □

5.3

Example 5 In [CdlORV00], Candelas, de la Ossa, and Rodriguez-Villegas have studied the family of quintic Calabi-Yau 3-folds in $\mathbb{P}^4$ given by

$$\sum_{i=1}^{5} x_i^5 - 5\psi x_1 x_2 x_3 x_4 x_5 = 0.$$

They have shown that

$$\frac{1}{(5\psi)^5} = w + 154w^2 + 179139w^3 + \cdots \in \mathbb{Z}[w]], w = e^{2\pi i t},$$

where

$$t = \frac{1}{2\pi i} \frac{\bar{\omega}_1(\psi)}{\bar{\omega}_0(\psi)}$$

is the ratio of two particular solutions of the Picard-Fuchs equations arising from the quintic family. By the above lemma, the Fourier coefficients of ψ have unbounded denominators.

5.4

Example 6 Let Γ_n be as in Example 2. A Hauptmodul of Γ_n is $t_n = \sqrt[n]{t}$ where the expansion of t is as in (4.2). By Lemma 2, when $n \neq 1, 5$, the coefficients of t_n have unbounded denominators. Therefore Γ_n is noncongruence.

Lemma 3 *Let $f_{n,k}$ be the weight k cuspform for $\Gamma^1(5)$ as in Lemma 1; then the Fourier coefficients of $1/f_{n,k}$ have bounded denominators.*

Proof By our choice of $f_{n,k}$ whose poles are supported at the cusps, we know $f_{n,k}$ satisfies the bounded denominator property. Take any integral weight normalized newform g for $\Gamma^1(5)$. Clearly for any positive integer m, both g^m and g^{-m} satisfy the bounded denominator property. When m is large enough, $g^m(f_{n,k})^{-1}$ is also a holomorphic cuspform h for $\Gamma^1(5)$, hence $(f_{n,k})^{-1} = hg^{-m}$ also satisfies the bounded denominator property. □

Proposition 1 *Let f be a weight $k > 3$ cuspform for some Γ_n with coefficients in some algebraic number field. If it is not a form for Γ_1 or Γ_5, then the coefficients of f at ∞ have unbounded denominators.*

Proof By Lemma 1, $f = f_{n,k} \cdot (\sum_{j=1}^{n} a_j t_n^j)$ for some $a_j \in \mathbb{C}$. Let $F(x) \in \mathbb{C}[x]$ be the nonconstant polynomial with minimal degree such that the coefficients of $F(t_n)$ have bounded denominators. This is $F(x) = x^n$ if $5 \nmid n$ or $F(x) = x^{n/5}$ otherwise. In the first case, if $f \notin S_k(\Gamma^1(5))$, then $a_j \neq 0$ for some j not divisible by n, hence $\sum a_j x^j$ is not a polynomial in x^n, i.e. the Fourier coefficients of $\sum a_j t_n^j$ have unbounded denominators. If f satisfies the bounded denominator property so does $f(f_{n,k})^{-1} = \sum a_j t_n^j$ which contradicts our assumption above. The second case can be handled in a similar manner. □

5.5 The Fermat curves and modular curves. In the next example, we give one of the realizations of the Fermat curves in terms of modular curves. It is motivated by a question raised by Koblitz (through private conversation) regarding the relations between the Fermat curves and noncongruence subgroups.

For any positive integer $n \geq 2$,

$$x^n + y^n = 1$$

is the Fermat curve and its properties have been studied extensively (cf. [GR78, KR78, Yui80, Yan96, etc]). Now let G_1 and G_2 be two genus 0 subgroups with Hauptmoduls

$$x = \sqrt[n]{\lambda}, \quad y = \sqrt[n]{1-\lambda}$$

respectively, where the lambda function

$$\begin{aligned}\lambda(q) &= 16q \prod_{n \geq 1} [(1+q^{2n})^2(1-q^{2n-1})]^8 \\ &= 16q(1 - 8q + 44q^2 - 192q^3 + 718q^4 - 2400q^5 + 7352q^6 + \cdots)\end{aligned}$$

and $\lambda(q^2)$ is a Hauptmodul for $\Gamma(2)$. (Not just any arbitrary algebraic function of λ is another modular function. These two functions can be Hauptmoduls due to the special locations of the zeros and poles of λ and $1-\lambda$). Clearly x and y satisfy $x^n + y^n = 1$. It is not difficult to see that the Fermat curve above is biholomorphic to the modular curve for $\Phi(n) = G_1 \cap G_2$. In the literature, $\Phi(n)$ is the so-called Fermat group. By Lemma 2, it is easy to check that when $n \neq 1, 2, 4, 8$ the Fourier coefficients of the expansions of x and y in q at infinity will have unbounded denominators, which implies $\Phi(n)$ is noncongruence. By using the idea used in the proof of Proposition 1, we can show that when $n \neq 1, 2, 4, 8$, the Fermat group $\Phi(n)$ satisfies the condition (UBD) (cf. Section 5.1).

5.6 Recently, Kurth has verified that for each of the following K3 configurations [Kur06]

$$[1,1,1,1,2,18], [1,1,1,1,10,10], [1,1,2,2,6,12], [1,1,2,2,9,9], [1,1,2,5,5,10]$$

$$[1,1,4,6,6,6], [2,2,2,3,3,12], [2,2,5,5,5,5], [2,3,3,4,6,6], [3,3,3,3,6,6]$$

arising as double covers of rational surfaces mentioned in subsection 2.4 and [TY07], the Fourier coefficients of every possible Hauptmodul of these monodromy groups have unbounded denominators.

5.7 In general, it is relatively easy to construct modular forms for type I character groups whose corresponding homomorphisms send all elliptic and hyperbolic generators to 1 (cf. Example 2 and Lemma 1). To handle all type II or II(A) character groups of a fixed congruence group is more challenging. The following theorem gives a partial positive answer to the open question mentioned at the beginning of this section for type II(A) character groups.

Theorem 6 ([KL08, Theorem 3]) *Let Γ^0 be any genus 1 congruence subgroup. If there exists a prime p such that every index-p type II(A) character group of Γ^0 satisfies (UBD), then there exists $c(\Gamma^0) > 0$ such that for any $X \gg 0$,*

$$\#\{\Gamma < \Gamma^0 \mid \textit{type II(A)}, [\Gamma^0 : \Gamma] < X, \Gamma \textit{ satisfies (UBD)}\} > c(\Gamma^0) \cdot X^2. \qquad (5.2)$$

It is computationally feasible to check the conditions above. For example $\Gamma^0 = \Gamma^0(11)$ is considered in [KL08] via some results of Atkin in [Atk67].

6 Modularity

6.1 l-adic Scholl representations. Let Γ be a noncongruence subgroup with X_Γ defined over $\mathbb{Q}$ and $k > 2$ be an integer. Let $d = \dim S_k(\Gamma)$. For every prime l, Scholl has constructed a $2d$-dimensional $\mathbb{Q}_l$-vector space W from weight k cuspforms for Γ such that there is a continuous homomorphism

$$\rho : \mathrm{Gal}(\overline{\mathbb{Q}}/\mathbb{Q}) \to \mathrm{Aut}(W),$$

(cf. [Sch85]). These representations satisfy good properties: They ramify only at a finite set of primes. For each prime p such that ρ is unramified, the characteristic polynomial of the Frobenius element at p has integral coefficients. It has been verified that Scholl's results also hold for odd weight cases (cf. [F$^+$05]). In [Sch88], Scholl investigated the 2-dimensional l-adic representation attached to the unique weight 4 genuine cuspform for an index 9 noncongruence subgroup. He showed that it is isomorphic to the l-adic representation attached to a weight 4 level 14 congruence newform. Scholl has given a few more unpublished examples of this kind [Sch93].

6.2 In [SB85], Stienstra and Beukers investigated several K3 surfaces with Picard number 20 over $\mathbb{C}$. These K3 surfaces are constructed as double covers of $\mathbb{C}P^2$ with branching points along degree 6 polynomials. According to a result of Shioda and Inose [SI77], the L-function of each algebraic K3 surface with Picard number 20 mainly consists of the L-function of a weight 3 congruence Hecke eigenform. In other words, the l-adic representations attached to the nontrivial part of the l-adic second étale cohomology of such a K3 surface are modular. Stienstra and Beukers have identified the Hecke newforms corresponding to each K3 surface in their consideration and have constructed elliptic fibers for those K3 surfaces including the one with configuration $[1, 1, 1, 1, 10, 10]$. However, using Sebbar's classification [Seb01], one sees that several monodromy groups attached to these elliptic fibrations are noncongruence. Combining the above discussions with the theory of elliptic modular surfaces by Shioda [Shi72], we have

Theorem 7 *Let Γ be a finite index subgroup of the modular group which is conjugate to a monodromy group of a K3 surface over $\mathbb{Q}$ with Picard number 20. Then the l-adic Scholl representation attached to a $S_3(\Gamma)$ is isomorphic to the l-adic representation attached to certain Hecke eigenform.*

This modularity result has significant influence on [LLY05b] and it explains why weight 3 cuspforms are given more attention. As a special case of the above theorem, we know the l-adic Scholl representation attached to $S_3(\Gamma_2)$ is modular.

6.3 Starting from [LLY05b], investigations have been focused on higher dimensional l-adic Scholl representations. In particular, l-adic Scholl representations attached to $S_3(\Gamma_n)$ have been examined extensively. To study these representations, it is natural to decompose the representation spaces into subrepresentation spaces by using available operators. Normalizers of Γ in $PSL_2(\mathbb{R})$ become important as they give rise to involutions on the corresponding l-adic representation spaces. Use Γ^N to denote the group of normalizers of Γ in $PSL_2(\mathbb{R})$. Clearly $\Gamma^0 \subset \Gamma^N$. (M.L. Lang gave an algorithm for finding Γ^N for a given Γ [Lan02]). Moreover, Γ^N/Γ

acts on the finite-dimensional spaces $S_k(\Gamma)$. We assume the action of any element in Γ^N/Γ is defined over a number field F. Consequently, the l-adic Scholl representation above, when restricted to the subgroup $\mathrm{Gal}(\overline{\mathbb{Q}}/F)$, can be decomposed into a direct sum of subrepresentations. In some ideal cases, these subrepresentations are 2-dimensional and are shown to relate to some l-adic Deligne representations attached to congruence newforms [LLY05b, F$^+$05, ALL08, Lon08, etc].

6.4 With the new exciting progress on major conjectures like the Fontaine-Mazur conjecture [Kis06], and the Serre conjecture [KW06a, KW06b, Kis07], it is promising that one can establish the modularity for a large class of l-adic Scholl representations.

7 Atkin and Swinnerton-Dyer congruences

7.1 When Γ is a congruence subgroup, we have the following classical result due to Hecke and Petersson. For any positive integer k, the space $S_k(\Gamma)$ has a basis consisting of Hecke eigenforms. Namely, for each of the basis elements $\sum_{n\geq 1} a(n)w^n$ there exists a suitable Dirichlet character χ such that for any prime p,

$$a(np) - a(p)a(n) + \chi(p)p^{k-1}a(n/p) = 0, \quad \forall\, n \in \mathbb{N}. \tag{7.1}$$

However, when Γ is a noncongruence subgroup, the Hecke operator theory is unsatisfactory as mentioned in Section 4.3. Consequently, such a result no longer holds for noncongruence cuspforms. While working with explicit examples, Atkin and Swinnerton-Dyer have found that three-term recursions like (7.1) still hold in a p-adic sense. What Atkin and Swinnerton-Dyer have observed is the following: given $S_k(\Gamma)$, then for almost all primes p, the space $S_k(\Gamma)$ has a basis (depending on p) consisting of forms of the form $\sum a(n)w^n$ satisfying the ASD congruence relations

$$a(np) - A(p)a(n) + B(p)a(n/p) = 0 \mod p^{(k-1)(1+\mathrm{ord}_p n)}, \quad \forall n \geq 1$$

where $A(p)$ and $B(p)$ are p-adic and in general such bases are different for each p.

7.2 To examine the ASD congruence relations computationally, Atkin uses "p-adic" Hecke operators. For any "good" prime p, define the p-adic Hecke operator $T_p = T^2 + s(p)\cdot p^{k-1}$ for $f = \sum a(n)w^n$ as follows:

$$T_p(f) = \sum_n (a(np) + s(p)p^{k-1}a(n/p))w^n.$$

Here, $s(p)$ is an algebraic number with absolute value 1. In general, it is totally nontrivial to guess the right $s(p)$ values since they are rarely Dirichlet characters. It is called a "p-adic" Hecke operator as the nth coefficient of $f|T_p$ will be determined modulo some suitable power of p depending on n. Unlike the classical Hecke operators which can be diagonalized simultaneously, these operators are a priori defined over various p-adic fields. It is clearly exceptional that the space $S_k(\Gamma)$ can be decomposed simultaneously for almost all "p-adic" Hecke operators.

7.3 In [Sch85], Scholl has proven a collective version of Atkin and Swinnerton-Dyer's observation. Basically, he showed that for any form f with rational coefficients in a d-dimensional space of even weight noncongruence cuspforms, the coefficients of f satisfy a $(2d+1)$-term congruence recursion. In [F$^+$05], the authors have verified the necessary conditions under which Scholl's results are valid for odd weight cases.

7.4 The available operators remain the key players when one attempts to reduce Scholl's result to a 3-term ASD congruence. In [LLY05b, ALL08, Lon08] significant efforts have been made to study the relations between these operators and elements in the Galois groups. With the aid of these operators, stronger results such as the existence of simultaneous (resp. semi-simultaneous) eigenbases for Atkin and Swinnerton-Dyer congruences at almost all primes p can be achieved cf. [LLY05b] (resp. [ALL08, Lon08]).

Example 7 ([LLY05b]) There exists a basis $f_\pm = \sum a_\pm(n)w^n$ with coefficients in $\mathbb{Z}[1/3](i)$ for $S_3(\Gamma_3)$ and a pair of newforms $g_\pm = \sum b_\pm(n)q^n$ with coefficients in $\mathbb{Z}[i]$ such that for any $n \geq 1$ and any odd prime p different from 3

$$\frac{a_\pm(np) - b_\pm(p)a_\pm(n) + \left(\frac{-3}{p}\right) p^2 a_\pm(n/p)}{(np)^2}$$

is integral for some prime in $\mathbb{Z}[i]$ above p.

In general, the symmetries inherited from Γ^N/Γ are not sufficient to derive 3-term Atkin and Swinnerton-Dyer congruences, especially when the dimension d is large [ALo07].

References

[AL70] A. O. L. Atkin and J. Lehner, *Hecke operators on $\Gamma_0(m)$*, Math. Ann. **185** (1970), 134–160.

[ALo07] A. O. L. Atkin and L. Long, *On Atkin and Swinnerton-Dyer congruence relations for some noncongruence cuspforms*, preprint (2007)

[ALL08] A. O. L. Atkin, W.C. Li, and L. Long, *On Atkin and Swinnerton-Dyer congruence relations (2)*, Math. Ann. **340**(2) (2008), 335–358

[ASD71] A. O. L. Atkin and H. P. F. Swinnerton-Dyer, *Modular forms on noncongruence subgroups*, Combinatorics (Proc. Sympos. Pure Math., Vol. XIX, Univ. California, Los Angeles, Calif., 1968), Amer. Math. Soc., Providence, R.I., 1971, pp. 1–25.

[Atk67] A. O. L. Atkin, *Proof of a conjecture of Ramanujan*, Glasgow Math. J. **8** (1967), 14–32.

[BCDT01] C. Breuil, B. Conrad, F. Diamond, and R. Taylor, *On the modularity of elliptic curves over* **Q**: *wild 3-adic exercises*, J. Amer. Math. Soc. (2001), (14) no. 4, 843–939.

[Bea82] A. Beauville, *Les familles stables de courbes elliptiques sur* $\mathbf{p}^1$ *admettant quatre fibres singulières*, C. R. Acad. Sci. Paris Sér. I Math. **294** (1982), no. 19, 657–660.

[Bel79] G. V. Belyĭ, *Galois extensions of a maximal cyclotomic field*, Izv. Akad. Nauk SSSR Ser. Mat. **43** (1979), no. 2, 267–276, 479.

[Ber94] G. Berger, *Hecke operators on noncongruence subgroups*, C. R. Acad. Sci. Paris Sér. I Math. **319** (1994), no. 9, 915–919.

[Ber00] ———, *Relations between cusp forms on congruence and noncongruence groups*, Proc. Amer. Math. Soc. **128** (2000), no. 10, 2869–2874.

[Bir94] B. Birch, *Noncongruence subgroups, covers and drawings*, The Grothendieck theory of dessins d'enfants (Luminy, 1993), London Math. Soc. Lecture Note Ser., vol. 200, Cambridge Univ. Press, Cambridge, 1994, pp. 25–46.

[BLS64] H. Bass, M. Lazard, and J.-P. Serre, *Sous-groupes d'indice fini dans* $\mathbf{SL}(n, \mathbf{Z})$, Bull. Amer. Math. Soc. **70** (1964), 385–392.

[BPVdV84] W. Barth, C. Peters, and A. Van de Ven, *Compact complex surfaces*, Ergebnisse der Mathematik und ihrer Grenzgebiete (3) [Results in Mathematics and Related Areas (3)], vol. 4, Springer-Verlag, Berlin, 1984.

[Car71] P. Cartier, *Groupes formels, fonctions automorphes et fonctions zêta des courbes elliptiques*, Actes du Congrès International des Mathématiciens (Nice, 1970), Tome 2, Gauthier-Villars, Paris, 1971, pp. 291–299.

[CdlORV00] P. Candelas, X. de la Ossa, and F. Rodriguez-Villegas, *Calabi-Yau manifolds over finite fields. I*, hep-th/0012233, 2000.

[CLY04] K. S. Chua, M. L. Lang, and Y. Yang, *On Rademacher's conjecture: congruence subgroups of genus zero of the modular group*, J. Algebra **277** (2004), no. 1, 408–428.

[CP03] C. J. Cummins and S. Pauli, *Congruence subgroups of* $\mathrm{PSL}(2,\mathbb{Z})$ *of genus less than or equal to 24*, Experiment. Math. http://www.math.tu-berlin.de/ pauli/congruence/ **12** (2003), no. 2, 243–255.

[Den75] J. B. Dennin, Jr., *The genus of subfields of* $K(n)$, Proc. Amer. Math. Soc. **51** (1975), 282–288.

[Dit76] B. Ditters, *Sur les congruences d'Atkin et de Swinnerton-Dyer*, C. R. Acad. Sci. Paris Sér. A-B **282** (1976), no. 19, Ai, A1131–A1134.

[DS05] F. Diamond and J. Shurman, *A first course in modular forms*, Graduate Texts in Mathematics, vol. 228, Springer-Verlag, New York, 2005.

[F+05] L. Fang, J. W. Hoffman, B. Linowitz, A. Rupinski, and Verrill H., *Modular forms on noncongruence subgroups and Atkin-Swinnerton-Dyer relations*, preprint (2005).

[Fri86] R. Fricke, *Über die Substitutionsgruppen, welche zu den aus dem Legendre'schen Integralmodul k2(w) gezogenen Wurzeln gehören. (Mit einer Figurentafel)*, Math. Ann. **28** (1886), 99–118.

[Gel84] S. Gelbart, *An elementary introduction to the Langlands program*, Bull. Amer. Math. Soc. (N.S.) **10** (1984), no. 2, 177–219.

[GR78] B. H. Gross and D. E. Rohrlich, *Some results on the Mordell-Weil group of the Jacobian of the Fermat curve*, Invent. Math. **44** (1978), no. 3, 201–224.

[Hec37] E. Hecke, *Über Modulfunktionen und die Dirichletschen Reihen mit Eulerscher Produktentwicklung. I and II*, Math. Ann. **114** (1937), no. 1, 1–28,316–351.

[HL05] P. Hubert and S. Lelièvre, *Noncongruence subgroups in* $\mathcal{H}(2)$, Int. Math. Res. Not. (2005), no. 1, 47–64.

[Hon70] T. Honda, *On the theory of commutative formal groups*, J. Math. Soc. Japan **22** (1970), 213–246.

[Hsu96] T. Hsu, *Identifying congruence subgroups of the modular group*, Proc. Amer. Math. Soc. **124** (1996), no. 5, 1351–1359.

[Jon79] G. A. Jones, *Triangular maps and noncongruence subgroups of the modular group*, Bull. London Math. Soc. **11** (1979), no. 2, 117–123.

[Jon86] ———, *Congruence and noncongruence subgroups of the modular group: a survey*, Proceedings of groups—St. Andrews 1985 (Cambridge), London Math. Soc. Lecture Note Ser., vol. 121, Cambridge Univ. Press, 1986, pp. 223–234.

[JS94] G. Jones and D. Singerman, *Maps, hypermaps and triangle groups*, The Grothendieck theory of dessins d'enfants (Luminy, 1993), London Math. Soc. Lecture Note Ser., vol. 200, Cambridge Univ. Press, Cambridge, 1994, pp. 115–145.

[KW06a] C. Khare and J.-P. Wintenberger, *Serre's modularity conjecture (I)*, Preprint, available at http://www.math.utah.edu/~shekhar/papers.html

[KW06b] ———, *Serre's modularity conjecture (II)*,Preprint, available at http://www.math.utah.edu/~shekhar/papers.html

[Kat81] N. M. Katz, *Crystalline cohomology, Dieudonné modules, and Jacobi sums*, Automorphic forms, representation theory and arithmetic (Bombay, 1979), Tata Inst. Fund. Res. Studies in Math., vol. 10, Tata Inst. Fundamental Res., Bombay, 1981, pp. 165–246.

[Kis06] M. Kisin, *The Fontaine-Mazur conjecture for* Gl_2, preprint, available at http://www.math.uchicago.edu/~kisin/preprints.html (2006).

[Kis07] ———, *Modularity of 2-adic Barsotti-Tate representations*, Preprint, available at http://www.math.uchicago.edu/~kisin/preprints.html (2007).

[KN65] M. I. Knopp and M. Newman, *Congruence subgroups of positive genus of the modular group*, Illinois J. Math. **9** (1965), 577–583.

[Kob93] N. Koblitz, *Introduction to elliptic curves and modular forms*, second ed., Springer-Verlag, New York, 1993.

[Kod63] K. Kodaira, *On compact analytic surfaces. II, III*, Ann. of Math. (2) 77 (1963), 563–626; ibid. **78** (1963), 1–40.

[KR78] N. Koblitz and D. Rohrlich, *Simple factors in the Jacobian of a Fermat curve*, Canad. J. Math. **30** (1978), no. 6, 1183–1205.

[Kul85] R. S. Kulkarni, *A new proof and an extension of a theorem of Millington on the modular group*, Bull. London Math. Soc. **17** (1985), no. 5, 458–462.

[Kur06] C. Kurth, available at http://orion.math.iastate.edu/linglong/Noncong/noncong.html (2006).

[KL08] C. A. Kurth and L. Long, *On modular forms for some noncongruence subgroups of* $SL_2(\mathbb{Z})$, J. Number Theory (2008), doi:10.1016/j.jnt.2007.10.007.

[KL07] ______, *Computations with finite index subgroups of* $PSL_2(\mathbb{Z})$ *using Farey Symbols*, arXiv:0710.1835 (2007)

[Lan02] M. L. Lang, *Normalisers of subgroups of the modular group*, J. Algebra **248** (2002), no. 1, 202–218.

[Li75] W.C. Li, *Newforms and functional equations*, Math. Ann. **212** (1975), 285–315.

[Li96] ______, *Number theory with applications*, World Scientific Publishing Co. Inc., River Edge, NJ, 1996.

[LLY05a] W.C Li, L. Long, and Z. Yang, *Modular forms for noncongruence subgroups*, Quarterly Journal of Pure and Applied Mathematics **1** (2005), no. 1, 205–221.

[LLY05b] ______, *On Atkin and Swinnerton-Dyer congruence relations*, J. of Number Theory **113** (2005), no. 1, 117–148.

[Lon08] L. Long, *On Atkin and Swinnerton-Dyer congruence relations (3)*, J. Number Theory (2008), doi:10.1016/J.jnt.2008.02.014.

[Mil69a] M. H. Millington, *On cycloidal subgroups of the modular group*, Proc. London Math. Soc. (3) **19** (1969), 164–176.

[Mil69b] ______, *Subgroups of the classical modular group*, J. London Math. Soc. (2) **1** (1969), 351–357.

[Miy71] T. Miyake, *On automorphic forms on* GL_2 *and Hecke operators*, Ann. of Math. (2) **94** (1971), 174–189.

[Mor17] L. J. Mordell, *On Mr. Ramanujan's empirical expansions of modular functions*, Proc. Cambridge Phil. Soc. **19** (1917), 117–124.

[MP89] R. Miranda and U. Persson, *Configurations of* I_n *fibers on elliptic* $K3$ *surfaces*, Math. Z. **201** (1989), no. 3, 339–361.

[Mur93] M. R. Murty, *A motivated introduction to the Langlands program*, Advances in number theory (Kingston, ON, 1991), Oxford Sci. Publ., Oxford Univ. Press, New York, 1993, pp. 37–66.

[Nér64] A. Néron, *Modèles minimaux des variétés abéliennes sur les corps locaux et globaux*, Inst. Hautes Études Sci. Publ.Math. No. **21** (1964), 128.

[New65] M. Newman, *Normal subgroups of the modular group which are not congruence subgroups*, Proc. Amer. Math. Soc. **16** (1965), 831–832.

[Nor85] M. Nori, *On certain elliptic surfaces with maximal Picard number*, Topology **24** (1985), no. 2, 175–186.

[Ogg69] A. P. Ogg, *Modular forms and Dirichlet series*, W. A. Benjamin, Inc., New York-Amsterdam, 1969.

[Pet39] H. Petersson, *Konstruktion der sämtlichen Lösungen einer Riemannschen Funktionalgleichung durch Dirichletreihen mit Eulerscher Produktentwicklung I*, Math. Ann. **116** (1939), no. 1, 401–412.

[Pic86] G. Pick, *Über gewisse ganzzahlige lineare Substitutionen, welche sich nicht durch algebraische Congruenzen erklären lassen*, Math. Ann. **28** (1886), 119–124.

[Ran67] R. A. Rankin, *Lattice subgroups of free congruence groups*, Invent. Math. **2** (1967), 215–221.

[Ran77] ______, *Modular forms and functions*, Cambridge University Press, Cambridge, 1977.

[Rei58] I. Reiner, *Normal subgroups of the unimodular group*, Illinois J. Math. **2** (1958), 142–144.

[Sar90] P. Sarnak, *Some applications of modular forms*, Cambridge Tracts in Mathematics, vol. 99, Cambridge University Press, Cambridge, 1990.

[SB85] J. Stienstra and F. Beukers, *On the Picard-Fuchs equation and the formal Brauer group of certain elliptic* $K3$*-surfaces*, Math. Ann. **271** (1985), no. 2, 269–304.

[Sch85] A. J. Scholl, *Modular forms and de Rham cohomology; Atkin-Swinnerton-Dyer congruences*, Invent. Math. **79** (1985), no. 1, 49–77.

[Sch88] ———, *The l-adic representations attached to a certain noncongruence subgroup*, J. Reine Angew. Math. **392** (1988), 1–15.

[Sch93] ———, *The l-adic representations attached to non-congruence subgroups II*, preprint (1993).

[Sch97] ———, *On the Hecke algebra of a noncongruence subgroup*, Bull. London Math. Soc. **29** (1997), no. 4, 395–399.

[Sch06] ———, *On some l-adic representations of galois group attached to noncongruence subgroups*, Bull. London Math. Soc. **38**, (2006), no. 4, 561–567.

[Seb01] A. Sebbar, *Classification of torsion-free genus zero congruence groups*, Proc. Amer. Math. Soc. **129** (2001), no. 9, 2517–2527 (electronic).

[Seb02] ———, *Modular subgroups, forms, curves and surfaces*, Canad. Math. Bull. **45** (2002), no. 2, 294–308.

[Sel65] A. Selberg, *On the estimation of Fourier coefficients of modular forms*, Proc. Sympos. Pure Math., Vol. VIII, Amer. Math. Soc., Providence, R.I., 1965, pp. 1–15.

[Shi71] G. Shimura, *Introduction to the arithmetic theory of automorphic functions*, Publications of the Mathematical Society of Japan, No. 11. Iwanami Shoten, Publishers, Tokyo, 1971, Kanô Memorial Lectures, No. 1.

[Shi72] T. Shioda, *On elliptic modular surfaces*, J. Math. Soc. Japan **24** (1972), 20–59.

[SI77] T. Shioda and H. Inose, *On singular K3 surfaces*, Complex analysis and algebraic geometry, Iwanami Shoten, Tokyo, 1977, pp. 119–136.

[Sti85] J. Stienstra, *Les groupes formels d'Artin-Mazur et les congruences d'Atkin-Swinnerton-Dyer*, Study group on ultrametric analysis, 12th year, 1984/85, No. 2, Secrétariat Math., Paris, 1985, pp. Exp. No. 18, 13.

[Sti06] ———, *Jan Mahler measure variations, Eisenstein series and instanton expansions.* Mirror symmetry. V, 139–150, AMS/IP Stud. Adv. Math., 38, Amer. Math. Soc., Providence, RI, 2006.

[Sto84] W. W. Stothers, *Level and index in the modular group*, Proc. Roy. Soc. Edinburgh Sect. A **99** (1984), no. 1-2, 115–126.

[Tho80] J. G. Thompson, *A finiteness theorem for subgroups of* PSL(2, **R**) *which are commensurable with* PSL(2, **Z**), The Santa Cruz Conference on Finite Groups (Univ. California, Santa Cruz, Calif., 1979), Proc. Sympos. Pure Math., vol. 37, Amer. Math. Soc., Providence, R.I., 1980, pp. 533–555.

[Tho89] ———, *Hecke operators and noncongruence subgroups*, Group theory (Singapore, 1987), de Gruyter, Berlin, 1989, Including a letter from J.-P. Serre, pp. 215–224.

[TW95] R. Taylor and A. Wiles, *Ring-theoretic properties of certain Hecke algebras*, Ann. of Math. (2) **141** (1995), no. 3, 553–572.

[TY07] J. Top and N. Yui, *Explicit equations of some elliptic modular surfaces*, Rocky Mountain J. Math, **37**, (2007) no. 2, 663–687.

[Wil95] A. Wiles, *Modular elliptic curves and Fermat's last theorem*, Ann. of Math. (2) **141** (1995), no. 3, 443–551.

[Woh64] K. Wohlfahrt, *An extension of F. Klein's level concept*, Illinois J. Math. **8** (1964), 529–535.

[Yan96] T. Yang, *Cusp forms of weight* 1 *associated to Fermat curves*, Duke Math. J. **83** (1996), no. 1, 141–156.

[Yui80] N. Yui, *On the Jacobian variety of the Fermat curve*, J. Algebra **65** (1980), no. 1, 1–35.

[Zog91] P. Zograf, *A spectral proof of Rademacher's conjecture for congruence subgroups of the modular group*, J. Reine Angew. Math. **414** (1991), 113–116.

Aspects of Geometric and Differential Equations

Fields Institute Communications
Volume **54**, 2008

Apéry Limits of Differential Equations of Order 4 and 5

Gert Almkvist
Matematikcentrum
Lunds Universitet
Matematik MNF, Box 118
SE-221 00 Lund, Sweden
gert@maths.lth.se

Duco van Straten
Fachbereich 08, Institut für Mathematik
AG Algebraische Geometrie
Johannes Gutenberg-Universität
D-55099 Mainz, Germany
straten@mathematik.uni-mainz.de

Wadim Zudilin
Department of Mechanics and Mathematics
Moscow Lomonosov State University
Vorobiovy Gory, GSP-1, 119991 Moscow, Russia
and
Steklov Mathematical Institute
Russian Academy of Sciences
Gubkina str. 8, 119991 Moscow, Russia
wadim@mi.ras.ru

Abstract. The concept of *Apéry limit* for second and third order differential equations is extended to fourth and fifth order equations, mainly of Calabi–Yau type. For those equations obtained from Hadamard products of second and third order equations we can prove that the limits are determined in terms of the factors by a certain formula. Otherwise the limits are found by using PSLQ in Maple and are only conjectural. All identified limits are rational linear combinations of the following numbers: π^2, Catalan's constant G, $\sum_{n=1}^{\infty}\frac{\left(\frac{n}{3}\right)}{n^2}$, π^3, $\zeta(3)$, $\pi^3\sqrt{3}$, π^4.

1 Introduction

In 1978 Apéry proved that $\zeta(3)$ is irrational by considering the recursion

$$(n+2)^3A_{n+2}-(2n+3)(17n^2+51n+39)A_{n+1}+(n+1)^3A_n=0$$

2000 *Mathematics Subject Classification.* Primary 14J32, 11Y60, 14D05.
The work of the third-named author was supported by a fellowship of the Max Planck Institute for Mathematics (Bonn).

with $A_{-1} = 0$, $A_0 = 1$. Let B_n satisfy the same recursion with $B_0 = 0$, $B_1 = 1$. Then

$$\frac{B_n}{A_n} \to \frac{\zeta(3)}{6} \qquad \text{as } n \to \infty.$$

He also considered the recursion

$$(n+2)^2 A_{n+2} - (11n^2 + 33n + 25)A_{n+1} - (n+1)^2 A_n = 0$$

with $A_{-1} = 0$, $A_0 = 1$. Let B_n satisfy the same recursion with $B_0 = 0$, $B_0 = 1$. Then

$$\frac{B_n}{A_n} \to \frac{\pi^2}{30} \qquad \text{as } n \to \infty.$$

Note that $y_0 = \sum_{n=0}^{\infty} A_n x^n$ is annihilated by the differential operator

$$\theta^2 - x(11\theta^2 + 11\theta + 3) - x^2(\theta+1)^2, \qquad \text{where} \quad \theta = x\frac{d}{dx}.$$

This is denoted by (b) in [1] and we note that

$$A_n = \sum_k \binom{n}{k}^2 \binom{n+k}{n}.$$

Later Zudilin [8] considered the recursion

$$(n+2)^5 A_{n+2} - 3(2n+3)(3n^2+9n+7)(15n^2+45n+34)A_{n+1} \\ -3(n+1)^3(3n+2)(3n+4)A_n = 0$$

with $A_{-1} = 0$, $A_0 = 1$. Let $B_0 = 0$, $B_1 = 1$. Then

$$\frac{B_n}{A_n} \to \frac{\pi^4}{1170} = \frac{\zeta(4)}{13}.$$

It turns out that

$$w_0 = \sum_{n=0}^{\infty} A_n x^n$$

is the Wronskian

$$w_0 = x \begin{vmatrix} y_0 & y_1 \\ y_0' & y_1' \end{vmatrix},$$

where y_0 and y_1 are solutions of a fourth order differential equation (this was the start of [1]). It is equation #32 in the “big” table [2] of Calabi–Yau differential equations (for a definition see Section 2). Further cases of connections between Diophantine approximations of di- and trilogarithms and Catalan's constant G led to Calabi–Yau equations #195, #209 and #257. These will be treated in section 2.1 (see Zudilin [9, 10]).

In an unpublished paper [7], Zagier considered the recursion

$$(n+2)^2 A_{n+2} - (a(n+1)^2 + a(n+1) + b)A_{n+1} + c(n+1)^2 A_n = 0$$

trying to find integer solutions A_n. He finds the cases denoted by (a), (b), (c), (d), (e), (f), (g) and (h) in [1]. In [1] we find two more cases (i) and (j); e.g., in case (j) we have

$$A_n = 432^n \sum_k (-1)^k \binom{-5/6}{k} \binom{-1/6}{n-k}^2$$

which satisfies Zagier's recursion with $a = 864$, $b = 372$, $c = 186624$. The sequence (A_n^2), the Hadamard square of (A_n), satisfies a recursion of order 3 and $y_0 = \sum_{n=0}^{\infty} A_n^2 x^n$ satisfies a fourth order Calabi–Yau equation. If (A_n') and (A_n'') are two sequences from the list (a)–(j), then $(A'_n A''_n)$, the Hadamard product of (A'_n) and (A''_n), also results in a Calabi–Yau equation. In Section 3 we find formulas for how the Apéry limits are transformed under various Hadamard products.

There are also ten cases of third order differential equations (α), (β), (γ), (δ), (ϵ), (ζ), (η), (ϑ), (ι), (κ) with recursions similar to those of the second order:

$$(n+2)^3 A_{n+2} - (2n+3)(a(n+1)^2 + a(n+1) + b)A_{n+1} + c(n+1)^3 A_n = 0.$$

For instance, in case (κ)

$$A_n = 432^n \sum_k \binom{-1/6}{k}^2 \binom{-5/6}{n-k}^2$$

satisfies the above recursion with $a = 432$, $b = 312$, $c = 186624$. Let B_n satisfy the same recursion with $B_0 = 0$, $B_1 = 1$. Then

$$\frac{B_n}{A_n} \to \frac{1}{432} \sum_{j=1}^{\infty} \frac{1}{(j-\frac{1}{6})^3} = \frac{91}{432}\zeta(3) - \frac{1}{216}\pi^3\sqrt{3}.$$

This will be proved in Section 2.3.

In Section 4 we list all the conjectured Apéry limits identified by PSLQ. We have also some limits which we cannot identify (usually listed with 50 digits).

2 Known results

2.1 Calabi–Yau differential equations with known Apéry limits. Consider a fourth order differential equation

$$y^{(4)} + a_3 y''' + a_2 y'' + a_1 y' + a_0 y = 0,$$

where the a_j are polynomials. We assume it is MUM (*maximal unipotent monodromy*) at the origin, so we have the Frobenius solutions

$$y_0 = 1 + A_1 x + \cdots, \qquad y_1 = y_0 \log x + B_1 x + \cdots,$$
$$y_2 = \frac{1}{2} y_0 \log^2 x + \cdots, \qquad y_3 = \frac{1}{6} y_0 \log^3 x + \cdots.$$

We also assume that the coefficients satisfy the relation

$$a_1 = \frac{1}{2}a_2 a_3 - \frac{1}{8}a_3^3 + a_2' - \frac{3}{4}a_3 a_3' - \frac{1}{2}a_3''.$$

Let

$$q = \exp\left(\frac{y_1}{y_0}\right) = x + c_2 x^2 + \cdots.$$

Solving for x we get the *mirror map*

$$x = q - c_2 q^2 + \cdots.$$

Then we have the Yukawa coupling

$$K(q) = \left(q\frac{d}{dq}\right)^2 \left(\frac{y_2}{y_0}\right) = 1 + \sum_{n=1}^{\infty} \frac{n^3 N_n q^n}{1-q^n}.$$

The N_n are called the instanton numbers. If there exists a fixed integer c such that cN_n are integers for all n we call the differential equation Calabi–Yau.

The maximal degree of the a_j is called the degree of the equation. Observe that the degree of the equation becomes the order of the corresponding recursion for the coefficients A_n. When computing the Apéry limit we usually try to use the recursion of lowest order.

The following two equations are in Zudilin [9].

Example (#209) Let

$$A_n = \binom{2n}{n} \sum_k \binom{n}{k}^2 \binom{n+k}{n} \binom{n+2k}{n}.$$

It satisfies the recursion

$$\begin{aligned}
&(n+3)^4(946n^2+3053n+2475)A_{n+3}\\
&\quad -(208120n^6+2752860n^5+1506697n^4+43651558n^3\\
&\qquad +70564960n^2+60321212n+21297414)A_{n+2}\\
&\quad +4(2n+3)(3784n^5+36808n^4+141179n^3\\
&\qquad +267255n^2+250336n+93060)A_{n+1}\\
&\quad -4(n+1)^2(2n+1)(2n+3)(946n^2+4945n+6474)A_n=0
\end{aligned}$$

with $A_{-2}=0$, $A_{-1}=0$, $A_0=1$. Let B_n satisfy the same recursion with $B_{-1}=0$, $B_0=0$, $B_1=1$; then

$$\frac{B_n}{A_n} \to \frac{\pi^2}{138}.$$

But if we start with $B_0=0$, $B_1=17$, $B_2=\frac{9405}{8}$ then

$$\frac{B_n}{A_n} \to \zeta(3),$$

so we have simultaneous approximation of $\zeta(2)$ and $\zeta(3)$.

Example (#195) Let $H_n=\sum_{j=1}^n \frac{1}{j}$ if $n>0$ and 0 otherwise be the harmonic numbers. Let

$$\begin{aligned}
A_n &= \binom{2n}{n}^2 \binom{3n}{n}^2 \sum_k \binom{n}{k}^5 \binom{3n}{n+k}^{-2}\\
&\qquad \times \{1+k(-5H_k+5H_{n-k}+2H_{n+k}-2H_{2n-k})\}\\
&= \sum_{i,j} \binom{n}{i}^2 \binom{n}{j}^2 \binom{i+j}{j} \binom{n+i}{n}
\end{aligned}$$

which satisfies

$$\begin{aligned}
&(n+3)^4(1457n^2+4465n+3450)A_{n+3}\\
&\quad +(148614n^6+1941570n^5+10489565n^4+29970066n^3\\
&\qquad +47717965n^2+40114368n+13906092)A_{n+2}\\
&\quad +(97619n^6+1080107n^5+4934487n^4+11925999n^3\\
&\qquad +16102866n^2+115316n+3425652)A_{n+1}\\
&\quad +3(n+1)^2(3n+2)(3n+4)(1457n^2+7379n+9372)A_n=0
\end{aligned}$$

with $A_{-2}=0$, $A_{-1}=0$, $A_0=1$. Let B_n satisfy the same recursion with $B_{-1}=0$, $B_0=0$, $B_1=1$; then

$$\frac{B_n}{A_n} \to -\frac{\pi^2}{78}.$$

But if we take $B_0 = 0$, $B_1 = \frac{29}{13}$, $B_2 = -\frac{7617}{104}$, then

$$\frac{B_n}{A_n} \to -\frac{3}{13}\zeta(3).$$

Example (#257) Let

$$A_n = (-1)^n \binom{2n}{n}^2 \sum_k \binom{n}{k}\binom{n+k}{n-k}\binom{2n-k}{k}\binom{2n+2n}{n+k}\binom{4n-2k}{2n-k}$$
$$\times \{1 + k(-2H_k + 2H_{n-k} - H_{n+k} + H_{2n-k} - 2H_{2k} + 2H_{2n-2k} + 2H_{2n+2k} - 2H_{4n-2k})\}.$$

Then

$$\begin{aligned}(n+2)^4&(20n^2+32n+13)A_{n+2}\\ &- (56320n^6 + 428032n^5 + 1328384n^4 + 2153472n^3\\ &\quad + 1923136n^2 + 897408n + 171184)A_{n+1}\\ &- 2^{12}(2n+1)^4(20n^2+72n+65)A_n = 0.\end{aligned}$$

Then with $A_{-1} = 0$, $A_0 = 1$ and $B_0 = 0$, $B_1 = 1$ we have

$$\frac{B_n}{A_n} \to \frac{G}{108}$$

where G is Catalan's constant,

$$G = \sum_{n=0}^{\infty} \frac{(-1)^n}{(2n-1)^2}.$$

This is proved in Zudilin [10].

2.2 Differential equations of order 2. In [1] we considered the following differential equations of order 2 as building blocks.

Hypergeometric:

	Coefficients	Differential operator
(A)	$A_n = \binom{2n}{n}^2$	$\theta^2 - 4x(2\theta+1)^2$
(B)	$A_n = \frac{(3n)!}{n!^3}$	$\theta^2 - 3x(3\theta+1)(3\theta+2)$
(C)	$A_n = \frac{(4n)!}{(2n)!n!^2}$	$\theta^2 - 4x(4\theta+1)(4\theta+3)$
(D)	$A_n = \frac{(6n)!}{(3n)!(2n)!n!}$	$\theta^2 - 12x(6\theta+1)(6\theta+5)$

Sporadic:

	Coefficients	Differential operator
(a)	$A_n = \sum_k \binom{n}{k}^3$	$\theta^2 - x(7\theta^2 + 7\theta + 2) - 8x^2(\theta+1)^2$
(b)	$A_n = \sum_k \binom{n}{k}^2 \binom{n+k}{n}$	$\theta^2 - x(11\theta^2 + 11\theta + 3) - x^2(\theta+1)^2$
(c)	$A_n = \sum_k \binom{n}{k}^2 \binom{2k}{k}$	$\theta^2 - x(10\theta^2 + 10\theta + 3) + 9x^2(\theta+1)^2$
(d)	$A_n = \sum_k \binom{n}{k}\binom{2k}{k}\binom{2n-2k}{n-k}$	$\theta^2 - 4x(3\theta^2 + 3\theta + 1) + 32x^2(\theta+1)^2$
(e)	$A_n = \sum_k 4^{n-k}\binom{2k}{k}^2\binom{2n-2k}{n-k}$	$\theta^2 - 4x(8\theta^2 + 8\theta + 3) + 256x^2(\theta+1)^2$
(f)	$A_n = \sum_k (-1)^k 3^{n-3k}\binom{n}{3k}\frac{(3k)!}{k!^3}$	$\theta^2 - 3x(3\theta^2 + 3\theta + 1) + 27x^2(\theta+1)^2$
(g)	$A_n = \sum_{i,j} 8^{n-i}(-1)^i \binom{n}{i}\binom{i}{j}^3$	$\theta^2 - x(17\theta^2 + 17\theta + 6) + 72x^2(\theta+1)^2$
(h)	$A_n = 27^n \sum_k (-1)^k \binom{-2/3}{k}\binom{-1/3}{n-k}^2$	$\theta^2 - 3x(18\theta^2 + 18\theta + 7) + 3^6x^2(\theta+1)^2$
(i)	$A_n = 64^n \sum_k (-1)^k \binom{-3/4}{k}\binom{-1/4}{n-k}^2$	$\theta^2 - 4x(32\theta^2 + 32\theta + 13)$ $+ 2^{12}x^2(\theta+1)^2$
(j)	$A_n = 432^n \sum_k (-1)^k \binom{-5/6}{k}\binom{-1/6}{n-k}^2$	$\theta^2 - 12x(72\theta^2 + 72\theta + 31)$ $+ 2^8 3^6 x^2(\theta+1)^2$

Then we have the following limits:

	Limit
(a)	$\pi^2/24$
(b)	$\pi^2/30$
(c)	$L(\chi_3, 2)/2$
(d)	$G/2$
(e)	slow convergence
(f)	not convergent
(g)	$5L(\chi_3, 2)/8$
(h)	slow convergence
(i)	slow convergence
(j)	slow convergence

The cases with four singular points (a), (b), (c), (d), (f) and (g) are treated in Zagier [7]. In the cases with three singular points convergence is slow and the limit hard to guess. In case (e) we found some evidence for the limit being $G/2$.

For case (h) Arne Meurman conjectures that the limit is $\frac{1}{2}L(\chi_3, 2) - \frac{2}{81}\pi^2$. His computations suggest the following Ramanujan-like conjecture

$$\sum_{n=1}^{\infty} \frac{\left(\frac{n}{3}\right)(-\exp(-\pi/\sqrt{3}))^n}{n^2(1-(-\exp(-\pi/\sqrt{3}))^n)} = \frac{2}{81}\pi^2 - \frac{1}{2}\sum_{n=1}^{\infty}\frac{\left(\frac{n}{3}\right)}{n^2}.$$

This was proved subsequently by Yifan Yang [6], who also settled case (e). He uses, as Beukers [4] and Zagier [7], the fact that these equations admit modular parametrisations. In [1] there are three more second order differential equations which give rise to some fourth order Calabi–Yau equations by Hadamard product:

	Coefficients	Differential operator
(k)	$A_n = 9^n \sum_k (-1)^k \binom{n}{k}\binom{-1/3}{n-k}\binom{-2/3}{k}$	$\theta^2 - 3x(2\theta+1) - 81x^2(\theta+1)^2$
(l)	$A_n = 8^n \sum_k (-1)^k \binom{n}{k}\binom{-1/4}{n-k}\binom{-3/4}{k}$	$\theta^2 - 4x(2\theta+1) - 64x^2(\theta+1)^2$
(m)	$A_n = 36^n \sum_k (-1)^k \binom{n}{k}\binom{-1/6}{n-k}\binom{-5/6}{k}$	$\theta^2 - 24x(2\theta+1) - 1296x^2(\theta+1)^2$

Then we have the following limits:

	Limit
(k)	$\frac{\pi^2}{27} - \frac{1}{4}L(\chi_3, 2)$
(l)	$\frac{\pi^2}{16} - \frac{G}{2}$
(m)	$\frac{\pi^2}{36} - \frac{5}{16}L(\chi_3, 2)$

Proof of case (l) Let

$$R_n(t) = (-1)^n \frac{n! \prod_{j=0}^{n-1}(t+\frac{1}{4}+j)}{\prod_{j=0}^{n}(t+j)^2} = \sum_{k=0}^{n} \left\{ \frac{U_k}{(t+k)^2} + \frac{\widetilde{U}_k}{t+k} \right\},$$

where

$$\begin{aligned} U_k = (R_n(t)(t+k)^2)_{t=-k} &= (-1)^{n-k} n! \frac{(1/4)_{n-k}(3/4)_k}{(k!(n-k)!)^2} \\ &= (-1)^k \binom{n}{k}\binom{-1/4}{n-k}\binom{-3/4}{k}. \end{aligned}$$

Let

$$\begin{aligned} r_n &= \sum_{j=1}^{\infty} (R_n(t))_{t=j-1/4} = \sum_{j=1}^{\infty}\sum_{k=0}^{n} \left\{ \frac{U_k}{(j+k-1/4)^2} + \frac{\widetilde{U}_k}{j+k-1/4} \right\} \\ &= a_n \sum_{j=1}^{\infty} \frac{1}{(j-\frac{1}{4})^2} - b_n, \end{aligned}$$

where

$$a_n = \sum_{k=0}^{n} U_k = \sum_{k=0}^{n} (-1)^k \binom{n}{k}\binom{-1/4}{n-k}\binom{-3/4}{k} = \frac{1}{8^n} A_n$$

and

$$b_n = \sum_{k=0}^{n}\sum_{j=1}^{k} \frac{U_k}{(j-1/4)^2} + \sum_{k=0}^{n}\sum_{j=1}^{k} \frac{\widetilde{U}_k}{j-1/4}$$

as in the proof of Theorem 1 in [10].

Here we used

$$\sum_{k=0}^{n} \widetilde{U}_k = \sum_{k=0}^{n} \operatorname{Res}_{t=-k} R_n(t) = -\operatorname{Res}_{t=\infty} R_n(t) = 0,$$

since

$$R_n(t) = O(t^{-(n+2)}) \qquad \text{as } t \to \infty.$$

Using for instance Maple's `Zeilberger` we find that a_n and $\sum_t R_n(t)$ both satisfy the recursion

$$2(n+2)^2N^2 - (2n+3)N - 2(n+1)^2.$$

(Here N is the shift operator, that is: $Nf(n) := f(n+1)$.) It follows that r_n and hence b_n also satisfy this recursion and we have the start values

$$a_0 = 1, \quad a_1 = \frac{1}{2}, \qquad b_0 = 0, \quad b_1 = -2.$$

Now we show that

$$\frac{r_n}{a_n} \to 0 \qquad \text{as} n \to \infty.$$

It follows from general considerations that the coefficients a_n behave asymptotically as

$$a_n \sim Cn^{-d}\left\{1 + \frac{e_1}{n} + \frac{e_2}{n^2} + \cdots\right\}$$

as $n \to \infty$. Substituting this in the recursion formula above we obtain

$$d = \frac{1}{2}, \qquad e_1 = -\frac{1}{4}, \quad e_2 = \frac{1}{16}.$$

Numerical evidence suggests $C = 2/3$, but we will not need this.

Consider

$$F_n(t) = R_n(t)\frac{\pi}{\sin(\pi t)} = (-1)^n\Gamma(t)^2\Gamma\Big(\frac{3}{4} - t\Big)n^{-3/4-t}(1 + O(n^{-1})).$$

Then

$$\operatorname{Res}_{t=k-1/4} F_n(t) = R_n(t)_{t=k-1/4}$$

and

$$r_n = \frac{1}{2\pi i}\int_{\mathscr{L}} F_n(t)\,dt,$$

where $\mathscr{L}$ is a rectangle with vertices in $\varepsilon + iM$, $\varepsilon - iM$, $M + \frac{1}{4} + iM$, $M + \frac{1}{4} - iM$, and M is a positive integer $> n$. It follows that

$$\frac{r_n}{a_n} \le Cn^{-1/4} \to 0$$

as $n \to \infty$. □

2.3 Third order differential equations. Here we also have ten differential equations suitable for Hadamard products:

	Coefficients	Differential operator
(α)	$A_n=\sum_k \binom{n}{k}^2\binom{2k}{k}\binom{2n-2k}{n-k}$	$\theta^3-2x(2\theta+1)(5\theta^2+5\theta+2)$ $+64x^2(\theta+1)^3$
(β)	$A_n=\sum_k \binom{2k}{k}^2\binom{2n-2k}{n-k}^2$	$\theta^3-8x(2\theta+1)(2\theta^2+2\theta+1)$ $+256x^2(\theta+1)^3$
(γ)	$A_n=\sum_k \binom{n}{k}^2\binom{n+k}{n}^2$	$\theta^3-x(2\theta+1)(17\theta^2+17\theta+5)$ $+x^2(\theta+1)^3$
(δ)	$A_n=\sum_k(-1)^k 3^{n-3k}\binom{n}{3k}\binom{n+k}{n}\frac{(3k)!}{k!^3}$	$\theta^3-x(2\theta+1)(7\theta^2+7\theta+3)$ $+81x^2(\theta+1)^3$
(ϵ)	$A_n=\sum_k \binom{n}{k}^2\binom{2k}{n}^2$	$\theta^3-4x(2\theta+1)(3\theta^2+3\theta+1)$ $+16x^2(\theta+1)^3$
(ζ)	$A_n=\sum_{i,j} \binom{n}{i}^2\binom{n}{j}\binom{i}{j}\binom{i+j}{j}$	$\theta^3-3x(2\theta+1)(3\theta^2+3\theta+1)$ $-27x^2(\theta+1)^3$
(η)	A_n is not known	$\theta^3-x(2\theta+1)(11\theta^2+11\theta+5)$ $+125x^2(\theta+1)^3$
(θ)	$A_n=64^n\sum_k \binom{-1/4}{k}^2\binom{-3/4}{n-k}^2$	$\theta^3-8x(2\theta+1)(8\theta^2+8\theta+5)$ $+4096x^2(\theta+1)^3$
(ι)	$A_n=27^n\sum_k \binom{-1/3}{k}^2\binom{-2/3}{n-k}^2$	$\theta^3-3x(2\theta+1)(9\theta^2+9\theta+5)$ $+729x^2(\theta+1)^3$
(κ)	$A_n=432^n\sum_k \binom{-1/6}{k}^2\binom{-5/6}{n-k}$	$\theta^3-24x(2\theta+1)(18\theta^2+18\theta+13)$ $+186624x^2(\theta+1)^3$

Then we find the following limits:

	Limit
(α)	$\frac{7}{24}\zeta(3)$
(β)	slow convergence
(γ)	$\frac{1}{6}\zeta(3)$
(δ)	not convergent
(ϵ)	$\frac{7}{32}\zeta(3)$
(ζ)	$\frac{1}{3}L(\chi_3,3)$
(η)	not convergent
(θ)	$\frac{7}{16}\zeta(3)-\frac{1}{64}\pi^3$
(ι)	$\frac{13}{27}\zeta(3)-\frac{2}{243}\pi^3\sqrt{3}$
(κ)	$\frac{91}{432}\zeta(3)-\frac{1}{216}\pi^3\sqrt{3}$

where

$$\frac{1}{3}L(\chi_3,3)=\sum_{j=0}^{\infty}\left\{\frac{1}{(3j+1)^3}-\frac{1}{(3j+2)^3}\right\}=\frac{4}{729}\pi^3\sqrt{3}.$$

Proof of case (θ) Consider

$$R_n(t)=\frac{\prod_{j=0}^{n-1}(t+\frac14+j)^2}{\prod_{j=0}^{n}(t+j)^2}=\sum_{k=0}^{n}\left\{\frac{U_k}{(t+k)^2}+\frac{\widetilde{U}_k}{t+k}\right\},$$

where

$$U_k=(R_n(t)(t+k)^2)_{t=-k}=\left\{\frac{(3/4)_k(1/4)_{n-k}}{k!(n-k)!}\right\}^2$$

and

$$\widetilde{U}_k=\frac{d}{dt}(R_n(t)(t+k)^2)_{t=-k}\in\mathbb{Q}.$$

Let

$$\begin{aligned}
r_n&=-\frac12\sum_{j=1}^{\infty}\left(\frac{d}{dt}R_n(t)\right)_{t=j-1/4}\\
&=\sum_{j=1}^{\infty}\sum_{k=0}^{n}\left\{\frac{U_k}{(j+k-\frac14)^3}+\frac12\frac{\widetilde{U}_k}{(j+k-\frac14)^2}\right\}\\
&=\sum_{k=0}^{n}U_k\left\{\sum_{j=1}^{\infty}\frac{1}{(j-\frac14)^3}-\sum_{j=1}^{k}\frac{1}{(j-\frac14)^3}\right\}\\
&\quad+\frac12\sum_{k=0}^{n}\widetilde{U}_k\left\{\sum_{j=1}^{\infty}\frac{1}{(j-\frac14)^2}-\sum_{j=1}^{n}\frac{1}{(j-\frac14)^2}\right\}\\
&=a_n\sum_{j=1}^{\infty}\frac{1}{(j-\frac14)^3}-b_n,
\end{aligned}$$

where

$$a_n=\sum_{k=0}^{n}U_k=\sum_{k=0}^{n}\binom{-1/4}{k}^2\binom{-3/4}{n-k}^2=\frac{1}{64^n}A_n$$

and

$$b_n=\sum_{k=0}^{n}\sum_{j=1}^{k}\frac{U_k}{(j-\frac14)^3}+\frac12\sum_{k=0}^{n}\sum_{j=1}^{k}\frac{\widetilde{U}_k}{(j-\frac14)^2}.$$

It follows that

$$\frac12\sum_{k=0}^{n}\widetilde{U}_k=\frac12\sum_{k=0}^{n}\operatorname{Res}_{t=-k}R_n(t)=-\frac12\operatorname{Res}_{t=\infty}R_n(t)=0,$$

since

$$R_n(t)=O(t^{-2})\qquad\text{as } t\to\infty.$$

Using Maple's `Zeilberger` we find that a_n and $\sum_t R_n(t)$ are annihilated by the difference operator

$$8(n+2)^3N^2-(16n^3+72n^2+114n+63)N+8(n+1)^3.$$

It follows that r_n, and hence also b_n, satisfy the same difference equation. We also have the start values

$$a_0=1,\quad a_1=\frac58,\quad \dots,\qquad b_0=0,\quad b_0=1,\quad \dots .$$

We want to show that
$$\frac{r_n}{a_n} \to 0 \qquad \text{as } n \to \infty.$$
First we note that
$$a_n > \binom{-3/4}{n}^2 = \Gamma(3/4)^{-2} n^{-1/2} + O(n^{-3/2}).$$
Consider the function
$$\begin{aligned} F_n(t) = R_n(t)\left\{\frac{\pi}{\sin(\pi(t+1/4))}\right\}^2 &= \left\{\frac{\Gamma(t)\Gamma(3/4-t)\Gamma(t+n+1/4)}{\Gamma(t+n+1)}\right\}^2 \\ &= \Gamma(t)^2\Gamma(3/4-t)^2 n^{-3/2} + O(n^{-5/2}) \qquad \text{as } n \to \infty. \end{aligned}$$
We note that
$$\mathrm{Res}_{t=k-1/4} F_n(t) = \left(\frac{d}{dt} R_n(t)\right)_{t=k-1/4}$$
and hence
$$-2r_n = \frac{1}{2\pi i}\int_{\mathscr{L}} F_n(t)\, dt,$$
where the path of integration $\mathscr{L}$ is taken as in the proof of case (l) above.

It follows that
$$|r_n| < Cn^{-3/2} \qquad \text{as } n \to \infty$$
and
$$\left|\frac{r_n}{a_n}\right| = O(n^{-1}).$$
Finally we note that
$$\sum_{j=0}^{\infty} \frac{1}{(4j+3)^3} = \frac{1}{2}\left\{\sum_{j=0}^{\infty}\frac{1}{(2j+1)^3} - \sum_{j=0}^{\infty}\frac{(-1)^j}{(2j+1)^3}\right\} = \frac{1}{2}\left(\frac{7}{8}\zeta(3) - \frac{\pi^3}{32}\right)$$
and, since $A_n = 64^n a_n$, we have to divide the result by 64. □

Remark: As was pointed out by the referee, some of the cases in the table can be proven using modular parametrisations as in Beukers [4]. For instance, the function $A(x) = \sum_{n=1}^{\infty} a_n x^n$ for case (ϵ) has the parametrisation
$$x(\tau) = \left(\frac{\eta(\tau)\eta(8\tau)}{\eta(2\tau)\eta(4\tau)}\right)^8, \quad A(\tau) = \frac{(\eta(2\tau)\eta(4\tau))^6}{(\eta(\tau)\eta(8\tau))^4},$$
modular forms for $\Gamma_0(8) + w_8$. Case (β) is proven along these lines in [6].

3 Hadamard products

It is well known that the Hadamard product of D-finite sequences is D-finite (see [5, p. 194] for a proof), i.e., given two sequences A'_n and A''_n that satisfy difference equations with polynomial coefficients then the same is the case for the sequence $A'_n A''_n$. Here we mostly consider the case when
$$\begin{aligned} A'_{n+2} &= P_1(n)A'_{n+1} + Q_1(n)A'_n, \\ A''_{n+2} &= P_2(n)A''_{n+1} + Q_2(n)A''_n, \end{aligned}$$
where the P, Q are rational functions of degree 2.

Proposition 1 *The coefficients*

$$C_n = A'_n A''_n$$

satisfy the recursion

$$T_4 C_{n+4} + T_3 C_{n+3} + T_2 C_{n+2} - W_1 C_{n+1} - W_0 C_n = 0,$$

where for $j = 1, 2$

$$\begin{aligned} R_j(n) &= P_j(n+1)P_j(n) + Q_j(n+1), \\ S_j(n) &= P_j(n+1)Q_j(n), \\ U_j(n) &= P_j(n+2)R_j(n) + Q_j(n+2)P_j(n), \\ V_j(n) &= P_j(n+2)S_j(n) + Q_j(n+2)Q_j(n), \end{aligned}$$

and (here $P_j = P_j(n)$ *etc.)*

$$\begin{aligned} T_2 &= R_2 S_1 U_1 V_2 - R_1 S_2 U_2 V_1, \\ T_3 &= P_1 Q_2 U_2 V_1 - P_2 Q_1 U_1 V_2, \\ T_4 &= R_1 S_2 P_2 Q_1 - R_2 S_1 P_1 Q_2, \\ W_0 &= Q_1 Q_2 T_2 + S_1 S_2 T_3 + V_1 V_2 T_4, \\ W_1 &= P_1 P_2 T_2 + R_1 R_2 T_3 + U_1 U_2 T_4. \end{aligned}$$

Proof We have

$$A'_{n+3} = R_1 A'_{n+1} + S_1 A'_n, \qquad A'_{n+4} = U_1 A'_{n+1} + V_1 A'_n,$$

and similarly formulas for A''_{n+3} and A''_{n+4}.

Multiplying the recursion formulas for A'_{n+2} and A''_{n+2} (respectively, for A'_{n+3} and A''_{n+3}, and for A'_{n+4} and A''_{n+4}) we have

$$\begin{aligned} C_{n+2} &= P_1 P_2 C_{n+1} + Q_1 Q_2 C_n + P_1 Q_2 A'_{n+1} A''_n + P_2 Q_1 A'_n A''_{n+1}, \\ C_{n+3} &= R_1 R_2 C_{n+1} + S_1 S_2 C_n + R_1 S_2 A'_{n+1} A''_n + R_2 S_1 A'_n A''_{n+1}, \\ C_{n+4} &= U_1 U_2 C_{n+1} + V_1 V_2 C_n + U_1 V_2 A'_{n+1} A''_n + U_2 V_1 A'_n A''_{n+1}. \end{aligned}$$

Considering $A'_{n+1} A''_n$ and $A'_n A''_{n+1}$ as unknowns we find

$$\begin{vmatrix} P_1 Q_2 & P_2 Q_1 & -C_{n+2} + P_1 P_2 C_{n+1} + Q_1 Q_2 C_n \\ R_1 S_2 & R_2 S_1 & -C_{n+3} + R_1 R_2 C_{n+1} + S_1 S_2 C_n \\ U_1 V_2 & U_2 V_1 & -C_{n+4} + U_1 U_2 C_{n+1} + V_1 V_2 C_n \end{vmatrix} = 0.$$

Expanding the determinant we get the wanted recursion formula. □

Let A'_n be the solution to

$$A'_{n+2} = P_1(n) A'_{n+1} + Q_1(n) A'_n$$

with

$$A_{-1} = 0 \quad \text{and} \quad A_0 = 1$$

(and similarly for A''_n). Let further B'_n be the solution to the same recursion but with

$$B_0 = 0 \quad \text{and} \quad B_1 = 1$$

(and similarly for B''_n). Let $C_n = A'_n A''_n$ as above and let D_n be the solution to the recursion in Proposition 1 with

$$D_n = 0 \qquad \text{for} \quad n \leq 0 \quad \text{and} \quad D_1 = 1.$$

Assume that

$$P_j(n) = \frac{a_j(n+1)^2 + a_j(n+1) + b_j}{(n+2)^2}, \qquad Q_j(n) = \frac{c_j(n+1)^2}{(n+2)^2}$$

for $j = 1, 2$. Define

$$f_1 = \frac{b_2c_1}{b_2^2c_1 - b_1^2c_2}, \qquad f_2 = \frac{b_1c_2}{b_1^2c_2 - b_2^2c_1}.$$

Theorem *We have for* $n \geq 1$

$$\frac{D_n}{C_n} = f_1\frac{B'_n}{A'_n} + f_2\frac{B''_n}{A''_n}.$$

Proof We use induction on n. First we let Maple check the identity for $n = 0, 1, 2, 3$. Assume that

$$\frac{D_k}{C_k} = f_1\frac{B'_k}{A'_k} + f_2\frac{B''_k}{A''_k},$$

i.e., $D_k = f_1A''_kB'_k + f_2A'_kB''_k$ for $k = n, n+1, n+2, n+3$. We have

$$\begin{aligned}
\frac{D_{n+4}}{C_{n+4}} &= -\frac{1}{T_4}\frac{T_3D_{n+3} + T_2D_{n+2} - W_1D_{n+1} - W_0D_n}{A'_{n+4}A''_{n+4}}, \\
&= -f_1\frac{1}{T_4A'_{n+4}A''_{n+4}}\left\{\begin{array}{c} T_3A''_{n+3}B'_{n+3} + T_2A''_{n+2}B'_{n+2} \\ -W_1A''_{n+1}B'_{n+1} - W_0A''_nB'_n \end{array}\right\} \\
&\quad - f_2\left\{\text{similar formula with } {}' \text{ replaced by } {}''\right\} \\
&= -f_1\frac{1}{T_4A'_{n+4}A''_{n+4}}\left\{\begin{array}{l} T_3(R_2A''_{n+1} + S_2A''_n)(R_1B'_{n+1} + S_1B'_n) \\ \quad +T_2(P_2A''_{n+1} + Q_2A''_n)(P_1B'_{n+1} + Q_1B'_n) \\ \quad -W_1A''_{n+1}B'_{n+1} - W_0A''_nB'_n \end{array}\right\} \\
&\quad - f_2\left\{\text{similar formula}\right\} \\
&= -f_1\frac{1}{T_4A'_{n+4}A''_{n+4}}\left\{\begin{array}{l} (T_3R_2R_1 + T_2P_1P_2 - W_1)A''_{n+1}B'_{n+1} \\ \quad +(T_3S_1S_2 + T_2Q_1Q_2 - W_0)A''_nB'_n \\ \quad +(T_3S_1R_2 + T_2P_2Q_1)A''_{n+1}B'_n \\ \quad +(T_3R_1S_2 + T_2P_1Q_2)A''_nB'_{n+1} \end{array}\right\} \\
&\quad - f_2\left\{\text{similar formula}\right\} \\
&= -f_1\frac{1}{T_4A'_{n+4}A''_{n+4}}\left\{-T_4(U_2A''_{n+1} + V_2A''_n)(U_1B'_{n+1} + V_1B'_n)\right\} \\
&\quad - f_2\left\{\text{similar formula}\right\} \\
&= f_1\frac{B'_{n+4}}{A'_{n+4}} + f_2\frac{B''_{n+4}}{A''_{n+4}}.
\end{aligned}$$

□

In [1] we consider the second order differential operators

$$\theta^2 - z(a\theta^2 + a\theta + b) - cz^2(\theta+1)^2, \qquad \text{where} \quad \theta = z\frac{d}{dz}.$$

This corresponds to the recursion formula

$$(n+2)^2A_{n+2} = (a(n+1)^2 + a(n+1) + b)A_{n+1} + c(n+1)^2A_n$$

We consider the following cases:

#	a	b	c
(a)	7	2	8
(b)	11	3	1
(c)	10	3	−9
(d)	12	4	−32
(e)	32	12	−256
(f)	9	3	−27
(g)	17	6	−72
(h)	54	21	−729
(i)	128	52	−4096
(j)	864	372	−186624

We get the following table for (f_1, f_2):

	(b)	(c)	(d)	(e)	(f)
(a)	$(\frac{6}{17}, -\frac{1}{34})$	$(\frac{2}{9}, \frac{1}{6})$	$(\frac{1}{8}, \frac{1}{4})$	$(\frac{3}{68}, \frac{4}{17})$	$(\frac{2}{15}, \frac{3}{10})$
(b)		$(\frac{1}{30}, \frac{3}{10})$	$(\frac{1}{76}, \frac{6}{19})$	$(\frac{1}{204}, \frac{16}{51})$	$(\frac{1}{84}, \frac{9}{28})$
(c)			$(-\frac{1}{4}, \frac{2}{3})$	$(-\frac{3}{28}, \frac{16}{21})$	$(-\frac{1}{6}, \frac{1}{2})$
(d)				$(\frac{3}{4}, -2)$	$(-\frac{2}{3}, \frac{3}{4})$
(e)					$(-\frac{16}{33}, \frac{9}{44})$

	(g)	(h)	(i)	(j)
(a)	$(\frac{1}{12}, \frac{1}{4})$	$(\frac{14}{537}, \frac{81}{358})$	$(\frac{13}{1188}, \frac{64}{297})$	$(\frac{31}{19308}, \frac{324}{1609})$
(b)	$(\frac{1}{114}, \frac{6}{19})$	$(\frac{7}{2334}, \frac{243}{778})$	$(\frac{13}{9892}, \frac{768}{2473})$	$(\frac{31}{151500}, \frac{3888}{12625})$
(c)	$(-\frac{1}{6}, \frac{2}{3})$	$(-\frac{7}{96}, \frac{27}{32})$	$(-\frac{13}{348}, \frac{256}{261})$	$(-\frac{31}{4020}, \frac{432}{335})$
(d)	∞	$(\frac{14}{51}, -\frac{81}{68})$	$(\frac{13}{164}, -\frac{32}{41})$	$(\frac{31}{3756}, -\frac{162}{313})$
(e)	$(-\frac{4}{3}, \frac{3}{4})$	$(\frac{112}{165}, -\frac{243}{220})$	$(\frac{13}{100}, -\frac{12}{25})$	$(\frac{31}{2784}, -\frac{243}{928})$
(f)	$(\frac{1}{2}, -\frac{2}{3})$	$(\frac{7}{66}, -\frac{9}{22})$	$(\frac{39}{1004}, -\frac{256}{753})$	$(\frac{31}{6348}, -\frac{144}{529})$
(g)		$(\frac{14}{51}, -\frac{27}{34})$	$(\frac{13}{164}, -\frac{64}{123})$	$(\frac{31}{3756}, -\frac{108}{313})$
(h)			$(\frac{1053}{4580}, -\frac{1792}{3445})$	$(\frac{31}{2124}, -\frac{112}{531})$
(i)				$(\frac{124}{5061}, -\frac{1053}{6748})$

We can also consider Hadamard products of third order differential equations, i.e., we have recursions for $j = 1, 2$

$$A'_{n+2} = \frac{(2(n+1)+1)(a_1(n+1)^2 + a_1(n+1) + b_1)}{(n+2)^3} A'_{n+1} + c_1 \frac{(n+1)^3}{(n+2)^3} A'_n$$

and similarly for A''_n. Then we can imitate everything we did for second order equations and the results are the same except that the degree of the difference equation is higher. In particular, the weights f_1 and f_2 are the same as before.

For a Hadamard square we must modify the construction of the recurrence. Let

$$A_{n+2} = P(n)A_{n+1} + Q(n)A_n$$

with $A_{-1} = 0$, $A_0 = 1$. Let B_n satisfy the same recursion with $B_0 = 0$, $B_1 = 1$. Then

$$A_{n+3} = R(n)A_{n+1} + S(n)A_n,$$

where

$$R(n) = P(n+1)P(n) + Q(n+1), \qquad S(n) = P(n+1)Q(n+1).$$

Put

$$T(n) = R(n)Q(n) - S(n)P(n);$$

then $C_n = A_n^2$ satisfies

$$PQC_{n+3} - RSC_{n+2} - PRTC_{n+1} + QSTC_n = 0.$$

Proposition 2 *Let $C_{-2} = 0$, $C_{-1} = 0$, $C_0 = 1$ and let D_n satisfy the same recursion with $D_{-1} = 0$, $D_0 = 0$, $D_1 = 1$. Then if*

$$P(n) = \frac{a(n+1)^2 + a(n+1) + b}{(n+2)^2}, \qquad Q(n) = c\frac{(n+1)^2}{(n+2)^2},$$

we have

$$\frac{D_n}{C_n} = \frac{1}{b}\frac{B_n}{A_n} \qquad \textit{for} \quad n > 0.$$

Proof Assume that

$$\frac{D_k}{C_k} = \frac{1}{b}\frac{B_k}{A_k} \qquad \text{for} \quad k = n, n+1, n+2.$$

Then

$$\begin{aligned}\frac{D_{n+3}}{C_{n+3}} &= \frac{RSA_{n+2}B_{n+2} + PRTA_{n+1}B_{n+1} - QSTA_nB_n}{bPQA_{n+3}^2} \\ &= \frac{1}{bPQA_{n+3}^2}\left\{\begin{array}{c} RS(PA_{n+1} + QA_n)(PB_{n+1} + QB_n) \\ +PRTA_{n+1}B_{n+1} - QSTA_nB_n \end{array}\right\} \\ &= \frac{PQ(RA_{n+1} + SA_n)(RB_{n+1} + SB_n)}{bPQA_{n+3}^2} = \frac{1}{b}\frac{B_{n+3}}{A_{n+3}}\end{aligned}$$

Maple checks that the formula is true for $n = 0, 1, 2$. □

We still have another case of Hadamard products, namely, the product of the second order differential equations (a), (b), (c), (d), (e), (f), (g), (h), (i), (j) (respectively, third order (α), (β), (γ), (δ), (ε), (ζ), (η), (θ), (ι), (κ)) with the second order differential equations (A), (B), (C), (D), i.e., multiplying the coefficients with

$$\binom{2n}{n}, \quad \frac{(3n)!}{n!^3}, \quad \frac{(4n)!}{(2n)!n!^2}, \quad \frac{(6n)!}{(3n)!(2n)!n!}.$$

The corresponding differential operators are $\theta^2 - zQ(\theta)$, where $Q(\theta)$ is given by

(A)	$(2\theta+1)^2$
(B)	$3(3\theta+1)(3\theta+2)$
(C)	$4(4\theta+1)(4\theta+3)$
(D)	$12(6\theta+1)(6\theta+5)$

We have the formulas

$$\begin{aligned}&\{\theta^2 - zP(\theta) - cz^2(\theta+1)^2\} * \{\theta^2 - zQ(\theta)\} \\ &\quad = \theta^4 - zP(\theta)Q(\theta) - cz^2Q(\theta)Q(\theta+1)\end{aligned}$$

$$\begin{aligned}&\{\theta^3 - z(2\theta+1)P(\theta) - cz^2(\theta+1)^3\} * \{\theta^2 - zQ(\theta)\} \\ &\quad = \theta^5 - z(2\theta+1)P(\theta)Q(\theta) - cz^2(\theta+1)Q(\theta)Q(\theta+1).\end{aligned}$$

Let now A_n solve the equation

$$\theta^2 - zP(\theta) - cz^2(\theta+1)^2$$

with $A_{-1}=0, A_0=1$. Then A_n satisfies the recursion

$$(n+2)^2A_{n+2} = P(n+1)A_{n+1} + c(n+1)^2A_n.$$

Let B_n satisfy the same recurrence but with $B_0=0, B_1=1$. Let further C_n satisfy the recursion

$$(n+2)^4C_{n+2} = P(n+1)Q(n+1)C_{n+1} + cQ(n)Q(n+1)C_n$$

with $C_{-1}=0, C_0=1$. Similarly, let D_n satisfy the same recursion but with $D_0=0, D_1=1$.

Proposition 3 *We have*

$$\frac{D_n}{C_n} = \frac{1}{Q(0)}\frac{B_n}{A_n}.$$

Proof We first show that

$$\frac{C_{n+1}}{C_n} = \frac{Q(n)}{(n+1)^2}\frac{A_{n+1}}{A_n}.$$

Assume this is true for n. Then

$$\begin{aligned}(n+2)^4\frac{C_{n+2}}{C_{n+1}} &= P(n+1)Q(n+1) + cQ(n)Q(n+1)\frac{C_n}{C_{n+1}}\\ &= P(n+1)Q(n+1) + cQ(n)Q(n+1)\frac{(n+1)^2}{Q(n)}\frac{A_n}{A_{n+1}}\\ &= \frac{Q(n+1)}{A_{n+1}}\{P(n+1)A_{n+1} + Q(n)A_n\}\\ &= \frac{Q(n+1)}{A_{n+1}}(n+2)^2A_{n+2}.\end{aligned}$$

We get, assuming the formula true for n and $n+1$,

$$\begin{aligned}\frac{D_{n+2}}{C_{n+2}} &= \frac{P(n+1)Q(n+1)D_{n+1} + cQ(n)Q(n+1)D_n}{P(n+1)Q(n+1)C_{n+1} + Q(n)Q(n+1)C_n}\\ &= \frac{1}{Q(0)}\frac{P(n+1)B_{n+1}C_{n+1}/A_{n+1} + cQ(n)B_nC_n/A_n}{P(n+1)A_{n+1}C_{n+1}/A_{n+1} + cQ(n)A_nC_n/A_n}\\ &= \frac{1}{Q(0)}\frac{P(n+1)B_{n+1} + c(n+1)^2B_n}{P(n+1)A_{n+1} + c(n+1)^2A_n} = \frac{1}{Q(0)}\frac{B_{n+2}}{A_{n+2}}.\end{aligned}$$

□

4 Conjectured limits

Using Maple's PSLQ we have found many limits, where we have no proof.

4.1 Limits of form $c\pi^2$.

#	15	18	19	20	21	22	23	24	25	26	27	33	45
c	$\frac{1}{144}$	$\frac{1}{60}$	$\frac{1}{56}$	$\frac{5}{216}$	$\frac{1}{48}$	$\frac{1}{36}$	$\frac{1}{32}$	$\frac{1}{180}$	$\frac{1}{120}$	$\frac{1}{84}$	$-\frac{1}{56}$	$\frac{1}{384}$	$\frac{1}{96}$

#	51	55	62	63	68	99	109	117	118	119
c	$\frac{1}{360}$	$-\frac{1}{128}$	$\frac{1}{1440}$	$-\frac{1}{1800}$	$\frac{1}{288}$	$-\frac{1}{198}$	$-\frac{1}{474}$	$\frac{1}{64}$	$\frac{1}{160}$	$\frac{3}{224}$

#	193	195	198	208	209	210	211	212	222	224
c	$\frac{1}{108}$	$\frac{1}{78}$	$-\frac{1}{72}$	$\frac{7}{1119}$	$\frac{1}{138}$	$-\frac{1}{276}$	$-\frac{1}{2048}$	$-\frac{1}{48}$	$\frac{1}{128}$	$\frac{1}{64}$

#	232	234	235	238	241	243	248	250	256	260
c	$\frac{1}{256}$	$\frac{2}{81}$	$\frac{1}{162}$	$\frac{1}{162}$	$-\frac{1}{512}$	$-\frac{1}{168}$	$\frac{1}{40}$	$\frac{1}{80}$	$-\frac{1}{320}$	$-\frac{25}{3096}$

#	261	262	279	282	297	338	341
c	$-\frac{1}{81}$	$-\frac{1}{54}$	$\frac{1}{36}$	$-\frac{4}{729}$	$\frac{67}{3870}$	$-\frac{1}{160}$	$\frac{29}{1842}$

4.2 Limits of the form cG.

#	36	38	48	65	231	233	237	239	257	258
c	$\frac{2}{2}$	$\frac{1}{24}$	$\frac{1}{12}$	$\frac{1}{120}$	$\frac{3}{76}$	$\frac{1}{8}$	$\frac{1}{40}$	$\frac{1}{152}$	$\frac{1}{104}$	$\frac{1}{56}$

#	264	277	288	289	300	329	331
c	$\frac{1}{1144}$	$\frac{1}{248}$	$\frac{1}{200}$	$\frac{1}{472}$	$-\frac{1}{168}$	$\frac{1}{16}$	$\frac{1}{24}$

4.3 Limits of the form $L(\chi_3, 2)$.

#	58	64	69	70	137	138	139	140	183	274	278
c	$\frac{1}{8}$	$\frac{1}{12}$	$\frac{1}{24}$	$\frac{1}{12}$	$\frac{5}{32}$	$\frac{5}{48}$	$\frac{5}{96}$	$\frac{1}{96}$	$\frac{1}{4}$	$-\frac{5}{19}$	$\frac{1}{30}$

4.4 Limits of the form $c\zeta(3)$.

#	16	28	29	37	42	44	50	52	53	60
c	$\frac{7}{48}$	$\frac{1}{7}$	$\frac{1}{12}$	$\frac{7}{288}$	$\frac{7}{64}$	$\frac{1}{24}$	$\frac{7}{144}$	$\frac{1}{72}$	$\frac{1}{76}$	$\frac{3}{23}$

#	66	67	149	182	189	205
c	$\frac{7}{1440}$	$\frac{7}{1440}$	$-\frac{1}{360}$	$\frac{3}{11}$	$\frac{1}{14}$	$\frac{21}{80}$

4.5 Limits of the form $cL(\chi_3, 3)$.

#	185
c	$\frac{1}{2}$

4.6 Limits of the form $c\pi^4$.

#	32	244
c	$\frac{1}{1170}$	$-\frac{1}{630}$

4.7 Limits of mixed form.

#	59	116	214
Limit	$-\frac{1}{208}\pi^2 + \frac{7}{26}G$	$-\frac{1}{320}\pi^2 + \frac{1}{10}G$	$-\frac{1}{144}\pi^2 + \frac{3}{16}L(\chi_3, 2)$

#	215	216 & 217	218
Limit	$-\frac{1}{736}\pi^2 + \frac{3}{46}G$	$\frac{1}{54}\pi^2 - \frac{1}{24}L(\chi_3, 2)$	$-\frac{1}{450}\pi^2 + \frac{7}{50}L(\chi_3, 2)$

#	219	220	221
Limit	$\frac{1}{18}\pi^2 - \frac{5}{8}L(\chi_3, 2)$	$-\frac{1}{128}\pi^2 + \frac{1}{8}G$	$-\frac{5}{4104}\pi^2 + \frac{10}{171}G$

#	226	240	242
Limit	$-\frac{1}{216}\pi^2 + \frac{5}{24}L(\chi_3, 2)$	$\frac{1}{6072}\pi^2 + \frac{26}{253}G$	$-\frac{1}{54}\pi^2 + \frac{1}{6}L(\chi_3, 2)$

#	251	252
Limit	$-\frac{1}{945}\pi^2 + \frac{2}{45}G + \frac{3}{28}L(\chi_3, 2)$	$\frac{1}{39888}\pi^2 - \frac{20}{277}G$

#	328	340
Limit	$\frac{1}{84}\pi^2 - \frac{1}{12}G$	$-\frac{1}{7712}\pi^2 + \frac{9}{241}G$

Remark Equation #251 is remarkable. The recursion is of order 4 and degree 8. The corresponding differential operator (of order 8 and degree 4) factors into two of order 4 (of degree 12 and 8, respectively). It is also the only case we have found which is a linear combination of values of three L-functions.

4.8 Unidentified limits with many digits. The following limits converge very fast but we have not been able to identify them:

#	Limit
17	0.3097538790467859180612009276621093227187570662744
34	0.3328799000999687141578669041535332990987481910397
147	−0.18294864738225619912816208889600198974512074131046
206	0.034533768895242744048924842293799661180236416
207	−0.00050462505145900474057831709244307528529622730007723
$\widetilde{214}$	0.078351361063584306378644830113068845245731042453985
229	0.017891639973294587136164078700439361783026615410295

For $\#\widetilde{214}$ see [3].

Acknowledgements. We thank Vasily Golyshev and Arne Meurman for several interesting discussions about this work. We also want to thank the referee for useful comments and spotting several mistakes in the first version of this paper.

References

[1] G. Almkvist and W. Zudilin, 'Differential equations, mirror maps and zeta values', in: *Mirror Symmetry V*, N. Yui, S.-T. Yau, and J. D. Lewis (eds.), AMS/IP Stud. Adv. Math. **38** (International Press & Amer. Math. Soc., Providence, RI, 2007) 481–515; `arXiv:math/0402386 [math.NT]` (2004).

[2] G. Almkvist, C. van Enckevort, D. van Straten, and W. Zudilin, 'Tables of Calabi–Yau equations', `arXiv:math/0507430 [math.AG]` (2005).

[3] G. Almkvist, 'Calabi–Yau differential equations of degree 2 and 3 and Yifan Yang's pullback', `arXiv:math/0612215 [math.AG]` (2006).

[4] F. Beukers, 'Irrationality proofs using modular forms', Astérisque,Vol.**147-148**, (271-283).

[5] R. Stanley, *Enumerative combinatorics, vol.* 2, Cambridge Studies in Advanced Mathematics **62** (Cambridge Univ. Press, Cambridge, 1999).

[6] Yifan Yang, 'Apéry limits and special values of L-functions', `arXiv:math/0709.1968[math.NT]` (2007).

[7] D. ZAGIER, 'Integral solutions of Apéry-like recurrence equations', Manuscript (2003).
[8] W. ZUDILIN, 'Binomial sums related to rational approximations of $\zeta(4)$', *Mat. Zametki* **75**:4 (2004) 637–640; English transl., *Math. Notes* **75**:4 (2004) 594–597; `arXiv:math/0311196 [math.CA]` (2003).
[9] W. ZUDILIN, 'Approximations to -, di- and tri- logarithms', J. Comput. Appl. Math. **202**:2 (2007) 450–459; `arXiv:math/0409023 [math.CA]` (2004).
[10] W. ZUDILIN, 'An Apéry-like difference equation for Catalan's constant', *Electronic J. Combinatorics* **10**:1 (2003) #R14, 10 pp.; `arXiv:math/0201024 [math.NT]` (2002).

Fields Institute Communications
Volume 54, 2008

Hypergeometric Systems in Two Variables, Quivers, Dimers and Dessins d'Enfants

Jan Stienstra
Mathematisch Instituut
Universiteit Utrecht
Postbus 80.010, 3508 TA Utrecht, The Netherlands
J.Stienstra@uu.nl

Abstract. This paper presents some parallel developments in Quiver/Dimer Models, Hypergeometric Systems and Dessins d'Enfants. It demonstrates that the setting in which Gelfand, Kapranov and Zelevinsky have formulated the theory of hypergeometric systems also provides a natural setting for dimer models. The Fast Inverse Algorithm of [14] and the untwisting procedure of [4] are recast in this more natural setting and then immediately produce from the quiver data the Kasteleyn matrix for dimer models, which is best viewed as the bi-adjacency matrix for the untwisted model. Some perfect matchings in the dimer models are direct reformulations of the triangulations in GKZ theory, and the rule which maps triangulations to the vertices of the secondary polygon extends to a rule for mapping perfect matchings to lattice points in the secondary polygon. Finally, it is observed in many examples, and then conjectured to hold in general, that the determinant of the Kasteleyn matrix with suitable weights becomes after a simple transformation equal to the principal $\mathcal{A}$-determinant in GKZ theory[1]. Illustrative examples are distributed throughout the text.

1 Introduction

In the last decade interesting correspondences were discovered relating Quiver Gauge Theories, lattice polygons and Calabi-Yau singularities. The motivation and evolution of these ideas in physics are well-documented in many articles; e.g. [1, 3, 4, 5, 6, 12, 13, 14, 15, 16, 17, 19, 20]. In the present paper we want to put some aspects of these correspondences from physics alongside the hypergeometric systems in two variables of Gelfand, Kapranov, Zelevinsky [8, 9, 10] and the dessins d'enfants of Grothendieck et al. [18, 22, 24]. This reveals intriguing connections between these fields.

2000 *Mathematics Subject Classification.* Primary 52B20; Secondary 33C70, 14M25, 14H55, 05C90.

[1] *Added in proof:* Meanwhile this conjecture has been proven; see [26].

The beautiful insight of Gelfand, Kapranov and Zelevinsky was that hypergeometric structures greatly simplify if one introduces extra variables and balances this with an appropriate torus action [8, 9, 10, 11, 25]. In order to profit from the simplification they developed tools like the *secondary fan, secondary polytope* and *principal $\mathcal{A}$-determinant.* This paper demonstrates that these are also very practical tools for studying quivers and dimer models.

A similar beautiful insight of simplification by going to higher dimensions appeared in De Bruijn's construction [2] of Penrose tilings and developed into the well-known projection method in the theory of quasi-crystals; e.g. [23]. We apply the same method to construct periodic rhombus tilings of the plane, which the physicists call *brane tilings* and *dimer models.*

From a geometric perspective this paper deals with embeddings of quivers into compact oriented surfaces without boundary. More specifically, the initial combinatorial data for a quiver Q are two finite sets E (arrows) and V (nodes) and two maps $s, t : E \to V$ (source and target). Embedding Q into a compact oriented surface without boundary $\mathcal{M}$ means that V becomes a subset of $\mathcal{M}$ and an arrow $e \in E$ becomes a path p_e in $\mathcal{M}$ from the point $s(e)$ to the point $t(e)$. It is required that the boundary of every *face of* $(\mathcal{M}, \mathsf{Q})$ – i.e. connected component of $\mathcal{M} \setminus \bigcup_{e \in E} p_e$ – is formed by a sequence of paths $(p_{e_1}, \dots, p_{e_n})$ with $p_{e_i} \cap p_{e_{i+1}} = t(e_i) = s(e_{i+1})$ for $1 \leq i \leq n$; $e_{n+1} = e_1$. It is evident from this requirement that the face boundaries receive an orientation from Q. When this is combined with the orientation on $\mathcal{M}$, faces lie either on the positive or on the negative side of their boundary. In pictures we mark faces which lie on the positive (resp. negative) side of their boundary with $\bullet$ (resp. $\circ$). It is obvious that adjacent faces get different colors. The dual graph of Q w.r.t. $\mathcal{M}$ is thus a bipartite graph $\Gamma^{\bullet\circ}$, embedded in $\mathcal{M}$: the nodes of $\Gamma^{\bullet\circ}$ correspond with the faces of $(\mathcal{M}, \mathsf{Q})$; the edges of $\Gamma^{\bullet\circ}$ connect nodes coming from adjacent faces and correspond bijectively with the edges of Q. A pair $(\mathcal{M}, \mathsf{Q})$ consisting of a compact oriented surface without boundary and an oriented graph embedded in it is called a *dessin d'enfants* or just *dessin*, in one of the various equivalent definitions of "dessin (d'enfants)"; see [22, 18, 24]. Other definitions refer to $(\mathcal{M}, \Gamma^{\bullet\circ})$ as dessin d'enfants. In case $\mathcal{M}$ has genus 1 one often calls $\Gamma^{\bullet\circ}$ and its lifting to the plane (i.e. the universal covering of the torus $\mathcal{M}$) a *dimer model* or *brane tiling*; see for instance [4, 5, 12, 13, 14, 15]. Yet another presentation of the same structure gives the dessin as the triangulation of $\mathcal{M}$ with vertex set the union of the nodes of Q and $\Gamma^{\bullet\circ}$; the triangles are given by the triplets of vertices consisting of two nodes of $\Gamma^{\bullet\circ}$ connected by an edge e^* of $\Gamma^{\bullet\circ}$ and one node v of Q incident to the edge e of Q which is dual to e^*. Figure 1 shows these three get-ups of the dessin d'enfants of the $\mathbb{P}^2$-quiver (case B_1 in Figure 3)

From an algebraic perspective this paper deals with *superpotentials for quivers.* There is a simple equivalence between the geometric and algebraic perspectives: the superpotential is a convenient way of writing the list of oriented boundaries $(p_{e_1}, \dots, p_{e_n})$ of the faces of $(\mathcal{M}, \mathsf{Q})$ with $\pm$-signs indicating how the orientation matches with that of $\mathcal{M}$. It can also be read as the instruction for building $\mathcal{M}$ by glueing polygons. For the dessin d'enfants in Figure 1 the superpotential (for some numbering of the arrows of Q) is

$$X_1X_2X_3 + X_4X_5X_6 + X_7X_8X_9 - X_2X_8X_5 - X_3X_9X_6 - X_4X_1X_7 .$$

Another algebraic perspective, equivalent to the previous one, is given by what we want to call *the bi-adjacency matrix of the dessin d'enfants* $(\mathcal{M}, \mathsf{Q})$ *with weight*

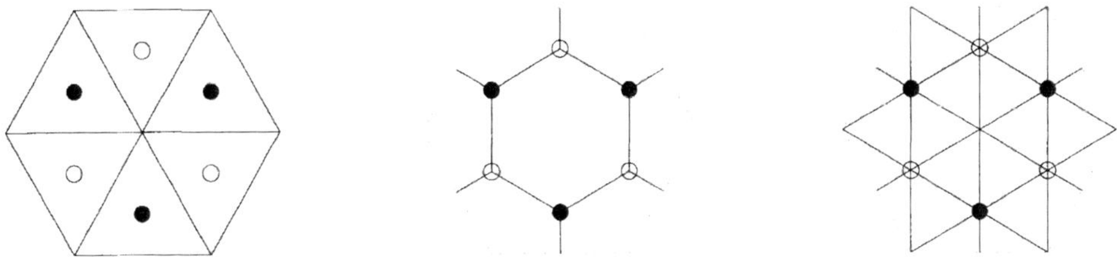

Figure 1 Three versions of the dessin d'enfants for $\mathbb{P}^2$ (= case B_1 in Fig. 3). The surface is obtained by identifying opposite sides of the hexagon.

ϖ. It is defined as follows. The *weight* is a function $\varpi : E \to \mathbb{C}$. The dessins in the present paper have as many black faces as white faces and every edge $e \in E$ lies in the boundary of a unique black face, denoted $\mathbf{b}(e)$, and a unique white face, denoted $\mathbf{w}(e)$. The bi-adjacency matrix of $(\mathcal{M}, \mathsf{Q})$ with weight ϖ is a square matrix $\mathbb{K}^{\varpi}$ with rows corresponding with the black faces, columns corresponding with the white faces and entries in the polynomial ring $\mathbb{C}[u_v \mid v \in V]$ which has one variable for every node v of the quiver Q: *the* $(\mathbf{b}, \mathbf{w})$*-entry of* $\mathbb{K}^{\varpi}$ *is*

$$\mathbb{K}^{\varpi}_{\mathbf{b},\mathbf{w}} = \sum_{e \in E:\, \mathbf{b}(e)=\mathbf{b},\, \mathbf{w}(e)=\mathbf{w}} \varpi(e)\, u_{s(e)} u_{t(e)}\,. \tag{1}$$

The bi-adjacency matrix is in fact the *Kasteleyn matrix* of a twist of the dimer model $(\mathcal{M}, \Gamma^{\bullet\circ})$; see Section 8. For the dessin in Figure 1 the bi-adjacency matrix (for some numbering of the nodes, edges and faces) is

$$\begin{bmatrix} \varpi_1 u_1 u_3 & \varpi_2 u_1 u_2 & \varpi_3 u_2 u_3 \\ \varpi_4 u_2 u_3 & \varpi_5 u_1 u_3 & \varpi_6 u_1 u_2 \\ \varpi_7 u_1 u_2 & \varpi_8 u_2 u_3 & \varpi_9 u_1 u_3 \end{bmatrix}.$$

In Section 2 we describe how the quivers for which we can solve the embedding problem are associated with certain rank 2 subgroups $\mathbb{L}$ of $\mathbb{Z}^N$; here N is the number of nodes of the quiver. Such subgroups $\mathbb{L} \subset \mathbb{Z}^N$ are the foundations for the theory of Gelfand, Kapranov and Zelevinsky. In Section 3 we describe the *secondary fan* and *the secondary polygon* associated with $\mathbb{L} \subset \mathbb{Z}^N$. These important structures in GKZ-theory surprisingly turn out to be also quite relevant for the dimer models. We show for instance in Theorem 9.3 that the secondary polygon is the *Newton polygon* of the determinant of the bi-adjacency matrix with non-zero weights. In Section 10 we report the observation, based on examples, that for the *critical weight*

$$\mathrm{crit} : E \to \mathbb{Z}_{>0}\,, \qquad \mathrm{crit}(e) = \sharp\{e' \in E \mid s(e') = s(e),\, t(e') = t(e)\} \tag{2}$$

the determinant of the bi-adjacency matrix $\mathbb{K}^{\mathrm{crit}}$, after a simple transformation, becomes equal to the *principal $\mathcal{A}$-determinant* of Gelfand, Kapranov, Zelevinsky [11]. In Sections 4 and 5 we recall some topics in the theory of GKZ-hypergeometric systems of differential equations.

Section 6 presents the algorithm to solve the quiver embedding problem. It is a combination of the *Fast Inverse Algorithm* of [14] and the *untwisting procedure* of [4]. In the cited papers, however, methods are mainly presented via visual inspection of pictures in some concrete examples. So an extrapolation to general situations was needed. The first impression that the Fast Inverse Algorithm of [14] is more or less De Bruijn's construction [2] of Penrose tilings did not quite yield the sought after embedding of the quiver; the untwisting procedure of [4] is also needed. The algorithm became a smoothly operating algebraic-combinatorial tool by consistently working from the philosophy that things look simpler from a higher dimensional viewpoint (with group action). For doing the computer experiments behind this paper I implemented the algorithm in MATLAB. The reader can find in Section 6 sufficient details for making a computer version of the algorithm.

Sections 7–10 demonstrate that consistently working from the higher dimensional viewpoint leads to new insights in the dimer technology tools *perfect matchings, Kasteleyn matrix and its determinant* and shows their close relation with the GKZ tools *secondary polytope and principal $\mathcal{A}$-determinant.*

Besides problems such as proving that the algorithm of Section 6 yields at least one superpotential for every quiver which satisfies the conditions in Theorem 2.10 or proving Conjecture 10.5, our work raises some interesting questions like:

Q1. The Calabi-Yau singularities in the background of this work are constructed from toric diagrams, which here are interpreted as secondary polygons. In [19] §6 the singularity is obtained from the toric diagram by a symplectic quotient construction. In [12] §2 the singularity is given as a toric variety constructed from a fan with one maximal cone, namely the cone over the toric diagram (polygon). For this the polygon is put into the plane of points with first coordinate 1 in $\mathbb{R}^3$. In the work of Gelfand, Kapranov and Zelevinsky the secondary polygon is put into a 2-dimensional plane in $\mathbb{R}^N$ which does not pass through $\mathbf{0}$; here N is the number of nodes of the quiver. One can perform the standard toric variety construction with the 3-dimensional cone over the secondary polygon in $\mathbb{R}^N$. *Does this toric variety give the appropriate view in the GKZ philosophy (higher dimension compensated by group action) on the singular Calabi-Yau 3-space?* Since this toric variety is a natural domain for GKZ hypergeometric functions (see 5.2) one may wonder: *Do hypergeometric functions provide new useful tools for investigating the geometry of 3-dimensional Calabi-Yau singularities?* Possibly positive indications are the fact that the dimension of the solution space of the GKZ system of hypergeometric differential equations equals the size of the bi-adjacency matrix $\mathbb{K}^{\mathrm{crit}}$ (see Theorem 5.3 and Corollary 7.3) and the observation (Conjecture 10.5 in the present paper = [26] Theorem 3) that the determinant of $\mathbb{K}^{\mathrm{crit}}$ equals, up to a simple transformation, the principal $\mathcal{A}$-determinant, which describes the singularities of the GKZ system.

Q2. The surface $\mathcal{M}$ with the pair of dual graphs Q and $\Gamma^{\bullet\circ}$ embedded in it has been constructed in a purely combinatorial topological way from the superpotential. It is a general fact (see [18] §1.2) that there is then a continuous map f from $\mathcal{M}$ to the 2-sphere S^2 which is a ramified covering with exactly three ramification points such that the set of vertices of Q, the set of black vertices of $\Gamma^{\bullet\circ}$ and the set of white vertices of $\Gamma^{\bullet\circ}$ are the fibers over the three ramification points. A highlight in the theory of dessins d'enfants is *Belyi's Theorem* (see [18] Theorem 2.1.1). It states that in the above situation the surface $\mathcal{M}$ admits a model $\mathcal{M}_K$ over a number field K such that f becomes a morphism $\mathcal{M}_K \to \mathbb{P}^1_K$ of algebraic curves over K which is unramified outside $\{0, 1, \infty\}$. The labeling of the ramification points can be taken such that $f^{-1}(\infty)$, $f^{-1}(0)$ and $f^{-1}(1)$ are the sets of white and black vertices of $\Gamma^{\bullet\circ}$ and the vertices of Q, respectively. The bipartite graph $\Gamma^{\bullet\circ}$ in $\mathcal{M}$ is then the inverse image of the negative real axis $[-\infty, 0]$ in the Riemann sphere $\mathbb{P}^1_{\mathbb{C}} = S^2$ and the quiver Q is the inverse image of the positively oriented unit circle

$\{z \in \mathbb{C} \mid |z| = 1\}$. It is usually difficult to find explicit algebraic equations for $\mathcal{M}_K$ and the *Belyi function* f.

On the other hand, the authors of [4] write in footnote [6]: .... *we have produced a dimer model that is defined on its own spectral curve* $\det \mathbb{K}^{\varpi} = 0$. Unfortunately their arguments are not (yet) sufficiently refined to yield the weight ϖ that is to be used in this equation.

It seems an interesting challenge to tackle the two problems simultaneously and look for a weight ϖ and a Belyi function f on the algebraic curve with equation $\det \mathbb{K}^{\varpi} = 0$ that realize $(\mathcal{M}, \mathsf{Q}, \Gamma^{\bullet\circ})$ as described above.

2 Rank 2 subgroups of $\mathbb{Z}^N$ and Quivers

2.1. The most basic object in this paper is a rank 2 subgroup $\mathbb{L} \subset \mathbb{Z}^N$ which is not contained in any of the standard coordinate hyperplanes of $\mathbb{Z}^N$ and is perpendicular to the vector $(1, \ldots, 1)$, with all coordinates 1. So, the elements of $\mathbb{L}$ are vectors $(\ell_1, \ldots, \ell_N) \in \mathbb{Z}^N$ with $\ell_1 + \ldots + \ell_N = 0$ and for every $i \in \{1, \ldots, N\}$ there is an $(\ell_1, \ldots, \ell_N) \in \mathbb{L}$ such that $\ell_i \neq 0$.

2.2. Notation. Throughout this paper $\mathbf{e}_1, \ldots, \mathbf{e}_N$ is the standard basis of $\mathbb{Z}^N$ and J denotes the matrix $\begin{pmatrix} 0 & 1 \\ -1 & 0 \end{pmatrix}$.

2.3. Definition/construction. Taking the second exterior powers one finds an inclusion $\bigwedge^2 \mathbb{L} \hookrightarrow \bigwedge^2 \mathbb{Z}^N$. The group $\bigwedge^2 \mathbb{L}$ is a free $\mathbb{Z}$-module of rank 1. After fixing the orientation on $\mathbb{L}$, i.e. choosing one of the two possible isomorphisms $\bigwedge^2 \mathbb{L} \simeq \mathbb{Z}$, one gets an inclusion $\mathbb{Z} \hookrightarrow \bigwedge^2 \mathbb{Z}^N$. The coordinates of $1 \in \mathbb{Z}$ with respect to the standard basis $\{\mathbf{e}_i \wedge \mathbf{e}_j\}_{1 \leq i < j \leq N}$ of $\bigwedge^2 \mathbb{Z}^N$ are called the *Plücker coordinates of* $\mathbb{L} \subset \mathbb{Z}^N$.

Dual to the above inclusion one finds the map $\bigwedge^2 \mathbb{Z}^N \rightarrow \mathbb{Z}$, i.e. an anti-symmetric bilinear form on $\mathbb{Z}^N$. In view of its relation with the Plücker coordinates we call this the *Plücker form of* $\mathbb{L}$.

To make this look more explicit we take a basis for $\mathbb{L}$ compatible with the chosen orientation. One can represent the two basis vectors as the rows of a $2 \times N$-matrix B. Let $\mathbf{b}_1, \ldots, \mathbf{b}_N$ be the columns of this matrix. Then the Plücker coordinates are $(\det(\mathbf{b}_i, \mathbf{b}_j))_{1 \leq i < j \leq N}$ and the (anti-symmetric) matrix for the Plücker form is $(\det(\mathbf{b}_i, \mathbf{b}_j))_{1 \leq i,j \leq N}$. The Plücker coordinates and the Plücker form do, of course, not change if one takes another basis for $\mathbb{L}$ with the same orientation. Note that the matrix for the Plücker form has rank 2:

$$\operatorname{rank}\left((\det(\mathbf{b}_i, \mathbf{b}_j))_{1 \leq i,j \leq N}\right) = 2. \tag{3}$$

2.4. Definition. A *quiver* Q is a finite directed graph. It can be given combinatorially by two finite sets E and V and two maps $s, t : E \rightarrow V$; in shorthand notation $\mathsf{Q} = (E, V, s, t)$. Each element of V is a node of the graph and an element $e \in E$ is an arrow from node $s(e)$ to node $t(e)$. The *adjacency matrix* of the quiver is the matrix $(q_{ij})_{i,j \in V}$ with entry q_{ij} equal to the number of arrows from node i to node j. If the quiver has no directed loops of length ≤ 2 (i.e. $q_{ij}q_{ji} = 0$ for all $i, j \in V$), it is faithfully described by the *anti-symmetrized adjacency matrix* $Q = (q_{ij} - q_{ji})_{i,j \in V}$. If the (i, j)-entry of Q is positive it equals the number of arrows from node i to node j. If it is negative it is the number of arrows from j to i.

2.5. Definition. We call the quiver without directed loops of length ≤ 2 with anti-symmetrized adjacency matrix equal to the matrix of the Plücker form of $\mathbb{L}$ the *Plücker quiver of* $\mathbb{L}$.

2.6. Examples of groups $\mathbb{L} \subset \mathbb{Z}^N$ and their Plücker quivers are shown in Figures 3 and 4.

2.7. Some easily visible geometric properties of a quiver without directed loops of length ≤ 2 are equivalent to easily seen algebraic properties of its anti-symmetrized adjacency matrix Q. For instance, the graph has no isolated nodes if and only if no column of Q is zero. Also, in every node of the quiver the number of incoming arrows equals the number of outgoing arrows precisely if in each row of Q the sum of the entries is 0. For Plücker quivers these properties correspond to the conditions that $\mathbb{L}$ is not contained in any of the standard coordinate hyperplanes of $\mathbb{Z}^N$ and is perpendicular to the vector $(1, \ldots, 1)$. One property of a Plücker quiver which one cannot easily read off from the graph is $\operatorname{rank} Q = 2$.

In the remainder of this section we show that these properties characterize Plücker quivers of rank 2 subgroups $\mathbb{L} \subset \mathbb{Z}^N$ as in 2.1.

2.8. Proposition. *An $N \times N$-matrix C with entries in $\mathbb{Z}$ is anti-symmetric and* $\operatorname{rank} C = 2$ *if and only if there is a $2 \times N$-matrix B with entries in $\mathbb{Z}$ such that* $\operatorname{rank} B = 2$ *and* $C = B^t \mathsf{J} B$.

Proof The "if"-statement is trivial. So let us consider the "only if" and assume that C is anti-symmetric and $\operatorname{rank} C = 2$. Let d denote the greatest common divisor of the entries of C and $C' = \frac{1}{d}C$. Choose an $N \times 2$-matrix D whose columns form a $\mathbb{Z}$-basis for the column space of C'. The equality of column spaces of C' and D means that there are an $N \times 2$-matrix E and a $2 \times N$-matrix F, both with entries in $\mathbb{Z}$ and of rank 2, such that $D = C'E$ and $C' = DF$. Then $C' = DF = C'EF = -C'^t EF = -F^t D^t EF$. So $D^t E$ is an anti-symmetric 2×2-matrix; say $D^t E = f\mathsf{J}$ with $f \in \mathbb{Z}$. In fact $f = \pm 1$ since it divides all entries of C'. If $f = 1$ we replace F by $\mathsf{J}F$. We then always have $C' = F^t \mathsf{J} F$. Let G be any 2×2-matrix with entries in $\mathbb{Z}$ and $\det G = d$. Let $B = GF$. Then $C = B^t \mathsf{J} B$ as wanted. □

2.9. Multiplying in Proposition 2.8 the matrix B from the left by a matrix from $SL_2(\mathbb{Z})$ does not change the matrix $B^t \mathsf{J} B$. So it is more natural to interpret C as the matrix of the Plücker form of the $\mathbb{Z}$-row space of B. Note that it follows from the proof of Proposition 2.8 that this space is uniquely determined if the greatest common divisor of the entries of C is 1. If on the other hand the greatest common divisor of the entries of C is $d > 1$ this space depends on the additional choice of a 2×2-matrix with entries in $\mathbb{Z}$ and determinant d.

Let us summarize the above discussion:

2.10. Theorem. *For every $\mathbb{L} \subset \mathbb{Z}^N$ as in 2.1 one has its Plücker quiver. This quiver has no isolated nodes or directed loops of length ≤ 2 and in every node the number of incoming arrows equals the number of outgoing arrows. Moreover the rank of its anti-symmetrized adjacency matrix is 2. Conversely, every quiver with these properties is the Plücker quiver of some $\mathbb{L} \subset \mathbb{Z}^N$ as in 2.1. The correspondence between such quivers and such $\mathbb{L} \subset \mathbb{Z}^N$ is one-to-one for those quivers for which the greatest common divisor of the entries of the anti-symmetrized adjacency matrix is 1 and those $\mathbb{L} \subset \mathbb{Z}^N$ for which $\mathbb{Z}^N/\mathbb{L}$ has no torsion.*

3 The secondary fan and polygon of $\mathbb{L} \subset \mathbb{Z}^N$

3.1. For $\mathbb{L} \subset \mathbb{Z}^N$ as in 2.1 we now define the *secondary fan* and the *secondary polytope*. In 4.12 we will compare this with the original definitions by Gelfand, Kapranov and Zelevinsky. The term "secondary" refers to the fact that in their theory of hypergeometric systems another polytope appears first, which is therefore called the *primary polytope*. Nonetheless both the secondary fan and the secondary polytope can most conveniently and directly be described using the lattice $\mathbb{L}$. In case the rank of $\mathbb{L}$ is 2, the constructions become particularly simple.

3.2. Definition. Let $\mathbb{L} \subset \mathbb{Z}^N$ be as in 2.1. Let $\mathbb{L}^\vee_{\mathbb{R}} := \mathrm{Hom}(\mathbb{L}, \mathbb{R})$ denote the real dual space of $\mathbb{L}$. Let $\mathbf{e}_1, \ldots, \mathbf{e}_N$ be the standard basis of $\mathbb{Z}^N$. Let $\mathbf{b}_i \in \mathbb{L}^\vee_{\mathbb{R}}$ be the image of $\mathbf{e}_i$ under the map $\mathbb{R}^N \to \mathbb{L}^\vee_{\mathbb{R}}$ dual to the inclusion $\mathbb{L} \hookrightarrow \mathbb{Z}^N$. Here and henceforth we identify $\mathbb{R}^N$ with the real dual space of $\mathbb{Z}^N$ by means of the standard dot product.

By definition, the *secondary fan of* $\mathbb{L}$ is the following collection of cones in $\mathbb{L}^\vee_{\mathbb{R}}$: the 0-dimensional cone $\{\mathbf{0}\}$, the 1-dimensional cones $\mathbb{R}_{\geq 0}\mathbf{b}_i$ $(i = 1, \ldots, N)$ and the 2-dimensional cones which are the closures of the connected components of $\mathbb{L}^\vee_{\mathbb{R}} \setminus \bigcup_{i=1}^N \mathbb{R}_{\geq 0}\mathbf{b}_i$.

3.3. Definition. With a 2-dimensional cone C in the secondary fan one associates the set

$$L_C := \{\{i,j\} \subset \{1, \ldots, N\} \mid C \subset (\mathbb{R}_{\geq 0}\mathbf{b}_i + \mathbb{R}_{\geq 0}\mathbf{b}_j)\,\} \tag{4}$$

of 2-element subsets of $\{1, \ldots, N\}$ and the vector

$$\psi_C := \sum_{\{i,j\} \in L_C} |\det(\mathbf{b}_i, \mathbf{b}_j)|\, (\mathbf{e}_i + \mathbf{e}_j)\,. \tag{5}$$

The *secondary polytope* $\Sigma(\mathbb{L})$ of $\mathbb{L} \subset \mathbb{Z}^N$ is then defined as

$$\Sigma(\mathbb{L}) := \textsf{convex hull}(\{\psi_C \mid C \text{ a 2-dim cone of the secondary fan}\})\,.$$

3.4. Remark. In §4.12 we give an interpretation of the above L_C as a triangulation of the primary polytope. In Equation (27) we associate with L_C a perfect matching in a bipartite graph.

3.5. Example. Figure 2 shows the secondary fan with the lists L_C, the secondary polytope $\Sigma(\mathbb{L})$ with coordinates for the vertices and the primary polytope for $\mathbb{L} =$

$\mathbb{Z}(0,1,1,-2) \oplus \mathbb{Z}(-1,0,2,-1) \subset \mathbb{Z}^4$. This is case B_2 of Figure 3; see also Example 4.6 and §4.12.

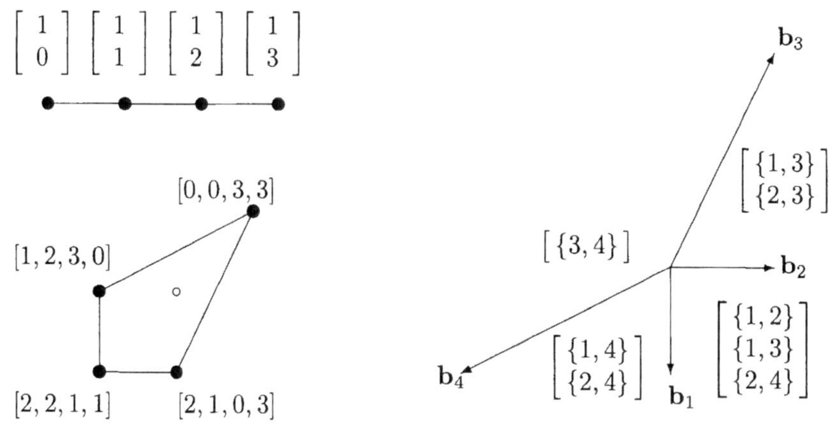

Figure 2 Secondary fan (right), primary polytope (top left) and secondary polytope (bottom left) for $\mathbb{L} = \mathbb{Z}(0,1,1,-2) \oplus \mathbb{Z}(-1,0,2,-1) \subset \mathbb{Z}^4$.

3.6. The geometric formulation of the construction of the secondary fan and polytope is quite attractive. Nonetheless in practical algorithms everything can be most easily obtained from the matrix of the Plücker form of $\mathbb{L} \subset \mathbb{Z}^N$. Indeed, a 2-dimensional cone in the secondary fan is bounded by two half-lines $\mathbb{R}_{\geq 0}\mathbf{b}_i$ and $\mathbb{R}_{\geq 0}\mathbf{b}_j$ with the property that $\det(\mathbf{b}_i, \mathbf{b}_j) > 0$ and $\det(\mathbf{b}_i, \mathbf{b}_k)\det(\mathbf{b}_j, \mathbf{b}_k) \geq 0$ for $k = 1, \ldots, N$. Thus to find the secondary fan one just needs to find all pairs $\mathbf{b}_i, \mathbf{b}_j$ with these properties.

For a cone C bounded by $\mathbb{R}_{\geq 0}\mathbf{b}_i$ and $\mathbb{R}_{\geq 0}\mathbf{b}_j$ the list L_C is then

$$L_C = \{\, \{k, l\} \mid \det(\mathbf{b}_k, \mathbf{b}_i) \geq 0 \text{ and } \det(\mathbf{b}_j, \mathbf{b}_l) \geq 0 \,\}.$$

Now consider two adjacent 2-dimensional cones in the secondary fan, say C and C'. Let $\mathbf{b}_{i_1}, \ldots, \mathbf{b}_{i_s}$ be those vectors from the set $\{\mathbf{b}_1, \ldots, \mathbf{b}_N\}$ that lie on the half-line $C \cap C'$. Without loss of generality we may assume C lies to the right of $\mathbf{b}_{i_1}$ and C' to the left. Then

$$\begin{aligned} L_C \setminus (L_C \cap L_{C'}) &= \{\, \{k, l\} \mid l \in \{i_1, \ldots, i_s\},\ \det(\mathbf{b}_l, \mathbf{b}_k) < 0 \,\}, \\ L_{C'} \setminus (L_C \cap L_{C'}) &= \{\, \{k, l\} \mid l \in \{i_1, \ldots, i_s\},\ \det(\mathbf{b}_l, \mathbf{b}_k) > 0 \,\}. \end{aligned}$$

From this we see, using $\sum_{k=1}^{N} \det(\mathbf{b}_i, \mathbf{b}_k) = 0$ for all i, that

$$\psi_{C'} - \psi_C = \sum_{r=1}^{s} \sum_{k=1}^{N} \det(\mathbf{b}_{i_r}, \mathbf{b}_k)\mathbf{e}_k\,. \tag{6}$$

In other words, $\psi_{C'} - \psi_C$ is the sum of rows $i_1, \ldots, i_s$ of the matrix of the Plücker form.

3.7. To make this even more explicit and simple looking we take a basis for $\mathbb{L}$ compatible with the chosen orientation. One can represent the two basis vectors as the rows of a $2 \times N$-matrix B. Let $\mathbf{b}_1, \ldots, \mathbf{b}_N$ be the columns of this matrix. The elements of $\mathbb{L}$ should now be written as row vectors with two components and the embedding $\mathbb{L} \hookrightarrow \mathbb{Z}^N$ is given by $\mathbf{v} \mapsto \sum_{k=1}^{N}(\mathbf{v} \cdot \mathbf{b}_k)\mathbf{e}_k$. Since $\det(\mathbf{b}_i, \mathbf{b}_k) = \mathbf{b}_i^t \mathsf{J} \mathbf{b}_k$, we can reformulate Equation (6) as

$$\psi_{C'} - \psi_C = \text{image of} \sum_{r=1}^{s} \mathbf{b}_{i_r}^t \mathsf{J} \text{ under the embedding } \mathbb{L} \hookrightarrow \mathbb{Z}^N . \tag{7}$$

In order to obtain the simplest formulation for the construction we order the vectors $\mathbf{b}_1, \ldots, \mathbf{b}_N$ so that the points $\mathbf{p}_k = \sum_{i=1}^{k} \mathbf{b}_i^t$ for $k = 1, \ldots, N$ lie ordered counterclockwise on the boundary of the polygon

$$\Delta = \mathsf{convex\ hull}\{\mathbf{p}_1, \ldots, \mathbf{p}_N\} . \tag{8}$$

Then the secondary polygon $\Sigma(\mathbb{L})$ is obtained by first rotating Δ clockwise by $90°$, next embedding it along with $\mathbb{L}$ into $\mathbb{Z}^N$ and finally translating it over the vector ψ_C, where C is the cone in the secondary fan with left hand boundary $\mathbb{R}_{\geq 0}\mathbf{b}_1$.

Note that while the secondary polytope $\Sigma(\mathbb{L})$ depends only on the embedding $\mathbb{L} \hookrightarrow \mathbb{Z}^N$, the polygon Δ usually changes when one puts another vector $\mathbf{b}_i$ in first position by a cyclic permutation or when one multiplies the vectors $\mathbf{b}_1, \ldots, \mathbf{b}_N$ by a matrix from $SL_2(\mathbb{Z})$.

3.8. For a converse to the above construction we start from a convex polygon Δ in $\mathbb{R}^2$ with vertices in $\mathbb{Z}^2$ and non-empty interior. Let $\partial\Delta$ denote its boundary. Next we choose a collection of points $\mathbf{p}_1, \ldots, \mathbf{p}_N$ in $\partial\Delta \cap \mathbb{Z}^2$ which includes all vertices of Δ. We number these points so that $\mathbf{p}_i$ and $\mathbf{p}_{i+1}$ are consecutive points in the counterclockwise orientation of $\partial\Delta$. Thinking of the elements of $\mathbb{Z}^2$ as row vectors we define column vectors $\mathbf{b}_1 = \mathbf{p}_1^t$ and $\mathbf{b}_i = \mathbf{p}_i^t - \mathbf{p}_{i-1}^t$ for $i = 2, \ldots, N$. Finally we define $\mathbb{L}$ to be the $\mathbb{Z}$-row space of the $2 \times N$-matrix B with columns $\mathbf{b}_1, \ldots, \mathbf{b}_N$.

Thus every convex polygon Δ in $\mathbb{R}^2$ with vertices in $\mathbb{Z}^2$ and non-empty interior can be viewed as the secondary polygon of some rank 2 subgroup $\mathbb{L} \subset \mathbb{Z}^N$ satisfying the conditions in 2.1.

Note however that in general there are several possible choices for the points $\mathbf{p}_1, \ldots, \mathbf{p}_N$ in $\partial\Delta \cap \mathbb{Z}^2$. The minimal choice takes only the vertices of Δ, while the maximal choice takes all points of $\partial\Delta \cap \mathbb{Z}^2$.

3.9. Theorem. *The (Euclidean) area of the polygon Δ in (8) is*

$$\text{area}\,\Delta = \tfrac{1}{2}\sum_{1\leq i<j\leq N} |\det(\mathbf{b}_i, \mathbf{b}_j)| - \sum_{\{i,j\}\in L_C} |\det(\mathbf{b}_i, \mathbf{b}_j)| \tag{9}$$

for every 2-dimensional cone C in the secondary fan.

Proof After applying, if necessary, a cyclic permutation to $\mathbf{b}_1, \ldots, \mathbf{b}_N$ we may assume, without loss of generality, that $C = \mathbb{R}_{\geq 0}\mathbf{b}_1 + \mathbb{R}_{\geq 0}\mathbf{b}_N$.

From the triangulation of Δ by the diagonals between the vertex $\mathbf{p}_N = \mathbf{0}$ and the other vertices of Δ one sees that area $\Delta = \frac{1}{2}\sum_{1\leq i<j\leq M} \det(\mathbf{b}_i, \mathbf{b}_j)$ where M is such that $\mathbf{b}_j \notin \mathbb{R}_{\geq 0}\mathbf{b}_N$ for $1 \leq j \leq M$ and $\mathbf{b}_j \in \mathbb{R}_{\geq 0}\mathbf{b}_N$ for $M+1 \leq j \leq N$. The fact that $\sum_{i=1}^{N} \mathbf{b}_i = \mathbf{0}$ implies

$$\sum_{1\leq i\leq M<j\leq N} \det(\mathbf{b}_i, \mathbf{b}_j) = \det(\sum_{1\leq i\leq M} \mathbf{b}_i, \sum_{M<j\leq N} \mathbf{b}_j) = 0 .$$

Thus we see

$$\text{area}\,\Delta = \tfrac{1}{2}\sum_{1\leq i<j\leq N} \det(\mathbf{b}_i, \mathbf{b}_j) .$$

Next note that for $1 \leq i < j \leq N$ the inequality $\det(\mathbf{b}_i, \mathbf{b}_j) < 0$ holds if and only if $C \subset \mathbb{R}_{\geq 0}\mathbf{b}_i + \mathbb{R}_{\geq 0}\mathbf{b}_j$, if and only if $\{i,j\} \in L_C$. Equality (9) now follows immediately. □

3.10. Corollary. *The number of lattice points in the interior of the polygon* Δ *in (8) is*

$$\tfrac{1}{2}\left(\sum_{1\leq i<j\leq N} |\det(\mathbf{b}_i, \mathbf{b}_j)| - N\right) - \sum_{\{i,j\}\in L_C} |\det(\mathbf{b}_i, \mathbf{b}_j)| + 1 .$$

Proof This follows from Theorem 3.9 in combination with *Pick's Formula* ([7] p.113): $2\text{area}\,\Delta = 2\sharp\left(\mathbb{Z}^2 \cap \text{interior}\,\Delta\right) + \sharp\left(\mathbb{Z}^2 \cap \text{boundary}\,\Delta\right) - 2.$ □

3.11. Example. Figures 3 and 4 show some examples of the above relation between quivers, rank 2 subgroups $\mathbb{L}$ of $\mathbb{Z}^N$ as in 2.1 and polygons. The quivers are taken from [3]. We name our examples neutrally as $B_1, \ldots, B_{10}$, but also mention the names given to these quiver models in [3]. The latter names refer to del Pezzo surfaces dP_k, i.e. the projective plane $\mathbb{P}^2$ blown up in k points, some of which, on the present occasion, may happen to be "infinitely near" and therefore require repeated blow-ups. With the latter somewhat liberal use of the name *del Pezzo surface*, dP_k matches well with the toric geometry of the fan whose 1-dimensional rays are the half-lines from the interior lattice point $\circ$ through a lattice point $\bullet$ at the boundary. Note however that this fan is in general not the same as the secondary fan and that therefore the singularity associated with the polygon (cf. [12, 19]) is not the singularity obtained by contracting to a point the zero-section of the canonical bundle of the del Pezzo surface.

The polygons in Figures 3 and 4, however, are in general different from the polygons associated with the same quiver in [3] Figure 8. We will clarify this issue in 6.14, 6.15, 6.16.

It is remarkable that the polygons in Figures 3 and 4 are exactly the lattice polygons with one interior lattice point and ≤ 6 lattice points on the boundary. There are exactly 16 lattice polygons with one interior lattice point. The remaining 6 are shown in Figure 5. A nice review of these so-called *reflexive polygons* can be found in [21].

In cases B_1, B_3 and B_4 the greatest common divisor d of the entries in the quiver's adjacency matrix is > 1 and the polygon in these cases depends on an integer matrix with determinant d. We have chosen this matrix such that the polygon is a reflexive polygon.

4 GKZ hypergeometric systems

4.1. In the late 1980's Gelfand, Kapranov and Zelevinsky discovered fascinating generalizations of the classical hypergeometric structures of Euler, Gauss, Appell, Lauricella, Horn [8, 9, 10, 11, 25]. The main ingredient for these new hypergeometric structures is a finite sequence $\mathcal{A} = (\mathbf{a}_1, \ldots, \mathbf{a}_N)$ of vectors in $\mathbb{Z}^{k+1}$ which generates $\mathbb{Z}^{k+1}$ as an abelian group and for which there exists a group homomorphism $h : \mathbb{Z}^{k+1} \to \mathbb{Z}$ such that $h(\mathbf{a}_i) = 1$ for all i. The latter condition means that $\mathcal{A}$ lies in a k-dimensional affine hyperplane in $\mathbb{Z}^{k+1}$. Figure 6 shows $\mathcal{A}$ (each black dot represents one vector) sitting in this hyperplane for some classical hypergeometric structures. Gelfand, Kapranov and Zelevinsky called these new structures *$\mathcal{A}$-hypergeometric systems*, but nowadays most authors call them *GKZ hypergeometric systems*.

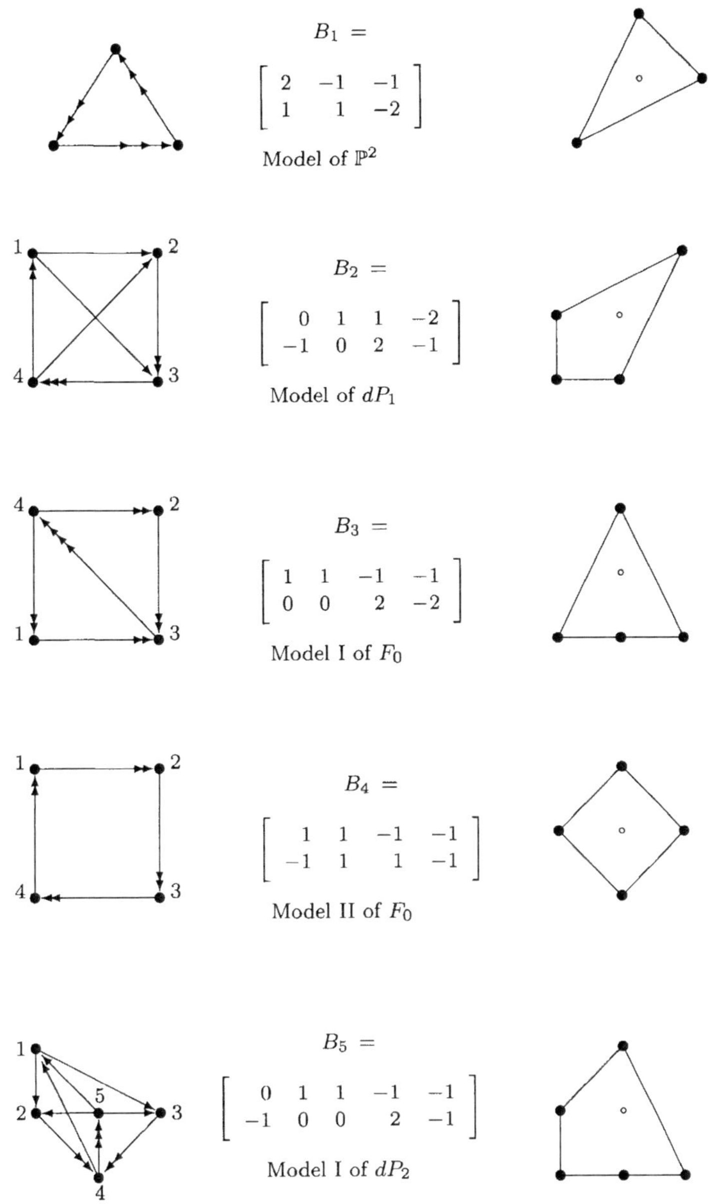

Figure 3 Quivers from [3] Figures 10, 11, 4, 12 and corresponding polygons.

Let $\mathbb{L}$ denote the lattice (= free abelian group) of linear relations in $\mathcal{A}$:

$$\mathbb{L} := \{(\ell_1, \ldots, \ell_N) \in \mathbb{Z}^N \mid \ell_1 \mathbf{a}_1 + \cdots + \ell_N \mathbf{a}_N = 0\}. \tag{10}$$

It follows from the above construction and assumptions that the quotient group $\mathbb{Z}^N/\mathbb{L}$ is $\mathbb{Z}^{k+1}$, and a fortiori it is torsion-free.

In this paper we consider only the case when the rank of $\mathbb{L}$ is 2; i.e. $k+1 = N-2$. Since we assumed the existence of a linear map $h : \mathbb{Z}^{N-2} \to \mathbb{Z}$ such that $h(\mathbf{a}_i) = 1$

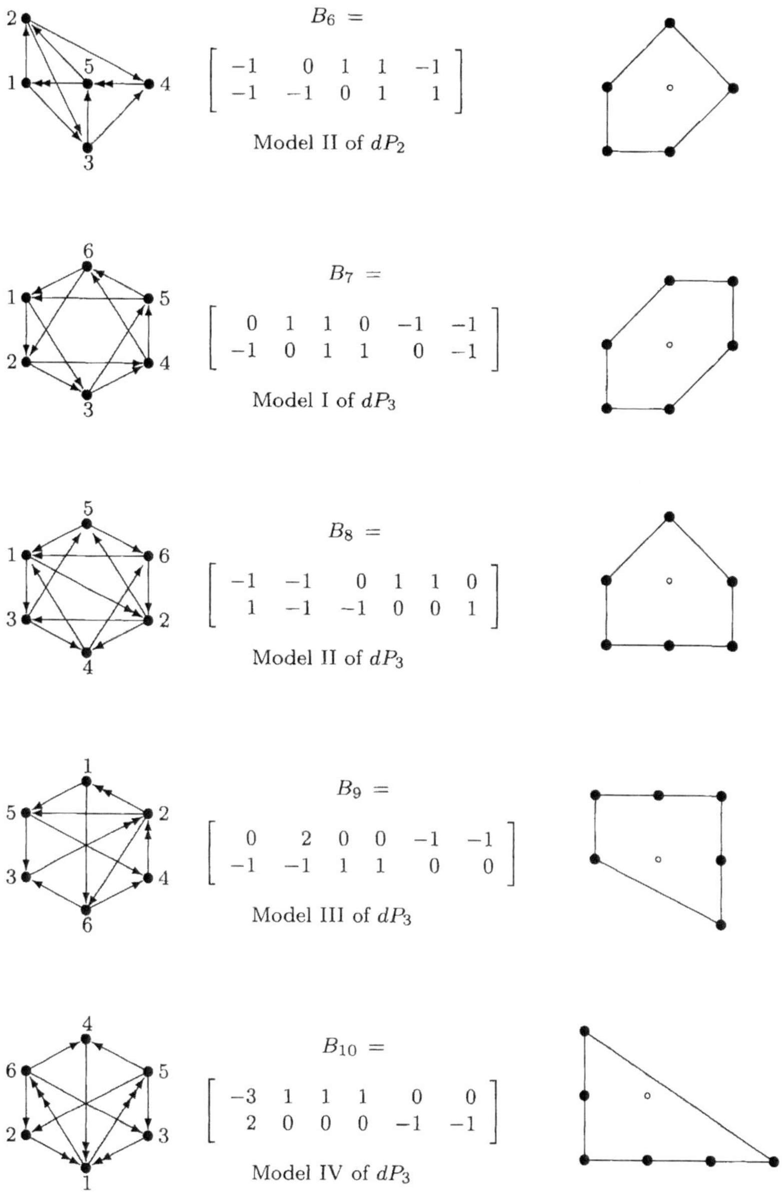

$$B_6 = \begin{bmatrix} -1 & 0 & 1 & 1 & -1 \\ -1 & -1 & 0 & 1 & 1 \end{bmatrix}$$

Model II of dP_2

$$B_7 = \begin{bmatrix} 0 & 1 & 1 & 0 & -1 & -1 \\ -1 & 0 & 1 & 1 & 0 & -1 \end{bmatrix}$$

Model I of dP_3

$$B_8 = \begin{bmatrix} -1 & -1 & 0 & 1 & 1 & 0 \\ 1 & -1 & -1 & 0 & 0 & 1 \end{bmatrix}$$

Model II of dP_3

$$B_9 = \begin{bmatrix} 0 & 2 & 0 & 0 & -1 & -1 \\ -1 & -1 & 1 & 1 & 0 & 0 \end{bmatrix}$$

Model III of dP_3

$$B_{10} = \begin{bmatrix} -3 & 1 & 1 & 1 & 0 & 0 \\ 2 & 0 & 0 & 0 & -1 & -1 \end{bmatrix}$$

Model IV of dP_3

Figure 4 Quivers from [3] Figures 12, 9 and corresponding polygons.

for all i, $\mathbb{L}$ lies in the kernel of the map $\mathbf{s}$ defined by:

$$\mathbf{s} : \mathbb{R}^N \to \mathbb{R}, \qquad \mathbf{s}(z_1, \ldots, z_N) = z_1 + \cdots + z_N . \tag{11}$$

The following lemma shows what the other condition for $\mathbb{L}$ in 2.1 means in terms of $\mathcal{A}$.

4.2. Proposition. *In the situation of 4.1 the following statements are equivalent:*

1. *$\mathbb{L}$ is not contained in any of the standard coordinate hyperplanes of $\mathbb{Z}^N$.*

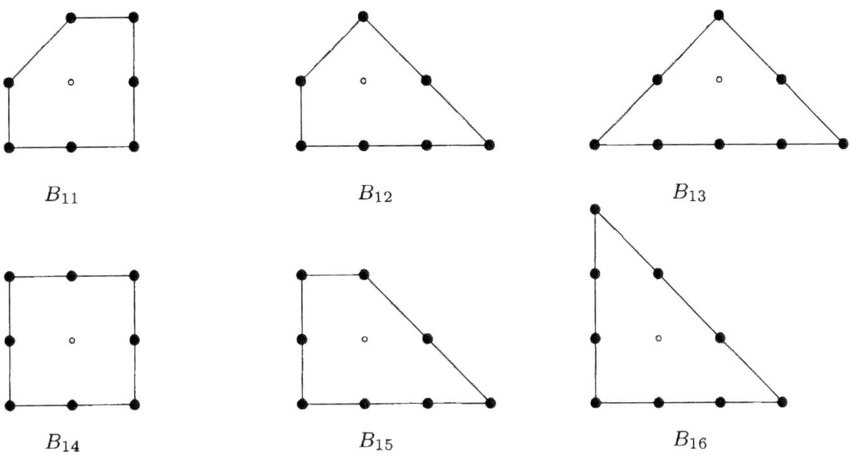

Figure 5 Remaining reflexive polygons.

2. *No vector in the sequence $\mathcal{A}$ is linearly independent of the other vectors.*
3. *For every i the lattice of linear relations between the vectors of $\mathcal{A} \setminus \{\mathbf{a}_i\}$ has rank 1.*

Proof Assume $\mathbb{L}$ is contained in the i-th coordinate hyperplane. Then the i-th component of every element of $\mathbb{L}$ is 0 and hence $\mathbf{a}_i$ occurs in no linear relation for the vectors in $\mathcal{A}$. In other words, $\mathbb{L}$ is the lattice of linear relations between the vectors of $\mathcal{A} \setminus \{\mathbf{a}_i\}$. Since $\mathbb{L}$ has rank 2, we conclude that statement 3 implies statement 1.

Conversely, assume that the lattice $\mathbb{L}_i$ of linear relations between the vectors of $\mathcal{A} \setminus \{\mathbf{a}_i\}$ has rank > 1. Since $\mathbb{L}_i$ is contained in $\mathbb{L}$ and the latter has rank 2, we see that $\mathbb{L}_i \otimes \mathbb{R} = \mathbb{L} \otimes \mathbb{R}$. Therefore, since $\mathbb{L}_i$ is contained in the i-th coordinate hyperplane, so is $\mathbb{L}$. Hence statement 1 implies statement 3.

It is obvious that statements 1 and 2 are equivalent. □

4.3. Definition. A subsequence $\mathcal{A}'$ of $\mathcal{A}$ is said to be *minimally dependent* if the vectors in $\mathcal{A}'$ are linearly dependent and the vectors in every subsequence $\mathcal{A}''$ of $\mathcal{A}'$ with $\mathcal{A}'' \neq \mathcal{A}'$ are linearly independent. A linear dependence relation $\sum_j \alpha_j \mathbf{a}_j = 0$ is a *minimal linear dependence relation* in $\mathcal{A}$ if the subsequence $(\mathbf{a}_j \mid \alpha_j \neq 0)$ is minimally dependent.

4.4. Proposition.

1. *Let B be a $2 \times N$-matrix such that its rows are a $\mathbb{Z}$-basis for $\mathbb{L}$. Let $\mathbf{b}_1, \ldots, \mathbf{b}_N$ be the columns of this matrix. Then one has for every i*

$$\sum_{j=1}^{N} \det(\mathbf{b}_i, \mathbf{b}_j)\, \mathbf{a}_j \;=\; \mathbf{0}\,. \tag{12}$$

2. *Assume the conditions in Proposition 4.2 are satisfied. Then every linear relation (12) is minimal and every minimal linear dependence relation in $\mathcal{A}$ is a non-zero scalar multiple of some relation (12).*

Proof For statement 1, let A be the matrix with columns $\mathbf{a}_1, \ldots, \mathbf{a}_N$. Then Equation (10) can be rewritten as $BA^t = 0$. This implies $B^t \mathsf{J} BA^t = 0$, which is just a compressed form of the relations we wanted to prove.

For statement 2 note that condition 3 in Proposition 4.2 implies that every linear relation (12) is minimal. Conversely, consider a minimal linear relation $\sum_j \alpha_j \mathbf{a}_j = 0$. Then the subsequence $(\mathbf{a}_j \mid \alpha_j \neq 0)$ has at most $N-1$ elements. The given minimal relation is therefore a non-zero scalar multiple of at least one relation (12). □

4.5. Gelfand, Kapranov and Zelevinsky [8, 9, 10] associate with a set $\mathcal{A}$ as in 4.1 and a vector $\mathbf{c} \in \mathbb{C}^{k+1}$ the following system of partial differential equations for functions Φ of N variables $u_1, \ldots, u_N$:

- for every $(\ell_1, \ldots, \ell_N) \in \mathbb{L}$ one differential equation

$$\prod_{\ell_i<0} \left(\frac{\partial}{\partial u_i}\right)^{-\ell_i} \Phi = \prod_{\ell_i>0} \left(\frac{\partial}{\partial u_i}\right)^{\ell_i} \Phi\,, \tag{13}$$

- the system of $k+1$ differential equations

$$\mathbf{a}_1\, u_1 \frac{\partial \Phi}{\partial u_1} + \ldots + \mathbf{a}_N\, u_N \frac{\partial \Phi}{\partial u_N} = \mathbf{c}\Phi\,. \tag{14}$$

4.6. Example. The sequence $\mathcal{A} = (\mathbf{a}_1, \mathbf{a}_2, \mathbf{a}_3, \mathbf{a}_4) \in \mathbb{Z}^2$ for Example B_2 in Figure 3 can be taken to be

$$\mathbf{a}_1 = \begin{bmatrix} 1 \\ 0 \end{bmatrix}, \quad \mathbf{a}_3 = \begin{bmatrix} 1 \\ 1 \end{bmatrix}, \quad \mathbf{a}_4 = \begin{bmatrix} 1 \\ 2 \end{bmatrix}, \quad \mathbf{a}_2 = \begin{bmatrix} 1 \\ 3 \end{bmatrix}.$$

It is a classical result of K. Mayr that the roots of the 1-variable cubic polynomial $P(x) = u_1 + u_3 x + u_4 x^2 + u_2 x^3$ as functions of the coefficients $u_1, \ldots, u_4$ satisfy the GKZ system of differential equations for this $\mathcal{A}$ and with $\mathbf{c} = \begin{bmatrix} 0 \\ -1 \end{bmatrix}$; see e.g. [25] §2.1.

4.7. Example. An important example in the toric geometry constructions of Sasaki-Einstein manifolds is known under the name $L^{a,b,c}$; see e.g. [6]. Here a, b, c are integers with $c \leq b$ and $0 < a \leq b$. The polygon for this example, displayed in [6] Figure 2, is a quadrangle with vertices $(0,0)$, $(1,0)$, (ak, b), $(-al, c)$ where k and l are integers such that $ck + bl = 1$. The method explained in 3.8 yields $\mathbb{L} = \mathbb{Z}(1, ak-1, -al-ak, al) \oplus \mathbb{Z}(0, b, c-b, -c)$. From this we see that we can take $\mathcal{A} = (\mathbf{a}_1, \mathbf{a}_2, \mathbf{a}_3, \mathbf{a}_4) \in \mathbb{Z}^2$ with

$$\mathbf{a}_1 = \begin{bmatrix} 1 \\ c-a \end{bmatrix}, \quad \mathbf{a}_2 = \begin{bmatrix} 1 \\ c \end{bmatrix}, \quad \mathbf{a}_3 = \begin{bmatrix} 1 \\ 0 \end{bmatrix}, \quad \mathbf{a}_4 = \begin{bmatrix} 1 \\ b \end{bmatrix}.$$

As in Example 4.6 this means that the corresponding GKZ system deals with the roots of the "four-nomial" $u_1 x^{c-a} + u_2 x^c + u_3 + u_4 x^b$ as functions of the coefficients u_1, u_2, u_3, u_4.

According to [6] §3 the examples $Y^{p,q}$ of [1, 19] are special cases of the above: $Y^{p,q} = L^{p-q,p+q,p}$. So, these correspond to the "four-nomials" $u_1 x^q + u_2 x^p + u_3 + u_4 x^{p+q}$. Thus Example 4.6 is in fact $Y^{2,1} = L^{1,3,2}$.

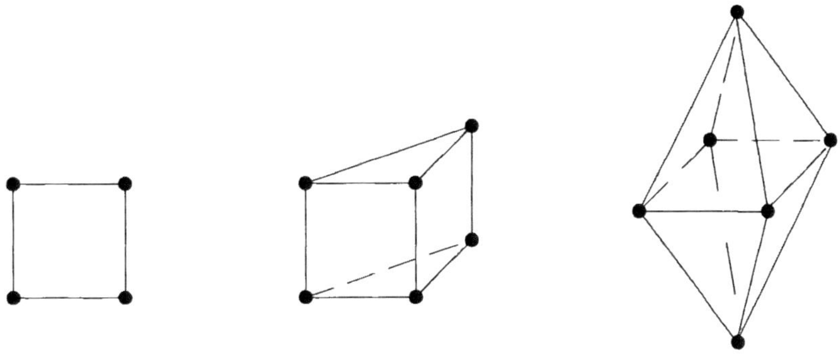

Figure 6 The sets $\mathcal{A}$ for Gauss's hypergeometric functions (left), Appell's F_1 (middle) and Appell's F_4 (right).

4.8. Example. From the pictures of $\mathcal{A}$ shown in Figure 6 one easily sees that for the Gauss system the lattice $\mathbb{L}$ is generated by the vector $(1,1,-1,-1)$ and thus has rank 1 and is not of interest here. For Appell's F_1 and F_4 the lattice $\mathbb{L}$ has rank 2 and the corresponding B-matrices are

$$F_1 : \begin{bmatrix} 1 & -1 & 0 & -1 & 1 & 0 \\ 1 & 0 & -1 & -1 & 0 & 1 \end{bmatrix}, \quad F_4 : \begin{bmatrix} -1 & -1 & 1 & 1 & 0 & 0 \\ -1 & -1 & 0 & 0 & 1 & 1 \end{bmatrix}.$$

Thus Appell's F_1 corresponds to B_7 in Figure 4. For Appell's F_4 the quiver and the polygon are

In [8] §3.2 Gelfand, Kapranov and Zelevinsky discuss how the 14 complete hypergeometric Horn series in two variables fit their theory. It is an amusing simple exercise to now find the corresponding quivers and polygons. It turns out that, with the exception of F_4, all Horn series give a quiver listed in Example 3.11:

$$\{G_3\} \leftrightarrow B_2\,, \quad \{H_5\} \leftrightarrow B_5\,, \quad \{G_1,\, H_3,\, H_6\} \leftrightarrow B_6\,, \quad \{F_1,\, G_2\} \leftrightarrow B_7\,,$$
$$\{H_1\} \leftrightarrow B_8\,, \quad \{H_4,\, H_7\} \leftrightarrow B_9\,, \quad \{F_2,\, F_3,\, H_2\} \leftrightarrow B_{11}\,.$$

4.9. In 4.1 we started from the sequence of vectors $\mathcal{A}$ and then defined $\mathbb{L}$ via Equation (10). Let us reverse the procedure and start with a rank 2 subgroup $\mathbb{L} \subset \mathbb{Z}^N$ as in 2.1. We can now define

$$\begin{aligned} \mathcal{A} &= (\mathbf{a}_1, \ldots, \mathbf{a}_N) \qquad \text{with} \\ \mathbf{a}_i &= \text{the class } \mathbf{e}_i \bmod \mathbb{L} \text{ in the quotient group } \mathbb{Z}^N/\mathbb{L}\,. \end{aligned} \tag{15}$$

If the quotient group $\mathbb{Z}^N/\mathbb{L}$ is torsion free, it is isomorphic to $\mathbb{Z}^{N-2}$ and the sequence of vectors $\mathcal{A}$ in (15) satisfies the requirements of 4.1. *Thus (10) and (15) give equivalences, inverse to each other, between the data $\mathcal{A}$ and $\mathbb{L}$.*

In combination with Theorem 2.10 this gives

4.10. Theorem. *There is an equivalence of data between on the one hand sequences $\mathcal{A}$ as in 4.1 which satisfy also the conditions in Proposition 4.2 and on the other hand quivers without isolated nodes or directed loops of length ≤ 2, such that in every node the number of incoming arrows equals the number of outgoing arrows and such that for its anti-symmetrized adjacency matrix the rank is 2 and the greatest common divisor of the entries is 1.*

4.11. Remark. We refer to 5.5 for a discussion of how to extend the definition of GKZ systems so as to allow for torsion in the group $\mathbb{Z}^N/\mathbb{L}$.

4.12. In order to have an efficient method to construct bases for the solution space of the hypergeometric system (13)-(14) Gelfand, Kapranov and Zelevinsky developed the theory of the *secondary fan* and the *secondary polytope*. The term "secondary" refers to the habit of considering the set $\mathcal{A}$ as primary data and to call the convex hull of $\mathcal{A}$ the *primary polytope*.

In case the lattice $\mathbb{L}$ of relations in $\mathcal{A}$ has rank 2, the constructions become particularly simple. The secondary fan is the one in Definition 3.2. With a 2-dimensional cone C in the secondary fan one associates a set L_C of 2-element subsets of $\{1,\ldots,N\}$ as in Equation (4). One can interpret this L_C as a triangulation of the primary polytope $\mathsf{convex\ hull}(\mathcal{A})$ as follows. Recall that $\mathcal{A} = (\mathbf{a}_1,\ldots,\mathbf{a}_N)$ and assume for convenience of imagining pictures that all $\mathbf{a}_i$'s are different. For $\{i,j\} \in L_C$ set $T_{\{i,j\}} := \mathsf{convex\ hull}(\{\mathbf{a}_k \mid k \neq i,j\})$. This $T_{\{i,j\}}$ is an $(N-3)$-dimensional simplex and the simplices $T_{\{i,j\}}$ with $\{i,j\} \in L_C$ together constitute a *triangulation of the primary polytope.*

In [11] p.220 the *secondary polytope* is constructed (and defined) in terms of these triangulations as follows. For a 2-element subset $\{i,j\} \subset \{1,\ldots,N\}$ with $i<j$ let A_{ij} denote the $(N-2)\times(N-2)$-matrix with columns $\mathbf{a}_k$ for $k \neq i,j$ in the natural order of increasing indices. To a 2-dimensional cone C in the secondary fan one then associates the vector

$$\varphi_C := \sum_{\{i,j\}\in L_C} |\det A_{ij}| \sum_{k\neq i,j} \mathbf{e}_k\,. \tag{16}$$

Then in [11] p.220 the *secondary polytope* $\Sigma(\mathcal{A})$ is defined as

$$\Sigma(\mathcal{A}) := \mathsf{convex\ hull}(\{\varphi_C \mid C \text{ a 2-dim cone of the secondary fan}\})\,. \tag{17}$$

As $|\det A_{ij}|$ is $(N-3)!$ times the Euclidean volume of the simplex $T_{\{i,j\}}$, the number $\mathrm{vol}_{\mathcal{A}} := \sum_{\{i,j\}\in L_C} |\det A_{ij}|$ is for every C equal to $(N-3)!$ times the Euclidean volume of the primary polytope $\mathsf{convex\ hull}(\mathcal{A})$ and the point

$$\frac{1}{(N-2)\mathrm{vol}_{\mathcal{A}}} \sum_{\{i,j\}\in L_C} |\det A_{ij}| \sum_{k\neq i,j} \mathbf{a}_k$$

coincides for every C with the barycentre of the primary polytope. The secondary polytope therefore lies in a 2-dimensional plane parallel to $\mathbb{L}_{\mathbb{R}}$.

4.13. Let us compare $\Sigma(\mathcal{A})$ with the polytope $\Sigma(\mathbb{L})$ of Definition 3.3. It is a well-known and easy to prove fact that $|\det A_{ij}| = |\det(\mathbf{b}_i,\mathbf{b}_j)|$ for all $\{i,j\} \in L_C$ (see e.g. [25] Eq. (62)). Thus Equations (16) and (5) yield

$$\varphi_C = -\psi_C + \mathrm{vol}_{\mathcal{A}} \cdot \sum_{k=1}^{N} \mathbf{e}_k \,, \tag{18}$$

with

$$\mathrm{vol}_{\mathcal{A}} = \sum_{\{i,j\}\in L_C} |\det A_{ij}| = \sum_{\{i,j\}\in L_C} |\det(\mathbf{b}_i, \mathbf{b}_j)| \,. \tag{19}$$

This means that the two secondary polytopes $\Sigma(\mathcal{A})$ and $\Sigma(\mathbb{L})$ are related by a point symmetry with centre $\frac{1}{2}\mathrm{vol}_{\mathcal{A}} \cdot \sum_{k=1}^{N} \mathbf{e}_k$:

$$\Sigma(\mathcal{A}) = -\Sigma(\mathbb{L}) + \mathrm{vol}_{\mathcal{A}} \cdot \sum_{k=1}^{N} \mathbf{e}_k \,. \tag{20}$$

5 About the solutions to GKZ systems

5.1. The variables in GKZ theory (see Section 4) are the natural coordinates on the space $\mathbb{C}^{\mathcal{A}} := \mathrm{Maps}(\mathcal{A}, \mathbb{C})$ of maps from $\mathcal{A}$ to $\mathbb{C}$. The torus $\mathbb{T}^{k+1} := \mathrm{Hom}(\mathbb{Z}^{k+1}, \mathbb{C}^*)$ of group homomorphisms from $\mathbb{Z}^{k+1}$ to $\mathbb{C}^*$ acts naturally on $\mathbb{C}^{\mathcal{A}}$ and on the functions on this space: for $\sigma \in \mathbb{T}^{k+1}$, $\mathbf{u} \in \mathbb{C}^{\mathcal{A}}$, $\mathbf{a} \in \mathcal{A}$ and $\Phi : \mathbb{C}^{\mathcal{A}} \to \mathbb{C}$:

$$(\sigma \cdot \mathbf{u})(\mathbf{a}) = \sigma(\mathbf{a})\mathbf{u}(\mathbf{a}) \,, \qquad (\Phi \cdot \sigma)(\mathbf{u}) = \Phi(\sigma \cdot \mathbf{u}) \,. \tag{21}$$

The GKZ hypergeometric functions associated with $\mathcal{A}$ and $\mathbf{c}$ are defined on open domains in $\mathbb{C}^{\mathcal{A}}$. One easily sees that if a function Φ on $\mathbb{C}^{\mathcal{A}}$ satisfies the differential equations (13) then for every $\sigma \in \mathbb{T}^{k+1}$ the function $\Phi \cdot \sigma$ also satisfies these differential equations. So, the torus $\mathbb{T}^{k+1}$ acts on the solution space of the system of differential equations (13).

On the other hand, if Φ_1 and Φ_2 are two functions which satisfy the differential equations (14) with the same $\mathbf{c}$, then their quotient $\Psi = \frac{\Phi_1}{\Phi_2}$ satisfies

$$\mathbf{a}_1 \, u_1 \frac{\partial \Psi}{\partial u_1} + \cdots + \mathbf{a}_N \, u_N \frac{\partial \Psi}{\partial u_N} = \mathbf{0} \,.$$

The latter equation is equivalent to Ψ being $\mathbb{T}^{k+1}$-invariant. Thus we find:

All quotients of pairs of solutions of the system (13)-(14) are functions on simply connected open subsets of the orbit space $\mathbb{C}^{\mathcal{A}}/\mathbb{T}^{k+1}$. In the case of interest in the present paper the dimension of this orbit space is $N - k - 1 = 2$.

For $\mathbf{c} \in \mathbb{Z}^{k+1}$ the differential equations (14) are equivalent with Φ transforming under the action of $\mathbb{T}^{k+1}$ according to the character given by $\mathbf{c}$:

$$\Phi \cdot \sigma = \sigma(\mathbf{c})\Phi \,. \tag{22}$$

In particular for $\mathbf{c} = \mathbf{0}$ all solutions of (14) are $\mathbb{T}^{k+1}$-invariant.

5.2. The space $(\mathbb{C}^*)^{\mathcal{A}} := \mathrm{Maps}(\mathcal{A}, \mathbb{C}^*)$ of maps from $\mathcal{A}$ to $\mathbb{C}^*$ is a torus of dimension N which contains $\mathbb{T}^{k+1}$ as a subtorus. The action (21) of $\sigma \in \mathbb{T}^{k+1}$ on $\mathbb{C}^{\mathcal{A}}$ obviously restricts to the action of $\mathbb{T}^{k+1}$ on $(\mathbb{C}^*)^{\mathcal{A}}$ by multiplication. The quotient space $(\mathbb{C}^*)^{\mathcal{A}}/\mathbb{T}^{k+1}$, which is a subspace of $\mathbb{C}^{\mathcal{A}}/\mathbb{T}^{k+1}$, is the torus $\mathrm{Hom}(\mathbb{L}, \mathbb{C}^*)$ of group homomorphisms from $\mathbb{L}$ to $\mathbb{C}^*$. The toric variety associated with the secondary fan (cf. Definition 3.2) gives a compactification of the torus $\mathrm{Hom}(\mathbb{L}, \mathbb{C}^*)$; see [7] for the general theory of toric varieties. It is an essential part of the GKZ philosophy that quotients of hypergeometric functions should be viewed as being defined on open subsets of this toric variety.

5.3. Theorem. (cf. [8] Theorems 2 and 5, [8]′, [27] Prop. 13.5) *Let*

$$\begin{aligned}\mathbb{N}\mathcal{A} &= \{x_1\mathbf{a}_1 + \cdots + x_N\mathbf{a}_N \in \mathbb{R}^{k+1} \mid \forall x_i \in \mathbb{Z}_{\geq 0}\}, \\ \mathbb{Z}\mathcal{A} &= \{x_1\mathbf{a}_1 + \cdots + x_N\mathbf{a}_N \in \mathbb{R}^{k+1} \mid \forall x_i \in \mathbb{Z}\}, \\ \mathrm{pos}(\mathcal{A}) &= \{x_1\mathbf{a}_1 + \cdots + x_N\mathbf{a}_N \in \mathbb{R}^{k+1} \mid \forall x_i \in \mathbb{R}_{\geq 0}\}.\end{aligned}$$

Assume $\mathbb{N}\mathcal{A} = \mathbb{Z}\mathcal{A} \cap \mathrm{pos}(\mathcal{A})$, *then the dimension of the space of solutions of the system of differential equations (13)-(14) at a general point of* $\mathbb{C}^{\mathcal{A}}$ *equals the number* $\mathrm{vol}_{\mathcal{A}}$ *in Equation (19).*

5.4. Remark. As shown in the proof of [27] Prop. 13.15 the condition $\mathbb{N}\mathcal{A} = \mathbb{Z}\mathcal{A}\cap\mathrm{pos}(\mathcal{A})$ in the above theorem is satisfied if $\mathcal{A}$ admits a *unimodular triangulation.* The latter condition is equivalent to: there is a cone C in the secondary fan such that $|\det(\mathbf{b}_i, \mathbf{b}_j)| = 1$ for all $\{i, j\} \in L_C$ (see Definitions 3.2 and 3.3).

5.5. Remark. From the discussion in 5.1 one may get an idea about extending the GKZ system (13)-(14) to the situation in which $\mathbb{Z}^N/\mathbb{L}$ has a non-trivial torsion subgroup $\left(\mathbb{Z}^N/\mathbb{L}\right)_{\mathrm{tors}}$. Equation (13) still makes sense in this more general situation, but Equation (14) must be adapted.

Recall from (15) a definition of $\mathcal{A}$ which also works in the torsion case. Let $\mathcal{G}_{\mathcal{A}} := \mathrm{Hom}(\mathbb{Z}^N/\mathbb{L}, \mathbb{C}^*)$. This is a commutative algebraic group of which the connected component of the identity is $\mathcal{G}^\circ_{\mathcal{A}} = \mathbb{T}^{k+1} = \mathrm{Hom}(\mathbb{Z}^{k+1}, \mathbb{C}^*)$ and the group of connected components is the finite abelian group $\mathcal{G}_{\mathcal{A}}/\mathcal{G}^\circ_{\mathcal{A}} = \mathrm{Hom}(\left(\mathbb{Z}^N/\mathbb{L}\right)_{\mathrm{tors}}, \mathbb{C}^*) \simeq \left(\mathbb{Z}^N/\mathbb{L}\right)_{\mathrm{tors}}$. Formula (21) defines an action of $\mathcal{G}_{\mathcal{A}}$ on $\mathbb{C}^{\mathcal{A}}$ and on the functions on $\mathbb{C}^{\mathcal{A}}$. It is now clear how to adapt Equation (14): let $\bar{\mathbf{a}}_i$ denote the projection of $\mathbf{a}_i$ on the free part $\mathbb{Z}^{k+1}$ of $\mathbb{Z}^N/\mathbb{L}$ and replace (14) by

$$\bar{\mathbf{a}}_1\, u_1 \frac{\partial \Phi}{\partial u_1} + \cdots + \bar{\mathbf{a}}_N\, u_N \frac{\partial \Phi}{\partial u_N} = \mathbf{c}\Phi$$

plus the requirement that the solution should transform according to some character of the finite abelian group $\left(\mathbb{Z}^N/\mathbb{L}\right)_{\mathrm{tors}}$.

5.6. Example. In case B_1 in Figure 3 one has $\mathbb{L} = \mathbb{Z}(2, -1, -1) \oplus \mathbb{Z}(1, 1, -2)$ and $\mathbb{Z}^3/\mathbb{L} = \mathbb{Z} \oplus \mathbb{Z}/3\mathbb{Z}$. As a generator for the torsion subgroup we take $\mathbf{g} = (1, -1, 0) \bmod \mathbb{L}$. Then the polynomials $\Phi_0 = \frac{1}{2}u_3^2 + u_1u_2$, $\Phi_1 = \frac{1}{2}u_2^2 + u_1u_3$ and $\Phi_2 = \frac{1}{2}u_1^2 + u_2u_3$ satisfy the differential equations (13) for $\mathbb{L}$, the differential equations (14) for $\mathbf{c} = 2$, while $\mathbf{g} \cdot \Phi_r = e^{2\pi i r/3}\Phi_r$ for $r = 0, 1, 2$.

6 From rank 2 subgroups of $\mathbb{Z}^N$ to Dessins

We are going to describe a construction which associates with a rank 2 subgroup $\mathbb{L}$ of $\mathbb{Z}^N$, as in 2.1, *dessins d'enfants*, i.e. bipartite graphs embedded in oriented Riemann surfaces.

The construction is given in [14] §5 and §6 as *Fast Inverse Algorithm.* In op. cit., however, this algorithm is only presented via explicit visual inspection of pictures in some concrete examples. In this section we want to present a general principle behind the Fast Inverse Algorithm of [14] which uses only linear algebra and can be performed by computer.

The main part of our construction produces rhombus tilings of the plane and is the same as N.G. de Bruijn's [2] construction of Penrose tilings. His work also led to effective methods for making *quasi-crystals* [23]. Since $\mathbb{L}$ is defined over $\mathbb{Z}$ the construction yields in our situation periodic tilings of the plane $\mathbb{L}_{\mathbb{R}}^{\vee} = \mathbb{R}^2$, not just quasi-periodic ones. We can therefore pass to $\mathbb{R}^2$ modulo the period lattice and find a tiling of the two-dimensional torus.

6.1. In this section $\mathbb{L}$ is as in 2.1. The quotient group $\mathbb{Z}^N/\mathbb{L}$ is allowed to have torsion. Let $\mathbb{L}_{\mathbb{R}}$ denote the real 2-plane in $\mathbb{R}^N$ which contains $\mathbb{L}$. We write $\lambda + \mathbb{L}_{\mathbb{R}}$ for the real 2-plane in $\mathbb{R}^N$ obtained by translating $\mathbb{L}_{\mathbb{R}}$ by a vector $\lambda \in \mathbb{R}^N$. On the other hand one has in $\mathbb{R}^N$ the *standard N-grid* consisting of the hyperplanes

$$\mathcal{H}_{i,k} := \{(z_1, \ldots, z_N) \in \mathbb{R}^N \mid z_i = k\} \qquad \text{for } i = 1, \ldots, N \text{ and } k \in \mathbb{Z}\,.$$

By intersecting with this standard N-grid we obtain in the 2-plane $\lambda+\mathbb{L}_{\mathbb{R}}$ an N-grid of lines:

$$\mathcal{L}_{i,k}^{\lambda} := \mathcal{H}_{i,k} \cap (\lambda + \mathbb{L}_{\mathbb{R}}) \qquad \text{for } i = 1, \ldots, N \text{ and } k \in \mathbb{Z}\,. \tag{23}$$

Note that this crucially uses the assumption that $\mathbb{L}$ is not contained in any of the standard coordinate hyperplanes in $\mathbb{Z}^N$.

The grid lines are naturally oriented: the orientation on $\mathcal{L}_{i,k}^{\lambda}$ is such that the points in $\lambda + \mathbb{L}_{\mathbb{R}}$ which are to the left of $\mathcal{L}_{i,k}^{\lambda}$ have a larger i-th coordinate than the points to the right of $\mathcal{L}_{i,k}^{\lambda}$.

6.2. Definition. We say that λ is *non-resonant* if no three grid lines pass through one point.

6.3. First step. Let $\mathbb{L}$ and λ be as in 6.1 with λ non-resonant as in 6.2. Consider the map

$$F : \mathbb{R}^N \to \mathbb{Z}^N\,, \qquad F(z_1, \ldots, z_N) = (\lfloor z_1 \rfloor, \ldots, \lfloor z_N \rfloor)\,,$$

where $\lfloor z \rfloor$ for a real number z denotes the largest integer $\leq z$. The map F contracts an open N-cube in the $\mathbb{Z}^N$ structure on $\mathbb{R}^N$ (and part of its boundary) onto one of its corners. F commutes with the translation action of $\mathbb{Z}^N$ on $\mathbb{R}^N$ and $\mathbb{Z}^N$. Suppressing $\mathbb{L}$ and λ from the notation we define

$$\mathcal{S}^0 := F(\lambda + \mathbb{L}_{\mathbb{R}})\,. \tag{24}$$

The map F is constant on each 2-cell (= connected component) of the grid complement in $\lambda + \mathbb{L}_{\mathbb{R}}$. Every point of $\mathcal{S}^0$ is in fact the image of a unique such 2-cell. The distance between two points of $\mathcal{S}^0$ is 1 if and only if the corresponding 2-cells are separated by exactly one grid line.

Because λ is non-resonant, an intersection point of grid lines is in the closure of exactly four 2-cells. If $\mathbf{x} = \mathcal{L}_{i,k}^{\lambda} \cap \mathcal{L}_{j,m}^{\lambda}$ the four points of $\mathcal{S}^0$ corresponding to these cells, i.e. $F(\mathbf{x})$, $F(\mathbf{x}) - \mathbf{e}_i$, $F(\mathbf{x}) - \mathbf{e}_j$, $F(\mathbf{x}) - \mathbf{e}_i - \mathbf{e}_j$, are the vertices of a square $\Box_{\mathbf{x}}$. When $\mathbf{x}$ runs through the set of all intersection points of grid lines, these squares fit together to form a connected surface $\mathcal{S} := \bigcup_{\mathbf{x}} \Box_{\mathbf{x}}$ embedded in $\mathbb{R}^N$. One can characterize the surface $\mathcal{S}$ also as

$$\mathcal{S} := \textit{union of all unit squares in } \mathbb{R}^N \textit{ with vertices in the set } \mathcal{S}^0\,. \tag{25}$$

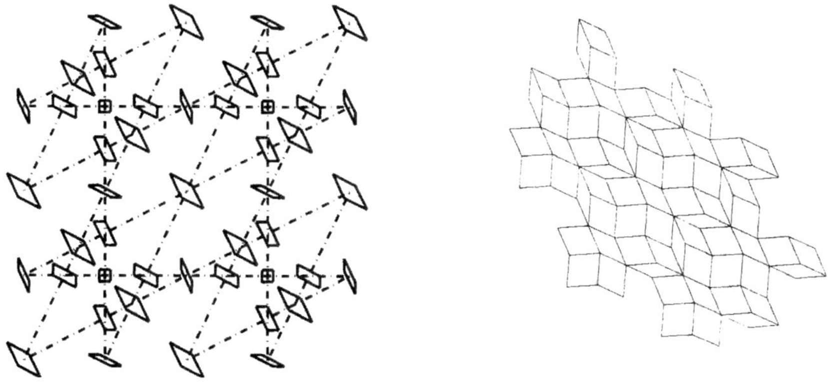

Figure 7 Converting grid to tiling for B_2 from Figure 3.

Here one may define *unit square* as the convex hull of four points $\mathbf{p}_1, \mathbf{p}_2, \mathbf{p}_3, \mathbf{p}_4$ in $\mathbb{R}^N$ with Euclidean distances $\|\mathbf{p}_1 - \mathbf{p}_2\| = \|\mathbf{p}_2 - \mathbf{p}_3\| = \|\mathbf{p}_3 - \mathbf{p}_4\| = \|\mathbf{p}_4 - \mathbf{p}_1\| = 1$ and $\|\mathbf{p}_1 - \mathbf{p}_3\| = \|\mathbf{p}_2 - \mathbf{p}_4\| = \sqrt{2}$. A unit square with vertices in $\mathbb{Z}^N$ is necessarily of the form $\mathbf{p} + \text{convex hull}(\mathbf{0}, \mathbf{e}_i, \mathbf{e}_j, \mathbf{e}_i + \mathbf{e}_j)$ for some $\mathbf{p} \in \mathbb{Z}^N$ and some i, j. We say that a unit square in $\mathbb{R}^N$ has *type* (i, j) if its sides are parallel to the vectors $\mathbf{e}_i$ or $\mathbf{e}_j$. Similarly, a side of a unit square in $\mathbb{R}^N$ has *type* i if it is parallel to the vector $\mathbf{e}_i$.

6.4. Proposition. *Let* $\mathbb{L}^\vee_\mathbb{R} = \mathrm{Hom}(\mathbb{L}, \mathbb{R})$ *denote the real dual of* $\mathbb{L}$. *Identify the real dual* $\mathrm{Hom}(\mathbb{Z}^N, \mathbb{R})$ *of* $\mathbb{Z}^N$ *with* $\mathbb{R}^N$ *by means of the standard dot product. Then the linear map* $\mathbb{R}^N \to \mathbb{L}^\vee_\mathbb{R}$ *which is dual to the inclusion* $\mathbb{L} \to \mathbb{Z}^N$ *restricts to a homeomorphism* $\mathcal{S} \xrightarrow{\sim} \mathbb{L}^\vee_\mathbb{R}$.

Proof Let $\mathbf{b}_1, \ldots, \mathbf{b}_N \in \mathbb{L}^\vee_\mathbb{R}$ denote the images of the standard basis vectors of $\mathbb{R}^N$. After identifying in the obvious way, the real plane $\mathbb{L}_\mathbb{R}$ with the real 2-plane $\lambda + \mathbb{L}_\mathbb{R}$ in $\mathbb{R}^N$ one obtains in $\mathbb{L}_\mathbb{R}$ an N-grid in which the grid lines $\mathcal{L}^\lambda_{i,k}$ are perpendicular to $\mathbf{b}_i$. After choosing a basis for $\mathbb{L}$ and taking the dual basis for $\mathbb{L}^\vee_\mathbb{R}$ we may identify $\mathbb{L}_\mathbb{R}$ and $\mathbb{L}^\vee_\mathbb{R}$ with $\mathbb{R}^2$. The term "perpendicular" then means perpendicular with respect to the standard inner product on $\mathbb{R}^2$.

Next draw for each intersection point of grid lines, say $\mathbf{x} = \mathcal{L}^\lambda_{i,k} \cap \mathcal{L}^\lambda_{j,m}$, a parallelogram with centre $\mathbf{x}$ and sides $\epsilon\mathbf{b}_i$ and $\epsilon\mathbf{b}_j$. Here the positive real number ϵ is so small that the parallelograms obtained from all grid intersection points are disjoint; see Figure 7 for an example. It is clear from this figure how one can glue the parallelograms of two consecutive intersection points on a grid line along their sides perpendicular to the grid line. The result of this glueing is then the same as the image (scaled with a factor ϵ) of the surface $\mathcal{S}$ under the projection $\mathbb{R}^N \to \mathbb{L}^\vee_\mathbb{R}$. □

It is obvious from the construction that the group $\mathbb{L}$ acts by translations on the point set $\mathcal{S}^0$ and on the surface $\mathcal{S}$, preserving the types of the squares and their sides. The cell structure on $\mathcal{S}$ given by the squares, their sides and vertices therefore induces on $\mathcal{S}/\mathbb{L}$ a cell structure. The following proposition counts the cells on $\mathcal{S}/\mathbb{L}$.

6.5. Proposition. *Let B be a $2 \times N$-matrix such that its rows are a $\mathbb{Z}$-basis for $\mathbb{L}$. Let $\mathbf{b}_1, \ldots, \mathbf{b}_N$ be the columns of this matrix. Then on $\mathcal{S}/\mathbb{L}$*

1. *the number of 2-cells of type (i,j) is $|\det(\mathbf{b}_i, \mathbf{b}_j)|$.*
2. *the number of 1-cells of type i is $\sum_{j=1}^{N} |\det(\mathbf{b}_i, \mathbf{b}_j)|$.*
3. *the number of 0-cells is $\sum_{1 \le i < j \le N} |\det(\mathbf{b}_i, \mathbf{b}_j)|$.*

Proof The counts can be made on the dual structure, which is given by the intersecting grid lines on $\mathbb{L}_{\mathbb{R}}$. Let B_{ij} denote the 2×2-matrix with columns $\mathbf{b}_i$ and $\mathbf{b}_j$. Then 1. amounts to counting for $\lambda_i, \lambda_j \in \mathbb{R}$ the number of elements in the set $\{\mathbf{x} = (x_1, x_2) \in \mathbb{R}^2 \mid 0 \le x_1, x_2 < 1, \mathbf{x} B_{ij} + (\lambda_i, \lambda_j) \in \mathbb{Z}^2\}$. Since there is a bijection between this set and the coset space $\mathbb{Z}^2/\mathbb{Z}^2 B_{ij}$, the number of elements is $|\det B_{ij}|$. The number of 1-cells of type i is equal to the number of intersection points of the i-th grid with the other grids. In view of 1. this number is therefore as stated in 2. Knowing the numbers of 0- and 1-cells on the 2-torus $\mathbb{L}_{\mathbb{R}}/\mathbb{L}$ one computes the number of 2-cells from the fact that the Euler characteristic of a 2-torus is 0. □

6.6. Second step. For the constructions in 6.1 and 6.3 we may randomly choose a non-resonant $\lambda \in \mathbb{R}^N$. Once we have constructed the set $\mathcal{S}^0$ and surface $\mathcal{S}$ we modify it by *elementary transformations*. For such an elementary transformation we need a point $\mathbf{p}_0 \in \mathcal{S}^0$ such that there are exactly three points $\mathbf{p}_1$, $\mathbf{p}_2$, $\mathbf{p}_3$ in $\mathcal{S}^0$ at distance 1 from $\mathbf{p}_0$ and such that the three points $\mathbf{p}_1 + \mathbf{p}_2 - \mathbf{p}_0$, $\mathbf{p}_1 + \mathbf{p}_3 - \mathbf{p}_0$ and $\mathbf{p}_2 + \mathbf{p}_3 - \mathbf{p}_0$ also lie in $\mathcal{S}^0$. These seven points are the vertices of three squares in $\mathcal{S}$. The elementary transformation now replaces $\mathbf{p}_0$ by $\mathbf{p}_0' := \mathbf{p}_1 + \mathbf{p}_2 + \mathbf{p}_3 - 2\mathbf{p}_0$ and replaces the three squares by the three squares in $\mathbb{R}^N$ with vertex configuration $\{\mathbf{p}_1, \mathbf{p}_2, \mathbf{p}_3, \mathbf{p}_1 + \mathbf{p}_2 - \mathbf{p}_0, \mathbf{p}_1 + \mathbf{p}_3 - \mathbf{p}_0, \mathbf{p}_2 + \mathbf{p}_3 - \mathbf{p}_0, \mathbf{p}_0'\}$; see Figure 8. In

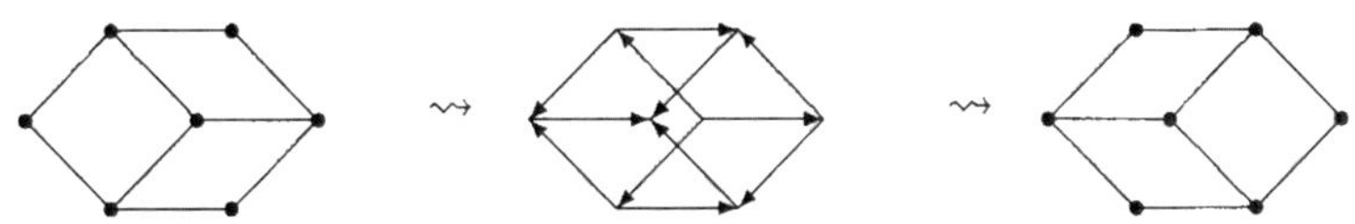

Figure 8 An elementary transformation.

order to preserve $\mathbb{L}$-periodicity we perform this transformation simultaneously at all configurations $\{\mathbf{p}_0 + \ell, \mathbf{p}_1 + \ell, \mathbf{p}_2 + \ell, \mathbf{p}_3 + \ell\}$ with $\ell \in \mathbb{L}$.

After this elementary transformation we obtain a surface $\mathcal{S}'$ in $\mathbb{R}^N$ which is the union of all unit squares with vertices in the set $\mathcal{S}'^0$, obtained from $\mathcal{S}^0$ by replacing the points $\mathbf{p}_0 + \ell$ by $\mathbf{p}_1 + \mathbf{p}_2 + \mathbf{p}_3 - 2\mathbf{p}_0 + \ell$ for all $\ell \in \mathbb{L}$.

The surface $\mathcal{S}'$, in turn, can be further modified by elementary transformations: just replace in the above construction $\mathcal{S}^0$ by $\mathcal{S}'^0$ and $\mathbf{p}_0$ by an appropriate point of $\mathcal{S}'^0$.

6.7. Third step. After starting the construction in 6.3 with a randomly chosen non-resonant $\lambda \in \mathbb{R}^N$ one can create by repeated elementary transformations many surfaces $\mathcal{S}$ each of which is the union of unit squares with vertices in an $\mathbb{L}$-invariant subset $\mathcal{S}^0$ of $\mathbb{Z}^N$ and which is mapped homeomorphically onto $\mathbb{L}_{\mathbb{R}}^{\vee}$ by the projection

$\mathbb{R}^N \to \mathbb{L}_{\mathbb{R}}^{\vee}$. Moreover the numbers of cells on each of these surfaces are still the same as in Proposition 6.5.

The number of surfaces one can make in this way is finite and it is possible to generate (by computer) a complete list.

6.8. Remark. The list of surfaces produced in 6.7 is, up to permutation of its entries, independent of the choice of $\lambda \in \mathbb{R}^N$ at the start of the algorithm. This can be seen as follows. Fix i, $1 \leq i \leq N$. Take $\lambda \in \mathbb{R}^N$ and let $\lambda_t = \lambda + t\mathbf{e}_i$ for $t \in \mathbb{R}$. Set $t_1 = \min\{t \in \mathbb{R}_{>0} \mid \lambda_t \text{ is not non-resonant}\}$ and $t_2 = \min\{t \in \mathbb{R}_{>t_1} \mid \lambda_t \text{ is not non-resonant}\}$. Then the surface $\mathcal{S}_t$ constructed from λ_t in 6.3 is equal to the surface $\mathcal{S}_0$ for $0 < t < t_1$. Figure 9 shows how the grid locally changes as t passes through t_1. Comparing Figures 9 and 8 one sees that such a local change

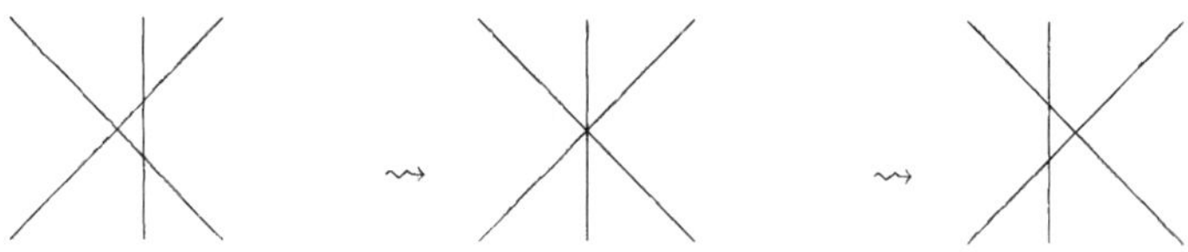

Figure 9 Local change in grid at resonance.

of the grid corresponds to an elementary transformation of the surface $\mathcal{S}_0$. Thus the surfaces $\mathcal{S}_t$ with $t_1 < t < t_2$ are obtained from $\mathcal{S}_0$ by a number of elementary transformations.

Since any two non-resonant λ, $\lambda' \in \mathbb{R}^N$ can be moved to a common value by coordinatewise changes as above we conclude that the surfaces $\mathcal{S}$ and $\mathcal{S}'$ obtained in 6.3 from λ and λ', respectively, are related by elementary transformations, and that, hence, the lists of surfaces which the algorithm produces from start values λ and λ' are the same, up to possibly a permutation of the entries.

6.9. Definition. We say that a surface $\mathcal{S}$ on this list is *perfect* if the function $\mathsf{s} : \mathbb{R}^N \to \mathbb{R}$, $\mathsf{s}(z_1, \ldots, z_N) = z_1 + \cdots + z_N$, takes only three values on the set $\mathcal{S}^0$ of vertices in $\mathcal{S}$. In that case these three values are consecutive integers, say $a+1, a, a-1$. We denote by $\mathcal{S}^{\bullet}$ (resp. $\mathcal{S}^*$ resp. $\mathcal{S}^{\circ}$) the set of vertices where s takes the value $a+1$ (resp. a resp. $a-1$) and say that the vertices in $\mathcal{S}^{\bullet}$ are *black*, those in $\mathcal{S}^{\circ}$ are *white* and those in $\mathcal{S}^*$ are *grey*. Each of these sets is invariant under the translation action of $\mathbb{L}$.

6.10. Remark. I have at present no proof that for every rank 2 subgroup $\mathbb{L} \subset \mathbb{Z}^N$ as in 2.1 the above construction indeed yields at least one perfect surface. On the other hand, in all examples I investigated the computer produced at least one perfect surface. It would also be very interesting to know which perfect surfaces can arise already in the first step 6.3 of the construction, by choosing λ appropriately (instead of at random).

6.11. For every surface $\mathcal{S}$ in the list of 6.7 one has the graph $\widehat{\Gamma}$ with set of vertices $\mathcal{S}^0$ and arrows given by the diagonals in the squares, oriented such that in a square of type (i, j) the oriented diagonals are $\mathbf{e}_i + \mathbf{e}_j$ and $\operatorname{sign}(\det(\mathbf{b}_i, \mathbf{b}_j))(\mathbf{e}_i - \mathbf{e}_j)$. The advantage of having $\mathcal{S}$ embedded in $\mathbb{R}^N$ is that the vertices and the arrows in these graphs are actual points and vectors in $\mathbb{R}^N$. If the surface $\mathcal{S}$ is perfect, every square in $\mathcal{S}$ has one vertex in $\mathcal{S}^\bullet$, one in $\mathcal{S}^\circ$ and two vertices in $\mathcal{S}^*$. One of its diagonals goes from the white to the black vertex and the other diagonal connects the two grey vertices. The graph $\widehat{\Gamma}$ therefore is the disjoint union of two oriented graphs $\widehat{\Gamma}^{\bullet\circ}$ and $\widehat{\Gamma}^*$, with vertex sets $\mathcal{S}^\bullet \cup \mathcal{S}^\circ$ and $\mathcal{S}^*$, respectively. The graph $\widehat{\Gamma}^{\bullet\circ}$ is a *bipartite graph*, i.e. its vertex set is the disjoint union of two sets (the black, resp. white vertices) and with edges connecting only vertices of different colors.

There is a natural duality between $\widehat{\Gamma}^{\bullet\circ}$ and $\widehat{\Gamma}^*$: every vertex of $\widehat{\Gamma}^{\bullet\circ}$ lies in a unique connected component of $\mathcal{S} \setminus \widehat{\Gamma}^*$ and vice versa. Moreover the boundaries of these connected components are polygons and an arrow between two nodes in one graph is a common side of the corresponding polygons for the other graph.

6.12. The perfect surface $\mathcal{S}$ is mapped homeomorphically onto $\mathbb{L}^\vee_\mathbb{R}$ by the projection $\mathbb{R}^N \to \mathbb{L}^\vee_\mathbb{R}$, which is dual to the inclusion $\mathbb{L} \hookrightarrow \mathbb{Z}^N$. This projection maps $\mathbf{e}_i$ to $\mathbf{b}_i$, for $i = 1, \ldots, N$. One may however pre-compose this projection with the linear map $\mathbb{R}^N \to \mathbb{R}^N$, $\mathbf{e}_i \mapsto \frac{1}{\|\mathbf{b}_i\|}\mathbf{e}_i$, $1 \le i \le N$. The composite linear map $\mathbb{R}^N \to \mathbb{L}^\vee_\mathbb{R}$ maps the tiling by squares on the surface $\mathcal{S}$ piecewise linearly and homeomorphically onto a tiling of $\mathbb{L}^\vee_\mathbb{R}$ by rhombi. It maps $\mathbb{L}$ isomorphically onto a lattice $\overline{\mathbb{L}}$ in $\mathbb{L}^\vee_\mathbb{R}$ and it maps the graphs $\widehat{\Gamma}^{\bullet\circ}$ and $\widehat{\Gamma}^*$ 'isomorphically' onto $\overline{\mathbb{L}}$-periodic graphs $\overline{\Gamma}^{\bullet\circ}$ and $\overline{\Gamma}^*$ in the plane $\mathbb{L}^\vee_\mathbb{R}$. In the literature $\overline{\Gamma}^{\bullet\circ} \subset \mathbb{L}^\vee_\mathbb{R}$ is called a *periodic dimer model* or *periodic brane tiling*; see for instance [4, 5, 12, 13, 14, 15].

6.13. Definition. All structures in 6.11 are invariant under translation by vectors in $\mathbb{L}$. Passing to the orbit space we obtain from a perfect surface $\mathcal{S}$ the 2-dimensional torus $\mathcal{T} = \mathcal{S}/\mathbb{L}$ (i.e. a compact oriented surface of genus 1 without boundary). Embedded in this torus are the two graphs $\Gamma^{\bullet\circ} = \widehat{\Gamma}^{\bullet\circ}/\mathbb{L}$ and $\Gamma^* = \widehat{\Gamma}^*/\mathbb{L}$, which are dual to each other. Note that 6.12 also yields $\mathcal{T} = \mathbb{L}^\vee_\mathbb{R}/\overline{\mathbb{L}}$, $\Gamma^{\bullet\circ} = \overline{\Gamma}^{\bullet\circ}/\overline{\mathbb{L}}$ and $\Gamma^* = \overline{\Gamma}^*/\overline{\mathbb{L}}$.

After Grothendieck one calls $(\mathcal{T}, \Gamma^*, \Gamma^{\bullet\circ})$ a *dessin d'enfants* [18, 22].

6.14. Example. Figure 10 shows two drawings of the dessin $(\mathcal{T}, \Gamma^*, \Gamma^{\bullet\circ})$ for B_{10} from Figure 3; or, rather, it shows a lifting of this dessin to the plane $\mathbb{L}^\vee_\mathbb{R}$ with basis so that the period lattice $\overline{\mathbb{L}}$ becomes just $\mathbb{Z}^2$. It also shows Γ^* as an unembedded quiver. *Note however that this is not the same as the quiver* Q *in case B_{10} in Figure 3.*

The quiver in Figure 10 is in fact the quiver for the singularity $\mathbb{C}^3/\mathbb{Z}_6$ where a generator of the cyclic group $\mathbb{Z}_6$ acts on $\mathbb{C}^3$ as multiplication by the diagonal matrix $\mathrm{diag}(e^{2\pi i/6}, e^{2\pi i/3}, e^{2\pi i/2})$. Also, the brane tiling picture in Figure 10 appears in [13] §3 in connection with the singularity $\mathbb{C}^3/\mathbb{Z}_6$.

This difference between [13] and our approach arises, because we still have to perform the untwist, as is explained below.

6.15. The quiver Γ^* in 6.13 is not the Plücker quiver of $\mathbb{L}$, defined in 2.5. The latter appears in the current setting through the *zigzag loops*, as follows. Let $\mathcal{S}$ be a perfect surface. Fix $i \in \{1, \ldots, N\}$. In every square on $\mathcal{S}$ which has two sides parallel to $\mathbf{e}_i$ draw the line segment connecting the midpoints of these two sides. If the square has type (i, j) this line segment is oriented so that it is a translate

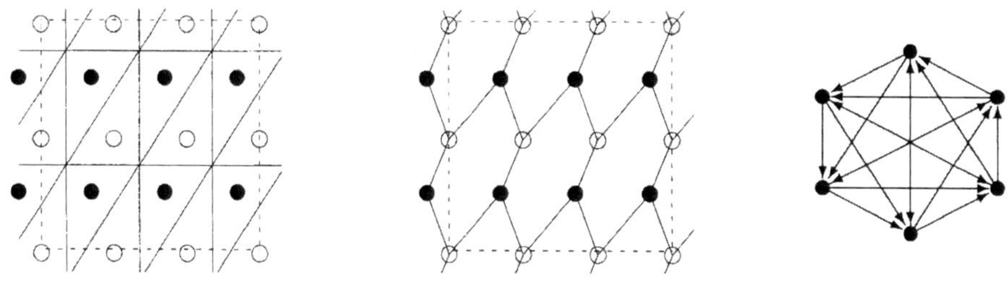

Figure 10 Two versions of the dessin $(\mathcal{T}, \Gamma^*, \Gamma^{\bullet\circ})$ for B_{10} from Figure 3. The dashed square is the period parallelogram. The picture on the left shows how the quiver on the right is embedded in a 2-torus.

of the vector $\text{sign}(\det(\mathbf{b}_i, \mathbf{b}_j))\mathbf{e}_j$. The union of these line segments projects to an oriented closed curve on $\mathcal{T}$, called the *i-th zigzag loop.* Note that this zigzag loop may consist of several connected components.

According to Proposition 6.5 the i-th and j-th zigzag loops on $\mathcal{T}$ intersect in exactly $|\det(\mathbf{b}_i, \mathbf{b}_j)|$ points. *Thus we see that the nodes of the Plücker quiver can be identified with the zigzag loops on $\mathcal{T}$ and the arrows between two nodes are the intersection points of the corresponding zigzag loops.* The orientation of the arrows follows from the orientation of the zigzag loops and can in pictures be indicated as an over/undercrossing of the zigzag loops.

Figure 11 shows the zigzag loops for the dessin in Figure 10.

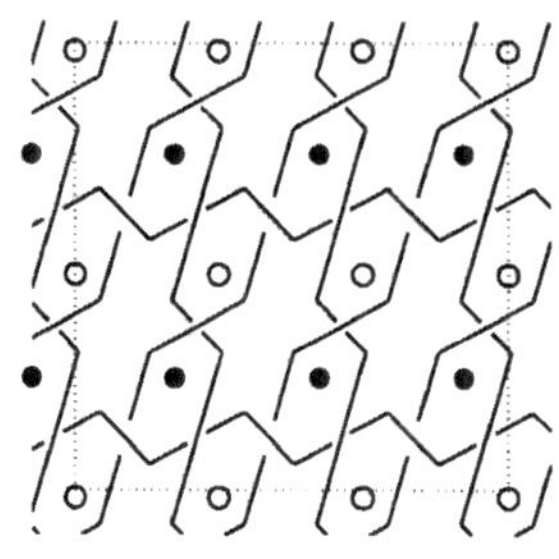

Figure 11 The zigzag loops for the dessin in Figure 10.

6.16. Final untwist. Pictures like Figure 11 can be viewed as showing an oriented surface with boundary, in which the connected components of the zigzag loops are the connected components of the boundary and which is embedded in three space in a twisted way so that the black dots are on one side of the surface and the white dots are on the other side. This surface comes with a tiling by helices as shown in Figure 12.

One can 'untwist' the helix tiles and complete the untwisted helix tiles to squares as indicated in Figure 12. The effect is that the boundary cycles are being capped off by discs. The result is an oriented surface $\mathcal{M}$ without boundary which is tiled by squares with one black, one white and two red vertices. The black-white diagonals give an embedding of the bipartite graph $\Gamma^{\bullet\circ}$ into $\mathcal{M}$. The red-red diagonals give an embedding of the quiver Q into $\mathcal{M}$ (if all zigzag loops have just

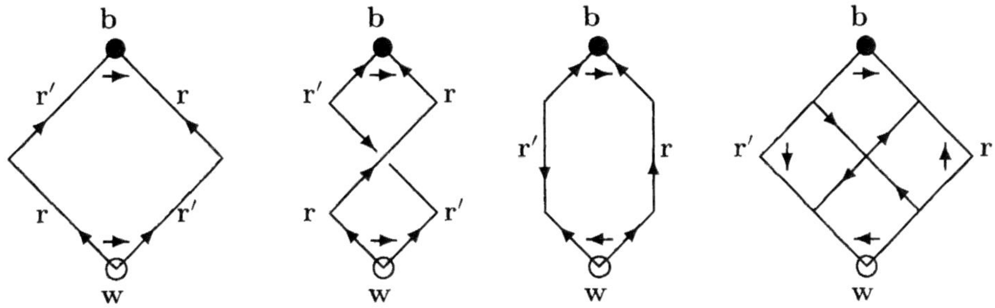

Figure 12 Untwisting: from left to right: original square tile, helix tile, untwisted helix tile, untwisted helix completed to square tile.

one connected component; otherwise some points on the surface must be pinched together).

The untwisting procedure is described in [4] §5 via examples, pictures and visual inspection. Here we want to build the 'untwisted surface' by simply reinterpreting the combinatorial data of the tiling of the torus $\mathcal{T}$, i.e. the $\mathbb{L}$-periodic tiling of the perfect surface $\mathcal{S}$ by unit squares taken modulo $\mathbb{L}$ (see 6.9, 6.13). Let B (resp. W) denote the set of black (resp. white) vertices. There is a bijection between the set of tiles and the set E of arrows of the quiver Q and we can refer to tiles as $e \in E$. A tile e has one black vertex $\mathbf{b}(e)$ and one white vertex $\mathbf{w}(e)$. A tile e is the image of a unit square with sides parallel to two of the basis vectors $\mathbf{e}_1, \ldots, \mathbf{e}_N$; let us denote the indices of this (unordered) pair of basis vectors as a 2-element subset $\{\mathbf{r}(e), \mathbf{r}'(e)\}$ of $\{1, \ldots, N\}$. Thus the tiling on $\mathcal{T}$ yields the list of quadruples

$$\mathbb{M} = \{ \, (\mathbf{b}(e), \mathbf{w}(e), \mathbf{r}(e), \mathbf{r}'(e)) \, \}_{e \in E} \, . \tag{26}$$

Now, for every $e \in E$ we take a unit square $\square_e$ and attach labels $\mathbf{b}(e)$, $\mathbf{w}(e)$, $\mathbf{r}(e)$, $\mathbf{r}'(e)$ and colors black, white, red, red, respectively, to its vertices such that the vertices labeled $\mathbf{b}(e)$ and $\mathbf{w}(e)$ are not adjacent. If two such squares $\square_e$ and $\square_{e'}$ have equal labels on two adjacent vertices, we glue $\square_e$ and $\square_{e'}$ along the corresponding sides. The result of this glueing is a surface $\widetilde{\mathcal{M}}$ tiled with squares.

On the surface $\widetilde{\mathcal{M}}$ there is for every $\mathbf{b} \in \mathsf{B}$ one black point with label $\mathbf{b}$ and for every $\mathbf{w} \in \mathsf{W}$ there is one white point with label $\mathbf{w}$. However for $i \in \{1, \ldots, N\}$ there may be several red points with label i. We finally obtain the desired surface $\mathcal{M}$ by identifying for every $i \in \{1, \ldots, N\}$ the red points with label i.

6.17. Proposition. *The surface $\widetilde{\mathcal{M}}$, and hence also $\mathcal{M}$, is oriented.*

Proof Cut the square $\square_e$ into four pieces like in the right-hand picture of Figure 12. Then for fixed $\mathbf{b} \in \mathsf{B}$ the small squares with black vertex $\mathbf{b}$ are glued together to form a polygon isomorphic to the polygon in $\mathcal{T}$ formed by the unit squares with one vertex $\mathbf{b}$. Similarly, for fixed $\mathbf{w} \in \mathsf{W}$ the small squares with white vertex $\mathbf{w}$ are glued together to form a polygon isomorphic to the polygon in $\mathcal{T}$ formed by the unit squares with one vertex $\mathbf{w}$ but with the orientation reversed. For $i \in \{1, \ldots, N\}$ the small squares with red vertex i are glued together to form a number of disjoint polygons, one for every connected component of the i-th zigzag loop and with oriented boundary 'equal to' that component. It is now clear from Figure 12 that these polygons are oriented consistently so that $\widetilde{\mathcal{M}}$ is oriented. $\square$

6.18. Conclusion. *The surface $\mathcal{M}$ is tiled with squares so that the black-white diagonals form a bipartite graph isomorphic to the bipartite graph $\Gamma^{\bullet\circ}$ in 6.13, while the red-red diagonals form a graph isomorphic to the quiver* Q.

6.19. Example. The list $\mathbb{M}$ (26) which our algorithm produces for model IV of dP_3, i.e. example B_{10} from Figure 3, is

e	:	1	2	3	4	5	6	7	8	9	10	11	12	13	14	15	16	17	18
$\mathbf{b}(e)$	:	6	5	1	2	4	3	1	5	3	6	4	2	5	6	1	2	3	4
$\mathbf{w}(e)$	:	5	6	2	1	3	4	3	1	5	2	6	4	5	6	1	2	3	4
$\mathbf{r}(e)$	:	2	2	3	3	4	4	1	1	1	1	1	1	5	6	5	6	5	6
$\mathbf{r}'(e)$	:	1	1	1	1	1	1	5	5	5	6	6	6	2	2	3	3	4	4

The surface $\mathcal{M}$ with the corresponding tiling by kites (instead of squares) is shown in the left-hand picture in Figure 13. The right-hand picture in Figure 13 shows the bipartite graph $\Gamma^{\bullet\circ}$ and the quiver Q embedded in $\mathcal{M}$. Note that this quiver is indeed the same as the one for example B_{10} in Figure 13.

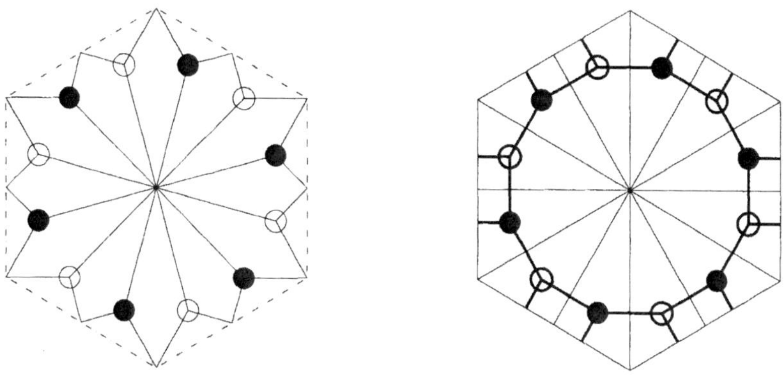

Figure 13 Surface $\mathcal{M}$ with kite tiling (left) and quiver Q and bipartite graph $\Gamma^{\bullet\circ}$ (right) for case B_{10} in Figure 4 (= model IV of dP_3). The surface $\mathcal{M}$ is obtained by identifying opposite sides of the hexagon.

7 Perfect matchings and the secondary fan

7.1. Definition. (see e.g. [17, 16, 20]) A *perfect matching* P on a bipartite graph $\Gamma^{\bullet\circ}$ (also known as a *dimer configuration*) is a subset of the edges of $\Gamma^{\bullet\circ}$ such that every node of $\Gamma^{\bullet\circ}$ is incident to exactly one edge in P.

7.2. Theorem. *Consider a subgroup $\mathbb{L} \subset \mathbb{Z}^N$ as in 2.1. Let $\Gamma^{\bullet\circ}$ be a bipartite graph which is obtained from $\mathbb{L} \subset \mathbb{Z}^N$ as explained in Section 6, and in Definition 6.13 in particular. Let C be a 2-dimensional cone in the secondary fan of $\mathbb{L} \subset \mathbb{Z}^N$ and let L_C be the corresponding collection of 2-element subsets of $\{1, \ldots, N\}$ as in 3.2 and 3.3. Identify E with the set of edges of $\Gamma^{\bullet\circ}$.*
Then, with the notation as in (26), the set

$$P_C = \{e \in E \mid \{\mathbf{r}(e), \mathbf{r}'(e)\} \in L_C\} \tag{27}$$

is a perfect matching on $\Gamma^{\bullet\circ}$.

Proof Consider a perfect surface $\mathcal{S} \subset \mathbb{R}^N$ as in 6.9. The map $\pi : \mathbb{R}^N \to \mathbb{L}^\vee_\mathbb{R}$ which is dual to the inclusion $\mathbb{L} \hookrightarrow \mathbb{Z}^N$ projects $\mathcal{S}$ homeomorphically onto the plane $\mathbb{L}^\vee_\mathbb{R}$. The tiling of $\mathcal{S}$ by unit squares gives a tiling of $\mathbb{L}^\vee_\mathbb{R}$ by parallelograms. A unit square with edges $\mathbf{e}_i$ and $\mathbf{e}_j$ projects onto a parallelogram with edges $\mathbf{b}_i$ and $\mathbf{b}_j$. Now let C be a 2-dimensional cone of the secondary fan (see 3.2) and let $\mathbf{w}$ be a white vertex of the tiling. By translating the origin of the secondary fan to $\mathbf{w}$ one sees that there is exactly one parallelogram with vertex $\mathbf{w}$ and with edges $\mathbf{b}_i$ and $\mathbf{b}_j$ such that $C \subset \mathbb{R}_{\geq 0}\mathbf{b}_i + \mathbb{R}_{\geq 0}\mathbf{b}_j$, i.e. such that $\{i,j\} \in L_C$. This shows that there is exactly one element of P_C incident to $\mathbf{w}$, namely the black-white diagonal of this parallelogram. A similar argument works for the black vertices. □

7.3. Corollary. *In the situation of Theorem 7.2 the number of black vertices and the number of white vertices are both equal to the number* $\mathrm{vol}_\mathcal{A}$ *in (19).*

Proof The number of black (resp. white) vertices of $\Gamma^{\bullet\circ}$ is equal to the number of edges in any perfect matching. This holds in particular for the perfect matching P_C corresponding to a 2-dimensional cone C in the secondary fan. The result now follows from (27), (4) and (19). □

The following result was also derived in [4] §5.1.

7.4. Corollary. *The genus of the surface $\mathcal{M}$ in 6.16 is equal to the number of lattice points in the interior of the secondary polytope Δ in (8).*

Proof The number of 1-cells in the tiling of $\mathcal{M}$ is twice the number of 2-cells. According to Proposition 6.5 the number of 2-cells is $\sum_{1\leq i<j\leq N} |\det(\mathbf{b}_i, \mathbf{b}_j)|$. The number of 0-cells is, according to Corollary 7.3 and Formula (19) equal to $N + 2\sum_{\{i,j\}\in L_C} |\det(\mathbf{b}_i, \mathbf{b}_j)|$. The desired result now follows from Corollary 3.10 and a calculation of the Euler characteristic $2 - 2\,\mathrm{genus}(\mathcal{M})$. □

8 Kasteleyn matrix, bi-adjacency matrix, 3-constellation and superpotential

8.1. In 6.12 we found the $\mathbb{L}$-periodic bipartite graph $\overline{\Gamma}^{\bullet\circ}$ in the plane $\mathbb{L}^\vee_\mathbb{R}$. Periodic bipartite graphs in the plane are also known as *periodic dimer models.* One of the main tools for studying periodic dimer models is the *Kasteleyn matrix.* It is usually defined as an adjacency matrix with extra factors x, x^{-1}, y, y^{-1} for edges crossing the sides of a fundamental parallelogram and determined by visual inspection of a picture of the periodic dimer model; see e.g. [17, 13, 15]. The purpose of the extra factors is to keep track of the homology class of a closed path on the bipartite graph $\Gamma^{\bullet\circ}$ in the torus $\mathcal{T}$ (see 6.13). We obtained $\overline{\Gamma}^{\bullet\circ} \subset \mathbb{L}^\vee_\mathbb{R}$ as the projection of the black-white diagonals in the tiling of the perfect surface $\mathcal{S}$ by unit squares. The advantage of having $\mathcal{S}$ embedded in $\mathbb{R}^N$ is that these diagonals are actual vectors in $\mathbb{Z}^N$ and that points lying on paths formed by these diagonals are simply given by vector addition. Thus all edges become "equally responsible" for the periodicity. An edge which is the diagonal in a square of type (i,j) contributes $\mathbf{e}_i + \mathbf{e}_j$.

Thus we are led to a reformulation of the Kasteleyn matrix of the dimer model $(\mathcal{T}, \Gamma^{\bullet\circ})$ in terms of the combinatorial data given as the list $\mathbb{M}$ in Equation (26).

As this same list is used in 6.16 for building the surface $\mathcal{M}$ we call the same matrix also the bi-adjacency matrix of the dessin $(\mathcal{M}, \mathsf{Q})$.

8.2. Definition. From the list $\mathbb{M} = \{\, (\,\mathbf{b}(e),\, \mathbf{w}(e),\, \mathbf{r}(e),\, \mathbf{r}'(e)\,)\,\}_{e\in E}$ in (26) and a function $\varpi : E \to \mathbb{C}$ on the set of edges of the quiver Q we construct a matrix $\mathbb{K}^{\varpi}$ with rows indexed by the set of black vertices B, columns indexed by the set of white vertices W and with entries in the polynomial ring $\mathbb{C}[u_1, \ldots, u_N]$:

$$(\mathbf{b}, \mathbf{w})\text{-entry of } \mathbb{K}^{\varpi} \quad = \sum_{e\in E:\, \mathbf{b}(e)=\mathbf{b},\, \mathbf{w}(e)=\mathbf{w}} \varpi(e)\, u_{\mathbf{r}(e)} u_{\mathbf{r}'(e)}\,. \tag{28}$$

We use for $\mathbb{K}^{\varpi}$ the name *Kasteleyn matrix of the dimer model* $(\mathcal{T}, \Gamma^{\bullet\circ})$ as well as the name *bi-adjacency matrix of the dessin* $(\mathcal{M}, \mathsf{Q})$.

8.3. Example. Although the relation between the list $\mathbb{M}$ and the matrix $\mathbb{K}^{\varpi}$ is so straightforward that there seems no point in writing here the matrix $\mathbb{K}^{\varpi}$ for Example 6.19, we do present this $\mathbb{K}^{\varpi}$ here to compare it with results in the literature.

$$\text{bi-adjacency matrix } \mathbb{K}^{\varpi} \quad \text{for model IV of } dP_3 \;=$$

$$\begin{bmatrix}
\varpi_{15}u_3u_5 & \varpi_3u_1u_3 & \varpi_7u_1u_5 & 0 & 0 & 0\\
\varpi_4u_1u_3 & \varpi_{16}u_3u_6 & 0 & \varpi_{12}u_1u_6 & 0 & 0\\
0 & 0 & \varpi_{17}u_4u_5 & \varpi_6u_1u_4 & \varpi_9u_1u_5 & 0\\
0 & 0 & \varpi_5u_1u_4 & \varpi_{18}u_4u_6 & 0 & \varpi_{11}u_1u_6\\
\varpi_8u_1u_5 & 0 & 0 & 0 & \varpi_{13}u_2u_5 & \varpi_2u_1u_2\\
0 & \varpi_{10}u_1u_6 & 0 & 0 & \varpi_1u_1u_2 & \varpi_{14}u_2u_6
\end{bmatrix}$$

(with ϖ_e denoting $\varpi(e)$). In view of the discussion in Example 6.14 this bi-adjacency matrix for model IV of dP_3 should be somehow the same as the Kasteleyn matrix for the singularity $\mathbb{C}^3/\mathbb{Z}_6$ in [13] Equation (3.5):

$$K(z,w) = \begin{bmatrix}
w & -1 & 0 & -w & 0 & 0\\
0 & 1 & -1 & 0 & -1 & 0\\
-1 & 0 & 1 & 0 & 0 & -1\\
-zw & 0 & 0 & w & -1 & 0\\
0 & -z & 0 & 0 & 1 & -1\\
0 & 0 & -z & -1 & 0 & 1
\end{bmatrix}.$$

Indeed, by dividing the first, third and fifth column of $\mathbb{K}^{\varpi}$ by u_5 and subsequently setting $u_5 = z^{-1}$, $u_3 = w$, $u_1 = u_2 = u_4 = u_6 = 1$, and appropriately setting $\varpi_e = \pm 1$ one obtains a matrix which up to a permutation of rows and columns is $K(z,w)$. This way of getting $K(z,w)$ from $\mathbb{K}^{\varpi}$ is nothing but passing from homogeneous to inhomogeneous coordinates.

8.4. Definition.(cf. [18] Definition 1.1.1.) A *3-constellation* (E, σ_0, σ_1) consists of a finite set E and two permutations σ_0, σ_1 of E such that the permutation group generated by σ_0, σ_1 acts transitively on E.

8.5. Notations. The cycle notation $(i_1\; i_2 \cdots i_{k-1}\; i_k)$ denotes the permutation ρ such that $\rho(i_j) = i_{j+1}$ for $j = 1, \ldots, k-1$ and $\rho(i_k) = i_1$. By a *cycle* of a permutation σ we mean an orbit of the group generated by σ. We denote the set of cycles of σ by E_σ. In the product $\sigma\tau$ of two permutations σ and τ one first applies τ ; so $(\sigma\tau)(i) = \sigma(\tau(i))$.

8.6. Definition. For our purposes we do not need the most general notion of a superpotential for a quiver. The superpotentials we need are just notational reformulations of 3-constellations. The *superpotential attached to a* 3-*constellation* (E, σ_0, σ_1) is the following polynomial W in non-commuting variables X_e $(e \in E)$:

$$W = \sum_{\gamma \in E_{\sigma_0}} \mathsf{X}_\gamma - \sum_{\gamma \in E_{\sigma_1}} \mathsf{X}_\gamma ,$$

where $\mathsf{X}_\gamma := X_{i_1} X_{i_2} \cdot \ldots \cdot X_{i_k}$ for a cyclic permutation $\gamma = (i_1\, i_2 \cdots i_k)$.

8.7. Definition. The 3-*constellation* (E, σ_0, σ_1) *associated with the dessin* $(\mathcal{M}, \mathsf{Q})$ in 6.18 is the following. E is the set of arrows of the quiver Q. Every $e \in E$ is a path p_e on $\mathcal{M}$. Every connected component of $\mathcal{M} \setminus \cup_{e \in E} p_e$ has an oriented boundary which can be viewed as a cyclic permutation of elements of E. Then σ_0 (resp. σ_1) is the composition of the cyclic permutations which are boundaries of connected components containing a black (resp. white) point.

8.8. Remark. In the same way one associates a 3-constellation with the dessin $(\mathcal{T}, \Gamma^*)$ in 6.13. It is obvious from the untwisting procedure in 6.16 that, if (E, σ_0, σ_1) is the 3-constellation associated with the dessin $(\mathcal{M}, \mathsf{Q})$, then $(E, \sigma_0, \sigma_1^{-1})$ is the 3-constellation associated with the dessin $(\mathcal{T}, \Gamma^*)$.

8.9. It is obvious that the 3-constellation (E, σ_0, σ_1) associated with the dessin $(\mathcal{M}, \mathsf{Q})$ contains the complete instructions for building $\mathcal{M}$: for every cycle of σ_0 and every cycle of σ_1 take a (convex planar) polygon with sides labeled by the elements in the cycle in their cyclic order. Next glue these polygons by identifying sides with the same label. The sides of these polygons correspond to arrows of the quiver Q and the vertices of these polygons correspond to vertices of Q. After the above procedure of glueing polygons along their sides one must still identify points which correspond to the same vertex of Q. The result is then $\mathcal{M}$ with Q embedded in it.

8.10. The 3-constellation (E, σ_0, σ_1) in 8.9 and the list $\mathbb{M}$ in 6.16 both are completely determined by and do completely determine the dessin $(\mathcal{M}, \mathsf{Q})$. One can, however, also describe the relation between (E, σ_0, σ_1) and $\mathbb{M}$ in a direct algebraic/combinatorial way.

For the construction of (E, σ_0, σ_1) from $\mathbb{M}$ one first forms for $\mathbf{b} \in \mathsf{B}$, $\mathbf{w} \in \mathsf{W}$ and $i \in \{1, \ldots, N\}$ the sets

$$\begin{aligned} |\sigma_{\mathbf{b}}| &= \{e \in E \mid \mathbf{b}(e) = \mathbf{b}\}, \qquad |\sigma_{\mathbf{w}}| = \{e \in E \mid \mathbf{w}(e) = \mathbf{w}\}, \\ |\mathbf{z}_i| &= \{e \in E \mid i \in \{\mathbf{r}(e), \mathbf{r}'(e)\}\}. \end{aligned}$$

These sets have an unoriented cyclic structure: $e, e' \in |\sigma_{\mathbf{b}}|$ (resp. $e, e' \in |\sigma_{\mathbf{w}}|$) are neighbors if and only if $\{\mathbf{r}(e), \mathbf{r}'(e)\} \cap \{\mathbf{r}(e'), \mathbf{r}'(e')\} \neq \emptyset$, while $e, e' \in |\mathbf{z}_i|$ are neighbors if and only if $\{\mathbf{b}(e), \mathbf{w}(e)\} \cap \{\mathbf{b}(e'), \mathbf{w}(e')\} \neq \emptyset$. In order to put a consistent orientation on these cyclic sets we choose one of the two possible orientations of $|\mathbf{z}_1|$. For every $\mathbf{b} \in \mathsf{B}$ we have $\sharp(|\mathbf{z}_1| \cap |\sigma_{\mathbf{b}}|) = 0$ or 2. In the latter case $|\mathbf{z}_1| \cap |\sigma_{\mathbf{b}}|$ consists of two elements, say e and e', which are neighbors both in $|\mathbf{z}_1|$ and in $|\sigma_{\mathbf{b}}|$. The orientation on $|\mathbf{z}_1|$ then induces an orientation on $|\sigma_{\mathbf{b}}|$ such that e is the successor of e' in $|\sigma_{\mathbf{b}}|$ if e is the successor of e' in $|\mathbf{z}_1|$. In this way the orientation on $|\mathbf{z}_1|$ induces an orientation on every $|\sigma_{\mathbf{b}}|$ and every $|\sigma_{\mathbf{w}}|$ which has a non-empty intersection with $|\mathbf{z}_1|$. Next this induces an orientation on every $|\mathbf{z}_i|$ which has a non-empty intersection with any of the already oriented sets $|\sigma_{\mathbf{b}}|$ or $|\sigma_{\mathbf{w}}|$. And so on. It is because of the geometric background of $\mathbb{M}$ (i.e. the orientations shown in the right-hand picture in Figure 12) that we can indeed go on. In the end all sets $|\sigma_{\mathbf{b}}|$ and $|\sigma_{\mathbf{w}}|$ have an oriented cyclic structure and can be identified with cyclic permutations $\sigma_{\mathbf{b}}$ and $\sigma_{\mathbf{w}}$. The construction of the 3-constellation (E, σ_0, σ_1) is finished by setting

$$\sigma_0 = \prod_{\mathsf{b}\in\mathsf{B}} \sigma_{\mathbf{b}}\,, \qquad \sigma_1 = \prod_{\mathsf{w}\in\mathsf{W}} \sigma_{\mathbf{w}}^{-1}\,;$$

the reason for inverting the white cycles is shown in Figure 14.

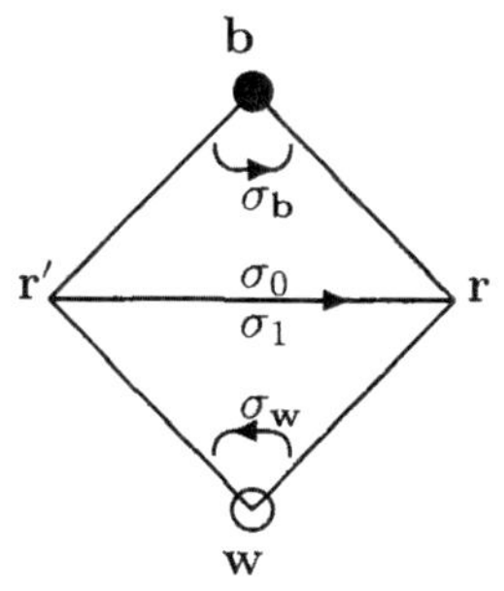

Figure 14 Orientation of cycles in 3-constellation and orientation of $\mathcal{M}$.

To go from (E, σ_0, σ_1) to $\mathbb{M}$ one views the cycles of the permutations σ_0, σ_1 and $\sigma_0\sigma_1^{-1}$ as elements of the sets E_{σ_0}, E_{σ_1} $E_{\sigma_0\sigma_1^{-1}}$, respectively, and then sets $\mathbb{M} = \{(\mathbf{b}(e), \mathbf{w}(e), \mathbf{r}(e), \mathbf{r}'(e))\}_{e\in E}$ with

$$\begin{aligned} \mathbf{b}(e) &= \text{ the cycle of } \sigma_0 \text{ which contains } e\,, \\ \mathbf{w}(e) &= \text{ the cycle of } \sigma_1 \text{ which contains } e\,, \\ \mathbf{r}(e) &= \text{ the cycle of } \sigma_0\sigma_1^{-1} \text{ which contains } e\,, \\ \mathbf{r}'(e) &= \mathbf{r}(\sigma_1(e))\,. \end{aligned}$$

8.11. Example. The procedure of 8.10 converts the list $\mathbb{M}$ in 6.19 into the 3-constellation (E, σ_0, σ_1) with

$$\begin{aligned} E &= \{1,2,3,4,5,6,7,8,9,10,11,12,13,14,15,16,17,18\}\,, \\ \sigma_0 &= (3,7,15)(4,12,16)(6,9,17)(5,11,18)(2,8,13)(1,10,14)\,, \\ \sigma_1 &= (4,8,15)(3,10,16)(5,7,17)(6,12,18)(1,9,13)(2,11,14)\,. \end{aligned}$$

Since in this example the cycle $\mathbf{z}_1 = (1,10,3,7,5,11,2,8,4,12,6,9)$ meets each of the above 3-cycles the conversion went especially fast.

The corresponding superpotential W is

$$\begin{aligned} &X_3X_7X_{15} + X_4X_{12}X_{16} + X_6X_9X_{17} + X_5X_{11}X_{18} + X_2X_8X_{13} + X_1X_{10}X_{14} \\ &-X_4X_8X_{15} - X_3X_{10}X_{16} - X_5X_7X_{17} - X_6X_{12}X_{18} - X_1X_9X_{13} - X_2X_{11}X_{14}. \end{aligned}$$

9 The determinant of the bi-adjacency matrix and the secondary polygon

9.1. Corollary 7.3 implies that the bi-adjacency matrix $\mathbb{K}^{\varpi} = (\kappa_{ij})$ of the dessin $(\mathcal{M}, \mathsf{Q})$ is a square matrix. Its determinant is by definition

$$\det \mathbb{K}^{\varpi} = \sum_{\tau} \operatorname{sign}(\tau)\kappa_{i\,\tau(i)}$$

where τ runs over the set of all bijections $\mathsf{B} \xrightarrow{\simeq} \mathsf{W}$ and $\operatorname{sign}(\tau)$ is defined as the sign of the permutation $\tau_0^{-1}\tau$ of B for some reference bijection $\tau_0 : \mathsf{B} \xrightarrow{\simeq} \mathsf{W}$. Of course, if we want to actually write $\mathbb{K}^{\varpi}$ as a matrix, we must choose bijections between B, W and the set of numbers $\{1, \ldots, \mathrm{vol}_{\mathcal{A}}\}$ and that fixes τ_0. Changing the reference bijection multiplies $\det \mathbb{K}^{\varpi}$ by ± 1, but that is for our purposes unimportant. Next note that the $(\mathbf{b}, \mathbf{w})$-entry of $\mathbb{K}^{\varpi}$ is non-zero if and only if there is an edge of $\Gamma^{\bullet\circ}$ connecting the nodes $\mathbf{b}$ and $\mathbf{w}$. So the only bijections τ that contribute to the determinant of $\mathbb{K}^{\varpi}$ are the *perfect matchings*; see Definition 7.1. Thus we find

$$\begin{aligned}\det \mathbb{K}^{\varpi} &= \pm \sum_{\text{perfect matchings } P} \operatorname{sign}(P) \prod_{e \in P} \varpi(e) u_{\mathbf{r}(e)} u_{\mathbf{r}'(e)} \\ &= \pm \sum_{\text{perfect matchings } P} \operatorname{sign}(P) \left(\prod_{e \in P} \varpi(e) \right) \mathbf{u}^{\widehat{P}} \end{aligned} \tag{29}$$

where

$$\widehat{P} = \sum_{e \in P} (\mathbf{e}_{\mathbf{r}(e)} + \mathbf{e}_{\mathbf{r}'(e)}) \in \mathbb{Z}^N, \tag{30}$$

and $\mathbf{u}^{\mathbf{p}} = u_1^{p_1} \cdot \ldots \cdot u_N^{p_N}$ for $\mathbf{p} = (p_1, \ldots, p_N) \in \mathbb{Z}^N$.

9.2. Definition. The *Newton polytope* of a Laurent polynomial

$$f = \sum_{(k_1, \ldots, k_N) \in \mathbb{Z}^N} c_{(k_1, \ldots, k_N)} u_1^{k_1} \cdot \ldots \cdot u_N^{k_N} \in \mathbb{C}[u_1^{\pm}, \ldots, u_N^{\pm}]$$

is the polytope

$$\mathsf{Newton}(f) := \mathsf{convex\ hull}\left(\{ (k_1, \ldots, k_N) \mid c_{(k_1, \ldots, k_N)} \neq 0 \} \right).$$

9.3. Theorem. *The Newton polytope of the determinant of the bi-adjacency matrix $\mathbb{K}^{\varpi}$ of the dessin $(\mathcal{M}, \mathsf{Q})$ is the same as the secondary polygon of $\mathbb{L} \subset \mathbb{Z}^N$ (see 3.3):*

$$\mathsf{Newton}(\det \mathbb{K}_{\varpi}) = \Sigma(\mathbb{L}). \tag{31}$$

The vertices of $\mathsf{Newton}(\det \mathbb{K}_{\varpi})$ *are the points $\widehat{P_C}$ (see (30)) given by the perfect matchings P_C with C a 2-dimensional cone in the secondary fan.*

Proof If P and P' are two perfect matchings then the edges in P oriented black-to-white together with the edges in P' oriented white-to-black form a collection of closed loops on the torus $\mathcal{T}$ in 6.13. The first homology group of this torus is $\mathbb{L}$. Therefore $\widehat{P} - \widehat{P'}$ lies in $\mathbb{L} \subset \mathbb{Z}^N$. This shows that $\mathsf{Newton}(\det \mathbb{K}^{\varpi})$ lies in a 2-dimensional plane in $\mathbb{R}^N$ parallel to $\mathbb{L}_{\mathbb{R}}$.

From Equation (30) one sees that for every perfect matching P

$$\widehat{P} = \sum_{i=1}^{N} \sharp\{ e \in P \mid i \in \{\mathbf{r}(e), \mathbf{r}'(e)\}\}\, \mathbf{e}_i \,,$$

i.e. the i-th coordinate of the vector $\widehat{P}$ equals the number of edges in the perfect matching P which intersect the i-th zigzag loop (cf. 6.15). It follows from Proposition 6.5 that the number of edges of the i-th zigzag loop is $\sum_{j=1}^{N} |\det(\mathbf{b}_i, \mathbf{b}_j)|$. Since consecutive edges in a zigzag loop can never intersect the same perfect matching we see that for every perfect matching P and every i

$$\sharp\{e \in P \mid i \in \{\mathbf{r}(e), \mathbf{r}'(e)\}\} \leq \tfrac{1}{2} \sum_{j=1}^{N} |\det(\mathbf{b}_i, \mathbf{b}_j)|. \tag{32}$$

Fix i and let C and C' be the two 2-dimensional cones in the secondary fan containing the half-line $\mathbb{R}_{\geq 0}\mathbf{b}_i$. Then $\{i,j\} \in L_C \Leftrightarrow \det(\mathbf{b}_i, \mathbf{b}_j) \geq 0$ and $\{i,j\} \in L_{C'} \Leftrightarrow \det(\mathbf{b}_i, \mathbf{b}_j) \leq 0$. Since $\sum_{j=1}^{N} \det(\mathbf{b}_i, \mathbf{b}_j) = 0$ we find

$$\sum_{j,\,\{i,j\}\in L_C} |\det(\mathbf{b}_i, \mathbf{b}_j)| = \sum_{j,\,\{i,j\}\in L_{C'}} |\det(\mathbf{b}_i, \mathbf{b}_j)| = \tfrac{1}{2} \sum_{j=1}^{N} |\det(\mathbf{b}_i, \mathbf{b}_j)|.$$

Thus we see that for i and the perfect matchings P_C and $P_{C'}$ Equation (32) is in fact an equality.

Fix a 2-dimensional cone C in the secondary fan. It is bounded by two half-lines $\mathbb{R}_{\geq 0}\mathbf{b}_i$ and $\mathbb{R}_{\geq 0}\mathbf{b}_{i'}$. The above argument now shows that Equation (32) is in fact an equality for the perfect matching P_C and i and for P_C and i' in place of i. Thus the point $\widehat{P_C}$ is a vertex of $\mathsf{Newton}(\det \mathbb{K}_w)$. This together with (32) proves:

$$\mathsf{Newton}(\det \mathbb{K}_{\varpi}) = \mathsf{convex\ hull}\left(\{\widehat{P_C} \mid C \text{ a 2-dim. cone of the secondary fan of } \mathbb{L}\}\right).$$

To complete the proof of (31) we note that Equations (4), (5), (27) and (30) together with Proposition 6.5 show $\widehat{P_C} = \psi_C$. □

9.4. Example. According to [5] §8 the quiver for model II of dP_3 (i.e. case B_8 in Figure 4) affords two different superpotentials. This is confirmed by our algorithm, which yields the following two bi-adjacency matrices

$$\mathbb{K}_1^{\varpi} = \begin{bmatrix} \varpi_2 u_1 u_2 & \varpi_8 u_2 u_4 & 0 & \varpi_4 u_1 u_4 \\ \varpi_{14} u_5 u_6 & \varpi_{10} u_2 u_6 & \varpi_9 u_2 u_5 & 0 \\ \varpi_6 u_1 u_6 + \varpi_7 u_2 u_3 & \varpi_{13} u_4 u_6 & \varpi_1 u_1 u_2 & \varpi_{11} u_3 u_4 \\ \varpi_{12} u_3 u_5 & 0 & \varpi_5 u_1 u_5 & \varpi_3 u_1 u_3 \end{bmatrix}$$

$$\mathbb{K}_2^{\varpi} = \begin{bmatrix} \varpi_2 u_1 u_2 & \varpi_9 u_2 u_5 & 0 & \varpi_5 u_1 u_5 \\ \varpi_4 u_1 u_4 & \varpi_1 u_1 u_2 & \varpi_8 u_2 u_4 & 0 \\ \varpi_7 u_2 u_3 & \varpi_{14} u_5 u_6 & \varpi_{10} u_2 u_6 & \varpi_{12} u_3 u_5 \\ \varpi_{11} u_3 u_4 & \varpi_6 u_1 u_6 & \varpi_{13} u_4 u_6 & \varpi_3 u_1 u_3 \end{bmatrix}$$

The corresponding superpotentials are:

$$\begin{aligned} W_1 &= X_2X_8X_4 + X_9X_{14}X_{10} + X_1X_7X_{11}X_{13}X_6 + X_3X_{12}X_5 \\ &\quad -X_2X_7X_{12}X_{14}X_6 - X_8X_{13}X_{10} - X_1X_9X_5 - X_3X_{11}X_4 \end{aligned}$$

$$\begin{aligned} W_2 &= X_2X_9X_5 + X_1X_8X_4 + X_7X_{12}X_{14}X_{10} + X_3X_{11}X_{13}X_6 \\ &\quad -X_2X_7X_{11}X_4 - X_1X_9X_{14}X_6 - X_8X_{13}X_{10} - X_3X_{12}X_5 \end{aligned}$$

After figuring out the rule for translating the edge labels one finds that the prepotentials in Eqs. (8.1) and (8.2) of [5] are $W_A = W_1$ and $W_B = W_2$.

For both superpotentials the surface $\mathcal{M}$ has genus 1. Figures 18 and 19 in [5] show pictures of the planar periodic bipartite graph which is the lifting of $\Gamma^{\bullet\circ}$ to the simply connected cover of $\mathcal{M}$.

One cannot expect the Kasteleyn matrices K_A and K_B in [5] Eq. (8.3) to be the same as the bi-adjacency matrices $\mathbb{K}_1^{\varpi}$ and $\mathbb{K}_2^{\varpi}$, respectively, because of the "untwist". $\mathbb{K}_1^{\varpi}$ and $\mathbb{K}_2^{\varpi}$ are Kasteleyn matrices of the dimer models before the untwisting (see 8.2). Moreover Eqs. (8.4) and (8.5) of [5] show that the Newton polygons of the determinants of the Kasteleyn matrices K_A and K_B are different. On the contrary, for both $\mathbb{K}_1^{\varpi}$ and $\mathbb{K}_2^{\varpi}$ the determinant is a linear combination of the monomials

$$\mathbf{u}^{[3,3,1,0,0,1]},\ \mathbf{u}^{[2,2,1,1,1,1]},\ \mathbf{u}^{[1,1,1,2,2,1]},\ \mathbf{u}^{[2,0,0,2,2,2]},\ \mathbf{u}^{[3,1,0,1,1,2]},\ \mathbf{u}^{[0,2,2,2,2,0]},\ \mathbf{u}^{[1,3,2,1,1,0]}.$$

So, in agreement with Theorem 9.3, the Newton polygons of $\det \mathbb{K}_1^{\varpi}$ and $\det \mathbb{K}_2^{\varpi}$ coincide with the polygon in Figure 4 case B_8 and 3.7. In spite of having the same Newton polygon $\det \mathbb{K}_1^{\varpi}$ and $\det \mathbb{K}_2^{\varpi}$ are not equal; indeed when all weights are 1 they differ in the coefficient of $\mathbf{u}^{[2,2,1,1,1,1]}$.

10 Bi-adjacency matrix with critical weights and the principal $\mathcal{A}$-determinant

10.1. In [11] p.297 Eq. (1.1) Gelfand, Kapranov and Zelevinsky define for a set $\mathcal{A} = (\mathbf{a}_1, \ldots, \mathbf{a}_N) \subset \mathbb{Z}^{k+1}$ as in 4.1 the *principal $\mathcal{A}$-determinant* $E_{\mathcal{A}}(f_{\mathcal{A}})$. The definition uses the Laurent polynomial

$$f_{\mathcal{A}} = \sum_{i=1}^{N} u_i \mathbf{x}^{\mathbf{a}_i}, \tag{33}$$

where $\mathbf{x}^{\mathbf{m}} = x_1^{m_1} x_2^{m_2} \cdot \ldots \cdot x_{k+1}^{m_{k+1}}$ for $\mathbf{m} = (m_1, \ldots, m_{k+1}) \in \mathbb{Z}^{k+1}$ and where the coefficients $u_1, \ldots, u_N$ are variables. The name "principal $\mathcal{A}$-determinant" refers to the fact ([11] p.298 Prop. 1.1) that $E_{\mathcal{A}}(f_{\mathcal{A}})$ can be written as the determinant of some exact complex, i.e.

$$E_{\mathcal{A}}(f_{\mathcal{A}}) = \prod_j (\det\, M_j)^{(-1)^j}$$

for certain matrices M_j. Another useful description of $E_{\mathcal{A}}(f_{\mathcal{A}})$ is given in [11] p.299 Prop. 1.2:

$$E_{\mathcal{A}}(f_{\mathcal{A}}) = \pm \prod_{\Gamma \subset \mathsf{convex\ hull}(\mathcal{A})} \Delta_{\mathcal{A}\cap\Gamma}(f_{\mathcal{A}\cap\Gamma})^{m(\Gamma)} \tag{34}$$

where the product runs over all faces Γ of the primary polytope $\mathsf{convex\ hull}(\mathcal{A})$ (cf. 4.12); $m(\Gamma)$ is a multiplicity and $\Delta_{\mathcal{A}\cap\Gamma}(f_{\mathcal{A}\cap\Gamma})$ is the *$(\mathcal{A}\cap\Gamma)$-discriminant* of the Laurent polynomial $f_{\mathcal{A}\cap\Gamma} = \sum_{i:\,\mathbf{a}_i\in\Gamma} u_i \mathbf{x}^{\mathbf{a}_i}$. The latter discriminant is a polynomial in the variables u_i with $\mathbf{a}_i \in \Gamma$. To define it (see [11] p.271) one needs the algebraic set $\nabla_{\mathcal{A}\cap\Gamma}$ which is the closure in $\mathbb{C}^{\mathcal{A}\cap\Gamma}$ of

$$\{\mathbf{u} \in \mathbb{C}^{\mathcal{A}\cap\Gamma} \mid \exists \mathbf{x}_0 \in (\mathbb{C}^*)^{k+1}\ s.t.\ f_{\mathcal{A}\cap\Gamma}(\mathbf{x}_0) = \frac{\partial f_{\mathcal{A}\cap\Gamma}}{\partial x_i}(\mathbf{x}_0) = 0,\ \forall i\}.$$

Then, by definition, $\Delta_{\mathcal{A}\cap\Gamma}(f_{\mathcal{A}\cap\Gamma}) = 1$ if $\mathrm{codim}_{\mathbb{C}^{\mathcal{A}\cap\Gamma}}(\nabla_{\mathcal{A}\cap\Gamma}) > 1$ and

$$\text{zero locus of } \Delta_{\mathcal{A}\cap\Gamma}(f_{\mathcal{A}\cap\Gamma}) = \nabla_{\mathcal{A}\cap\Gamma} \quad \text{if} \quad \mathrm{codim}_{\mathbb{C}^{\mathcal{A}\cap\Gamma}}(\nabla_{\mathcal{A}\cap\Gamma}) = 1.$$

So, $E_{\mathcal{A}}(f_{\mathcal{A}})$ gives the locus of the points $(u_1, \ldots, u_N) \in \mathbb{C}^N$ for which at least one of the Laurent polynomials $f_{\mathcal{A}\cap\Gamma}$ has a critical point with critical value 0.

10.2. Theorem.([11]p.302 Thm.1.4; cf.(17))
The Newton polytope of $E_{\mathcal{A}}(f_{\mathcal{A}})$ *coincides with the secondary polytope* $\Sigma(\mathcal{A})$.

10.3. Theorems 10.2 and 9.3 in combination with Formula (20) make one wonder whether for an appropriate choice of the weight ϖ the determinant of the bi-adjacency matrix $\mathbb{K}^{\varpi}$ is equal to the principal $\mathcal{A}$-determinant $E_{\mathcal{A}}(f_{\mathcal{A}})$, up to the simple transformation necessitated by (20). In order to formulate this transformation we must make the dependence on $u_1, \ldots, u_N$ visible by writing $\mathbb{K}^{\varpi}(u_1, \ldots, u_N)$ and $E_{\mathcal{A}}(f_{\mathcal{A}}(u_1, \ldots, u_N))$.

After some experimenting with examples I found a very natural and simple choice for the weight that does the job.

10.4. Definition. The *critical weight* for the arrows of the quiver Q is the function

$$\mathrm{crit} : E \to \mathbb{Z}_{>0}\,, \qquad \mathrm{crit}(e) = \sharp\{e' \in E \,|\, s(e') = s(e),\, t(e') = t(e)\} \tag{35}$$

10.5. Conjecture.[1] *For every set* $\mathcal{A}$ *as in 4.1 and every dessin* $(\mathcal{M}, \mathsf{Q})$ *(see 6.18) constructed from* $\mathcal{A}$ *by the algorithm in Section 6 the determinant of the bi-adjacency matrix with critical weight and the principal* $\mathcal{A}$*-determinant satisfy:*

$$(u_1 \cdot \ldots \cdot u_N)^{\mathrm{vol}_{\mathcal{A}}} \det \mathbb{K}^{\mathrm{crit}}(u_1^{-1}, \ldots, u_N^{-1}) \;=\; E_{\mathcal{A}}(f_{\mathcal{A}}(u_1, \ldots, u_N))\,. \tag{36}$$

10.6. Remark. *In support of the above conjecture we can point out that the coefficients of the monomials corresponding to the vertices of the secondary polygons are, up to sign, the same for the two sides of Equation (36).* Indeed, Theorem 1.4 of [11]p.302 gives the coefficient of the monomial in the principal $\mathcal{A}$-determinant with corresponds to a vertex of the secondary polytope. Such a vertex corresponds to a maximal cone C in the secondary fan. In the notations of Definition 3.3 the formula in loc. cit. for the coefficient of the monomial corresponding to C reads

$$\pm \prod_{\{i,j\} \in L_C} |\det(\mathbf{b}_i, \mathbf{b}_j)|^{|\det(\mathbf{b}_i, \mathbf{b}_j)|}\,. \tag{37}$$

On the other hand, Theorems 7.2 and 9.3 show that the same vertex corresponds to a perfect matching P_C. From the role of perfect matchings in the computation of the determinant of the bi-adjacency matrix (see 9.1) one now easily checks that the coefficient of the monomial in the determinant, which corresponds to C, is (possibly up to sign) the same as (37).

In the remainder of this section we explicitly verify Conjecture 10.5 in some examples.

10.7. Example. For case $B_2 = \begin{bmatrix} 0 & 1 & 1 & -2 \\ -1 & 0 & 2 & -1 \end{bmatrix}$ in Figure 3 the algorithm in Section 6 yields in the following bi-adjacency matrix with critical weights:

$$\mathbb{K}^{\mathrm{crit}} = \begin{bmatrix} 2u_1u_4 & 3u_3u_4 & u_1u_3 \\ 2u_2u_3 & u_2u_4 & 3u_3u_4 \\ u_1u_2 + 3u_3u_4 & 2u_2u_3 & 2u_1u_4 \end{bmatrix}.$$

[1] *Added in proof:* Meanwhile this conjecture has been proven; see [26].

One easily computes

$$\det \mathbb{K}^{\text{crit}} = 27\mathbf{u}^{[0,0,3,3]} + 4\mathbf{u}^{[1,2,3,0]} + 4\mathbf{u}^{[2,1,0,3]} - 18\mathbf{u}^{[1,1,2,2]} - \mathbf{u}^{[2,2,1,1]}$$

Note that the exponents are the same as the coordinates of the vertices and the interior point of the secondary polygon in 3.5. For the computation of the principal $\mathcal{A}$-determinant we note that in this case (see also Example 4.6)

$$\mathcal{A} = (\mathbf{a}_1, \mathbf{a}_2, \mathbf{a}_3, \mathbf{a}_4) = \left(\begin{bmatrix} 1 \\ 0 \end{bmatrix}, \begin{bmatrix} 1 \\ 3 \end{bmatrix}, \begin{bmatrix} 1 \\ 1 \end{bmatrix}, \begin{bmatrix} 1 \\ 2 \end{bmatrix}\right)$$

and hence $f_{\mathcal{A}} = x_1(u_1 + u_2x_2^3 + u_3x_2 + u_4x_2^2)$. The primary polytope is shown in Figure 2. The two boundary points of this primary polytope contribute factors u_1 and u_2 to the principal $\mathcal{A}$-determinant (cf. Equation (34)). The contribution from the full primary polytope is the very classical discriminant of the cubic polynomial $u_1 + u_3x_2 + u_4x_2^2 + u_2x_2^3$, which is (e.g. [11] p.405)

$$27u_1^2u_2^2 + 4u_1u_4^3 + 4u_3^3u_2 - u_3^2u_4^2 - 18u_1u_3u_4u_2\,.$$

Thus we find

$$E_{\mathcal{A}}(f_{\mathcal{A}}) = 27\mathbf{u}^{[3,3,0,0]} + 4\mathbf{u}^{[2,1,0,3]} + 4\mathbf{u}^{[1,2,3,0]} - \mathbf{u}^{[1,1,2,2]} - 18\mathbf{u}^{[2,2,1,1]}\,.$$

This shows that Conjecture 10.5 holds in this case.

10.8. Example. When critical weights are used the two bi-adjacency matrices in Example 9.4 become

$$\mathbb{K}_1^{\text{crit}} = \begin{bmatrix} 2u_1u_2 & u_2u_4 & 0 & u_1u_4 \\ u_5u_6 & u_2u_6 & u_2u_5 & 0 \\ u_1u_6 + u_2u_3 & u_4u_6 & 2u_1u_2 & u_3u_4 \\ u_3u_5 & 0 & u_1u_5 & u_1u_3 \end{bmatrix}$$

$$\mathbb{K}_2^{\text{crit}} = \begin{bmatrix} 2u_1u_2 & u_2u_5 & 0 & u_1u_5 \\ u_1u_4 & 2u_1u_2 & u_2u_4 & 0 \\ u_2u_3 & u_5u_6 & u_2u_6 & u_3u_5 \\ u_3u_4 & u_1u_6 & u_4u_6 & u_1u_3 \end{bmatrix}$$

One easily computes det $\mathbb{K}_1^{\text{crit}}$ and det $\mathbb{K}_2^{\text{crit}}$ and finds that both are equal to

$$4\mathbf{u}^{[3,3,1,0,0,1]} + 2\mathbf{u}^{[1,1,1,2,2,1]} + \mathbf{u}^{[1,3,2,1,1,0]} + \mathbf{u}^{[3,1,0,1,1,2]} - \\ -\mathbf{u}^{[2,0,0,2,2,2]} - \mathbf{u}^{[0,2,2,2,2,0]} - 6\mathbf{u}^{[2,2,1,1,1,1]}.$$

This gives a remarkable contrast with the last line of Example 9.4 and underlines the role of the critical weight in making Conjecture 10.5 reasonable for every dessin obtained from $\mathcal{A}$.

Transforming the above determinant as in the left-hand side of Equation (36) yields

$$4\mathbf{u}^{[1,1,3,4,4,3]} + 2\mathbf{u}^{[3,3,3,2,2,3]} + \mathbf{u}^{[3,1,2,3,3,4]} + \mathbf{u}^{[1,3,4,3,3,2]} - \mathbf{u}^{[2,4,4,2,2,2]} - \\ -\mathbf{u}^{[4,2,2,2,2,4]} - 6\mathbf{u}^{[2,2,3,3,3,3]}$$

$$= u_1u_2u_3^2u_4^2u_5^2u_6^2\,(u_4u_5 - u_1u_2)(4u_3u_4u_5u_6 + u_2^2u_3^2 - 2u_1u_2u_3u_6 + u_1^2u_6^2). \qquad (38)$$

From the matrix B_8 in Figure 4 one easily finds $\mathcal{A}$ and the corresponding polynomial

$$f_{\mathcal{A}} = u_1x_1 + u_2x_1x_2 + u_3x_1x_3 + u_4x_1x_4 + u_5x_1x_2x_4^{-1} + u_6x_1x_2x_3\,.$$

The primary polytope convex hull($\mathcal{A}$) is 3-dimensional and has seven 2-dimensional faces, six of which are triangles and the other is a quadrangle. Its only integer points are its vertices and these correspond to the monomials of $f_{\mathcal{A}}$. One can compute the discriminants for the polynomials supported by the faces of the primary polytope and multiply these to get the principal $\mathcal{A}$-determinant $E_{\mathcal{A}}(f_{\mathcal{A}})$ as in Equation (34). The result is exactly as in (38), with the 4-term factor coming from the primary polytope itself, the 2-term factor coming from the quadrangle face and the other factors with multiplicities coming from the vertices. *This shows that conjecture 10.5 holds in this case.*

10.9. Further examples. Exactly as in Example 10.7 one can verify Conjecture 10.5 for other "four-nomials" of degree ≤ 5 (see Example 4.7) by using the formulas in [11] for discriminants of quartic and quintic polynomials.

The method of Example 10.8 can be successfully employed to verify Conjecture 10.5 for, for instance, cases B_5, B_6, B_{10} in Figure 4. It also works for $B = \begin{bmatrix} -3 & 1 & 1 & 1 & 0 & 0 & 0 \\ -3 & 0 & 0 & 0 & 1 & 1 & 1 \end{bmatrix}$ and $B = \begin{bmatrix} 1 & 1 & 0 & -2 & 0 \\ 0 & 0 & 1 & 1 & -2 \end{bmatrix}$, which are closely related to B_{16} and B_9, respectively.

References

[1] Benvenuti, S., S. Franco, A. Hanany, D. Martelli, J. Sparks, *An Infinite Family of Superconformal Quiver Gauge Theories with Sasaki-Einstein Duals*, JHEP 0506 (2005) 064; also arXiv:hep-th/0411264.

[2] Bruijn, N.G. de, *Algebraic theory of Penrose's non-periodic tilings of the plane I*, Proceedings of the Koninklijke Nederlandse Akademie van Wetenschappen, Series A, vol 84 (= Indagationes Mathematicae, vol 43), 1981, 39-52.

[3] Feng, B., S. Franco, A. Hanany, Y-H. He, *Symmetries of Toric Duality* JHEP 0212 (2002) 076; also arXiv:hep-th/0205144.

[4] Feng, B., Y-H. He, K. Kennaway, C. Vafa, *Dimer Models from Mirror Symmetry and Quivering Amoebae*, arXiv:hep-th/0511287.

[5] Franco, S., A. Hanany, K. Kennaway, D. Vegh, B. Wecht, *Brane Dimers and Quiver Gauge Theories*, JHEP 0601 (2006) 096; also arXiv:hep-th/0504110.

[6] Franco, S., A. Hanany, D. Martelli, J. Sparks, D. Vegh, B. Wecht, *Gauge Theories from Toric geometry and Brane tilings*, JHEP 0601 (2006) 128; also arXiv:hep-th/0505211.

[7] Fulton, W., *Introduction to Toric Varieties*, Annals of Math. Studies, Study 131, Princeton University Press, 1993.

[8] Gelfand, I.M., A.V. Zelevinskii, M.M. Kapranov, *Hypergeometric functions and toral manifolds*, Functional Analysis and its Applications 23 (1989), 94-106.
[8]′ correction to [8], Funct. Analysis and its Appl. 27 (1993) 295.

[9] Gelfand, I.M., M.M. Kapranov, A.V. Zelevinsky, *Generalized Euler Integrals and A- Hypergeometric Functions*, Advances in Math. 84 (1990), 255-271.

[10] Gelfand, I.M., A.V. Zelevinsky, M.M. Kapranov, *Equations of Hypergeometric Type and Newton Polyhedra*, Soviet Math. Dokl. 37 (1988), 678-683.

[11] Gelfand, I.M., M.M. Kapranov, A.V. Zelevinsky, *Discriminants, Resultants and Multidimensional Determinants*, Birkhäuser Boston, 1994.

[12] Hanany, A., C. Herzog, D. Vegh, *Brane Tilings and Exceptional Collections*, JHEP 0607 (2006) 001; also arXiv:hep-th/0602041.

[13] Hanany, A., K. Kennaway, *Dimer models and toric diagrams*, JHEP (2005) 029; also arXiv:hep-th/0503149.

[14] Hanany, A., D. Vegh, *Quivers, Tilings, Branes and Rhombi*, JHEP 0710 (2007) 029; also arXiv:hep-th/0511063.

[15] Kennaway, K., *Brane Tilings*, Internat. J. Modern Phys. A 22 (2007), no. 18, 2977–3038.; also arXiv:0706.1660.

[16] Kenyon, R., *An introduction to the dimer model*, School and Conference on Probability Theory, 267–304, ICTP Lect. Notes XVII, Abdus Salam Int. Cent. Theoret. Phys., Trieste, 2004; also arXiv:math/0310326.

[17] Kenyon, R., A. Okounkov, S. Sheffield, *Dimers and Amoebae*, Annals of Math. (2) **163** (2006), no.3, 1019-1056; also arXiv:math-ph/0311005.

[18] Lando, S., Zvonkin, A., *Graphs on Surfaces and Their Applications*, Encyclopedia of Math. Sciences vol. 141, subseries Low-Dimensional Topology vol. 11, Springer-Verlag, Berlin-Heidelberg, 2004.

[19] Martelli, D., J. Sparks, *Toric Geometry, Sasaki-Einstein Manifolds and a New Infinite Class of AdS/CFT Duals*, Commun. Math. Phys. 262 (2006) 51-89; also arXiv:hep-th/0411238.

[20] Okounkov, A., N. Reshetikin, C. Vafa, *Quantum Calabi-Yau and Classical Crystals*, The unity of mathematics, Progress in Mathematics vol. 244, Birkhäser, Boston 2006, pp. 597–618; also arXiv:hep-th/0309208.

[21] Poonen, B., F. Rodriguez-Villegas, *Lattice polygons and the number 12*, Am. Math. Monthly 107 (2000) 238-250.

[22] *The Grothendieck Theory of Dessins d'Enfants*, L. Schneps (ed.) London Math. Soc. Lecture Note Series 200, Cambridge University Press, 1994.

[23] Senechal, M., *Quasicrystals and geometry*, Cambridge University Press, 1995.

[24] Shabat, G., V. Voevodsky, *Drawing Curves Over Number Fields*, in: *The Grothendieck festschrift Volume III*, P. Cartier et. al. (eds), Progress in Mathematics vol. 88, Birkhäuser, Boston, 1990, pp. 199-227.

[25] Stienstra, J., *GKZ Hypergeometric Structures*, in: *Arithmetic and Geometry Around Hypergeometric Functions*, R-P. Holzapfel, A. M. Uludağ, M. Yoshida (eds.), Progress in Mathematics vol. 260, Birkhäuser, Basel, 2007, pp. 313-371; see also math.AG/0511351.

[26] Stienstra, J., *Chow Forms, Chow Quotients and Quivers with Superpotential*, arXiv:0803.3908.

[27] Sturmfels, B., *Gröbner Bases and Convex Polytopes,* University Lecture Series vol. 8, American Math. Soc., 1996.

Fields Institute Communications
Volume **54**, 2008

Some Properties of Hypergeometric Series Associated with Mirror Symmetry

Don Zagier
Max-Planck-Institut für Mathematik
Vivatsgasse 7
53111 Bonn, Germany
zagier@mpim-bonn.mpg.de

Aleksey Zinger
Department of Mathematics
SUNY Stony Brook
NY 11794-3651 USA
azinger@math.sunysb.edu

Abstract. We show that certain hypergeometric series used to formulate mirror symmetry for Calabi-Yau hypersurfaces, in string theory and algebraic geometry, satisfy a number of interesting properties. Many of these properties are used in separate papers to verify the BCOV prediction for the genus one Gromov-Witten invariants of a quintic threefold and more generally to compute the genus one Gromov-Witten invariants of any Calabi-Yau projective hypersurface.

1 Introduction

An astounding prediction for the genus zero Gromov-Witten invariants of (counts of rational curves in) a quintic threefold was made in [CaDGP]. It was formulated in terms of the function $\mathcal{F}$ defined in (1.1) below and related objects. This 1991 mirror symmetry prediction was mathematically verified about five years later. The 1993 mirror symmetry prediction of [BCOV] for the genus one Gromov-Witten invariants of a quintic threefold was verified in [Z1], and a generalization to degree n hypersurfaces in $\mathbb{C}P^{n-1}$ for arbitrary n was formulated and proved in [Z2]. The proofs in these two papers make use of the properties of $\mathcal{F}$ described by Theorems 1.1, 1.2 and 1.4 below. Theorem 1.5 explores related properties of $\mathcal{F}$ which appear to be of interest in their own right and may also be of use in computation of higher genus Gromov-Witten invariants. Some further conjectural properties are stated in Section 3.

We denote by

$$\mathcal{P} \subset 1 + x\mathbb{Q}(w)[[x]]$$

2000 *Mathematics Subject Classification.* Primary 33C70; Secondary 14J32, 14N35, 33C90.

the subgroup of power series in x with constant term 1 whose coefficients are rational functions in w which are holomorphic at $w=0$. Thus, the evaluation map

$$\mathcal{P} \to 1 + x\mathbb{Q}[[x]], \qquad F(w,x) \mapsto F(0,x),$$

is well-defined. We define a map $\mathbf{M}: \mathcal{P} \to \mathcal{P}$ by

$$\mathbf{M}F(w,x) = \left\{1 + \frac{x}{w}\frac{\partial}{\partial x}\right\}\frac{F(w,x)}{F(0,x)}.$$

Our first result says that the hypergeometric functions arising in the mirror symmetry predictions are periodic fixed points of the map $\mathbf{M}$.

Theorem 1.1 *Let n be a positive integer and $\mathcal{F} \in \mathcal{P}$ the hypergeometric series*

$$\mathcal{F}(w,x) = \sum_{d=0}^{\infty} x^d \frac{\prod_{r=1}^{r=nd}(nw+r)}{\prod_{r=1}^{r=d}((w+r)^n - w^n)}. \tag{1.1}$$

Then $\mathbf{M}^n\mathcal{F} = \mathcal{F}$.

Note that we consider n as fixed and therefore omit it from the notations.

If we now define further power series $\mathcal{F}_p \in \mathcal{P}$ and $I_p \in 1 + x\mathbb{Q}[[x]]$ for all $p \geq 0$ by

$$\mathcal{F}_p(w,x) = \mathbf{M}^p\mathcal{F}(w,x), \qquad I_p(x) = \mathcal{F}_p(0,x),$$

so that $\mathcal{F}_{p+1} = (1 + w^{-1}x\,\partial/\partial x)(\mathcal{F}_p/I_p)$, then Theorem 1.1 says that $\mathcal{F}_{n+p} = \mathcal{F}_p$ and consequently $I_{n+p} = I_p$ for all $p \geq 0$. The next result gives further properties of the functions $\{I_p\}_{p\in\mathbb{Z}/n\mathbb{Z}}$.

Theorem 1.2 *The power series $I_p(x)$, $0 \leq p \leq n-1$, satisfy*

$$I_0(x)\,I_1(x)\cdots I_{n-1}(x) = (1 - n^n x)^{-1}, \tag{1.2}$$

$$I_0(x)^{n-1}I_1(x)^{n-2}\cdots I_{n-1}(x)^0 = (1 - n^n x)^{-(n-1)/2}, \tag{1.3}$$

$$I_p(x) = I_{n-1-p}(x) \qquad (0 \leq p \leq n-1). \tag{1.4}$$

We note that (1.2) and the symmetry property (1.4) imply (1.3). However, (1.3) is simpler to prove directly than (1.4) and will be verified together with (1.2) before we give the proof of (1.4).

The power series I_p describe the structure of $\mathcal{F}$ at $w=0$. We will also give information about its structure at $w=\infty$. We begin with the following observation, which will be proved in Subsection 2.3.

Lemma 1.3 *If $F \in \mathcal{P}$ and $\mathbf{M}^k F = F$ for some $k > 0$, then every coefficient of the power series $\log F(w,x) \in \mathbb{Q}(w)[[x]]$ is $\mathrm{O}(w)$ as $w \to \infty$.*

Applying this lemma to $F = \mathcal{F}$, which satisfies its hypothesis by Theorem 1.1, we find that $\log\mathcal{F}(w,x)$ has an asymptotic expansion $\sum_{j=-1}^{\infty}\mu_j(x)w^{-j}$ with $\mu_j(x) \in x\mathbb{Q}[[x]]$ for all $j \geq -1$ or equivalently, that $\mathcal{F}(w,x)$ itself has an asymptotic expansion

$$\mathcal{F}(w,x) \sim e^{\mu(x)w}\sum_{s=0}^{\infty}\Phi_s(x)\,w^{-s} \qquad (w \to \infty) \tag{1.5}$$

for some power series $\mu = \mu_{-1}$, $\Phi_0 = e^{\mu_0}$, $\Phi_1 = \Phi_0\mu_1, \ldots$ in $\mathbb{Q}[[x]]$.

Theorem 1.4 *The first three coefficients $\mu(x)$, $\Phi_0(x)$, and $\Phi_1(x)$ in the expansion (1.5) are given by*

$$\mu(x) = \int_0^x \frac{L(u)-1}{u}\,du\,, \qquad \Phi_0(x) = L(x)\,, \qquad \Phi_1(x) = \frac{(n-2)(n+1)}{24n}\,\bigl(L(x)-L(x)^n\bigr)\,, \tag{1.6}$$

where $L(x)$ denotes the power series $(1-n^n x)^{-1/n} \in \mathbb{Z}[[x]]$.

The proof of this theorem in Subsection 2.3 can be systematized and streamlined to obtain an algorithm for computing every Φ_s by a differential recursion, which we now state. For integers $m \ge j \ge 0$ (and for our fixed integer n) we define $\mathcal{H}_{m,j} = \mathcal{H}_{m,j}(X) \in \mathbb{Q}[X]$ inductively by

$$\mathcal{H}_{0,j} = \delta_{0,j}, \qquad \mathcal{H}_{m,j} = \mathcal{H}_{m-1,j} + (X-1)\Bigl(X\frac{d}{dX} + \frac{m-j}{n}\Bigr)\mathcal{H}_{m-1,j-1} \quad \text{for } m \ge 1 \tag{1.7}$$

(with the convention that $\mathcal{H}_{m-1,j-1} = 0$ if $j=0$). For example, for $0 \le j \le 2$ we find

$$\begin{aligned} \mathcal{H}_{m,0}(X) = 1, \qquad \mathcal{H}_{m,1}(X) &= \frac{1}{n}\binom{m}{2}(X-1), \\ \mathcal{H}_{m,2}(X) &= \frac{1}{n^2}\binom{m}{3}\bigl((n+1)X-1\bigr)(X-1) + \frac{3}{n^2}\binom{m}{4}(X-1)^2\,; \end{aligned} \tag{1.8}$$

more generally, $\mathcal{H}_{m,j}$ for fixed $j \ge 1$ and varying m has the form $\sum_{k=1}^{j}\binom{m}{j+k}Q_{j,k}(X)$ with $Q_{j,k} \in \mathbb{Z}[n^{-1}, X]$ defined inductively by

$$\begin{aligned} Q_{0,k} &= \delta_{0,k}, \\ Q_{j,k} &= (X-1)\bigl(XQ'_{j-1,k} + (k\,Q_{j-1,k} + (k+j-1)\,Q_{j-1,k-1})/n\bigr) \quad \text{for } j \ge 1. \end{aligned}$$

We now define differential operators $\mathfrak{L}_k$ $(0 \le k \le n)$ on $\mathbb{Q}[[x]]$ by

$$\mathfrak{L}_k = \sum_{i=0}^{k}\Biggl(\binom{n}{i}\mathcal{H}_{n-i,k-i}(L^n) - (L^n-1)\sum_{r=1}^{k-i}\binom{n-r}{i}\frac{S_r(n)}{n^r}\,\mathcal{H}_{n-i-r,k-i-r}(L^n)\Biggr)D^i\,, \tag{1.9}$$

where $D = x\,d/dx$ and $S_r(n)$ denotes the rth elementary symmetric function of $1, 2, \ldots, n$ (a Stirling number of the first kind). Using (1.8), we find that the first two of these operators are

$$\mathfrak{L}_1 = nD - (L^n - 1) = nLDL^{-1}\,, \tag{1.10}$$

$$\mathfrak{L}_2 = \binom{n}{2}D^2 - \frac{3(n-1)}{2}(L^n-1)D + \frac{n-1}{n}\Bigl(\frac{(n-2)(n-11)}{24}L^n - 1\Bigr)(L^n-1)\,. \tag{1.11}$$

Theorem 1.5 *(i) The power series $\Phi_s \in \mathbb{Q}[[x]]$, $s \ge 0$, are determined by the first-order ODEs*

$$\mathfrak{L}_1(\Phi_s) + \frac{1}{L}\,\mathfrak{L}_2(\Phi_{s-1}) + \frac{1}{L^2}\mathfrak{L}_3(\Phi_{s-2}) + \cdots + \frac{1}{L^{n-1}}\,\mathfrak{L}_n(\Phi_{s+1-n}) = 0, \quad s \ge 0, \tag{1.12}$$

(with the convention $\Phi_r = 0$ for $r < 0$) together with the initial condition $\Phi_s(0) = \delta_{0,s}$.
(ii) For fixed s and n, $\Phi_s(x)$ belongs to $L\,\mathbb{Q}[L]$.
(iii) For fixed s, $\Phi_s(x)$ belongs to $\mathbb{Q}(n)[L, L^{-1}, L^n]$.

The meaning of part *(iii)* in Theorem 1.5 is that for each $s \geq 0$ there exists

$$\Psi_s \equiv \Psi_s(a, X, Y, Z) \in \mathbb{Q}(a)[X, Y, Z]$$

such that the function $\Phi_s(x)$ defined by (1.1) and (1.5) is given by

$$\Phi_s(x) = \Psi_s\big(n, L(x), L(x)^{-1}, L(x)^n\big).$$

In particular, *(iii)* neither implies nor is implied by *(ii)*.

For example, from (1.12) for $s = 0$ and $s = 1$ together with equations (1.10) and (1.11) one finds the second and third identity in (1.6), and continuing the same way one obtains

$$\begin{aligned}
\Phi_2 &= \frac{(n+1)^2(n-2)^2}{2\,(24n)^2}\big(L - 2L^n + L^{2n-1}\big) = \Phi_1^2/2L\Phi_0\,,\\
\Phi_3 &= \frac{(n+1)(n-2)}{30\,(24n)^3}\bigg\{(1003n^4 - 2366n^3 + 3759n^2 - 1676n - 164)\,L^{3n-2}\\
&\qquad - 72\,(n-1)(3n-1)(7n^2 - 9n + 14)\,L^{2n-2}\\
&\qquad + 15\,(n+1)^2(n-2)^2\big(L^{2n-1} - L^n\big)\\
&\qquad + 72\,(n-1)(7n^3 - 17n^2 + 22n - 24)\,L^{n-2}\\
&\qquad + (5n^4 + 134n^3 - 447n^2 + 308n - 556)\,L\bigg\},
\end{aligned}$$

illustrating parts (ii) and (iii) of the theorem. These expressions, and the similar formulas obtained for $s \leq 7$, suggest that in fact Φ_s for s fixed and n varying is an element of $\mathbb{Q}[n, n^{-1}, L, L^{-1}, L^n]$, sharpening statement (iii), but we do not know how to prove this. Some further data and a further conjecture concerning the functions $\Phi_s(x)$ are given in Section 3.

Remark. Although the calculations in this paper are completely independent of the theory of Gromov-Witten invariants, this theory does form the principal motivation, so we give here a brief indication of how the function $\mathcal{F}$ defined in (1.1) is used in GW-theory. Mathematical formulations of mirror symmetry ordinarily involve the coefficients of $w^0, \ldots, w^{n-1}$ of the Taylor series expansion of the function $\widetilde{\mathcal{F}}_0$ in equation (2.13) at $w=0$. These coefficients are of course the same as the corresponding coefficients of $\mathcal{F}$. The power series $\widetilde{\mathcal{F}}_0$ and $\mathcal{F}$ are the evaluations at $\alpha_1 = \ldots = \alpha_n = 0$ and $\hbar = 1$ of the functions

$$\widetilde{\mathcal{Y}}_0 \equiv \sum_{d=0}^{\infty} x^d \frac{\prod_{r=1}^{r=nd}(nw+r\hbar)}{\prod_{r=1}^{r=d}\prod_{k=1}^{k=n}(w-\alpha_k+r\hbar)} \qquad \text{and}$$

$$\mathcal{Y} \equiv \sum_{d=0}^{\infty} x^d \frac{\prod_{r=1}^{r=nd}(nw+r\hbar)}{\prod_{r=1}^{r=d}\big(\prod_{k=1}^{k=n}(w-\alpha_k+r\hbar) - \prod_{k=1}^{k=n}(w-\alpha_k)\big)},$$

which appear prominently in the proofs of genus 0 mirror symmetry in [Gi] and of genus 1 mirror symmetry in [Z1], respectively; both proofs involve a localization computation on $\mathbb{P}^{n-1}$. As elements of

$$H_{\mathbb{T}}^*(\mathbb{P}^{n-1};\mathbb{Q})\big[\big[\hbar^{-1},x\big]\big] \approx \Big(\mathbb{Q}[\alpha_1,\ldots,\alpha_n]/\prod_{k=1}^{k=n}(w-\alpha_k)\Big)\big[\big[\hbar^{-1},x\big]\big],$$

where $\mathbb{T}$ is the n-torus acting in the usual way on $\mathbb{P}^{n-1}$, $\widetilde{\mathcal{Y}}_0$ and $\mathcal{Y}$ are the same. On the other hand,

$$\mathrm{Res}_{\hbar=0}\{\mathcal{Y}(w=\alpha_i)\} = \{\mathrm{Res}_{\hbar=0}\mathcal{Y}(w)\}\big|_{w=\alpha_i}.$$

This property, which clearly does not hold for $\widetilde{\mathcal{Y}}_0$ (viewed as a power series in rational functions), is used in an essential way in Subsection 3.4 in [Z1] and eventually reduces the problem of computing genus 1 GW-invariants of X_n to determining the first three coefficients in the expansion (1.5) of $\mathcal{F}$.

2 Proofs

2.1 Preliminaries. It will be convenient to introduce notations D and D_w for the first order differential operators $D = x\frac{d}{dx}$ and $D_w = D + w$ on $\mathbb{Q}(w)[[x]]$. (Here we think of w as a parameter rather than a variable and write simply $\frac{d}{dx}$ instead of $\frac{\partial}{\partial x}\cdot$.) The effect of D_w on a power series $\sum c_d(w)x^d \in \mathbb{Q}(w)[[x]]$ is to multiply each $c_d(w)$ by $w+d$, so D_w has an inverse operator D_w^{-1} which replaces each $c_d(w)$ by $(w+d)^{-1}c_d(w)$. The operator $\mathbf{M}$ defined above can be written in terms of D_w as $F(w,x) \mapsto w^{-1}D_w\big[F(w,x)/F(0,x)\big]$.

We remark that instead of working with the functions $\mathcal{F}_p(w,x)$, we could have worked with the functions $R_p(w,t) = e^{wt}\mathcal{F}_p(w,e^t)$, which are the objects that actually arise in the analysis of the mirror symmetry predictions for Gromov-Witten invariants. If we had done that, then the differential operator $D_w = w + x\,d/dx$ would have been replaced by the simpler differential operator d/dt, explaining why this operator plays such a ubiquitous role in our analysis. But it is easier, both in the calculations and for purposes of exposition, to work with power series over $\mathbb{Q}(w)$ in a single variable x rather than with objects in the less familiar space $e^{wt}\mathbb{Q}(w)[[e^t]]$.

The following lemma and its corollary are key to the proofs of the four theorems stated above.

Lemma 2.1 *Suppose $c_0, \ldots, c_m, f, g, a$ are functions of t (with f not identically 0) satisfying*

$$\begin{aligned} c_m\, f^{(m)} + c_{m-1}\, f^{(m-1)} + \ldots + c_0\, f &= 0, \\ c_m\, g^{(m)} + c_{m-1}\, g^{(m-1)} + \ldots + c_0\, g &= a, \end{aligned} \tag{2.1}$$

where $f^{(k)} = d^k f/dt^k$. Then the function $h := (g/f)'$ satisfies

$$\widetilde{c}_{m-1}\, h^{(m-1)} + \widetilde{c}_{m-2}\, h^{(m-2)} + \ldots + \widetilde{c}_0\, h = a, \tag{2.2}$$

where $\widetilde{c}_s(t) = \sum_{r=s+1}^{m}\binom{r}{s+1}c_r(t)\, f^{(r-1-s)}(t)$.

Proof Using Leibnitz's rule and (2.1), we find

$$\begin{aligned} a = \sum_{r=0}^{m} c_r\,(f\cdot g/f)^{(r)} &= \sum_{r=0}^{m} c_r\left(f^{(r)}g/f + \sum_{s=0}^{r-1}\binom{r}{s+1}f^{(r-1-s)}h^{(s)}\right) \\ &= \sum_{s=0}^{m-1}\widetilde{c}_s\, h^{(s)}. \end{aligned}$$

□

Corollary 2.2 *Suppose $F(w,x)\in\mathcal{P}$ satisfies*

$$\Big(\sum_{r=0}^{m} C_r(x)\,D_w^r\Big)F(w,x) \;=\; A(w,x) \tag{2.3}$$

for some power series $C_0(x),\,\ldots,\,C_m(x)\in\mathbb{Q}[[x]]$ and $A(w,x)\in\mathbb{Q}(w)[[x]]$ with $A(0,x)\equiv 0$. Then

$$\Big(\sum_{s=0}^{m-1} \widetilde{C}_s(x)\,D_w^s\Big)\mathbf{M}F(w,x) \;=\; \frac{1}{w}\,A(w,x)\,, \tag{2.4}$$

where $\widetilde{C}_s(x):=\sum_{r=s+1}^{m}\binom{r}{s+1}C_r(x)\,D^{r-1-s}F(0,x)$.

Proof Apply the lemma with $c_r(t)=C_r(e^t)$, $f(t)=F(0,e^t)$, $g(t)=e^{wt}F(w,e^t)$, $a(t)=e^{wt}A(w,e^t)$, noting that then $h(t)=we^{wt}\mathbf{M}F(w,e^t)$. □

2.2 Proof of Theorems 1.1 and 1.2. For the proof of (1.2) and (1.3), it is convenient to define $\mathcal{F}_p(w,x)$ also for $p=-1$. Set

$$\mathcal{F}_{-1}(w,x) \;=\; w\,D_w^{-1}\mathcal{F}(w,x) \;=\; \sum_{d=0}^{\infty} x^d\,\frac{\prod_{r=0}^{r=nd-1}(nw+r)}{\prod_{r=1}^{r=d}\big((w+r)^n-w^n\big)} \;\in\;\mathcal{P}. \tag{2.5}$$

We have $\mathcal{F}_{-1}(0,x)=1$ and $w^{-1}D_w\mathcal{F}_{-1}=\mathcal{F}$, so $\mathcal{F}_p=\mathbf{M}^{p+1}\mathcal{F}_{-1}$ for all $p\geq 0$, justifying the notation. It is straightforward to check that $\mathcal{F}_{-1}$ is a solution of the differential equation

$$\Big(D_w^n \;-\; x\prod_{j=0}^{n-1}(nD_w+j)\Big)\,\mathcal{F}_{-1} \;=\; w^n\,\mathcal{F}_{-1}\,. \tag{2.6}$$

This has the form of (2.3) with $F=\mathcal{F}_{-1}$, $A=w^n\mathcal{F}_{-1}$, $m=n$, and

$$C_n(x) \;=\; 1-n^nx\,,\quad C_r(x) \;=\; -n^r\,S_{n-r}(n-1)\,x \qquad (0<r<n), \qquad C_0(x) \;=\; 0\,, \tag{2.7}$$

where $S_{n-r}(n-1)$ as before denotes the $(n-r)$-th elementary symmetric function of $1,2,\ldots,n-1$. Applying Corollary 2.2 repeatedly, we obtain

$$\sum_{s=0}^{n-1-p} C_s^{(p)}(x)\,D_w^s\mathcal{F}_p(w,x) \;=\; w^{n-p-1}\,\mathcal{F}_{-1}(w,x) \qquad (0\leq p\leq n-1), \tag{2.8}$$

where $C_s^{(0)}(x)=C_{s+1}(x)$ with $C_r(x)$ as in (2.7) and $C_s^{(p)}$ for $p>0$ is given inductively by

$$C_s^{(p)} \;=\; \sum_{r=s+1}^{n-p}\binom{r}{s+1}\,C_r^{(p-1)}(x)\,D^{r-1-s}I_{p-1}(x)\,. \tag{2.9}$$

In particular, by induction on p we find that the first two coefficients in (2.8) are given by

$$C_{n-1-p}^{(p)} \;=\; (1-n^nx)\prod_{r=0}^{p-1}I_r(x)\,, \tag{2.10}$$

$$C_{n-2-p}^{(p)} \;=\; \Big(-\frac{n^n(n-1)}{2}\,x\;+\;(1-n^nx)\sum_{r=0}^{p-1}(n-r-1)\,\frac{I_r'(x)}{I_r(x)}\Big)\prod_{r=0}^{p-1}I_r(x)\,. \tag{2.11}$$

Equations (2.8) and (2.10) for $p = n - 1$ give

$$(1 - n^n x) \prod_{r=0}^{n-2} I_r(x)\, \mathcal{F}_{n-1}(w,x) \;=\; \mathcal{F}_{-1}(w,x)\,. \tag{2.12}$$

Setting $w = 0$ in this relation and using $\mathcal{F}_{-1}(0,x) = 1$ gives equation (1.2). Then substituting (1.2) back into (2.12) gives $\mathcal{F}_{n-1}/I_{n-1} = \mathcal{F}_{-1}$ and hence, applying $w^{-1}D_w$ to both sides, $\mathcal{F}_n = \mathcal{F}$, proving Theorem 1.1. Similarly, taking $p = n-2$ in equations (2.8), (2.10), and (2.11) and then setting $w = 0$ gives

$$\sum_{r=0}^{n-2} (n - r - 1)\, \frac{I_r'(x)}{I_r(x)} \;=\; \frac{n-1}{2}\, \frac{n^n x}{1 - n^n x}\,,$$

and integrating this and exponentiating gives (1.3).

Finally, we must prove the reflection symmetry (1.4). For this purpose, it is useful to construct the power series I_p in another way. Define a function $\widetilde{\mathcal{F}}_0 \in \mathcal{P}$ by

$$\widetilde{\mathcal{F}}_0(w,x) \;=\; \sum_{d=0}^{\infty} x^d\, \frac{\prod_{r=1}^{r=nd}(nw+r)}{\prod_{r=1}^{r=d}(w+r)^n} \tag{2.13}$$

and set $\widetilde{\mathcal{F}}_p(w,x) \;=\; \mathbf{M}^p \widetilde{\mathcal{F}}_0(w,x)$ for all $p \geq 0$. Since $\widetilde{\mathcal{F}}_0(w,x)$ is congruent to $\mathcal{F}(w,x)$ modulo w^n, we find by induction on p that $\widetilde{\mathcal{F}}_p(w,x)$ is congruent to $\mathcal{F}_p(w,x)$ modulo w^{n-p} for all $0 \le p \le n-1$ and hence that $I_p(x) = \widetilde{\mathcal{F}}_p(0,x)$ in this range. We now argue as above, using $\widetilde{\mathcal{F}}_0$ instead of $\mathcal{F}_{-1}$. This function satisfies the differential equation

$$\Bigl(D_w^{n-1} \;-\; nx \prod_{j=1}^{n-1} (nD_w + j)\Bigr)\, \widetilde{\mathcal{F}}_0 \;=\; w^{n-1}\,.$$

Applying Corollary 2.2 repeatedly, we obtain

$$\sum_{s=0}^{n-1-p} \widetilde{C}_s^{(p)}(x)\, D_w^s \widetilde{\mathcal{F}}_p(w,x) \;=\; w^{n-p-1}$$

for $0 \le p \le n-1$, where the coefficients $\widetilde{C}_s^{(p)}(x) \in \mathbb{Q}[[x]]$ can be calculated recursively, the top one being given by

$$\widetilde{C}_{n-1-p}^{(p)}(x) = (1 - n^n x) I_0(x) \cdots I_{p-1}(x).$$

Specializing to $p = n-1$ and using (1.2), we find that $\widetilde{\mathcal{F}}_{n-1}(w,x) = I_{n-1}(x)$ is independent of w. Now by downwards induction on p, using the equation $\widetilde{\mathcal{F}}_p = I_p w D_w^{-1} \widetilde{\mathcal{F}}_{p+1}$, we can "reconstruct" all of the power series $\widetilde{\mathcal{F}}_p(w,x)$ $(n-1 \ge p \ge 0)$ from their special values $I_p(x) = \widetilde{\mathcal{F}}_p(0,x)$ at $w = 0$, obtaining in particular the formula

$$w^{1-n}\, \widetilde{\mathcal{F}}_0(w,x) \;=\; I_0\, D_w^{-1}\, I_1\, D_w^{-1} \cdots I_{n-2}\, D_w^{-1}\, I_{n-1}$$

for the initial series $\widetilde{\mathcal{F}}_0$. Comparing the coefficients of x^d on both sides of this equation, we find

$$\frac{n^{-1}\prod_{r=0}^{nd}(nw+r)}{[w(w+1)\cdots(w+d)]^n} = \sum_{\substack{d_0,\ldots,d_{n-1}\geq 0\\ d_0+\cdots+d_{n-1}=d}} \frac{c_0(d_0)\,\cdots\,c_{n-1}(d_{n-1})}{(w+d_1+\cdots+d_{n-1})(w+d_2+\cdots+d_{n-1})\cdots(w+d_{n-1})}$$

for all $d\geq 0$, where $c_p(d)$ denotes the coefficient of x^d in $I_p(x)$. Splitting up the sum on the right into the subsum over n-tuples $(d_0,\ldots,d_{n-1})$ with $\max\{d_r\}\leq d-1$ and the sum over the n-tuples which are permutations of $(d,0,\ldots,0)$, and using that $c_p(0)=1$ for all p, we can rewrite this equation as

$$\sum_{p=0}^{n-1}\frac{c_p(d)}{w^{n-p-1}(w+d)^p} = \frac{\prod_{r=0}^{nd}(nw+r)}{n\prod_{r=0}^{d}(w+r)^n} - \sum_{\substack{0\leq d_0,\ldots,d_{n-1}<d\\ d_0+\cdots+d_{n-1}=d}} \frac{c_0(d_0)\,\cdots\,c_{n-1}(d_{n-1})}{(w+d_1+\cdots+d_{n-1})\cdots(w+d_{n-1})}$$

Now suppose by induction that $c_p(d')=c_{n-p-1}(d')$ for all $d'<d$ and all $0\leq p\leq n-1$. (Notice that this is true for $d'=0$ because $c_p(0)=I_p(0)=1$ for all p, providing the starting point for the induction.) Then both terms on the right are $(-1)^{n-1}$-invariant under the map $w\to -w-d$, as one sees for the second term by making the renumbering $d_r\to d_{n-1-r}$. It follows that the left-hand side has the same invariance and hence that $c_p(d)=c_{n-1-p}(d)$ for all $0\leq p\leq n-1$, completing the inductive proof of the desired symmetry $I_{n-1-p}=I_p$.

2.3 Proof of Theorem 1.4. We now turn to the expansion of $\mathcal{F}(w,x)$ near $w=\infty$. We first prove Lemma 1.3, which said that any periodic fixed point of the map $\mathbf{M}:\mathcal{P}\to\mathcal{P}$ has a logarithm which belongs to $w\,\mathbb{Q}[[x,w^{-1}]]$.

Proof of Lemma 1.3: The effect of $\mathbf{M}$ on logarithms is given by $\mathbf{M}\big(e^{H(w,x)}\big)=e^{H^*(w,x)}$, where

$$H^*(w,x) = H(w,x)-H(0,x)+\log\left(1+\frac{DH(w,x)-DH(0,x)}{w}\right); \tag{2.14}$$

here, as before, D denotes $x\frac{\partial}{\partial x}$. Suppose that $H(w,x):=\log F(w,x)$ is not $\mathrm{O}(w)$, and let e be the smallest integer such that the coefficient of x^e in $H(w,x)$ is not $\mathrm{O}(w)$ as $w\to\infty$. Then

$$H(w,x) = Cx^ew^N + x\mathrm{O}_w(w) + x^e\mathrm{O}_w(w^{N-1}) + \mathrm{O}(x^{e+1}) \qquad (w\to\infty) \tag{2.15}$$

for some $C\neq 0$ and $N\geq 2$, where $\mathrm{O}_w(w^\nu)$ denotes a polynomial in x with coefficients that grow at most like w^ν as $w\to\infty$ and $\mathrm{O}(x^{e+1})$ denotes an element of $x^{e+1}\,\mathbb{Q}(w)[[x]]$. From (2.14) and (2.15),

$$H^*(w,x) = H(w,x)+Cex^ew^{N-1}+x\mathrm{O}_w(1)+x^e\mathrm{O}_w(w^{N-2})+\mathrm{O}(x^{e+1}).$$

This has the same form as (2.15) with the same C, e, and N. Iterating, we find that

$$\log\big(\mathbf{M}^kF(w,x)\big) = H(w,x)+kCex^ew^{N-1}+x\mathrm{O}_w(1)+x^e\mathrm{O}_w(w^{N-2})+\mathrm{O}(x^{e+1}),$$

and this contradicts the assumption that $\mathbf{M}^kF=F$, since $C\neq 0$ and $N\geq 2$.

As already mentioned in the introduction, Lemma 1.3 together with Theorem 1.1 implies that $\mathcal{F}(w,x)$ has an asymptotic expansion of the form (1.5). From the proof of the lemma, we see that each $\mathcal{F}_p(w,x) = \mathbf{M}^p\mathcal{F}(w,x)$ has an asymptotic expansion

$$\mathcal{F}_p(w,x) \sim e^{\mu(x)w}\sum_{s=0}^{\infty}\Phi_{p,s}(x)\,w^{-s} \qquad (w\to\infty) \tag{2.16}$$

of the same form, with the same function $\mu(x)$ in the exponent. The equation $\mathcal{F}_{p+1} = \mathbf{M}\mathcal{F}_p$ gives

$$\Phi_{0,s} = \Phi_s, \quad \Phi_{p+1,s} = \frac{1+\mu'}{I_p}\,\Phi_{p,s} + \begin{cases}(\Phi_{p,s-1}/I_p)' & \text{if } s\ge 1,\\ 0 & \text{otherwise,}\end{cases} \tag{2.17}$$

where f' denotes $Df = x\,df/dx$. We want to solve these equations by induction on p for small s.

Before doing this, we begin with the following observation. Let $L(x) = (1 - n^n x)^{-1/n}$ as in Theorem 1.4. Then (1.2) says that the product of the functions $I_p(x)/L(x)$ ($p\in\mathbb{Z}/n\mathbb{Z}$) equals 1, so if we define

$$H_p(x) = \frac{L(x)^p}{I_0(x)\cdots I_{p-1}(x)} \qquad (p\ge 0), \tag{2.18}$$

then we have the properties

$$H_0 = 1, \quad H_p/H_{p+1} = I_p/L, \quad H_1H_2\cdots H_n = 1, \quad H_{p+n} = H_p, \quad H_{n-p} = H_p^{-1}, \tag{2.19}$$

where the last equality is originally true for $0\le p\le n$ but then, in view of the periodicity of $\{H_p\}$, holds for any $p\in\mathbb{Z}/n\mathbb{Z}$. A number of identities below are simpler to state in terms of the functions $H_p(x)$ than in terms of the original functions $I_p(x)$.

The case $s=0$ of (2.17) gives by induction the formula

$$\Phi_{p,0} = (1+\mu')^p/I_0\cdots I_{p-1}.$$

Combining this with the formulas $\mathcal{F}_n = \mathcal{F}$ and (1.2), we obtain $(1+\mu')^n = L^n$, from which the first equation in (1.6) follows since $\mu(x)$ is a power series in x with no constant term. This also gives us the formula

$$\Phi_{p,0}(x) = H_p(x)\,\Phi_0(x) \qquad \text{for all } p\ge 0,$$

with H_p as in (2.18). Now substituting this into the case $s=1$ of (2.17) we find inductively

$$\Phi_{p,1}(x) = H_p(x)\left(\Phi_1(x) + p\,\frac{\Phi_0' - L'}{L} + \frac{\Phi_0}{L}\sum_{r=1}^{p}\frac{H_r'}{H_r}\right) \qquad \text{for all } p\ge 0.$$

Setting $p=n$ in this relation and using the third and fourth of equations (2.19) and $\mathcal{F}_n = \mathcal{F}$, we deduce that $\Phi_0 = L$, which is the second assertion of Theorem 1.4. At the same time we can refine the last two equations to

$$\Phi_{p,0} = H_p\,L, \qquad \Phi_{p,1} = H_p\left(\Phi_1 + \sum_{r=1}^{p}\frac{H_r'}{H_r}\right) \qquad (p\ge 0). \tag{2.20}$$

The proof of the third identity in (1.6) is similar, but the calculations are more complicated. The case $s=2$ of (2.17) gives by induction the formula

$$\begin{aligned}\Phi_{p,2} \;=\; H_p\Bigg(\Phi_2 + p\Big(\frac{\Phi_1}{L}\Big)' + \Big(\sum_{r=1}^{p}\frac{H_r'}{H_r}\Big)\frac{\Phi_1}{L} \\ + \frac{1}{L}\sum_{s=2}^{p}\sum_{r=1}^{s-1}\frac{H_r'}{H_r}\frac{H_s'}{H_s} \\ + \Big(\frac{1}{L}\sum_{r=1}^{p-1}(p-r)\frac{H_r'}{H_r}\Big)'\Bigg)\end{aligned}$$

for all $p\ge 0$. Taking $p=n$, observing that

$$\sum_{s=2}^{n}\sum_{r=1}^{s-1}\frac{H_r'}{H_r}\frac{H_s'}{H_s} \equiv \frac{1}{2}\Bigg(\Big(\sum_{p=1}^{n}\frac{H_p'}{H_p}\Big)^2 - \sum_{p=1}^{n}\Big(\frac{H_p'}{H_p}\Big)^2\Bigg) = -\frac{1}{2}\sum_{p=1}^{n}\Big(\frac{H_p'}{H_p}\Big)^2$$

by the third equation in (2.19), and using $\mathcal{F}_n=\mathcal{F}$, we find that

$$n\Big(\frac{\Phi_1}{L}\Big)' = \frac{1}{2L}\sum_{p=1}^{n}\Big(\frac{H_p'}{H_p}\Big)^2 + \Big(\frac{1}{L}\sum_{p=0}^{n-1}p\,\frac{H_p'}{H_p}\Big)' = -\frac{(n+1)(n-2)}{24}\,(L^{n-1})',$$

the last equation being Lemma 2.3 below. Integrating and using $\Phi_1(0)=0$ gives the last identity in (1.6).

Lemma 2.3 *The functions* $\{H_p(x)\}_{p\in\mathbb{Z}/n\mathbb{Z}}$ *satisfy*

$$\frac{1}{2L}\sum_{p\ (\mathrm{mod}\ n)}\Big(\frac{H_p'}{H_p}\Big)^2 = -\Big(\frac{(n+1)(n-2)}{24}L^{n-1} + \frac{1}{L}\sum_{p=0}^{n-1}p\,\frac{H_p'}{H_p}\Big)'. \tag{2.21}$$

The proof consists of expressing the left-hand side of (2.21) in terms of the functions $I_0, I_1, \ldots, I_{n-1}$ and their derivatives, getting rid of all square terms via the product rule, and then eliminating I_{n-1}, I_{n-2}, and I_{n-3}. The last elimination is achieved by computing the coefficients $C_p^{(n-3-p)}$ inductively by (2.8), starting with

$$C_{n-3}^{(0)} = -n^{n-2}\,S_2(n-1)\,x = -\frac{(n-1)(n-2)(3n-1)}{24}L'/L^{n+1},$$

and then setting $p=n-3$, exactly as we did with $C_p^{(n-1-p)}$ and $C_p^{(n-2-p)}$ in Subsection 2.2 to prove eqs. (1.2) and (1.3). At this stage, all terms involving products of two functions I_p cancel, and the resulting expression can be integrated. We omit the details, which are somewhat tedious, since the last identity in (1.6) also follows easily from Theorem 1.5.

2.4 Proof of Theorem 1.5. We set $X=L^n$ and $Y=(L^n-1)/n$. Note that

$$D(\mu)=L-1, \qquad D(L)=LY, \qquad D(X)=X^2-X, \qquad D(Y)=XY. \tag{2.22}$$

The first identity implies that $D_w\, e^{\mu w} = e^{\mu w}\, \widetilde{D}_w$, where $\widetilde{D}_w = D + Lw$. By induction on k, the powers of the differential operator $\widetilde{D}_w$ are given by

$$\begin{aligned}\widetilde{D}_w^k &= \sum_{m=0}^{k} \binom{k}{m} \widetilde{D}_w^m(1)\, D^{k-m} \\ &= D^k + k\, Lw\, D^{k-1} + \frac{k(k-1)}{2}\left((Lw)^2 + Y(Lw)\right) D^{k-2} + \dots .\end{aligned} \tag{2.23}$$

A second induction gives the formula

$$\widetilde{D}_w^m(1) = \sum_{j=0}^{m} \mathcal{H}_{m,j}(X)\,(Lw)^{m-j}, \tag{2.24}$$

with $\mathcal{H}_{m,j} \in \mathbb{Z}[X,Y] \subseteq \mathbb{Q}[X]$ given by (1.7).

The function $\mathcal{F}(w,x)$ satisfies the ODE

$$\left(D_w^n - w^n - x\prod_{j=1}^{n}(nD_w + j)\right)\mathcal{F} = 0 .$$

Since $D_w\, e^{\mu w} = e^{\mu w}\, \widetilde{D}_w$, the function $\widetilde{\mathcal{F}}(w,x) = e^{-\mu(x)w}\mathcal{F}(w,x)$ satisfies the differential equation $\mathfrak{L}\widetilde{\mathcal{F}} = 0$, where $\mathfrak{L}$ is the differential operator

$$\begin{aligned}\mathfrak{L} &= L^n\left(\widetilde{D}_w^n - w^n - x\prod_{j=1}^{n}(n\widetilde{D}_w + j)\right) \\ &= \widetilde{D}_w^n - (Lw)^n - (L^n - 1)\sum_{r=1}^{n}\frac{S_r(n)}{n^r}\,\widetilde{D}_w^{n-r} .\end{aligned}$$

Using (2.23) and (2.24), we can expand $\mathfrak{L}$ as $\mathfrak{L} = \sum_{k=1}^{n}(Lw)^{n-k}\mathfrak{L}_k$, with $\mathfrak{L}_k$ defined by (1.9). Combining the differential equation $\mathfrak{L}\widetilde{\mathcal{F}} = 0$ with the asymptotic expansion $\widetilde{\mathcal{F}}(w,x) \sim \sum_{s\geq 0}\Phi_s(x)w^{-s}$ for large w, we obtain (1.12).

We will next use (1.12) to prove by induction that Φ_s belongs to $L\mathbb{Q}[L]$. Since $\mathfrak{L}_1(L\mathbb{Q}[L]) = L^2 Y\mathbb{Q}[L]$, it suffices to show that

$$\mathfrak{L}_k(L\mathbb{Q}[L]) \subseteq L^{k+1}Y\,\mathbb{Q}[L] \qquad (2 \leq k \leq n). \tag{2.25}$$

Let $\mathcal{I} \subset \mathbb{Q}[L]$ be the ideal generated by XY. Since D and Y commute modulo $\mathcal{I}$ by (2.22) and since $(D - rY)L^r = 0$, we have

$$(D-Y)(D-2Y)\cdots(D-kY)L^r \in \begin{cases} L^r\,\mathcal{I} & \text{if } 1 \leq r \leq k\,, \\ L^rY\,\mathbb{Q}[L] & \text{if } r \geq k+1\,. \end{cases}$$

Therefore (2.25) is a consequence of the following lemma.

Lemma 2.4 *For all $k > 1$,*

$$\mathfrak{L}_k \equiv \binom{n}{k}(D-Y)(D-2Y)\cdots(D-kY) \pmod{\mathcal{I}} .$$

Proof The recursion (1.7) for $H_{m,j}$ shows that $H_{m,j} \equiv h_{m,j}Y^j \pmod{\mathcal{I}}$, where $h_{m,j} \in \mathbb{Z}$ is given recursively by

$$h_{0,j} = \delta_{0,j}, \qquad h_{m,j} = h_{m-1,j} + (m-j)h_{m-1,j-1}\ \forall\, m \geq 1 \tag{2.26}$$

(with $h_{m-1,j-1} = 0$ for $j = 0$). Thus $h_{m,j} = \mathfrak{S}_m^{(m-j)}$, where $\mathfrak{S}_m^{(k)}$ denotes a Stirling number of the second kind (the number of ways of partitioning a set of m elements into k non-empty subsets). We also note that $(1-L^n)n^{-r} \equiv (-1)^r Y^r \pmod{\mathcal{I}}$ for all $r \geq 1$. Combining these facts with (1.9), we find that

$$\mathfrak{L}_k \equiv \sum_{i=0}^{k}\left(\sum_{r=0}^{k-i}(-1)^r\binom{n-r}{i} S_r(n)\,\mathfrak{S}_{n-r-i}^{(n-k)}\right) Y^{k-i}D^i \pmod{\mathcal{I}}.$$

The desired congruence for $\mathfrak{L}_k$ now follows from the generating series calculation

$$\begin{aligned}
&\sum_{i=0}^{k}\left(\sum_{r=0}^{k-i}(-1)^r\binom{n-r}{i} S_r(n)\,\mathfrak{S}_{n-r-i}^{(n-k)}\right) t^i \\
&= \sum_{i=0}^{n}\left(\sum_{r=0}^{n-i}(-1)^r\binom{n-r}{i} S_r(n)\left[\frac{1}{(n-k)!}\sum_{j=0}^{n-k}(-1)^{n-k-j}\binom{n-k}{j} j^{n-r-i}\right]\right) t^i \\
&= \frac{1}{(n-k)!}\sum_{j=0}^{n-k}(-1)^{n-k-j}\binom{n-k}{j}\sum_{r=0}^{n}(-1)^r S_r(n)\sum_{i=0}^{n-r}\binom{n-r}{i} j^{n-r-i}t^i \\
&= \frac{1}{(n-k)!}\sum_{j=0}^{n-k}(-1)^{n-k-j}\binom{n-k}{j}\sum_{r=0}^{n}(-1)^r S_r(n)(j+t)^{n-r} \\
&= \frac{n!}{(n-k)!}\sum_{j=0}^{n-k}(-1)^{n-k-j}\binom{n-k}{j}\binom{j+t-1}{n} \\
&= \frac{n!}{(n-k)!}\binom{t-1}{k} = \binom{n}{k}(t-1)(t-2)\cdots(t-k),
\end{aligned}$$

where the first equality follows from the well-known fact that the expression in square brackets equals $\mathfrak{S}_{n-r-i}^{(n-k)}$ if $i+r \leq k$ and 0 for $i+r > k$ and the second-to-last equality is obtained by expanding $(1+u)^{t-1}((1+u)-1)^{t-1}$ by the binomial theorem and equating coefficients of t^n. □

This completes the proof of part (ii) of Theorem 1.5. Part (iii) of Theorem 1.5 follows from the differential equation (1.12) by induction on s.

3 Further discussion of the large w expansion of $\mathcal{F}(w,x)$

In this final section we give some further information and conjectures about the power series $\Phi_s(x)$ defined by equation (1.5). We begin by giving the numerical values for $n \leq 5$ and $s \leq 4$. For this purpose it is convenient to divide Φ_s/L by $((n-2)(n+1)/24n)^s/s!$ and write the result as the sum of $(1-L^{n-1})^s$ and a correction term, because the formulas then become much simpler than without this renormalization:

Table 1 List of values of $s!\left(\frac{24n}{(n-2)(n+1)}\right)^s \Phi_s/L$ for $s = 1, 2, 3, 4$ and $n = 3, 4, 5$

$$
\begin{array}{lll}
n=3: & s=1: & 1-L^2 \\
 & s=2: & (1-L^2)^2 \\
 & s=3: & (1-L^2)^3 + 144\,(1-5L^3+4L^6) \\
 & s=4: & (1-L^2)^4 + 576\,(1-94L^2-5L^3+245L^5+4L^6-151L^8) \\
 & & \\
n=4: & s=1: & 1-L^3 \\
 & s=2: & (1-L^3)^2 \\
 & s=3: & (1-L^3)^3 + \frac{36}{25}\,(4+72L-297L^5+221L^9) \\
 & n=4 & (1-L^3)^4 + \frac{144}{125}\,(884+360L-20L^3-19584L^4-1485L^5 \\
 & & \qquad +44253L^8+1105L^9-25513L^{12}) \\
 & & \\
n=5: & s=1: & 1-L^4 \\
 & s=2 & (1-L^4)^2 \\
 & s=3: & (1-L^4)^3 + \frac{32}{45}\,(7+134L^2-504L^7+363L^{12}) \\
 & s=4: & (1-L^4)^4 + \frac{16}{135}\,(168+8576L+3216L^2-168L^4-127568L^6 \\
 & & \qquad -12096L^7+270144L^{11}+8712L^{12}-150984L^{16})
\end{array}
$$

This suggests that the series $\sum_s (\Phi_s/L)\, w^{-s}$ is given to a first approximation by a pure exponential $\exp\left(\frac{(n-2)(n+1)}{24n}(1-L^{n-1})/w\right)$ and hence that the formulas for the coefficients of the expansion (1.5) may become simpler if we take the logarithm. Doing this, we find an expansion which begins

$$\log \mathcal{F}(w,x) \;=\; \mu(x)w + \log L(x) + \frac{(n-2)(n+1)(1-L(x)^{n-1})}{24n}\, w^{-1} + 0\, w^{-2} + \cdots$$

and in which, at least experimentally, the coefficient of w^{-j} for $j \ge 1$ is the sum of a term independent of x and a term of the form L^{-j} times a polynomial (without constant term) in L^n. By applying the operator $w^{-1}D$ and adding 1, this can be stated more elegantly as follows.

Conjecture: If $\mathcal{F}$ is given by (1.1), then

$$1 + \frac{x}{w}\frac{\partial}{\partial x} \log \mathcal{F}(w,x) \;\stackrel{?}{=}\; L \sum_{k=0}^{\infty} \frac{P_k(n, L^n)}{(nLw)^k}\,, \tag{3.1}$$

where $P_k(n, X)$ is a polynomial in X of degree k with coefficients in $\mathbb{Q}[n]$.

We have verified this conjecture up to order $O(w^{-6})$, with the values of the corresponding coefficients P_k being given by

$$
\begin{aligned}
P_0(n,X) &= 1\,, \\
P_1(n,X) &= X-1\,, \\
P_2(n,X) &= -\frac{(n+1)(n-1)(n-2)}{24}\,(X-1)X\,, \\
P_3(n,X) &= 0\,, \\
P_4(n,X) &= \frac{(n+1)(n-1)(n-2)}{5760}\,(X-1)(A_3X^3+A_2X^2+A_1X)\,, \\
P_5(n,X) &= -\frac{(n+1)(n-1)(n-2)}{5760}\,(X-1)(B_4X^4+B_3X^3+B_2X^2+B_1X),
\end{aligned}
$$

where

$$
\begin{aligned}
A_1 &= (n-3)(7n^3-17n^2+22n-24)\,,\\
A_2 &= -(2n-3)(3n-1)(7n^2-9n+14)\,,\\
A_3 &= 3\,(14n^4-33n^3+52n^2-23n-2)\,,\\
B_1 &= -(n-3)(n-4)(7n^3-17n^2+22n-24)\,,\\
B_2 &= 2\,(n-1)(n-2)(49n^3-115n^2+152n-124)\,,\\
B_3 &= -4\,(n-1)(3n-1)(3n-4)(7n^2-9n+14)\,,\\
B_4 &= 8\,(n-1)(3n-2)(7n^3-11n^2+17n-1)\,.
\end{aligned}
$$

The coefficients of the polynomials P_k follow no apparent pattern apart from the divisibility by $(n+1)(n-1)(n-2)X(X-1)$: the common factors of A_1 and B_1 and of A_2 and B_3 are striking, but nothing similar occurs for the next two polynomials. On the other hand, there is a simple formula for the leading coefficient of $P_k(n,X)$ with respect to n, namely (at least up to $k=7$)

$$P_k(n,X) \;=\; \begin{cases} \alpha_j\, e_k(X)\, n^{4j-1} + \mathrm{O}(n^{4j-2}) & \text{if } k=2j>0,\\ (j-1)\alpha_j\, e_k(X)\, n^{4j} + \mathrm{O}(n^{4j-1}) & \text{if } k=2j+1, \end{cases}$$

where α_j denotes the coefficient of u^{2j} in $\frac{u/2}{\sinh u/2}$ ($\alpha_0=1$, $\alpha_1=-\frac{1}{24}$, $\alpha_2=\frac{7}{5760}$, $\alpha_3=-\frac{31}{967680}$, $\ldots$) and where $e_1=X-1$, $e_2=X^2-X$, $e_3=2X^3-3X^2+X$, $\ldots$ are the polynomials defined by

$$e_k(X) \;=\; \sum_{l=1}^{k}(-1)^{k-l}\,(l-1)!\,\mathfrak{S}_k^{(l)}\,X^l \quad\in\quad \mathbb{Z}[X]$$

with $\mathfrak{S}_k^{(l)}$ as before a Stirling number of the second kind. This is interesting because the argument $X=L^n$ of $P_k(n,X)$ in equation (3.1) is in fact $(1-n^n x)^{-1}$ and the functions $e_k((1-x)^{-1})$ have the basic property

$$e_k\left(\frac{1}{1-x}\right) \;=\; \sum_{d=1}^{\infty} d^{k-1}x^d \quad\in\; x\,\mathbb{Z}[[x]] \qquad (k\geq 1)\,.$$

There is also a possible intriguing connection with modular and elliptic functions since, for example, the power series in two variables $\sum \alpha_j e_{2j}\big(\frac{1}{1-x}\big)\,u^{2j-1}$ is closely related to the expansion of the Weierstrass $\wp$-function and related Jacobi forms. This suggests possible hidden modularity properties of the original function $\mathcal{F}(w,x)$.

As a final remark, we observe that (3.1), if it is true, defines the power series $\mathcal{F}(w,x)$ even for non-integral values of n and shows that this function is analytic in n as well as in w and x. This seems surprising since $\mathcal{F}$ is defined as a hypergeometric function of order n and we would usually not expect such series to have a reasonable interpolation with respect to the order of the differential equation which they satisfy.

References

[BCOV] M. Bershadsky, S. Cecotti, H. Ooguri, and C. Vafa, *Holomorphic Anomalies in Topological Field Theories*, Nucl. Phys. B405 (1993), 279–304.

[CaDGP] P. Candelas, X. de la Ossa, P. Green, L. Parkes, *A Pair of Calabi-Yau Manifolds as an Exactly Soluble Superconformal Theory*, Nuclear Phys. B359 (1991), 21–74.

[Gi] A. Givental, *The Mirror Formula for Quintic Threefolds*, Amer. Math. Soc. Transl. Ser. 2, 196 (1999), 49–62.

[Z1] A. Zinger, *The Reduced Genus-One Gromov-Witten Invariants of Calabi-Yau Hypersurfaces*, math/0705.2397.

[Z2] A. Zinger, *Standard vs. Reduced Genus-One Gromov-Witten Invariants*, math/0706.0715.

Fields Institute Communications
Volume **54**, 2008

Ramanujan-type Formulae for $1/\pi$: A Second Wind?

Wadim Zudilin
Department of Mechanics and Mathematics
Moscow Lomonosov State University
Vorobiovy Gory, GSP-1, 119991 Moscow, Russia
and
Steklov Mathematical Institute
Russian Academy of Sciences
Gubkina str. 8, 119991 Moscow, Russia
wadim@mi.ras.ru

Abstract. In 1914 S. Ramanujan recorded a list of 17 series for $1/\pi$. We survey the methods of proofs of Ramanujan's formulae and indicate recently discovered generalizations, some of which are not yet proven.

The twentieth century was full of mathematical discoveries. Here we present two significant contributions from that time, in reverse chronological order. At first glance, the stories might be thought to be of a different nature. But we will try to convince the reader that they have much in common.

1 Ramanujan and Apéry: $1/\pi$ and $\zeta(3)$

In 1978 R. Apéry showed the irrationality of $\zeta(3)$ (see [5] and [21]). His rational approximations to the number in question (known nowadays as *Apéry's constant*) have the form $v_n/u_n \in \mathbb{Q}$ for $n = 0, 1, 2, \dots$, where the denominators $\{u_n\} = \{u_n\}_{n=0,1,\dots}$ and numerators $\{v_n\} = \{v_n\}_{n=0,1,\dots}$ satisfy the same polynomial recurrence

$$(n+1)^3 u_{n+1} - (2n+1)(17n^2 + 17n + 5)u_n + n^3 u_{n-1} = 0 \tag{1.1}$$

with the initial data

$$u_0 = 1, \quad u_1 = 5, \qquad v_0 = 0, \quad v_1 = 6.$$

Then

$$\lim_{n\to\infty} \frac{v_n}{u_n} = \zeta(3)$$

2000 *Mathematics Subject Classification.* Primary 11F11, 11Y60, 33C20; Secondary 05A19, 11B65, 11J82, 11M06, 14H52, 14J32, 33C75, 33F10, 34M50, 40G99, 65B10, 65Q05.

The work was supported by a fellowship of the Max Planck Institute for Mathematics (Bonn) and supported in part by the INTAS foundation, grant no. 03-51-5070.

and, surprisingly, the denominators $\{u_n\}$ are integers:

$$u_n = \sum_{k=0}^{n} \binom{n}{k}^2 \binom{n+k}{k}^2 \in \mathbb{Z}, \qquad n = 0, 1, 2, \dots, \tag{1.2}$$

while the numerators $\{v_n\}$ are 'close' to being integers.

In 1914 S. Ramanujan [22], [7] recorded a list of 17 series for $1/\pi$, from which we indicate the simplest one

$$\sum_{n=0}^{\infty} \frac{(\frac{1}{2})_n^3}{n!^3}(4n+1) \cdot (-1)^n = \frac{2}{\pi} \tag{1.3}$$

and also two quite impressive examples

$$\sum_{n=0}^{\infty} \frac{(\frac{1}{4})_n(\frac{1}{2})_n(\frac{3}{4})_n}{n!^3}(21460n+1123) \cdot \frac{(-1)^n}{882^{2n+1}} = \frac{4}{\pi}, \tag{1.4}$$

$$\sum_{n=0}^{\infty} \frac{(\frac{1}{4})_n(\frac{1}{2})_n(\frac{3}{4})_n}{n!^3}(26390n+1103) \cdot \frac{1}{99^{4n+2}} = \frac{1}{2\pi\sqrt{2}} \tag{1.5}$$

which produce rapidly converging (rational) approximations to π. Here

$$(a)_n = \frac{\Gamma(a+n)}{\Gamma(a)} = \begin{cases} a(a+1)\cdots(a+n-1) & \text{for } n \ge 1, \\ 1 & \text{for } n = 0, \end{cases}$$

denotes the Pochhammer symbol (the rising factorial). The Pochhammer products occurring in all formulae of this type may be written in terms of binomial coefficients:

$$\frac{(\frac{1}{2})_n^3}{n!^3} = 2^{-6n}\binom{2n}{n}^3, \qquad \frac{(\frac{1}{3})_n(\frac{1}{2})_n(\frac{2}{3})_n}{n!^3} = 2^{-2n}3^{-3n}\binom{2n}{n}\frac{(3n)!}{n!^3},$$

$$\frac{(\frac{1}{4})_n(\frac{1}{2})_n(\frac{3}{4})_n}{n!^3} = 2^{-8n}\frac{(4n)!}{n!^4}, \qquad \frac{(\frac{1}{6})_n(\frac{1}{2})_n(\frac{5}{6})_n}{n!^3} = 12^{-3n}\frac{(6n)!}{n!^3(3n)!}.$$

Ramanujan's original list was subsequently extended to several other series which we plan to touch on later in the paper. For the moment we give two more celebrated examples:

$$\sum_{n=0}^{\infty} \frac{(\frac{1}{3})_n(\frac{1}{2})_n(\frac{2}{3})_n}{n!^3}(14151n+827) \cdot \frac{(-1)^n}{500^{2n+1}} = \frac{3\sqrt{3}}{\pi}, \tag{1.6}$$

$$\sum_{n=0}^{\infty} \frac{(\frac{1}{6})_n(\frac{1}{2})_n(\frac{5}{6})_n}{n!^3}(545140134n+13591409) \cdot \frac{(-1)^n}{53360^{3n+2}} = \frac{3}{2\pi\sqrt{10005}}. \tag{1.7}$$

Formula (1.6) was proven by H. H. Chan, W.-C. Liaw and V. Tan [13] and (1.7) is the Chudnovskys' famous formula [15] which enabled them to hold the record for the calculation of π in 1989–94. On the left-hand side of each formula (1.3)–(1.7) we have linear combinations of a (generalized) hypergeometric series

$${}_mF_{m-1}\left(\begin{matrix} a_1, a_2, \dots, a_m \\ b_2, \dots, b_m \end{matrix} \,\middle|\, z\right) = \sum_{n=0}^{\infty} \frac{(a_1)_n(a_2)_n\cdots(a_m)_n}{(b_2)_n\cdots(b_m)_n}\frac{z^n}{n!} \tag{1.8}$$

and its derivative at a point close to the origin. The rapid convergence of the series in (1.4)–(1.7) may be used for proving the quantitative irrationality of the numbers $\pi\sqrt{d}$ with $d \in \mathbb{N}$ (see [26] for details).

In both Ramanujan's and Apéry's cases, there were just hints on how the things might be proven. Rigorous proofs appeared somewhat later. We will not discuss proofs of Apéry's theorem and its further generalizations (see [17] for a review of the subject), just concentrating on the things around the remarkable Ramanujan-type series. But we will see that both Ramanujan's and Apéry's discoveries have several common grounds.

2 Elliptic proof of Ramanujan's formulae

Although Ramanujan did not indicate how he arrived at his series, he hinted that these series belong to what is now known as 'the theories of elliptic functions to alternative bases'. The first rigorous mathematical proofs of Ramanujan's series and their generalizations were given by the Borweins [11] and Chudnovskys [15]. Let us sketch, following [15], the basic ideas of those very first proofs.

One starts with an elliptic curve $y^2 = 4x^3 - g_2x - g_3$ over $\overline{\mathbb{Q}}$ with fundamental periods ω_1, ω_2 (where $\operatorname{Im}(\omega_2/\omega_1) > 0$) and corresponding quasi-periods η_1, η_2. Besides the Legendre relation

$$\eta_1\omega_2 - \eta_2\omega_1 = 2\pi i,$$

the following *linear* relations between $\omega_1, \omega_2, \eta_1, \eta_2$ over $\overline{\mathbb{Q}}$ are available in the complex multiplication case, i.e., when $\tau = \omega_2/\omega_1 \in \mathbb{Q}[\sqrt{-d}]$ for some $d \in \mathbb{N}$:

$$\omega_2 - \tau\omega_1 = 0, \qquad A\tau\eta_2 - C\eta_1 + (2A\tau + B)\alpha\omega_1 = 0, \tag{2.1}$$

where the integers A, B and C come from the equation $A\tau^2 + B\tau + C = 0$ defining the quadratic number τ and $\alpha \in \mathbb{Q}(\tau, g_2, g_3) \subset \overline{\mathbb{Q}}$. Equations (2.1) allow one to express ω_2, η_2 by means of ω_1, η_1 only. Substituting these expressions into (2.1), and using the hypergeometric formulae for ω_1, η_1 and also for $\omega_1^2, \omega_1\eta_1$ (which follow from Clausen's identity) one finally arrives at a formula of Ramanujan type. An important (and complicated) problem in the proof is computing the algebraic number

$$\alpha = \frac{8\pi^2}{81\omega_1^2}\biggl(E_2(\tau) - \frac{3}{\pi \operatorname{Im}\tau}\biggr), \qquad \text{where} \quad E_2(\tau) = 1 - 24\sum_{n=1}^{\infty} e^{2\pi i n\tau}\sum_{d|n} d.$$

Note that α viewed as a function of τ is a non-holomorphic modular form of weight 2. Although the Chudnovskys attribute the knowledge of the fact that $\alpha(\tau)$ takes values in the Hilbert class field $\mathbb{Q}(\tau, j(\tau))$ of $\mathbb{Q}(\tau)$ to Kronecker (in Weil's presentation [24]), we would refer the reader to the work [8] by B. C. Berndt and H. H. Chan.

3 Modular proof of Ramanujan's formulae

An understanding of the complication of the above proof came in 2002 with T. Sato's discovery of the formula

$$\sum_{n=0}^{\infty} u_n \cdot (20n + 10 - 3\sqrt{5})\biggl(\frac{\sqrt{5}-1}{2}\biggr)^{12n} = \frac{20\sqrt{3} + 9\sqrt{15}}{6\pi} \tag{3.1}$$

of Ramanujan type, involving Apéry's numbers (1.2). The modular argument was essentially simplified by H. H. Chan with his collaborators and later by Y. Yang to produce a lot of new identities like (3.1) based on a not necessarily hypergeometric series $F(z)=\sum_{n=0}^{\infty} u_n z^n$. Examples are

$$\sum_{n=0}^{\infty}\sum_{k=0}^{n}\binom{n}{k}^2\binom{2k}{k}\binom{2n-2k}{n-k}\cdot(5n+1)\frac{1}{64^n}=\frac{8}{\pi\sqrt{3}} \tag{3.2}$$

due to H. H. Chan, S. H. Chan and Z.-G. Liu [12];

$$\sum_{n=0}^{\infty}\sum_{k=0}^{[n/3]}(-1)^{n-k}3^{n-3k}\frac{(3k)!}{k!^3}\binom{n}{3k}\binom{n+k}{k}\cdot(4n+1)\frac{1}{81^n}=\frac{3\sqrt{3}}{2\pi} \tag{3.3}$$

due to H. H. Chan and H. Verrill (2005);

$$\sum_{n=0}^{\infty}\sum_{k=0}^{n}\binom{n}{k}^4\cdot(4n+1)\frac{1}{36^n}=\frac{18}{\pi\sqrt{15}} \tag{3.4}$$

due to Y. Yang (2005).

It should be mentioned that Picard–Fuchs differential equations (of order 3) satisfied by the series $F(z)$ always have very nice arithmetic properties [25]. Therefore, it is not surprising that $F(z)$ admits a modular parametrization: $f(\tau)=F(z(\tau))$ is a modular form of weight 2 for a modular (uniformizing) substitution $z=z(\tau)$.

Let us follow Yang's argument to show the basic ideas of the new proof in the example of (3.1). Our choice is

$$z(\tau)=\biggl(\frac{\eta(\tau)\eta(6\tau)}{\eta(2\tau)\eta(3\tau)}\biggr)^{12}, \qquad f(\tau)=\frac{\eta(2\tau)^7\eta(3\tau)^7}{\eta(\tau)^5\eta(6\tau)^5},$$

which are modular forms of level 6; the expressions were obtained by F. Beukers in his proof of Apéry's theorem using modular forms [9]. Here

$$\eta(\tau)=e^{\pi i\tau/12}\prod_{n=1}^{\infty}(1-e^{2\pi i n\tau})$$

is the Dedekind eta-function. The function $g(\tau)=(2\pi i)^{-1}f'(\tau)/f(\tau)$ satisfies the functional equation

$$g(\gamma\tau)=\frac{c(c\tau+d)}{\pi i}+(c\tau+d)^2g(\tau) \qquad \text{for} \quad \gamma=\begin{pmatrix} a & b\\ c & d\end{pmatrix}\in\Gamma_0(6)+w_6,$$

where w_6 denotes the Atkin–Lehner involution. Taking

$$\gamma=\frac{1}{\sqrt{6}}\begin{pmatrix} 0 & -1\\ 6 & 0\end{pmatrix}, \qquad \tau=\tau_0=\frac{i}{\sqrt{30}}$$

we obtain

$$g(\tau_0)+5g(5\tau_0)=\frac{\sqrt{30}}{\pi}. \tag{3.5}$$

On the other hand, $h(\tau)=g(\tau)-5g(5\tau)$ is a modular form of weight 2 and level 30 (on $\Gamma_0(30)+\langle w_5,w_6\rangle$). This implies that $h(\tau)/f(\tau)$ is an algebraic function of $z(\tau)$ and after explicit evaluations at $\tau=\tau_0$ we arrive at

$$g(\tau_0)-5g(5\tau_0)=h(\tau_0)=\frac{900\sqrt{2}-402\sqrt{10}}{5}f(\tau_0)=(900\sqrt{2}-402\sqrt{10})f(5\tau_0). \tag{3.6}$$

Combining (3.5) and (3.6) we deduce that

$$\frac{\sqrt{30}}{\pi} = (900\sqrt{2} - 402\sqrt{10})f(5\tau_0) + 10g(5\tau_0),$$

and it only remains to use the expansion

$$g(5\tau_0) = z\frac{\mathrm{d}f/\mathrm{d}z}{f} \cdot \frac{1}{2\pi i}\frac{z'(\tau)}{z(\tau)}\bigg|_{\tau=5\tau_0} = (108\sqrt{2} - 48\sqrt{10})\sum_{n=0}^{\infty} nu_n \cdot z(5\tau_0)^n$$

(since $z'(\tau)/(2\pi i)$ and $f(\tau)$ are modular forms of weight 2 on $\Gamma_0(6) + w_6$, the function

$$\frac{1}{f} \cdot \frac{1}{2\pi i}\frac{z'}{z}$$

is an algebraic function of z) and the evaluation

$$z(5\tau_0) = z(\tau_0) = 161 - 72\sqrt{5} = \left(\frac{\sqrt{5}-1}{2}\right)^{12}.$$

As pointed out to us by H. H. Chan, the main difficulty one meets in the above proof is to prove the algebraicity evaluations rigorously (cf. [8]).

4 Creative telescoping

There is yet another method of proof, but applicable only to a small number of Ramanujan-type series. It is based on the algorithm of creative telescoping, due to Gosper–Zeilberger. Note that an essential part of the first proof of Apéry's theorem [21], namely, the proof of the recurrence (1.1), was given by D. Zagier also using a telescoping argument. D. Zeilberger (and his automatic collaborator S. B. Ekhad) could prove the simplest Ramanujan's identity (1.3) in the following way [16]. One verifies the (terminating) identity

$$\sum_{n=0}^{\infty} \frac{(1/2)_n^2(-k)_n}{n!^2(3/2+k)_n}(4n+1)(-1)^n = \frac{\Gamma(3/2+k)}{\Gamma(3/2)\Gamma(1+k)} \tag{4.1}$$

for all *non-negative* integers k. To do this, divide both sides of (4.1) by the right-hand side and denote the summand on the left by $F(n,k)$:

$$F(n,k) = (4n+1)(-1)^n \frac{(1/2)_n^2(-k)_n}{n!^2(3/2+k)_n} \frac{\Gamma(3/2)\Gamma(1+k)}{\Gamma(3/2+k)};$$

then take

$$G(n,k) = \frac{(2n+1)^2}{(2n+2k+3)(4n+1)}F(n,k)$$

with the motive that $F(n,k+1) - F(n,k) = G(n,k) - G(n-1,k)$, hence $\sum_n F(n,k)$ is a constant, which is seen to be 1 by plugging in $k = 0$. Finally, to deduce (1.3) one takes $k = -1/2$, which is legitimate in view of Carlson's theorem [6, Section 5.3].

5 Guillera's series for $1/\pi^2$

If one wishes to use the latter method of proof for other Ramanujan-type formulae, ingenuity is required in order to put the new parameter k in the right place. This was done only recently by J. Guillera [18, 20], who used the method to prove some other identities of Ramanujan (in those cases when z has only 2 and 3 in its prime decomposition). If the reader doubts the applicability of the method, then take into account that the purely hypergeometric origin of the method and its independence from the elliptic and modular stuff allowed Guillera [18, 19, 20] to prove new generalizations of Ramanujan-type series, namely,

$$\sum_{n=0}^{\infty} \frac{(\frac12)_n^5}{n!^5}(20n^2+8n+1)\frac{(-1)^n}{2^{2n}} = \frac{8}{\pi^2}, \tag{5.1}$$

$$\sum_{n=0}^{\infty} \frac{(\frac12)_n^5}{n!^5}(820n^2+180n+13)\frac{(-1)^n}{2^{10n}} = \frac{128}{\pi^2}, \tag{5.2}$$

$$\sum_{n=0}^{\infty} \frac{(\frac12)_n^3(\frac14)_n(\frac34)_n}{n!^5}(120n^2+34n+3)\frac{1}{2^{4n}} = \frac{32}{\pi^2}, \tag{5.3}$$

and also to find experimentally [19] four additional formulae

$$\sum_{n=0}^{\infty} \frac{(\frac12)_n(\frac14)_n(\frac34)_n(\frac16)_n(\frac56)_n}{n!^5}(1640n^2+278n+15)\frac{(-1)^n}{2^{10n}} = \frac{256\sqrt{3}}{3\pi^2}, \tag{5.4}$$

$$\sum_{n=0}^{\infty} \frac{(\frac12)_n(\frac14)_n(\frac34)_n(\frac13)_n(\frac23)_n}{n!^5}(252n^2+63n+5)\frac{(-1)^n}{48^n} = \frac{48}{\pi^2}, \tag{5.5}$$

$$\sum_{n=0}^{\infty} \frac{(\frac12)_n(\frac13)_n(\frac23)_n(\frac16)_n(\frac56)_n}{n!^5}(5418n^2+693n+29)\frac{(-1)^n}{80^{3n}} = \frac{128\sqrt{5}}{\pi^2}, \tag{5.6}$$

$$\sum_{n=0}^{\infty} \frac{(\frac12)_n(\frac18)_n(\frac38)_n(\frac58)_n(\frac78)_n}{n!^5}(1920n^2+304n+15)\frac{1}{7^{4n}} = \frac{56\sqrt{7}}{\pi^2}. \tag{5.7}$$

As Guillera notices, the series in (5.5)–(5.7) are closely related to the series

$$\sum_{n=0}^{\infty} \frac{(\frac12)_n(\frac14)_n(\frac34)_n}{n!^3}(28n+3)\frac{(-1)^n}{48^n} = \frac{16}{\pi\sqrt{3}},$$

$$\sum_{n=0}^{\infty} \frac{(\frac12)_n(\frac16)_n(\frac56)_n}{n!^3}(5418n+263)\frac{(-1)^n}{80^{3n}} = \frac{640\sqrt{15}}{3\pi},$$

$$\sum_{n=0}^{\infty} \frac{(\frac12)_n(\frac14)_n(\frac34)_n}{n!^3}(40n+3)\frac{1}{7^{4n}} = \frac{49}{3\pi\sqrt{3}},$$

respectively, proven by the methods in Sections 2 and 3. However, there is no obvious way to deduce any of formulae (5.1)–(5.7) by modular means; the problem lies in the fact that the (Zariski closure of the) projective monodromy group for the corresponding series $F(z)=\sum_{n=0}^{\infty} u_n z^n$ is always $O_5(\mathbb{R})$ (this is an immediate consequence of a general result of F. Beukers and G. Heckman [10]), which is essentially 'richer' than $O_3(\mathbb{R})$ for classical Ramanujan series.

There exists also the higher-dimensional identity

$$\sum_{n=0}^{\infty} \frac{(\frac12)_n^7}{n!^7}(168n^3+76n^2+14n+1)\frac{1}{2^{6n}} = \frac{32}{\pi^3},$$

discovered by B. Gourevich in 2002 (using an integer relations algorithm). Guillera also found experimentally an analogue of Sato's series:

$$\sum_{n=0}^{\infty} v_n \cdot (36n^2 + 12n + 1)\frac{1}{2^{10n}} = \frac{32}{\pi^2},$$
$$\text{where} \quad v_n = \binom{2n}{n}^2 \sum_{k=0}^{n} \binom{2k}{k}^2 \binom{2n-2k}{n-k}^2. \tag{5.8}$$

6 Transformations of hypergeometric series

As we have seen, Ramanujan's original formulae as well as Guillera's formulae (5.1)–(5.7) involve classical hypergeometric series (1.8), while series like (3.1)–(3.4) and (5.8) are based on double hypergeometric series. A natural way to pass from one formula to another is by algebraic transformations of the hypergeometric series involved. For instance, formula (3.1) may be deduced from the transformation

$$\sum_{n=0}^{\infty} u_n z^n = \frac{1}{2 + 2z - \sqrt{1 - 34z + z^2}} \cdot {}_3F_2\left(\begin{matrix}\frac{1}{4}, \frac{1}{2}, \frac{3}{4} \\ 1, 1\end{matrix} \,\middle|\, 256t(z)\right)$$

where

$$t(z) = \frac{z}{2(1 + 14z + z^2)^4}\big(1 - 36z + 199z^2 + 184z^3 + 199z^4 - 36z^5 + z^6$$
$$+ (1 + z)(1 - z)^2(1 - 18z + z^2)\sqrt{1 - 34z + z^2}\,\big),$$

given by Y. Yang (2005), together with the Ramanujan-type formula for the ${}_3F_2$-series on the right-hand side specialized at the point

$$\left(\frac{5 + 4\sqrt{2}}{7\sqrt{3}}\right)^4 = 256t\left(\left(\frac{\sqrt{5} - 1}{2}\right)^{12}\right).$$

A similar argument is used by M. D. Rogers in [23] to deduce some further identities for $1/\pi$ of Ramanujan–Sato type.

Using the quadratic transformation $z \mapsto -4z/(1 - z)^2$ of the hypergeometric series, we were able to produce from (5.1), (5.2) two more series of the latter type [27]:

$$\sum_{n=0}^{\infty} w_n \frac{(4n)!}{n!^2 (2n)!}(18n^2 - 10n - 3)\frac{1}{(2^8 5^2)^n} = \frac{10\sqrt{5}}{\pi^2},$$

$$\sum_{n=0}^{\infty} w_n \frac{(4n)!}{n!^2 (2n)!}(1046529n^2 + 227104n + 16032)\frac{1}{(5^4 41^2)^n} = \frac{5^4 41\sqrt{41}}{\pi^2},$$

where the sequence of integers

$$w_n = \sum_{k=0}^{n} \binom{2k}{k}^3 \binom{2n-2k}{n-k} 2^{4(n-k)}, \qquad n = 0, 1, 2, \dots,$$

satisfies the recurrence relation

$$(n + 1)^3 w_{n+1} - 8(2n + 1)(8n^2 + 8n + 5)w_n + 4096n^3 w_{n-1} = 0, \qquad n = 1, 2, \dots.$$

In [28] we show that a huge family of formulae for $1/\pi^2$ (as well as for $1/\pi^3$, $1/\pi^4$, etc) can be derived by taking powers of Ramanujan-type formulae for $1/\pi$. For instance, the square of the Chudnovskys' formula (1.7) takes the monstrous form

$$\sum_{n=0}^{\infty} w_n \frac{(3n)!}{n!^3}(222883324273153467n^2 + 16670750677895547n + 415634396862086)\frac{(-1)^n}{640320^{3n+3}} = \frac{1}{64\pi^2}.$$

7 Further observations and open problems

It is worth mentioning that identities like (4.1) are valid for all non-negative *real* values of k. This fact has several other curious implications; for instance, the series

$$G(k) = \sum_{n=0}^{\infty} \frac{(1/2+k)_n^5}{(1+k)_n^5}(820(n+k)^2 + 180(n+k) + 13)\frac{(-1)^n}{2^{10n}} \tag{7.1}$$

has a closed-form evaluation at $k = 0$ and $k = 1/2$:

$$G(0) = \frac{128}{\pi^2} \qquad \text{and} \qquad G(\tfrac{1}{2}) = 256\zeta(3),$$

where the first formula follows from (5.2) while the second one was given by T. Amdeberhan and D. Zeilberger [4]. Guillera has conjectured (and proven) evaluations for series like (7.1) viewed as functions of the continuous (complex or real) parameter k.

It seems to be a challenge to develop a modular-like theory for proving Guillera's identities and finding a (more or less) general pattern of them. For the moment, we have only speculations in this respect on a relationship to mirror symmetry, namely, to the linear differential equations for the periods of certain Calabi–Yau threefolds. A standard example here is the hypergeometric series (cf. (5.1) and (5.2))

$$F(z) = {}_5F_4\left(\begin{matrix} \frac{1}{2}, \frac{1}{2}, \frac{1}{2}, \frac{1}{2}, \frac{1}{2} \\ 1, 1, 1, 1 \end{matrix} \,\middle|\, 2^{10}z\right) = \sum_{n=0}^{\infty} \binom{2n}{n}^5 z^n,$$

which satisfies the 5th-order linear differential equation

$$(\theta^5 - 32z(2\theta+1)^5)Y = 0, \qquad \text{where} \quad \theta = z\frac{\mathrm{d}}{\mathrm{d}z}.$$

If $G(z)$ is another solution of the latter equation of the form $F(z)\log z + F_1(z)$ with $F_1(z) \in z\mathbb{Q}[[z]]$, then

$$\widetilde{F}(z) = (1 - 2^{10}z)^{-1/2} \det \begin{pmatrix} F & G \\ \theta F & \theta G \end{pmatrix}^{1/2}$$

(the sharp normalization factor $(1-2^{10}z)^{-1/2}$ is due to Y. Yang) satisfies the 4th-order equation

$$(\theta^4 - 16z(128\theta^4 + 256\theta^3 + 304\theta^2 + 176\theta + 39) + 2^{20}z^2(\theta+1)^4)Y = 0 \tag{7.2}$$

(entry #204 in [3, Table A]). For a quadratic transformation of the new function $\widetilde{F}(z)$ we have the following explicit formula [2]:

$$\frac{1+z}{(1-z)^2}\widetilde{F}\left(\frac{-z}{(1-z)^2}\right) = \sum_{n=0}^{\infty}\left(\sum_{k=0}^{n} 4^{n-k}\binom{2k}{k}^2\binom{2n-2k}{n-k}\right)^2 z^n, \tag{7.3}$$

where the right-hand side is the Hadamard square of the series

$$\frac{1}{1-16z}\,{}_2F_1\left(\begin{matrix}\frac{1}{2},\frac{1}{2}\\1\end{matrix}\,\middle|\,\frac{-16z}{1-16z}\right)=\sum_{n=0}^{\infty}\binom{2n}{n}^2\frac{(-1)^nz^n}{(1-16z)^{n+1}}$$
$$=\sum_{n=0}^{\infty}\biggl(\sum_{k=0}^{n}4^{n-k}\binom{2k}{k}^2\binom{2n-2k}{n-k}\biggr)z^n$$

which admits a modular uniformization (cf. Section 3). It is worth mentioning that (7.2) and the differential equation of order 4 for the right-hand side of (7.3) are of Calabi–Yau type [1], [14], i.e., they imitate all properties of a differential equation for the periods of a Calabi–Yau threefold. Are there analogues of Hilbert class fields for this and similar situations? Can formulae (5.1)–(5.8) be deduced from formulae for $1/\pi$ by means of algebraic transformations of hypergeometric series? There is still some work to do in the subject originated by Ramanujan's note [22] almost 100 years ago.

Acknowledgements. It is a pleasure for me to thank Gert Almkvist, Heng Huat Chan, Jesús Guillera, Jonathan Sondow, Yifan Yang, Don Zagier, and the anonymous referee for helpful comments.

References

[1] G. ALMKVIST and W. ZUDILIN, *Differential equations, mirror maps and zeta values*, in: *Mirror Symmetry V*, N. Yui, S.-T. Yau, and J. D. Lewis (eds.), AMS/IP Stud. Adv. Math. **38** (International Press & Amer. Math. Soc., Providence, RI, 2007) 481–515.

[2] G. ALMKVIST and W. ZUDILIN, *A generalization of Clausen's formula*, in preparation (2007).

[3] G. ALMKVIST, C. VAN ENCKEVORT, D. VAN STRATEN, and W. ZUDILIN, *Tables of Calabi–Yau equations*, `arXiv: math/0507430 [math.AG]` (2005).

[4] T. AMDEBERHAN and D. ZEILBERGER, *Hypergeometric series acceleration via the WZ method*, The Wilf Festschrift (Philadelphia, PA, 1996), *Electron. J. Combin.* **4**:2 (1997) Research Paper 3, 4 pp.

[5] R. APÉRY, *Irrationalité de* $\zeta(2)$ *et* $\zeta(3)$, *Astérisque* **61** (1979) 11–13.

[6] W. N. BAILEY, *Generalized hypergeometric series*, Cambridge Math. Tracts **32** (Cambridge Univ. Press, Cambridge, 1935); 2nd reprinted edition (Stechert-Hafner, New York–London, 1964).

[7] B. C. BERNDT, *Ramanujan's notebooks. Part IV* (Springer-Verlag, New York, 1994).

[8] B. C. BERNDT and H. H. CHAN, *Eisenstein series and approximations to* π, *Illinois J. Math.* **45**:1 (2001) 75–90.

[9] F. BEUKERS, *Irrationality proofs using modular forms*, Journées arithmétiques (Besançon, 1985), *Astérisque* **147-148** (1987) 271–283.

[10] F. BEUKERS and G. HECKMAN, *Monodromy for the hypergeometric function* ${}_nF_{n-1}$, *Invent. Math.* **95**:2 (1989) 325–354.

[11] J. M. BORWEIN and P. B. BORWEIN, *Pi and the AGM* (Wiley, New York, 1987).

[12] H. H. CHAN, S. H. CHAN, and Z. LIU, *Domb's numbers and Ramanujan–Sato type series for* $1/\pi$, *Adv. Math.* **186**:2 (2004) 396–410.

[13] H. H. CHAN, W.-C. LIAW, and V. TAN, *Ramanujan's class invariant* λ_n *and a new class of series for* $1/\pi$, *J. London Math. Soc.* (2) **64**:1 (2001) 93–106.

[14] Y.-H. CHEN, Y. YANG, and N. YUI, *Monodromy of Picard–Fuchs differential equations for Calabi–Yau threefolds*, *J. Reine Angew. Math.* (to appear). `arXiv: math/0605675 [math.AG]` (2006).

[15] D. V. CHUDNOVSKY and G. V. CHUDNOVSKY, *Approximations and complex multiplication according to Ramanujan*, in: *Ramanujan revisited* (Urbana-Champaign, Ill., 1987) (Academic Press, Boston, MA, 1988) 375–472.

[16] S. B. EKHAD and D. ZEILBERGER, *A WZ proof of Ramanujan's formula for* π, in: *Geometry, Analysis, and Mechanics*, J. M. Rassias (ed.) (World Scientific, Singapore, 1994) 107–108.

[17] S. Fischler, *Irrationalité de valeurs de zêta (d'après Apéry, Rivoal, ...)*, *Astérisque* **294** (2004) 27–62.

[18] J. GUILLERA, *Some binomial series obtained by the WZ-method*, *Adv. in Appl. Math.* **29**:4 (2002) 599–603.

[19] J. GUILLERA, *About a new kind of Ramanujan-type series*, *Experiment. Math.* **12**:4 (2003) 507–510.

[20] J. GUILLERA, *Generators of some Ramanujan formulas*, *Ramanujan J.* **11**:1 (2006) 41–48.

[21] A. VAN DER POORTEN, *A proof that Euler missed.. Apéry's proof of the irrationality of* $\zeta(3)$, *Math. Intelligencer* **1**:4 (1978/79) 195–203.

[22] S. RAMANUJAN, *Modular equations and approximations to* π, *Quart. J. Math. Oxford Ser.* (2) **45** (1914) 350–372; reprinted in: *Collected papers of Srinivasa Ramanujan*, G. H. Hardy, P. V. Sechu Aiyar, and B. M. Wilson (eds.) (Cambridge University Press, Cambridge, 1927; Chelsea Publ., New York, 1962) 23–39.

[23] M. D. ROGERS, *New* ${}_5F_4$ *hypergeometric transformations, three-variable Mahler measures, and formulas for* $1/\pi$, *Ramanujan J.* (to appear). `arXiv: 0704.2438 [math.NT]` (2007).

[24] A. WEIL, *Elliptic functions according to Eisenstein and Kronecker*, Ergebnisse der Mathematik und ihrer Grenzgebiete **88** (Springer-Verlag, Berlin, 1976).

[25] Y. YANG, *On differential equations satisfied by modular forms*, *Math. Z.* **246**:1-2 (2004), 1–19.

[26] W. ZUDILIN, *Ramanujan-type formulae and irrationality measures of certain multiples of* π, *Russian Acad. Sci. Sb. Math.* **196**:7 (2005) 983–998.

[27] W. ZUDILIN, *Quadratic transformations and Guillera's formulas for* $1/\pi^2$, *Math. Notes* **81**:3 (2007) 297–301.

[28] W. ZUDILIN, *More Ramanujan-type formulae for* $1/\pi^2$, *Russian Math. Surveys* **62**:3 (2007) 634–636.

Aspects of Physics and String Theory

Fields Institute Communications
Volume 54, 2008

Meet Homological Mirror Symmetry

Matthew Robert Ballard
Department of Mathematics
University of Washington
Seattle, WA 98195 USA
ballard@math.washington.edu

Abstract. In this paper, we introduce the interested reader to homological mirror symmetry. After recalling a little background knowledge, we tackle the simplest cases of homological mirror symmetry: curves of genus zero and one. We close by outlining the current state of the field and mentioning what homological mirror symmetry has to say about other aspects of mirror symmetry.

CONTENTS

1 Introduction

As it stands today, mirror symmetry is a well-established field of mathematics. Seventeen years have passed since physicists correctly predicted the number of rational curves of given degrees on the Fermat quintic [9]. In the intervening years, the physical inspiration has energized the mathematics community. In many cases, including anti-canonical hypersurfaces and Calabi-Yau complete intersections in toric varieties, mathematical mirror constructions have been realised allowing one to see the mirror reflection in the Hodge diamond and establish the relation between curve counts and period integrals. Unfortunately, rigorous formulations of the physical

2000 *Mathematics Subject Classification.* Primary 14J32; Secondary 18E30, 53D40, 53D35.
Key words and phrases. homological mirror symmetry, A and B-branes, A_∞-algebra, Fukaya category, autoequivalence, elliptic curves, Calabi-Yau manifold.

arguments yielding these predictions have not arisen. Such rigorous formulations would allow mathematicians to tap into the deep physical understanding of the situation, but modern physical tools, such as the path integral, lie just beyond the current scope of mathematics. Consequently, to fully understand the phenomenon of mirror symmetry, mathematicians must follow the physical inspiration and develop their own overarching approaches.

At the International Congress of Mathematicians in 1994, M. Kontsevich proposed such an approach [34]. As is it commonly known now, homological mirror symmetry, or the homological mirror conjecture, reformulates mirror symmetry as an equivalence of triangulated categories built from different aspects of the Calabi-Yau geometry of two manifolds, called mirror manifolds. The Calabi-Yau metric is determined by two pieces of data — the complex structure and the Kähler form. Each piece lands in a different branch of mathematics. If we only care about the complex structure, we are naturally led into algebraic geometry. If we instead remember the Kähler form, we study symplectic geometry. Upon passage to the mirror manifold, we should see an exchange of these two geometries. Kontsevich sought to capture this exchange. Today, homological mirror symmetry remains an intriguing and daunting challenge to mathematics. The scope has expanded to encompass other manifolds beyond Calabi-Yau manifolds. It has become a powerful source of inspiration motivating insights in algebraic geometry, symplectic geometry, homological algebra, noncommutative geometry, and beyond.

Mixing all these fields, homological mirror symmetry is a very attractive conjecture, but it remains outside the common knowledge of the working mathematician. Why? Along with its relative youth, its proper formulation requires an imposing amount of unfamiliar technology. This article offers a introduction to homological mirror symmetry through two explicit examples. The cases of $\mathbb{P}^1$ and elliptic curves are very concrete. We give a holistic approach that treats both sides of mirror symmetry for $\mathbb{P}^1$, instead of speaking to one side without reference to the other as is often the case in the literature. Since homological mirror symmetry manifests itself in noticeably different ways whether one considers Fano or Calabi-Yau varieties, we also review the case of elliptic curves. Seeing both cases side by side will hopefully give the reader a deeper appreciation of the duality. To move into examples we need to review a little of the necessary formalism. This is done in section 2. With this over, we move on to the examples in sections 3 and 4. Algebraic curves of genus zero and one provide simple case studies. Within these, the reader can meet concrete incarnations of the relevant categories and appreciate the unexpected equivalences that homological mirror symmetry predicts. After the examples have been covered, we outline the current state of knowledge in the field in section 5 and then mention how homological mirror symmetry relates to other aspects of mirror symmetry in section 6.

The author heartily thanks his adviser, Charles Doran, for encouraging the creation of this survey, the referee for his careful reading, and Ursula Whitcher for her time and numerous suggestions to improve the article. All errors belong solely to the author.

2 Building some common knowledge

Part of the difficulty in dealing with homological mirror symmetry is the breadth of knowledge required for a proper formulation. Before we can dive into the promised simple examples, we recall some terminology and results from homological algebra, algebraic geometry, and symplectic geometry.

2.1 Homological algebra. I assume the reader has some familiarity with derived categories, at least in the case of modules over an associative algebra, Hochschild cohomology, and dg-algebras. A good reference for homological algebra is [20], a good reference for homological algebra and Hochschild cohomology is [60], and a good reference for dg-algebras is [33].

Our goal is here is to define a triangulated category which appears on each side of mirror symmetry. The main algebraic tool is the A_∞-algebra. A_∞-algebras are avoidable in algebraic geometry but not in symplectic geometry. (However, the A_∞-algebras that appear in homological mirror symmetry for $\mathbb{P}^1$ are honest associative algebras).

Definition 2.1 An A_∞-algebra over a base field k is a graded k-module A with k-linear maps $m_n : A^{\otimes n} \to A$ of degree $2 - n$ for each $n > 0$, satisfying the following quadratic relations for all $n > 0$:

$$\sum_{r,s}(-1)^{rs+(n-r-s)} m_l(\mathrm{id}^{\otimes r} \otimes m_s \otimes \mathrm{id}^{\otimes n-r-s}) = 0$$

Let us look at the first three of these relations. The first relation says $m_1^2 = 0$ so (A, m_1) is a chain complex. The second says m_2 is a chain map when we use the differential $1 \otimes m_1 + m_1 \otimes 1$ on $A^{\otimes 2}$. The third says that m_2 is associative up to a homotopy m_3. Thus, we can pass to the cohomology with respect to m_1 and get an associative algebra $H(A)$.

Familiar examples of A_∞-algebras are associative algebras, where $m_n = 0$ for $n \neq 2$, and dg-algebras, where $m_n = 0$ for $n \neq 1, 2$. An A_∞-algebra with $m_1 = 0$ is called minimal.

We can (and often have to) do something troublesome and add a degree 0 multiplication $m_0 : k \to A$ and continue to require the quadratic relations to hold. For example, the first two relations become

$$m_1(m_0) = 0 \quad m_1(m_1) + m_2(m_0, \mathrm{id}) + m_2(\mathrm{id}, m_0) = 0$$

This in general destroys the possibility of taking cohomology. An A_∞-algebra possessing such an m_0 is called curved or obstructed. If we assume, that m_0 is central (with respect to m_2) and $m_n(\mathrm{id}^{\otimes r} \otimes m_0 \otimes \mathrm{id}^{\otimes n-r-1}) = 0$ for all $n > 2$ we can once again take cohomology $H(A)$. In this case, A is called weakly obstructed. Below our A_∞-algebras will be unobstructed unless explicitly indicated.

Definition 2.2 A morphism $f : A \to B$ of A_∞-algebras over k is a collection of k-linear maps $f_n : A^{\otimes n} \to B$ of degree $1 - n$ satisfying

$$\sum(-1)^{\heartsuit} m_k(f_{i_1} \otimes \cdots \otimes f_{i_k}) = \sum(-1)^{sr+(n-s-r)} f_l(\mathrm{id}^{\otimes s} \otimes m_r \otimes \mathrm{id}^{\otimes n-r-s})$$

where $\heartsuit = (k-1)(i_1 - 1) + (k-2)(i_2 - 1) + \cdots + (i_{k-1} - 1)$.

The first relation says that f_1 commutes with the differential. The second says f_1 respects m_2 up to f_2. f is a called a quasi-isomorphism if $f_1 : H(A) \to H(B)$ is an isomorphism.

Remark 2.3 These definitions become more compact and perhaps clearer when one passes from A to the bar complex $B(A)$ on A. The A_∞-algebra structure on A is equivalent to a coderivation on $B(A)$. A_∞-algebra morphisms are then coalgebra morphisms commuting with the coderivations. For a reference, see [21].

Now we recall a result that allows us to pass to cohomology of an A_∞-algebra without losing information.

Theorem 2.4 [31, 40, 37] *Given an A_∞-algebra A over a field k, choose a splitting $A = H \oplus B \oplus D$ over k, where H is the cohomology and $m_1 : D \to B$ is an isomorphism. Then, there exists an A_∞-algebra structure on $H(A)$ with zero first order composition, second order composition induced by m_2, and quasi-isomorphisms $i : H(A) \to A, \pi : A \to H(A)$. Moreover, π_1 is the associated restriction map $A \to H$ and i_1 is the associated inclusion map $H \to A$.*

Consider for a moment a dg-algebra A. Applying this procedure yields a minimal A_∞-algebra structure on $H(A)$. Thus, the chain-level data lost when we just consider $H(A)$ has been transmogrified, returning as the higher compositions in the A_∞-structure.

How complex can A_∞-algebras be? A partial answer to this question is given by the following observation.

From theorem 2.4, there is an A_∞-structure on $H(A)$ which makes it quasi-isomorphic to the A_∞-algebra A. Given a minimal A_∞-algebra (A, m_n), take m_k to be the first non-zero operation with $k > 2$. Then, the first A_∞-relation involving m_k is

$$\sum_r (-1)^r m_k(\mathrm{id}^{\otimes r} \otimes m_2 \otimes \mathrm{id}^{\otimes k-1-r}) + m_2(m_k, \mathrm{id}) - m_2(\mathrm{id}, m_k) = 0$$

This equation states that m_k is a Hochschild cochain for the algebra (A, m_2). Suppose we wanted to find another minimal A_∞-structure m'_n on (A, m_2) which is isomorphic to our original one. Then, we have our collection $f_k : A^{\otimes k} \to A[1-k]$. We need f_1 to be an automorphism of the algebra (A, m_2), consequently we can inductively solve

$$\sum (-1)^{\heartsuit} m_k(f_{i_1} \otimes \cdots \otimes f_{i_k}) = \sum (-1)^{s+r(n-s-r)} f_l(\mathrm{id}^{\otimes s} \otimes m_r \otimes \mathrm{id}^{\otimes n-r-1})$$

for m'_n. Thus, the f_n and m_n uniquely determine m'_n. So, if we want to find an isomorphic (A, m'_n) with $m'_l = 0$ for $2 < l \leq k$, we just need to solve

$$\sum_s f_l(\mathrm{id}^{\otimes s} \otimes m_2 \,\mathrm{id}^{\otimes n-r-1}) + m_2(f_l, \mathrm{id}) + m_2(\mathrm{id}, f_l) = m_k(f_1^{\otimes k})$$

and then take $f_l = 0$ for $1 < l < k$ and f_l arbitrary for $l > k$. Applying an automorphism, we can set $f_1 = \mathrm{id}$. The previous equation says that m_k is a Hochschild coboundary for (A, m_2). Thus, we get the following result.

Lemma 2.5 *If the relevant pieces of the Hochschild cohomology of (A, m_2) are zero, we can trivialise up to any arbitrary order. From the construction, the limiting composition of these maps exists, giving a trivialisation for the A_∞-structure.*

As in the case of an associative algebra, we wish to form a category of modules over an A_∞-algebra A. We do so in a manner strongly analogous to forming chain complexes of modules over an associative algebra.

Definition 2.6 A (right) module M over an A_∞-algebra A is a graded module over k equipped with k-linear maps $m_n^M : A^{\otimes n-1} \otimes M \to M$ of degree $2-n$ for each $n > 0$, satisfying the following quadratic relations for all $n > 0$:

$$\sum_{r,s,n-r-s>0} (-1)^{rs+(n-r-s)} m_l^M(\mathrm{id}^{\otimes r} \otimes m_s \otimes \mathrm{id}^{\otimes n-r-s})$$
$$+ \sum_{u,v} (-1)^{uv} m_l^M(\mathrm{id}^{\otimes u} \otimes m_v^M) = 0$$

A morphism $g : M \to N$ of A-modules is a collection of k-linear maps $g_n : A^{\otimes n-1} \otimes M \to N$ satisfying quadratic relations similar to the case of morphisms of A_∞-algebras. For more details, see [21].

Let us repackage this definition. Instead of thinking of A as an A_∞-algebra, we shall think of it as an A_∞-category with one object $*$. The morphisms in this category are only the endomorphisms of $*$ and $\mathrm{Hom}(*, *) := A$. To give a category $\mathcal{A}$ an A_∞-structure, we need to define multi-compositions of morphisms

$$m_n : \mathrm{Hom}(X_0, X_1) \otimes \cdots \otimes \mathrm{Hom}(X_{n-1}, X_n) \to \mathrm{Hom}(X_0, X_n)$$

satisfying the A_∞-relations. Here X_i are objects of $\mathcal{A}$. In our case, we just use the operations $m_n : A^{\otimes n} \to A$ coming from the A_∞-algebra structure of A.

For an A_∞-category $\mathcal{A}$, each object X of $\mathcal{A}$ furnishes a functor $\mathrm{Hom}_{\mathcal{A}}(X, \cdot)$ from $\mathcal{A}$ to chain complexes over k, $Ch(k)$. These are naturally A_∞-functors (for more, see [32]).

Definition 2.7 Given two A_∞-categories $\mathcal{A}$ and $\mathcal{B}$. An A_∞-functor $\mathcal{F} : \mathcal{A} \to \mathcal{B}$ is an assignment of objects $X \mapsto \mathcal{F}(X)$ and a collection of maps

$$\mathcal{F}_n : \mathrm{Hom}_{\mathcal{A}}(X_0, X_1) \otimes \cdots \otimes \mathrm{Hom}_{\mathcal{B}}(X_{n-1}, X_n) \to \mathrm{Hom}_{\mathcal{B}}(\mathcal{F}(X_0), \mathcal{F}(X_n))$$

which satisfy quadratic relations similar to those given in the definition of a morphism of A_∞-algebras.

The A_∞-version of the Yoneda embedding is given by sending X to $\mathrm{Hom}_{\mathcal{A}}(X, -)$. This gives an A_∞-functor Y from $\mathcal{A}$ to the category of A_∞-functors from $\mathcal{A}^{op}$ to $Ch(k)$. Denote this category by $\mathrm{Mod}\,\mathcal{A}$. The standard Yoneda embedding is full and faithful. After taking H^0, Y reduces to the standard Yoneda embedding. Since $H^0(Y)$ is an equivalence onto its image, we say that Y is a quasi-equivalence onto its image.

If we consider the A_∞-category with a single object and morphisms algebra A, this gives another definition of A-module. It is a good exercise for the reader to translate between the two definitions of A-modules for an A_∞-algebra.

Since any A-module is, in particular, a chain complex, we have a notion of quasi-isomorphism in the category of A-modules. In analogy with the case of an associative algebra, we wish to invert these. When we do, we get $D(A)$, the derived category of A-modules. As in the case of associative algebras, the resulting category is triangulated. We let $D^\pi(A)$ be the smallest triangulated subcategory of $D(A)$ containing A and closed under taking triangles, direct sums, and direct summands. To finish, we recall the following useful result. For a proof, see [49], where the reader can find more details about the construction of $D(A)$.

Proposition 2.8 *If A and B are quasi-isomorphic A_∞-algebras, then $D(A)$ is triangle equivalent to $D(B)$ and $D^\pi(A)$ is triangle equivalent to $D^\pi(B)$.*

2.2 Algebraic geometry. Algebraic geometry is the most natural geometric field for the application of homological algebra. Indeed, homological algebra permeated algebraic geometry long ago. We shall only recall a small amount. For more details, see [27].

Let us recall some standard abelian categories associated to an algebraic variety X. Any algebraic variety X comes with its sheaf of functions $\mathcal{O}_X$. A sheaf $\mathcal{E}$ on X is called a quasi-coherent sheaf if there is an action of $\mathcal{O}_X$ on $\mathcal{E}$ and, locally in the Zariski topology, $\mathcal{E}$ is the cokernel of a morphism between free $\mathcal{O}_X$-modules. $\mathcal{E}$ is called coherent if locally, in the Zariski topology, $\mathcal{E}$ is the cokernel of a morphism between free, finite-rank $\mathcal{O}_X$-modules. Let us restrict ourselves to the case that X is quasi-projective over a field k. In this case, the category of coherent sheaves on X, $\mathrm{Coh}(X)$, is an abelian category. The category of quasi-coherent sheaves, $\mathrm{QCoh}(X)$, is also abelian and it possesses enough injective objects. Therefore, we can resolve any quasi-coherent sheaf by a bounded below complex of injective sheaves. Thus, we can form the derived category of $\mathrm{Coh}(X)$, or $\mathrm{QCoh}(X)$, by taking the homotopy category of the subcategory of all bounded below complexes of injectives with bounded coherent cohomology, or by taking the homotopy category of the subcategory of all injectives with quasi-coherent cohomology.

An important class of coherent sheaves is the locally free coherent sheaves, i.e. ones which are locally isomorphic to a finite-rank free $\mathcal{O}_X$-module. Given a locally free coherent sheaf $\mathcal{E}$, we can associate to it an algebraic vector bundle E whose sheaf of sections is the dual sheaf of $\mathcal{E}$, i.e. the sheaf whose sections over U are given by $\mathrm{Hom}_U(\mathcal{E}(U), \mathcal{O}_X(U))$. This gives a contravariant equivalence between the categories of locally free coherent sheaves and algebraic vector bundles.

2.3 Symplectic geometry background. In this section, we review some of the underlying notions of symplectic geometry. With this knowlege in hand, we proceed, in this section, to define Lagrangian intersection Floer homology. An excellent reference for all things symplecto-topological is [38] and, correspondingly, an excellent introduction to the uses of J-holomorphic curves in symplectic topology is [39]

Let M be a smooth manifold and ω an anti-symmetric two-form.

Definition 2.9 ω is a symplectic form if $d\omega = 0$ and ω is non-degenerate, i.e. the pairing on vectors in T_xM at all points $x \in M$ is non-degenerate.

Note that a symplectic manifold must have even dimension.

Example 2.10

1. The canonical example of a symplectic manifold is the cotangent space T^*X to any smooth manifold X. Let us look at an open chart U and denote the coordinates on U by p and denote the coordinates in the fiber direction by q. $T^*U \cong U \times \mathbb{R}^n$. We let $\sigma_{can} = q\ dp$. One can check that even though we have defined this locally, it is globally well-defined. We set $\omega_{can} = d\sigma_{can} = dq \wedge dp$. ω_{can} is clearly non-degenerate as vectors in the base pair with vectors in the fiber. (T^*X, ω_{can}) is also an important example of an exact symplectic manifold, that is, a symplectic manifold for which $\omega = d\sigma$ for some one-form σ.
2. Any smooth projective complex algebraic variety is a symplectic manifold. The Fubini-Study Kähler form on $\mathbb{P}^n_{\mathbb{C}}$ restricts to a symplectic form on any complex submanifold.

The following result shows that symplectic manifolds are all locally isomorphic.

Proposition 2.11 *(Darboux) Given a symplectic manifold* (M^{2n}, ω) *and a point* $x \in M$*, then there is a neighborhood* U *of* x *and a set of coordinates*

$$(x^1, \dots, x^n, y^1, \dots, y^n)$$

on U *such that* $\omega|_U = \sum_i dx^i \wedge dy^i$.

If ω is an anti-symmetric two-form on a vector space T, for any subspace S we can define the symplectic orthogonal $S^\perp = \{v \in T : \omega(s, v) = 0 \ \forall s \in S\}$. If ω is non-degenerate, then $S \oplus S^\perp = T$. A subspace S is isotropic if $S \subset S^\perp$ and coisotropic if $S^\perp \subset S$. It is called Lagrangian if it is both isotropic and coisotropic and symplectic if $S \cap S^\perp = \{0\}$. Given an symplectic manifold (M, ω), we say that an embedded submanifold S is isotropic (resp. coisotropic, Lagrangian, symplectic) if all the tangent spaces are isotropic (resp. coisotropic, Lagrangian, symplectic).

For T^*X with ω_{can}, the zero section X is a Lagrangian submanifold, as is each fiber. The general case is similar.

Proposition 2.12 *(Weinstein) Let* L *be a Lagrangian submanifold of a symplectic manifold* (M, ω)*. Then there is a tubular neighborhood* U *of* L *which is symplectomorphic to a neighborhood of* L *in* T^*L *with* ω_{can}.

Since a symplectic form is non-degenerate, it affords a way to turn vectors into one-forms and vice versa. If η is a one-form, we can define $\eta^\vee$ to be the vector field such that $\omega(\eta^\vee, -) = \eta$. We often start with the differential of a smooth function $f : M \to \mathbb{R}$, or of a time-varying function $f : M \times \mathbb{R} \to \mathbb{R}$. Given such a function, we define the Hamiltonian vector field X_{f_t} to be $df_t^\vee$. Given a vector field, we can often integrate it out to a diffeomorphism called a flow $\phi^t : M \to M$. The flow associated to X_{f_t} is called a Hamiltonian flow. One can check that since df_t is closed, Hamiltonian flows preserve the symplectic form. Note that given a submanifold N of M, $\phi^t(N)$ defines an isotopy between N and $\phi^1(N)$. This is called a Hamiltonian isotopy.

Definition 2.13 A bundle endomorphism $J : TX \to TX$ is called an almost complex structure if $J^2 = -1$.

The simplest example of an almost complex structure is multiplication by $\imath$ on $\mathbb{C} \cong \mathbb{R}^2$. An almost complex structure on a vector space is the same thing as a complex structure. Complex manifolds therefore have natural almost complex structures, namely multiplication by $\imath$. An almost complex structure J that makes X into a complex manifold is called a complex structure. However, not all almost complex structures are complex structures. The $\imath$-eigenspaces of J on $TX \otimes \mathbb{C}$ must be involutive. See [42].

Definition 2.14 If (M, ω) is a symplectic manifold, then an almost complex structure J is called compatible if $\omega(J-, -)$ is a Riemannian metric, i.e. symmetric, positive definite, and non-degenerate.

Given a symplectic manifold (M, ω), we can always locally find a compatible almost complex structure using Darboux's theorem. Since Riemannian metrics form a convex space (we can add them), we can find a partition of unity and patch together our local compatible almost complex structures into a global almost complex structure J. Given a complex manifold Y with complex structure j, we say that a map $\phi : Y \to M$ is J-holomorphic if $d\phi \circ j = J \circ d\phi$. If $Y = M = \mathbb{C}$ and $j = J = \imath$, then this equation is equivalent to the set of Cauchy-Riemann equations.

Solutions of the Cauchy-Riemann equations satisfy nice properties (all of complex algebraic geometry stems from this.) Gromov realised [22] that J-holomorphic maps have many of the same nice properties. He introduced the study of J-holomorphic curves into symplectic manifolds and gave a useful new tool to symplectic geometry.

2.4 Floer cohomology and Fukaya categories. Now we can outline Lagrangian intersection Floer cohomology. Let L and L' be two closed Lagrangian submanifolds of a compact symplectic manifold. If they are not transverse, replace L by a Hamiltonian isotopic Lagrangian. We can then assume that L and L' intersect transversely and thus $L \cap L'$ is a finite set. Given two intersection points p and q, consider the set of J-holomorphic maps $\phi : D \to M$ from the unit disc with two marked points at -1 and 1 such that $\phi(\partial D \cap \mathbb{H}) \subset L, \phi(\partial D \cap (\mathbb{C} - \mathbb{H})) \subset L', \phi(1) = q$, and $\phi(-1) = p$. The expected dimension of the space of solutions can be determined as follows. The pullback tangent bundle ϕ^*TM is trivial since we are working with a disc. There is a real sub-bundle along the boundary determined by the tangent spaces to the Lagrangians. We can change our trivialisation so that $\phi^*(T_pL') = \imath\phi^*(T_pL)$ and $\phi^*(T_qL) = \imath\phi^*(T_qL')$. If we rotate at p and q, we get a loop of Lagrangian subspaces in $\mathbb{R}^{2n}$. Let Λ_n denote the Lagrangian Grassmannian of $\mathbb{R}^{2n}$. $H^1(\Lambda_n) \cong \mathbb{Z}$ and there is a distinguished generator μ called the Maslov class. The index of the operator is given by applying μ to the loop of Lagrangian subspaces. Let us denote this index by $\mu_\phi(p,q)$, or by $\mu(p,q)$. Note that this a relative grading in the sense that we only know the difference between p and q. No absolute grading on the critical points is specified.

The space of such J-holomorphic discs is at least one-dimensional since we have a free action by conformal automorphisms of the unit disc $PSL(2,\mathbb{R})$ on the unit disc. If the dimension is one, taking the quotient by the free action we expect to get a zero-dimensional manifold, $\mathcal{M}(p,q)$, as the moduli space of solutions. A wonderful fact is that there are natural compactifications of spaces of J-holomorphic maps called Gromov compactifications $\bar{\mathcal{M}}(p,q)$ [39]. The codimension one component of the boundary of these compactifications, in the best situations, is

$$\partial\bar{\mathcal{M}}(p,q) = \coprod_{r \in L \cap L'} \mathcal{M}(p,r) \times \mathcal{M}(r,q)$$

This is analogous to finite-dimensional Morse theory where, to compactify the space of gradient trajectories, one adds in trajectories which are broken at an intermediate critical points. The codimension one piece is where we only have one intermediate critical point. In the case where $\mu(p,q)$ is one, there is no r we can squeeze in and hence $\mathcal{M}(p,q)$ is compact. Thus, $\mathcal{M}(p,q)$ is a finite set of points which we can count. Define the chain complex $CF(L,L')$ as the free $\mathbb{Z}/2\mathbb{Z}$-graded vector space with basis $L \cap L'$ and set

$$m_1(p) = \sum_{q:\mu(p,q)=1} n_{pq} q$$

where n_{pq} is the number of points in $\mathcal{M}(p,q)$.

Proposition 2.15 $m_1^2 = 0$.

To see this, let $\mu(p,q) = 2$ and consider the coefficient of the q-term in $m_1^2(p)$. It is exactly $\sum_r n_{pr} n_{rq}$, which is the number of points in $\coprod_{r \in L \cap L'} \mathcal{M}(p,r) \times \mathcal{M}(r,q)$. $\coprod_{r \in L \cap L'} \mathcal{M}(p,r) \times \mathcal{M}(r,q)$ is the boundary of $\bar{\mathcal{M}}(p,q)$, which is one-dimensional, hence must be an even number. We can improve the counts to lie in $\mathbb{Z}$, instead of

$\mathbb{Z}/2\mathbb{Z}$, when we can coherently orient the moduli spaces of solutions. This requires a Spin structure [15].

The cohomology of $CF(L, L')$ is called the (Lagrangian intersection) Floer cohomology of L and L' and is denoted by $HF(L, L')$.

What could go wrong with this situation? There are a few problems related to the Gromov compactification. The compactness relies on bounds on the energy $\int \phi^*\omega$ of a J-holomorphic disc ϕ. However, in general, we could have energies of a sequence of J-holomorphic discs tending to infinity. To remedy this, we can include a formal parameter which keeps track of the symplectic area of these discs. This introduces Novikov rings into the discussion [29]. If we are lucky, as we will be later in this paper, the formal series actually converges if we specialize it.

Another, more serious, problem is that if the symplectic form does not vanish on the second homotopy group of M, $\pi_2(M)$, we would have to include sphere bubbles in the Gromov compactification. Perturbing the almost complex structure and adding a zeroth order term are no longer enough to guarantee that the compactified moduli spaces have the proper structure. This issue was overcome by introducing a new, more general method of perturbation — virtual perturbation theory. For the details see [19, 48, 55, 36].

On the other hand, if ω does not vanish on $\pi_2(M, L)$, we could have disc bubbles which generally occupy a (real) codimension one subset of the compactified moduli spaces. Thus, if $\mu(p, q) = 2$, the boundary of $\bar{\mathcal{M}}(p, q)$ may now also include other components. Consequently, m_1^2 may no longer be zero. While this problem cannot always be fixed, there is a nice general framework developed in [18] which allows one to address the issue. For a simple and illuminating example of what could go wrong in this case, see [30].

Assume now that our Lagrangian L is sitting inside a Calabi-Yau manifold X. Since X is Calabi-Yau, it possesses a non-vanishing holomorphic volume form Ω. Given a Lagrangian submanifold L of X, we take a frame $\{v_1, \cdots, v_n\}$ for it at a point $x \in X$. This gives a complex number

$$\theta(x) = \frac{\Omega(v_1, \cdots, v_n)^2}{|\Omega(v_1, \cdots, v_n)|^2}$$

and consequently, a map $\theta : L \to S^1$ called the phase squared map. A graded Lagrangian submanifold is a Lagrangian L and a lifting of the phase map to $\alpha : L \to \mathbb{R}$. This induces a grading on the intersection points of graded Lagrangian submanifolds. One can check that this grading is compatible with the relative Maslov grading given in the Floer complex; see [51].

A choice of Spin structure, if it exists, and the grading allow one to work, in principle, with $\mathbb{Z}$-coefficients and $\mathbb{Z}$-gradings in Floer cohomology.

There is no reason to stop at J-holomorphic discs with two marked points. One can also consider J-holomorphic discs with $n > 2$ marked points. We outline the construction and refer the inquisitive reader to the references. First we consider the universal family of n-pointed discs, $\mathcal{A}_n$. By an n-pointed disc, we mean a disc with n distinguished points on the boundary considered up to equivalence given by Möbius transformations. The marked points $\xi_0, \cdots, \xi_{n-1}$ are ordered cyclically and we label the boundary segment between ξ_i and ξ_{i+1} as C_i. Given Lagrangians $L_0, \cdots, L_n$ and intersection points $p_{ii+1} \in L_i \cap L_{i+1}$ where i is taken modulo $n + 1$, we can look for J-holomorphic maps ϕ from an $(n + 1)$-pointed disc into our symplectic manifold M so that $\phi(C_i) \subset L_i$ and $\phi(\xi_i) = p_{ii+1}$. When $\tilde{p}_{n0} = 2 - n + \sum \tilde{p}_{ii+1}$, the space of such maps (when considering all possible $(n+1)$-pointed discs, assuming transversality) is zero-dimensional. As before, there are natural compactifications of these moduli spaces which add pieces of codimension larger or equal to one. Therefore, in the case where we have chosen the correct intersection points, we can count the number of such J-holomorphic maps. Denote this count by $n_{p_{01}\cdots p_{n-1n}p_{n0}}$ and define

$$m_i(p_{01},\cdots,p_{n-1n}) = \sum_{q\in L_0\cap L_n:\tilde{q}=2-n+\sum \tilde{p}_{ii+1}} n_{p_{01}\cdots p_{n-1n}q}q$$

Recall that m_1 satisfied $m_1^2=0$. The total collection of the m_i satisfy the A_∞-relations. Let us sketch why this is true. We first consider the case that M is simply a point and look at the natural compactifications of these moduli spaces. Any n-pointed disc provides a J-holomorphic map. Thus, our moduli space of such maps is all of $\mathcal{A}_n$. We need to compactify this space. To do this we add degenerations of the form

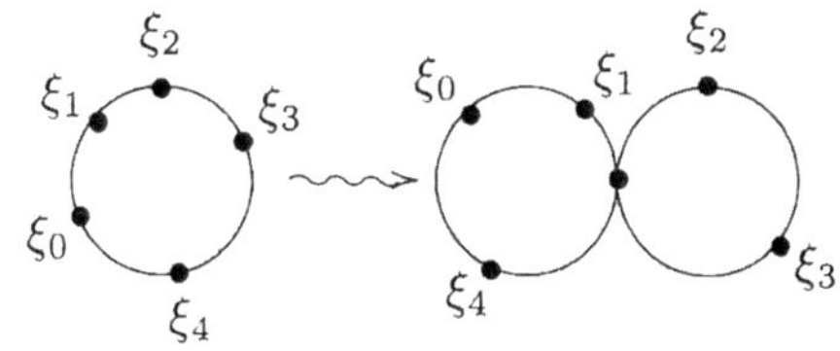

as the codimension one boundary strata and work our way down inductively. The codimension one boundary strata of $\overline{\mathcal{A}}_n$ are given by

$$\coprod_{n-1>j>2} \mathcal{A}_j \times \mathcal{A}_{n+1-j}$$

corresponding to the splitting of a single marked disc into two pointed discs, the first with j marked points and the second with $n+1-j$ marked points, joined at a marked point. If we view these as counting discs, we see that the relations in the codimension one boundary strata are exactly the A_∞-relations with vanishing m_1. The compactification in the general case works in a very similar way. When we generalise from a point, the compactifications include J-holomorphic maps with two marked points corresponding to the non-vanishing m_1. Thus, in the best of circumstances, we can form an A_∞-category called the Fukaya category of M, Fuk(M). Its objects are Lagrangian submanifolds (with some extra structure) and its morphisms are intersection points between two given Lagrangians. The multi-compositions are those outlined above.

We shall see a few variations on the theme of the Fukaya category as we proceed. Most variations correspond to restricting one's attention to pieces of a larger Fukaya category. Others correspond to slight modifications if we do not actually experience the best of circumstances.

Remark 2.16 If one wishes to examine a fuller treatment of the construction of the Fukaya category in the case of an exact symplectic manifold, see [49] and the references therein. For the reader interested in the most general results, see [18].

3 Homological mirror symmetry for the projective line

3.1 B-branes on the projective line. We begin by tackling B-branes before A-branes. The name "branes" is short for membranes and is a reflection of the subject's physical origin. In string theory, branes are boundary conditions for an open string proprogating through spacetime. The relative simplicity of the algebro-geometric side, or B-side, of mirror symmetry spurs this alphabetic rebellion. After defining the category of B-branes, we seek and find a concrete description of the category in terms of a directed graph, i.e. a quiver.

Let k be a field.

Definition 3.1 The category of B-branes on $\mathbb{P}^1_k$ is the bounded derived category of coherent sheaves on $\mathbb{P}^1_k$.

If we are seeking a category that reflects the algebraic geometry of $\mathbb{P}^1_k$, there is perhaps no better choice. Evidence is given by the following result of Bondal and Orlov [8]. In passing from $\mathbb{P}^1_k$ to $D^b(\mathrm{Coh}(\mathbb{P}^1_k))$, we lose nothing.

Theorem 3.2 *Let X and Y be smooth projective varieties and let ω_X or ω_X^{-1} be ample. If $D^b(\mathrm{Coh}(X))$ is equivalent to $D^b(\mathrm{Coh}(Y))$, then X is isomorphic to Y.*

In order to get a better handle on $D^b(\mathrm{Coh}(\mathbb{P}^1_k))$ we seek an alternative description. Consider the structure sheaf $\mathcal{O}$ and the twisting sheaf $\mathcal{O}(1)$. Let $\mathcal{E} = \mathcal{O} \oplus \mathcal{O}(1)$. The coherent sheaf $\mathcal{E}$ "touches everything" in $D^b(\mathrm{Coh}(\mathbb{P}^1_k))$. More precisely,

Lemma 3.3 *The smallest triangulated subcategory of $D^b(\mathrm{Coh}(\mathbb{P}^1_k))$ containing $\mathcal{E}$ and closed under taking direct summands is $D^b(\mathrm{Coh}(\mathbb{P}^1_k))$.*

Proof The set of objects $\{\mathcal{O}(n)\}_{n \in \mathbb{Z}}$ generates $D^b(\mathrm{Coh}(\mathbb{P}^1_k))$, since any coherent sheaf on $\mathbb{P}^1_k$ has a finite resolution by members of this collection ([28]). Consider the following exact sequence

$$0 \to \mathcal{O}(n-2) \stackrel{(y,-x)}{\to} \mathcal{O}(n-1)^{\oplus 2} \stackrel{(x,y)^t}{\to} \mathcal{O}(n) \to 0$$

Setting $n = 2$ shows that we can resolve $\mathcal{O}(2)$ by $\mathcal{O}(1)$ and $\mathcal{O}$ and consequently by $\mathcal{E}$. Letting $n = 0$ shows that $\mathcal{O}(-1)$ can also be resolved (in the other direction) by copies of $\mathcal{E}$. Thus, $\mathcal{O}(2), \mathcal{O}(-1)$ lie in the smallest triangulated subcategory of $D^b(\mathrm{Coh}(\mathbb{P}^1_k))$ generated by $\mathcal{E}$ and closed under taking idempotents. Iterating the argument shows inductively that $\mathcal{O}(n)$ also lies in the subcategory generated by $\mathcal{E}$, for all $n \in \mathbb{Z}$. Hence, the smallest triangulated category in $D^b(\mathrm{Coh}(\mathbb{P}^1_k))$ containing $\mathcal{E}$ is $D^b(\mathrm{Coh}(\mathbb{P}^1_k))$ itself. □

The next lemma is well known.

Lemma 3.4 $\mathrm{Ext}^i(\mathcal{E}, \mathcal{E}) = 0$ *for* $i > 0$.

The only interesting Ext-group is then $\mathrm{Ext}^0(\mathcal{E}, \mathcal{E}) = \mathrm{Hom}(\mathcal{E}, \mathcal{E})$. A simple graphical means of encoding the data of this algebra is the following quiver Q.

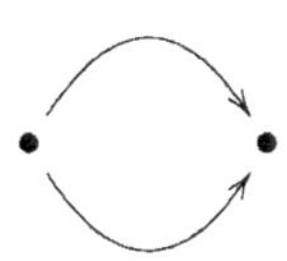

The path algebra kQ of Q is the algebra generated by oriented paths in Q with multiplication given by concatenation of paths. Note that the trivial paths at each of the vertices are idempotents.

Lemma 3.5 $\mathrm{Hom}(\mathcal{E}, \mathcal{E}) \cong kQ$.

Proof $\mathcal{O}$ corresponds to the first node and $\mathcal{O}(1)$ to the second. The two arrows are the morphisms $x, y : \mathcal{O} \to \mathcal{O}(1)$. □

From what we have just seen, the following should not be too surprising.

Proposition 3.6 *$D^b(\mathrm{Coh}(\mathbb{P}^1_k))$ is triangle equivalent to $D^b(kQ\text{-mod})$.*

The proof of this statement requires a little sophistication. Recall that, to get $D^b(\mathrm{Coh}(\mathbb{P}^1_k))$, we replaced all our bounded complexes of coherent sheaves with choices of injective resolutions and then took the homotopy category. Let us take an injective resolution of $\mathcal{E}$ and denote it by $I_{\mathcal{E}}$ and consider the category $I(\mathbb{P}^1_k)$ formed by injective complexes with bounded coherent cohomology. We then consider the functor $\mathrm{Hom}_{I(\mathbb{P}^1_k)}(I_{\mathcal{E}}, -)$ from $I(\mathbb{P}^1_k)$ to $Ch(k)$. In fact, the image of an object J in $Ch(k)$ has the structure of a left dg-module (and hence A_∞-module) over the dg-algebra $A = \mathrm{Hom}_{I(\mathbb{P}^1_k)}(I_{\mathcal{E}}, I_{\mathcal{E}})$ given by precomposition. Thus, we have a functor from $I(\mathbb{P}^1_k)$ to A-Mod. This induces a functor from $I(\mathbb{P}^1_k)$ to $D(A)$. The question then becomes: does it descend to a functor from $D^b(\mathrm{Coh}(P^1_k))$ to $D(A^{op})$? In other words, are all quasi-isomorphisms in $I(\mathbb{P}^1_k)$ inverted by the functor to $D(A^{op})$? The answer is yes. Since any quasi-isomorphism has an inverse up to homotopy in $I(\mathbb{P}^1_k)$, we just need to check that our functor kills all null-homotopic maps. This is equivalent to checking that the functor kills all null-homotopic complexes, i.e. those complexes for which the identity map is null-homotopic. But the contracting homotopy for the identity map of the complex, N, provides the contracting homotopy for any chain map in $\mathrm{Hom}_{I(\mathbb{P}^1_k)}(I_{\mathcal{E}}, N)$. Thus, the functor descends. Let us denote the resulting functor by $\mathrm{RHom}(\mathcal{E}, -)$. The dg-algebra A is quasi-isomorphic to kQ. Applying Proposition 2.8, we can assume that $\mathrm{RHom}(\mathcal{E}, -)$ maps from $D^b(\mathrm{Coh}(\mathbb{P}^1_k))$ to $D(kQ^{op})$. $\mathrm{RHom}(\mathcal{E}, -)$ is full and faithful on the smallest triangulated category containing $\mathcal{E}$ and closed under direct summands and hence is fully faithful. It is also essentially surjective onto $D^\pi(kQ^{op})$ since both have to be the smallest triangulated categories containing kQ and closed under direct summands. Now, as is well known, kQ has finite global dimension thus, $D^\pi(kQ^{op}) \cong D^b(kQ\text{-mod})$; see [47].

Thus, the structure of the seemingly complicated $D^b(\mathrm{Coh}(\mathbb{P}^1_k))$ is controlled by the relatively simple algebra kQ. We shall see the quiver Q again shortly.

3.2 A-branes on the mirror to the projective line. The mirror partner for $\mathbb{P}^1_k$ is something a little more exotic than a variety. We consider the function $W : \mathbb{C}^\times \to \mathbb{C}$ given by $W(z) = z + q/z$ with $q \in \mathbb{C}^\times$. Such a pair $(\mathbb{C}^\times, W)$ is called a Landau-Ginzburg model, or LG-model. Since the value of q does not affect the category of A-branes, we set $q = 1$. When dealing with LG-models, a useful moral to keep in mind is that we should study the geometry of the critical locus and how it relates to the ambient space. Now, on the A-side, we are interested in symplectic geometry, in particular Lagrangians. First, we need a symplectic form. We take it to be $\imath \dfrac{dz \wedge d\bar{z}}{z\bar{z}}$.

We could just take Lagrangian submanifolds of $\mathbb{C}^\times$ as our A-branes but that would not reflect the presence of W. Instead, we will look at non-closed Lagrangian submanifolds of $\mathbb{C}^\times$ whose boundaries lie on $W^{-1}(z)$ for some generic choice z. We could proceed as in [1] and investigate all such Lagrangians with some additional requirements. This approach seems much more natural, but for simplicity's sake we shall follow the ideas of Seidel in [53] or [4] and only consider a few special

Lagrangians. W has non-degenerate singularities, i.e., for each p such that $W'(p) = 0$ we have $W''(p) \neq 0$. Each such p is called a critical point; $W(p)$ is called the critical value. There is a natural connection on the complement of the singular fibers coming in this situation. Given a vector field X on $\mathbb{C}$ we can find a horizontal lift $\tilde{X}$ by using the symplectic form to split the tangent space of the domain of W into the tangent space of the fiber and its symplectic orthogonal. We take $\tilde{X}$ to lie in this orthogonal and project down to X. With a connection, we can parallel transport vectors. Thanks to our choice of connection, parallel transport preserves the symplectic form. This gives our Lagrangians. Fix a non-critical point q. If we take a path $\gamma : [0,1] \to \mathbb{C}$ so that $\gamma(0) = p$ and $\gamma(1) = q$, we can look at the set

$$\{x \in W^{-1}(q) : \lim_{t \to 0} T_{\gamma|_{[t,1]}}(x) = q\} \cup \{q\}$$

where $T_{\gamma|_{[t,1]}}(x)$ is the symplectomorphism coming from parallel transport along $\gamma|_{[t,1]}$. The resulting set is just an interval (or a one-dimensional disc). It is a Lagrangian submanifold with boundary on $W^{-1}(q)$. Stated in this language, the fact that the stable manifold is an n-dimensional Lagrangian disc holds in greater generality; see [53]. This Lagrangian is called the vanishing thimble associated with p. The boundary, which is an $(n-1)$-dimensional sphere, is called the vanishing cycle associated with p. Since our example is low-dimensional, we can actually draw the vanishing thimbles associated to the two critical points for $z + 1/z$.

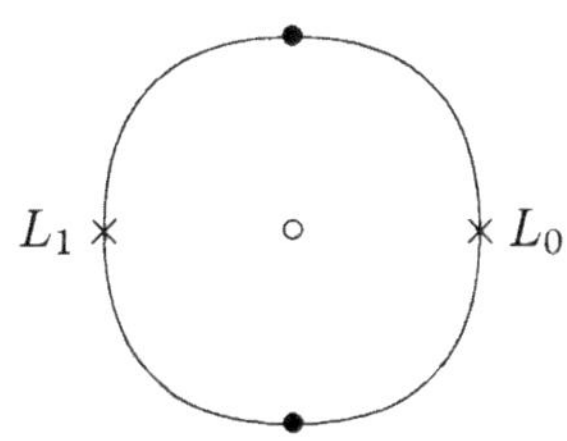

We can now define our category of A-branes associated with the LG-model.

Our two critical points give us two Lagrangian thimbles which intersect in two points. Order the Lagrangian thimbles L_0 and L_1.

Definition 3.7 The category of vanishing cycles, $FS(W)$ for the Landau-Ginzburg model W is an A_∞-category whose objects are collection of vanishing thimbles L_i, and whose morphisms are

$$\operatorname{Hom}(L_i, L_j) = \begin{cases} CF(\partial L_i, \partial L_j) & \text{if } i < j \\ k \operatorname{id}_{L_i} & \text{if } i = j \\ 0 & \text{otherwise} \end{cases}$$

We use the usual A_∞-structure on the vanishing cycles of the regular fiber $W = 0$. The category of A-branes is the bounded derived idempotent completion of the category of vanishing cycles, $D^\pi FS(W)$.

One should first notice that the A_∞-relations respect the (strict) ordering, in the sense that any degeneration of an ordered disc splits into two ordered discs. Thus, the A_∞-structure on the Fukaya category gives $FS(W)$ its A_∞ structure.

One should also note that, a priori, $D^\pi \mathcal{FS}(W)$ depends on the choice of paths. Different choices yield different Lagrangians. But, in fact, $D^\pi FS(W)$ is independent up to equivalence of the choice of paths [52].

Definition 3.8 The category of A-branes for the LG-model W is the category $D^{\pi}FS(W)$.

Now, let us take a closer look at the category of A-branes for $W(z) = z + 1/z$. We have two objects, two morphisms from the first object to the second, and the identity morphisms. Thus, it almost a tautology that the morphism algebra of $FS(W)$ is the path algebra of our quiver Q. $D^{\pi}(FS(W))$ is simply the bounded derived category of finite-dimensional representations of Q. Indeed, the category of B-branes on $\mathbb{P}^1_k$ is equivalent to the category of A-branes on W.

Remark 3.9 The simplicity of this example hides several important details. One of these details is that $\mathcal{O}, \cdots, \mathcal{O}(n)$ is the wrong exceptional collection to look at. One should instead consider $\Omega^n(n), \cdots, \Omega(1), \mathcal{O}$. The case of the mirror to $\mathbb{P}^2_k$ provides another illuminating example of homological mirror symmetry; see [4].

3.3 B-branes on the mirror to the projective line. With half of the homological mirror symmetry correspondence done, we now turn our attention to the other half. The naive guess for B-branes on the Landau-Ginzburg theory $W : \mathbb{C}^{\times} \to \mathbb{C}$ would simply be coherent sheaves on $\mathbb{C}^{\times}$ or equivalently modules over $\mathbb{C}[z, z^{-1}]$. What we should remember here is that, as in the case of A-branes on the Landau-Ginzburg model, we want to take the potential W into account. We must measure the singularities of W using complex geometry. This reflects the similar considerations on the A-side where the singular fibers of W gave rise to the vanishing thimbles. How do we measure singularities? By a classical result of Serre, a Noetherian commutative algebra A is regular if and only if every module over A has a projective resolution of uniformly bounded length. For a general A, modules with such bounded resolutions form a triangulated subcategory of $D^b(\text{mod-}A)$. Let us denote this subcategory by $\operatorname{Perf}(A)$. Then a measure of the singularity of A is the quotient $D^b(\text{mod-}A)/\operatorname{Perf}(A)$. This is often called the stable category of A or the category of singularities of A and it is indeed the desired definition of B-branes on W.

Definition 3.10 The category of B-branes of $W : Y \to \mathbb{C}$ is

$$D_{\text{sing}}(W) = \prod_{\lambda \in \mathbb{C}} D^b(\operatorname{Coh}(Y_\lambda))/\operatorname{Perf}(Y_\lambda)$$

where $\operatorname{Perf}(Y_\lambda)$ is the full triangulated subcategory of complexes of coherent sheaves admitting a bounded locally free resolution.

Note that, since we are working over $\mathbb{C}$, there are only a finite number of singular values and hence the product is finite. Of course, for this definition to be useful, one needs to know how to take the quotient of $D^b(\operatorname{Coh}(Y_\lambda))$ by $\operatorname{Perf}(Y_\lambda)$. We will not cover this, because there is a more computationally accessible version of the category of B-branes on W — the category of matrix factorizations of W, $MF(W)$.

Let A be an algebra and $f \in A$. A matrix factorization of f is a periodic complex

$$\cdots \to P_0 \overset{d_P^0}{\to} P_1 \overset{d_P^1}{\to} P_0 \overset{d_P^0}{\to} P_1 \to \cdots$$

with P_0, P_1 projective modules over A and $d_P^0 d_P^1 = d_P^1 d_P^0 = f \,\text{id}$. Notice that we could repackage the data by considering $P = P_0 \oplus P_1$ with $d_P = \begin{pmatrix} 0 & d_P^0 \\ d_P^1 & 0 \end{pmatrix}$ so that $d^2 = f\,\mathrm{id}$. Complexes of this form will be the objects in the category of matrix factorizations of f. Given two matrix factorizations $P^\bullet$ and $Q^\bullet$, we define the set of morphisms $\mathrm{Hom}_{CF(f)}(P^\bullet, Q^\bullet)$ to be $\mathrm{Hom}_A(P, Q)$. It is $\mathbb{Z}/2\mathbb{Z}$-graded, with the degree zero piece being given by $\mathrm{Hom}_A(P_0, Q_0) \oplus \mathrm{Hom}_A(P_1, Q_1)$ and the degree one piece being given by $\mathrm{Hom}_A(P_0, Q_1) \oplus \mathrm{Hom}_A(P_1, Q_0)$. One can view this as a periodic version of the ordinary Hom-complex. One nice feature is that while $P^\bullet$ and $Q^\bullet$ are not chain complexes, $\mathrm{Hom}(P^\bullet, Q^\bullet)$ is. The action of the differential $d_{P,Q}$ is simply super-commutation, namely $d_{P,Q}\phi = d_Q\phi - (-1)^{\tilde{\phi}}\phi d_P$. The proof of the following result is left as a simple exercise to the reader.

Lemma 3.11 *The composition $d^2_{P,Q}$ is 0. Thus it makes sense to speak of the cohomology of* $\mathrm{Hom}(P^\bullet, Q^\bullet)$.

We also have a shift functor [1] which sends P_i to $P_{i+1 \bmod 2}$ and d_i to $-d_{i+1 \bmod 2}$ and which acts on morphisms in the standard way. Notice that $[2] \cong \mathrm{Id}$. We can form cones over morphisms $\phi : P^\bullet \to Q^\bullet$ by taking $C(\phi) = P[1] \oplus Q$ with factorizations $d_{C(\phi)} = \begin{pmatrix} d_{P[1]} & \phi \\ 0 & d_Q \end{pmatrix}$.

Definition 3.12 The category of matrix factorizations of f, $MF(f)$, is the category whose objects are matrix factorizations and whose morphisms are

$$\mathrm{Hom}_{MF(f)}(P^\bullet, Q^\bullet) = H^0(\mathrm{Hom}_{CF(f)}(P^\bullet, Q^\bullet))$$

Proposition 3.13 *$MF(f)$ is a triangulated category where the shift is as above and triangles are of the form*

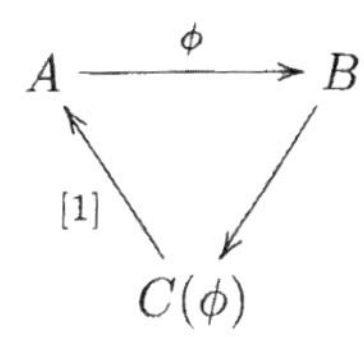

This proposition is proved in [44], where we find the following very useful theorem.

Theorem 3.14 *Consider a Landau-Ginzburg model $W : Y \to \mathbb{C}$ with Y affine. Then $D_{\mathrm{sing}}(W) = \prod_{\lambda \in \mathbb{C}} MF(W - \lambda) =: MF(W)$ as triangulated categories.*

Remark 3.15 This is an extension of the proof by Eisenbud that the category of matrix factorizations is equivalent to the category of Cohen-Macaulay modules [16].

Here is an elucidating example.

Example 3.16 Let $A = k[x]/(x^2)$. Then the minimal free resolution of $A/(x)$ is

$$\cdots A \xrightarrow{x} A \xrightarrow{x} A \to A/(x) \to 0$$

Lifting this resolution to $k[x]$ we get a matrix factorization of x^2.

In fact, our situation is quite similar to the previous example. In the case of the mirror to $\mathbb{P}^1_k$, $W = z + q/z$. The critical points of W are $\pm\sqrt{q}$. The critical values are $z = \pm 2\sqrt{q}$ and $W \pm 2\sqrt{q} = (1/z)(z \pm \sqrt{q})^2$. Let us take the matrix factorization, F, for $W - 2\sqrt{q}$, given by $P_0 = P_1 = \mathbb{C}[z, z^{-1}]$ and $d^0_P = (1/z)(z - \sqrt{q})$, $d^1_P = (z - \sqrt{q})$. First, let us compute the cohomology algebra $\mathrm{Hom}_{MF(W-2\sqrt{q})}(F, F)$.

Proposition 3.17 $\mathrm{Hom}_{MF(W-2\sqrt{q})}(F,F)$ *is isomorphic to the Clifford algebra on a single generator* ϕ *with* $\phi\cdot\phi = -(\partial_z^2 W)(\sqrt{q})$.

Proof A general morphism $\psi \in \mathrm{Hom}_{CF(W-2\sqrt{q})}(F,F)$ is of the form $\psi = \begin{pmatrix} A & B \\ C & D \end{pmatrix}$ and

$$d_{F,F}\psi =$$

$$\begin{pmatrix} 0 & (1/z)(z-\sqrt{q}) \\ (z-\sqrt{q}) & 0 \end{pmatrix}\begin{pmatrix} A & B \\ C & D \end{pmatrix} - \begin{pmatrix} A & -B \\ -C & D \end{pmatrix}\begin{pmatrix} 0 & (1/z)(z-\sqrt{q}) \\ (z-\sqrt{q}) & 0 \end{pmatrix}$$

$$= \begin{pmatrix} B/z+C & D-A \\ B/z+C & A-D \end{pmatrix}\begin{pmatrix} z-\sqrt{q} & 0 \\ 0 & z-\sqrt{q} \end{pmatrix}$$

Thus for ψ to lie in the kernel we must have $A = D$ and $B = -Cz$. Over $\mathbb{C}[z, z^{-1}]$, the cohomology is generated by the identity and by $\phi = \begin{pmatrix} 0 & 1/z \\ -1 & 0 \end{pmatrix}$. Note that $\phi^2 = \begin{pmatrix} -1/z & 0 \\ 0 & -1/z \end{pmatrix}$. Notice that $d_F^2 = W - \lambda$ implies that $d_F(\partial_z d_F) + (\partial_z d_F)d_F = \partial_z W$. Hence $\partial_z W\,\mathrm{id}$ is trivial in cohomology. Thus, the cohomology algebra becomes a module over the Jacobian ring $\mathbb{C}[z,z^{-1}]/(\partial_z W)$, which in particular is finite dimensional over $\mathbb{C}$. Now $\phi^2 = (-1/z)\,\mathrm{id}$. The image of $1/z$ in the Jacobian ring is $1/\sqrt{q}$ and hence $\phi^2 = -(\partial_z^2 W)(\sqrt{q})$. □

A similar result holds at $z = -\sqrt{q}$.

Usually, the cohomology algebra is not enough data. We need to know something about the underlying A_∞-structures. However, in this case, we actually have all the data we need. Hochschild cohomology of a graded associative algebra A can be computed as $\mathrm{Ext}^*_{A^e}(A,A)$.

Proposition 3.18 *A Clifford algebra* C*, considered as a* $\mathbb{Z}/2\mathbb{Z}$*-graded algebra, is projective over* C^e.

We shall not recall the proof here, but only refer the reader to [6] for satisfaction.

Appealing to the previous result and lemma 2.5, we can conclude that the dg-algebra $\mathrm{Hom}_{CF(W-2\sqrt{q})}(F,F)$ is quasi-isomorphic to its cohomology. Again a similar result holds at $z = -\sqrt{q}$.

Finally, we can appeal to Orlov's result, theorem 3.14, to conclude that the zero object and F generate $MF(W-2\sqrt{q})$, since the fiber over the singular points is isomorphic to $k[x]/(x^2)$. We can conclude that $MF(W-2\sqrt{q})$ is equivalent to the bounded derived category of modules over the Clifford algebra $\mathrm{Hom}_{MF(W-2\sqrt{q})}(F,F)$ via an argument analogous to that in subsection 3.1. Thus, the category of B-branes for $z + q/z$ is equivalent to two copies of the bounded derived category of modules over the Clifford algebras $k\langle x\rangle / \langle x^2 = \pm 1/\sqrt{q}\rangle$.

3.4 A-branes on the projective line, a.k.a. the sphere. Equip $\mathbb{P}^1_{\mathbb{C}}$ with the Fubini-Study metric and its associated Kähler form ω. In a Riemann surface, any curve is a Lagrangian. We shall choose to deal only with embedded and closed Lagrangians, i.e. simple closed curves in $\mathbb{P}^1_{\mathbb{C}}$. Any such curve divides the sphere into two, perhaps unequal (with respect to the Fubini-Study area form), halves. We first reproduce an illustrative calculation from [30]. Assume we have two Lagrangians γ and γ' that intersect in exactly two points. Let us denote the regions as in the image below.

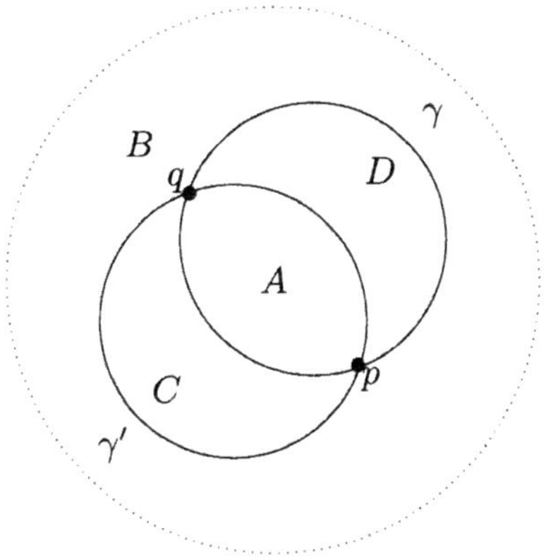

Let u be a formal parameter that we will use to keep track of the symplectic area of the discs of interest. We shall use the standard complex structure on the sphere. Before embarking on the computation, we must make sure that we have the signs associated to our Floer differential correct and we must make sure that we have the proper gradings on the intersection points. A thorough discussion of a coherent sign convention would take us off track, so we shall carry out our computations with $\mathbb{Z}/2\mathbb{Z}$ coefficients. This will initially eliminate considerations of the line bundles supported along the Lagrangians. The issue of grading, however, cannot be avoided. To simplify the situation, here we shall assume that one of our Lagrangians is a small Hamiltonian deformation associated to a Morse function. In this case, the intersection points are identified with the critical points of the Morse function and we can use the associated Morse index to grade them.

Now we compute the Floer differential.

$$m_1(q) = (u^{\omega(C)} + u^{\omega(D)})p$$
$$m_1(p) = (u^{\omega(A)} + u^{\omega(B)})q$$

Then,

$$m_1^2(p) = (u^{\omega(C)} + u^{\omega(D)})(u^{\omega(A)} + u^{\omega(B)})p =$$
$$u^{\omega(C)+\omega(B)}(u^{\omega(\mathrm{Int}(\gamma))} + u^{\omega(\mathrm{Int}(\gamma'))})(u^{\omega(\mathrm{Int}(\gamma'))} + u^{\omega(\mathrm{Out}(\gamma))})p$$

Since γ' is a Hamiltonian deformation of γ the areas of the interiors of γ and γ' are equal. Hence, $m_1^2(p) = 0$ and similarly $m_1^2(q) = 0$. Thus, the Floer cohomology between γ and γ' is well-defined. Note that the equality of the interior areas implies that $\omega(A) = \omega(B)$. Thus, $m_1(p) = 0$. Now consider $m_1(q)$.

$$m_1(q) = (u^{\omega(C)} + u^{\omega(D)})p = u^{\omega(C)}(1 + u^{\omega(D)-\omega(C)})p =$$
$$u^{\omega(C)}(1 + u^{\omega(\mathrm{Out}(\gamma))-\omega(\mathrm{Int}(\gamma'))})p =$$
$$u^{\omega(C)}(1 + u^{\omega(\mathrm{Out}(\gamma))-\omega(\mathrm{Int}(\gamma))})p$$

Thus, unless γ cuts the sphere in half, $m_1(q) \neq 0$ and the Floer cohomology of γ with itself vanishes. In the cohomological category of the Fukaya category of $\mathbb{P}^1_{\mathbb{C}}$, γ is isomorphic to the zero object if and only if γ separates the sphere into two unequal halves.

The most obvious choice of a γ which is cohomologically non-zero is a great circle. It is a classical result of Poincaré that any length minimizing area-bisecting curve in the sphere must be a great circle. The good news is that any area-bisecting curve is Hamiltonian isotopic to a great circle [43]. Thus, up to isomorphism the only nontrivial objects in the Fukaya category for $\mathbb{P}^1_{\mathbb{C}}$ are the great circles. Given two great circles, we can take the poles of the sphere to be their intersection points and take the height function h on the sphere. The associated Hamiltonian vector field is simply $J\nabla h$ where J is the complex structure on $\mathbb{P}^1_{\mathbb{C}}$. Thus, the associated Hamiltonian flow will simply rotate the sphere along the given axis and eventually take one great circle into another.

From our brusque analysis, we have concluded that with $\mathbb{Z}/2\mathbb{Z}$-coefficients there is only one (cohomologically) non-zero object in the Fukaya category — a choice of a great circle. If we keep trivial flat line bundles, this continues to hold with $\mathbb{Z}$ coefficients. We shall switch to $\mathbb{C}$ coefficients and outline the computation showing that the holonomy of our flat line bundle must be ± 1 in order for the A-brane to be non-zero. The computation showing that $m_1^2 = 0$ remains essentially the same. In the presence of a non-trivial flat line bundle, we get the following equation

$$m_1(q, v) = u^{\omega(C)}(v - e^{\int_{\gamma'} A'} v e^{\int_{\gamma} A} u^{\omega(\mathrm{Out}(\gamma)) - \omega(\mathrm{Int}(\gamma))}) p =$$

$$u^{\omega(C)}(v - (\mathrm{Hol}_{\gamma}(A))^2 v u^{\omega(\mathrm{Out}(\gamma)) - \omega(\mathrm{Int}(\gamma))} p$$

Thus, the holonomy of the flat line bundle must be ± 1 and as before γ must bisect $\mathbb{P}^1_{\mathbb{C}}$. Consequently, we have two non-trivial objects in the Fukaya category — a great circle with either trivial line bundle or a flat line bundle with holonomy -1. To determine the full structure of the category we must compute the morphism spaces. A computation similar to the above shows that $\mathrm{Hom}((\gamma, L_1), (\gamma, L_{-1}))$ has zero cohomology. Thus, there are two non-zero objects which are orthogonal to each other, at least cohomologically. The structure of the category of A-branes on $\mathbb{P}^1_{\mathbb{C}}$ now begins to resemble the structure of the category of B-branes on the mirror to $\mathbb{P}^1_{\mathbb{C}}$.

It is important to keep in mind that the above discussion sidestepped many of the crucial points in the construction of the Fukaya category, such as whether it honestly factors through Hamiltonian isotopy, orientations of the relevant moduli spaces, transversality, checking the A_∞-structure, etc. Many of these issues are resolved. For instance, the issues of Hamiltonian isotopy and transversality are taken care of in this case; see [12].

Now that we have seen that there are really only two non-zero objects of the Fukaya category of $\mathbb{P}^1_{\mathbb{C}}$ and that the only non-zero morphism sets are their endomorphisms, we will present another way of computing $HF(L, L)$ which was originally conceived in [18]. It is a natural adaptation of Bott's generalisation of Morse theory to allow non-isolated critical points. Naturally, it is called Bott-Morse-Floer cohomology. As mentioned above, bubbling holomorphic discs can occur in codimension one strata of the moduli space of holomorphic discs. The resolution of this problem was the central motivation of [18]. The presence of nontrivial discs with boundary on a given Lagrangian should deform the singular cohomology algebra structure in much the same way that the presence of nontrivial holomorphic spheres deforms the singular cohomology into the (small) quantum cohomology. Consider the singular chain complex of our great circle with its usual intersection product. The intersection product is not defined on all singular chains so we must restrict the allowed class of composable objects to those that are transversal. As the reader might appreciate now, the issue of transversality, or more precisely the lack thereof, crops up often in these cases. One means to sidestep the issue is the definition of a pre-category [35, 1]. While we shall not be so formal, it is important to keep this point in mind. We now outline the computation of Bott-Morse-Floer cohomology obtained in [10, 11].

Let $\mathcal{M}_k(\beta)$ be the moduli space of stable maps f from $(k+1)$-marked discs to $\mathbb{P}^1_{\mathbb{C}}$ with $f_*[\partial D] = \beta \in \pi_2(\mathbb{P}^1_{\mathbb{C}}, L)$. Here we have cyclically ordered the marked points. The markings provide us with evaluation morphisms $ev_i : \mathcal{M}(\beta) \to L$. Now, we deform the intersection product as follows:

$$m_2(S,T) = S \cap T + \sum_{\beta \in \pi_2(\mathbb{P},L)} ev_0(ev_1^* S \cap ev_2^* T) u^{\omega(\beta)}.$$

As the reader might have guessed by now, the deformed composition is no longer associative. We must add in compositions of all orders to obtain an A_∞-algebra. In fact, in this case, we must add in an m_0 also. We set

$$m_0 = \sum_{\beta \in \pi_2(\mathbb{P},L)} ev_0(\mathcal{M}_0(\beta)) u^{\omega(\beta)}$$

and

$$m_k(S_1, \cdots, S_k) = \sum_{\beta \in \pi_2(\mathbb{P}^1_{\mathbb{C}},L)} ev_0(\cap ev_i^* S_i) u^{\omega(\beta)}$$

In fact we only get a $\mathbb{Z}/2\mathbb{Z}$-grading on the resulting A_∞-algebra. If we added in another formal parameter counting the Maslov index of β, we could rectify this.

The proper way to view expressions such as $ev_0(\cap ev_i^* S_i)$ is as currents, i.e. distribution-valued differential forms. Thus, the image, $ev_0(\cap ev_i^* S_i)$, will represent the same current as its closure.

Proposition 3.19 *$m_k(x_1, \cdots, [L], \cdots, x_{k-1}) = 0$ for $k \neq 2$ and $m_2([L], x) = (-1)^{1-\tilde{x}} m_2(x, [L]) = x$.*

Proof It is clear that the fundamental class is closed since any quantum correction gives a chain with too large a degree. Since any class meets L, the class

$$ev_0(ev_1^* L \cap ev_i^* S_i)$$

lies inside $ev_0(\cap ev_i^* S_i)$. If $k > 2$, the dimension of $\mathcal{M}_{k-1}(\beta)$ is one less than the dimension of $\mathcal{M}_k(\beta)$ since we have lost a marked point and consequently a degree of freedom. Thus, the actual dimension of the current is smaller than the expected dimension and the current vanishes. The same reasoning tells us that in the case $k = 2$ any quantum corrections vanish leaving only the standard cap product. The sign arises from the choice of convention in [11]. □

Notice that m_0 is a multiple of the fundamental class. As we remarked previously, we still have an honest A_∞-algebra structure on cohomology. Since L is a strict unit, the only compositions that matter are those involving chains in the class of points. Now we check the previous calculation concerning the (Floer-cohomological) non-triviality of a given Lagrangian.

Proposition 3.20 *Unless the Lagrangian divides $\mathbb{P}^1_{\mathbb{C}}$ in half, the Floer cohomology vanishes.*

Proof $m_1(p) = [L](u^{\omega(\beta_1)} - u^{\omega(\beta_2)})$. The reversal in the sign comes from the different orientations induced on the boundary from each attaching disc. With a single marked point we trace out the whole of L. Now, if the symplectic area associated to β_1 is not equal to that β_2, we see that $m_1(p)$ is a nonzero multiple of the $[L]$. Note that, for any one-chain S, $m_1(S)$ contains a piece corresponding to the boundary of S as a chain. The rest is a singular chain of dimension one. Hence, for a singular chain of dimension one to be in the kernel of m_1, the chain needs to be the fundamental class or some multiple of it. Thus, if the Lagrangian does not cut $\mathbb{P}^1_{\mathbb{C}}$ in half, the Floer cohomology is zero. □

The argument in the previous proposition also shows that, if we take the equator as Lagrangian, the Floer cohomology is isomorphic to the standard cohomology as a module over our ring $\mathbb{C}[u]$. If we equip the Lagrangian with a flat line bundle $\mathcal{L}$, we incorporate the holonomy into the A_∞-structure as follows

$$m_k(S_1, \cdots, S_k) = \sum_{\beta \in \pi_2(\mathbb{P}^1_{\mathbb{C}}, L)} ev_0(\cap ev_i^* S_i) u^{\omega(\beta)} \operatorname{Hol}_{\partial\beta}(\mathcal{L})$$

As before, we have reduced to the two A-branes of interest: the equator equipped with the trivial line bundle and the equator equipped with the non-trivial flat line bundle. In these two cases, the Floer cohomology is non-vanishing and isomorphic as a module to the usual cohomology. The richness lies in the deformation of the standard algebra structure. In our case, the usual exterior algebra structure of the cohomology of a circle is deformed to a Clifford algebra $k\langle x\rangle / \left\langle x^2 = \pm u^{\omega(\mathbb{P}^1_{\mathbb{C}})/2}\right\rangle$.

Proposition 3.21 *Let p, q be distinct points thought of as transversal chains. Then*

$$m_2(p, q) = [L] e^{\omega(\beta)} \operatorname{Hol}_{\partial\beta}(\mathcal{L})$$

Proof When we consider a disc with three marked points two of which must lie on p and q, we can either orient the disc so that the zeroth marked point is before or after p. The two orientations are realized by the two discs attaching to L. Thus, we trace out all of L. The subtlety is in the signs. The moduli spaces of the discs all have natural orientations from auxiliary choices. In this case the induced orientations match up and we get all of the fundamental class. □

Stopping and taking stock of the situation, we see that on the A-side and B-side we have two special objects whose $\mathbb{Z}/2\mathbb{Z}$-graded endomorphism algebras are Clifford algebras.

If we specialize our formal parameter u to e, as is physically motivated, we end up in the same situation as before. Any A_∞-structure on $HF(L, L)$ is necessarily isomorphic to the trivial one. When we pass to $D^\pi(\operatorname{Fuk}(\mathbb{P}^1_{\mathbb{C}}))$ we will get a sum of the bounded derived category of modules over the two Clifford algebras $k\langle x\rangle / \left\langle x^2 = \pm e^{\omega(\mathbb{P}^1_{\mathbb{C}})/2}\right\rangle$.

"Proposition" 3.22 *The category of A-branes on $\mathbb{P}^1_C$ is equivalent to the category of B-branes on $W : \mathbb{C}^\times \to \mathbb{C}, W(z) = z + q/z$ where $q = e^{-\omega(\mathbb{P}^1_{\mathbb{C}})}$. Mathematically, the derived idempotent-completed Fukaya category of $\mathbb{P}^1_k$ is equivalent to the triangulated category of matrix factorizations of its mirror.*

Remark 3.23 We can interpret this correspondence of A-branes on $\mathbb{P}^1$ and B-branes on W as an instance of T-duality. T here is for the torus. $\mathbb{P}^1$ admits a torus fibration over the interval given by the momentum map of the $U(1)$-rotation along an axis. In T-duality, we replace these tori with their dual tori, i.e. $\operatorname{Hom}(T, U(1))$. Often, we think about the torus as having some metric structure which is a choice of length R. The T-dual torus will then have length $1/R$. Thus, if we T-dualize the momentum map torus fibration for $\mathbb{P}^1$, the two ends open up and we get the algebraic torus $\mathbb{C}^\times$. The two non-trivial A-branes map to 1 and -1 as determined by their holonomies. These are exactly the critical points of the superpotential on $\mathbb{C}^\times$.

Remark 3.24 As mentioned before, there are a few details missing from a complete proof of proposition 3.22, such as the complete definition of $\mathrm{Fuk}(\mathbb{P}^1_{\mathbb{C}})$! A rigorous reformulation of the previous proposition would just require an assumption that a $\mathrm{Fuk}(\mathbb{P}^1_{\mathbb{C}})$ exists such that the computations carried out above are valid.

4 Homological mirror symmetry for elliptic curves

The name "mirror symmetry" implies that two objects related by the correspondence should be similar enough for one to be a reflection of the other. In the previous section we saw that $\mathbb{P}^1_k$ had as mirror not another algebraic variety, not even another topological space, but a function. To the untrained eye, these two objects appear quite dissimilar. For something a little more symmetric, we move to the next case of homological mirror symmetry: elliptic curves.

For the purposes of mirror symmetry, an elliptic curve is a smooth Calabi-Yau variety of dimension one. We shall stick to $\mathbb{C}$ as a ground field. A smooth variety X is Calabi-Yau if its canonical bundle is trivial. For most Calabi-Yau varieties, mirror symmetry is a little more symmetric. Unlike the previous section where we had four different definitions of categories of branes, for Calabi-Yau's we only have two. Given a Calabi-Yau X, the category of B-branes on X is $D^b(\mathrm{Coh}(X))$ and the category of A-branes is $D^\pi(\mathrm{Fuk}(X))$. Homological mirror symmetry should simply exchange A-branes and B-branes of each ellipitic curve.

Testing this conjecture on the case of elliptic curves allows us to exploit some additional structure. We can uniformize elliptic curves. Namely, we can present any elliptic curve as $\mathbb{C}/(\mathbb{Z}+\tau\mathbb{Z})$ for some τ in the upper half of the complex plane. This determines its complex structure, as we simply take the complex structure provided by $\mathbb{C}$. This also determines the symplectic structure, since $dz \wedge d\bar{z}$ is invariant under translations. We can of course scale the volume of our elliptic curve without changing the complex structure. This makes sense for any positive real number, but we will want to complexify and allow scalings by complex numbers with positive imaginary components. The resulting complex-valued symplectic form is usually called the complexified Kähler form. As such, we can describe our elliptic curve E with complexified Kähler form by two numbers τ and ρ which lie in the upper half plane. Let us incorporate this observation by denoting our curve by E^ρ_τ. Then, mirror symmetry is simply the exchange $E^\rho_\tau \leftrightarrow E^\tau_\rho$.

Below, we generally follow the notation from [46].

4.1 A-branes on an elliptic curve. Another useful property of elliptic curves is that mean curvature flow behaves well on them. One can also note that mean curvature flow is a Hamiltonian deformation [57]. Thus, we can flow our (homologically non-trivial) Lagrangian (curve) into a geodesic and be none the worse off in terms of the Fukaya category. The result is that we only need to consider special Lagrangians when working with the Fukaya category. Of course, the standard means of computing $CF(L,L)$ is by using a slight Hamiltonian deformation so we need either to modify our definitions or forget about these types of computations. While one could attempt generalisations of Fukaya-type compositions using singular homology and incidence requirements [18], such a discussion would certainly take us beyond the scope of our introductory paper. We shall therefore restrict our attention to a subclass of morphisms when defining compositions.

We reduce to the following. The objects of our category will be special Lagrangian submanifolds of E_1^ρ. Our special Lagrangians are simply lines with rational (or infinite) slope. We will of course need to choose a grading and a spin structure as before. The choice of spin structure we shall suppress. Let us recall how a grading of a Lagrangian L is chosen. We need to lift the phase map $L \to S^1$. Our holomorphic volume form is simply dz. Restricting that volume form to the line $mx = ny$, we get $\frac{(n+\imath m)}{\sqrt{m^2+n^2}} dl$ where l is some coordinate and we have $l \to (1/\sqrt{m^2+n^2})(nl, ml)$. With the parametrization, the induced volume form on our line is simply dl. The phase map is constant and equal to $e^{2\pi\imath\alpha}$ for some choice of α so that $e^{\pi\imath\alpha}$ is equal to the argument of $n + \imath m$. Thus, to choose a grading for our Lagrangian L we simply need to make a choice of α. There are $\mathbb{Z}$-fold options and each choice gives a different object in our category of A-branes.

Now, the grading (which we take from [46]) on an intersection point $p \in L_i \cap L_j$ is then $\mu(p) = -[\alpha_j - \alpha_i]$ where $[r]$ is the greatest integer less than or equal to r. If we exclude vertical lines, we can choose a grading that satisfies $\mu(p) = 0$ if $s_i < s_j$ and $\mu(p) = 1$ if $s_i > s_j$ for all lines simultaneously.

We also include flat line bundles $\mathcal{E}$ on our special Lagrangians with monodromy of unit modulus (or flat $U(1)$-bundles). Then an A-brane on our elliptic curve is a triple $(L, \alpha, \mathcal{E})$ with L a special Lagrangian, α a lift of the argument of the slope, and $\mathcal{E}$ a flat line bundle on L. We shall denote the collection of this data as $\mathcal{L}$.

Let us set $\mathrm{Hom}(\mathcal{L}_i, \mathcal{L}_j) = CF(L_i, L_j; \mathrm{Hom}(\mathcal{E}_i, \mathcal{E}_j))$.

As discussed above, we restrict the class of allowed compositions. In our A_∞-category, we are only allowed to form compositions

$$m_n : \mathrm{Hom}(\mathcal{L}_1, \mathcal{L}_2) \otimes \cdots \otimes \mathrm{Hom}(\mathcal{L}_{n-1}, L_n) \to \mathrm{Hom}(\mathcal{L}_1, \mathcal{L}_n)$$

where all Lagrangians involved are distinct. This of course forces us to abandon identity morphisms.

We have the compositions.

$$m_n : \mathrm{Hom}(\mathcal{L}_1, \mathcal{L}_2) \otimes \cdots \otimes \mathrm{Hom}(\mathcal{L}_{n-1}, L_n) \to \mathrm{Hom}(\mathcal{L}_1, \mathcal{L}_n)$$

$$m_n(p_1, \cdots, p_n) = \sum_{p_0 \in L_1 \cap L_n} \sum_{\phi \in \mathcal{M}(p_0, \cdots, p_n)} \pm \exp(2\pi\imath \int_D \phi^*\omega) P\exp(\phi^*\beta)$$

when $\deg(p_0) = \sum_{i=1}^n \deg(p_i) + 2 - n$, $\mathcal{M}(p_0, \cdots, p_n)$ is the space of holomorphic maps from an $(n+1)$-pointed disc (modulo automorphisms when $n = 1$) with $\phi(\xi_i) = p_i$ and $\phi(S_i) \subset L_i$, and $P\exp(\phi^*\beta)$ is the path ordered exponential

$$P\exp(\phi^*\beta) = P\exp(\int_{S_n} \beta_n) t_n P\exp(\int_{S_{n-1}} \beta_n) t_{n-1} \cdots t_1 P\exp(\int_{S_1} \beta_n)$$

Here β is a connection form for the flat bundle $\mathcal{E}$.

Consider two A-branes $\mathcal{L}$ and $\mathcal{L}'$. Then as a set the morphisms between $\mathcal{L}$ and $\mathcal{L}'$ are the same — simply $L \cap L'$. Assume we have equipped these A-branes with a grading with phase within $(-1/2, 1/2)$. Then, all the morphisms between $\mathcal{L}$ and $\mathcal{L}'$ are either of degree zero or of degree one. Let us assume they are of degree zero. Assume that L has slope m and L' has slope n. Then there are $n - m$ intersection points on the torus given by

$$\left(\frac{k}{n-m}, \frac{mk}{n-m} \right)$$

for $k \in \mathbb{Z}/(n-m)\mathbb{Z}$. Assume now that the x-axis intercept of L is $(\alpha_1, 0)$ and the x-axis intercept for L' is α_2 for $0 \leq \alpha_i \leq 1$. Then the equation of L becomes $(\alpha_1 + t, (n-1)\alpha_1 + nt)$ and the equation of L' is $(\alpha_2 + t, (m-1)\alpha_2 + mt)$. Thus the intersection points are

$$\left(\frac{k + \alpha_2 - \alpha_1}{n-m}, \frac{mk + n\alpha_2 - n\alpha_1}{n-m}\right)$$

for $k \in \mathbb{Z}/(n-m)\mathbb{Z}$.

There is a natural non-degenerate pairing of degree one

$$\mathrm{Hom}(\mathcal{L}, \mathcal{L}') \times \mathrm{Hom}(\mathcal{L}', \mathcal{L}) \to \mathbb{C}$$

where we declare $p \in \mathrm{Hom}(\mathcal{L}, \mathcal{L}')$ dual to $p \in \mathrm{Hom}(\mathcal{L}', \mathcal{L})$. We then take the trace on the bundle components. Let us denote this by $(\cdot, \cdot)$. Then

$$(m_2(a,b), c) = (a, m_2(b,c))$$

If we set $\mathcal{L} = \mathcal{L}'$, then the pairing $(\cdot, \cdot)$ is the Floer manifestation of Poincaré duality. The cyclic symmetry and nondegeneracy of the pairing allows us to reconstruct two-fold compositions involving degree one morphisms from the two-fold compositions only involving degree zero morphisms. Thus, at the level of graded categories, if we formally add in identity morphisms to $H^0(\mathrm{Fuk}(E^\tau))$ then we can determine the proper two-fold composition rules for the degree one piece of $\mathrm{Hom}(\mathcal{L}, \mathcal{L})$.

4.2 B-branes on an elliptic curve. We now move onto the other side. If we quotient $\mathbb{C}$ out by $z \mapsto z + 1$ first, our elliptic curve E_τ can be viewed as $\mathbb{C}^\times/\mathbb{Z}$ where the action by $\mathbb{Z}$ is given by multiplication by $e^{2\pi\imath\tau}$. We can pull back any line bundle from E_τ to $\mathbb{C}^\times$. Any line bundle on $\mathbb{C}^\times$ is necessarily trivial (and consequently so is the pullback of any vector bundle). Since E_τ is projective and smooth, the collection of all line bundles on E_τ generates $D^b(\mathrm{Coh}(E_\tau))$. All line bundles, since they are trivial when pulled back to $\mathbb{C}^\times$, can be described as the quotient $(\mathbb{C}^\times \times \mathbb{C})/\mathbb{Z}$ where $\mathbb{Z}$ acts by $(u, v) \mapsto (qu, \phi(u)v)$ for some holomorphic map $\phi : \mathbb{C}^\times \to \mathbb{C}^\times$. The map ϕ is the choice of holomorphic trivialisation over $\mathbb{C}^\times$ of the pulled-back bundle. Let $L(\phi)$ denote the line bundle determined in this way.

With the similarity of the Fukaya-compositions to theta functions kept in mind, there are obvious choices for ϕ. Let $\phi_0(u) = \exp(-\pi\imath\tau)u^{-1}$. The sections of $L(\phi_0)$ are the classical theta functions. A general theta function is given as follows

$$\theta[c', c''](\tau, z) = \sum_{m \in \mathbb{Z}} \exp(2\pi\imath(\tau(m + c')^2/2 + (m + c')(z + c'')))$$

The classical theta function is simply $\theta[0,0](\tau, z)$. To check that this is actually a section of L, we simply need to check that it behaves properly when we translate by 1 and τ. $\theta[0,0](\tau, z)$ is clearly invariant under the shift $z \mapsto z + 1$. And

$$\theta[0,0](\tau, z + \tau) = \sum_{m=-\infty}^{\infty} \exp(2\pi\imath((m^2/2)\tau + mz + m\tau))$$

Letting $l = m + 1$, we have $(m^2/2)\tau + mz + m\tau = (l^2/2)\tau + lz - (1/2)\tau - z$. Thus,

$$\theta[0,0](\tau, z + \tau) = e^{-\pi\imath\tau} e^{-2\pi\imath z} \theta[0,0](\tau, z),$$

so it is indeed a section of L. One can similarly check that $\theta[a/n, 0](n\tau, nz)$ for $a \in \mathbb{Z}/n\mathbb{Z}$ are sections of L^n. By pulling back L^n via translation, we get sections for all possible line bundles.

The degree of $L(\phi_0) = L$ is 1. Any line bundle on an elliptic curve is the tensor product of a degree zero line bundle and L^n. It is well known that $\mathrm{Pic}^0(E_\tau)$ is isomorphic to E_τ. Thus, any line bundle is simply $t_x^* L \otimes L^{n-1}$ where $x \in E$ and n is the degree. If the degree of a line bundle is < 0, the corresponding divisor is not equivalent to an effective divisor. Hence, we can have no global sections. For $n > 0$, the sections of L^n are also given by theta functions: $\theta[a/n, 0](n\tau, nz)$ for $a \in \mathbb{Z}/n\mathbb{Z}$. The sections of $t_{\alpha+\imath\beta}^* L^n$ are then

$$t_{\alpha+\imath\beta}^* \theta[a/n, 0](n\tau, nz) = \theta[a/n, 0](n\tau, n(z + \alpha + \imath\beta))$$

Note that

$$\mathrm{Hom}(t_x^* L \otimes L^{n-1}, t_y^* L \otimes L^{m-1}) \cong \mathrm{Hom}(t_x^* L^{-1} \otimes t_y^* L \otimes L^{m-n})$$

Similarly, $t_x^* L^{-1} \otimes t_y^* L^{-1}$ can have no global sections if $x \neq y$. Thus, there are no morphisms from $t_x^* L \otimes L^{n-1}$ to $t_y^* L \otimes L^{m-1}$ unless $m > n$ or $m = n$ and $x = y$. If we let $x = \alpha_1 + \imath\beta_1$ and $y = \alpha_2 + \imath b_2$, then we can also rewrite $t_x^* L^{-1} \otimes t_y^* L^{-1}$ as $(t_{\alpha_{21}+\imath\beta_{21}}^* L)^{m-n}$ where

$$\alpha_{21} = \frac{\alpha_2 - \alpha_1}{m-n}, \beta_{21} = \frac{\beta_2 - \beta_1}{m-n}$$

The sections of $(t_{\alpha_{21}+\imath\beta_{21}}^* L)^{m-n}$ are given by

$$\theta[a/(m-n), 0]((m-n)\tau, (m-n)(z + \alpha_{21} + \imath\beta_{21}))$$

Since E_τ is a Calabi-Yau variety, the canonical bundle ω_{E_τ} is trivial. For a general smooth projective X, we have $\mathrm{Ext}^*(F, G) \cong \mathrm{Ext}^{n-*}(G, F \otimes \omega_X)^\vee$ [28]. The pairing is given by composing $a \in \mathrm{Ext}^*(F, G)$ and $b \in \mathrm{Ext}^{n-*}(G, F \otimes \omega_X)$, taking the trace on F giving an element of $H^n(\omega_X) \cong \mathbb{C}$. The functor $S : D^b(\mathrm{Coh}(X)) \to D^b(\mathrm{Coh}(X))$ given by $F \mapsto F \otimes \omega_X[-\dim X]$ is called the Serre functor [7]. For a Calabi-Yau, we get something rather special as $S \cong [-\dim X]$. We also have

$$(a \cdot b, c) = (-1)^{\tilde{a}(\tilde{b}+\tilde{c})}(b \cdot c, a)$$

Thus, as in the case of A-branes, one can then use Serre duality and the degree zero morphisms to compute the Ext-groups of these line bundles and reconstruct $D^b(\mathrm{Coh}(E_\tau))$ as a graded category.

In [46], Polishchuk and Zaslow treat all possible vector bundles on E_τ and torsion sheaves.

4.3 An outline of the equivalence. The first step in checking homological mirror symmetry for elliptic curves is to give a functor from the category of line bundles (with only honest morphisms) to the sub-category of the Fukaya category given by lines of non-vertical slope with flat $U(1)$-bundles on them. From the discussion above, one can see that the correspondence between theta functions and intersection points is pretty strong, so we make the obvious definition. Let

$$\Phi : \mathcal{L}(E_\tau) \to H^0(\mathrm{Fuk}(E^\tau))$$

$$\Phi(t_{\alpha\tau+\beta}^* L \otimes L^{n-1}) = (\Lambda, A)$$

where $\Lambda = (\alpha + t, (n-1)\alpha + nt)$ and $A = (-2\pi\imath\beta)dx$. Recall that morphisms between $t_{\alpha_1\tau+\beta_1}^* L \otimes L^{n_1-1}$ and $t_{\alpha_2\tau+\beta_2}^* L \otimes L^{n_2-1}$ are the same thing as sections of $t_{\alpha_{21}\tau+\beta_{21}}^* L^{n_2-n_1}$ where

$$\alpha_{21} = \frac{\alpha_2 - \alpha_1}{n_2 - n_1}, \beta_{21} = \frac{\beta_2 - \beta_1}{n_2 - n_1}.$$

$t^*_{\alpha_{21}\tau+\beta_{21}} L^{n_2-n_1}$ has sections given by

$$f_k\theta[k/(n_2-n_1),0]((n_2-n_1)\tau,(z+\alpha_{21}\tau+\beta_{21}))$$

for $k \in \mathbb{Z}/(n_2-n_1)\mathbb{Z}$. Correspondingly the intersection points of

$$\Phi(t^*_{\alpha_1\tau+\beta_1} L \otimes L^{n_1-1}) \text{ and } \Phi(t^*_{\alpha_2\tau+\beta_2} L \otimes L^{n_1-1})$$

are given by

$$e_k = \left(\frac{k+\alpha_2-\alpha_1}{n_2-n_1}, \frac{n_1k+n_1\alpha_2-n_1\alpha_1}{n_2-n_1}\right)$$

for $k \in \mathbb{Z}/(n_2-n_1)\mathbb{Z}$. We then set $\Phi(f_k) = \exp(-\pi\imath\tau\alpha_{21}^2(n_2-n_1)) \cdot e_k$.

Theorem 4.1 [46] *Φ is full and faithful. It gives an equivalence with the full subcategory of $H^0(\mathrm{Fuk}(E^\tau))$ consisting of special Lagrangians with non-vertical slope and the chosen special gradings paired with flat $U(1)$-bundles. Moreover, Φ naturally extends to an equivalence of the abelian category of coherent sheaves on E_τ and the subcategory $H^0(\mathrm{Fuk}(E^\tau))$ consisting of all Lagrangians with flat $U(N)$-bundles and the chosen special gradings.*

In lieu of reviewing the proof, we shall treat an illuminating example. The example is taken directly from [46], as the author feels he cannot improve upon the choice.

Under Φ we take the structure sheaf of E_τ to the x-axis with zero connection. L^n is mapped to a line passing through the origin with slope n and no connection. Let us consider the composition $\mathrm{Hom}(\mathcal{O},L) \times \mathrm{Hom}(L,L^2) \to \mathrm{Hom}(\mathcal{O},L^2)$. The vector space $\mathrm{Hom}(\mathcal{O},L)$ is one-dimensional and spanned by the classical theta function $\theta[0,0](\tau,z)$. As $\mathrm{Hom}(L,L^2) \cong \mathrm{Hom}(\mathcal{O},L)$, this space is also spanned by the classical theta function.

On the other side, we have $\Lambda_1 \cap \Lambda_2 = \{e_1\}$, $\Lambda_2 \cap \Lambda_3 = \{e_1\}$, and $\Lambda_1 \cap \Lambda_3 = \{e_2, e_3\}$ where e_1 and e_2 are the origin and $e_3 = (1/2, 0)$. Under Φ, we identify e_1 with $\theta[0,0](\tau,z)$ and e_2 with $\theta[0,0](2\tau,2z)$ and e_3 with $\theta[1/2,0](2\tau,2z)$.

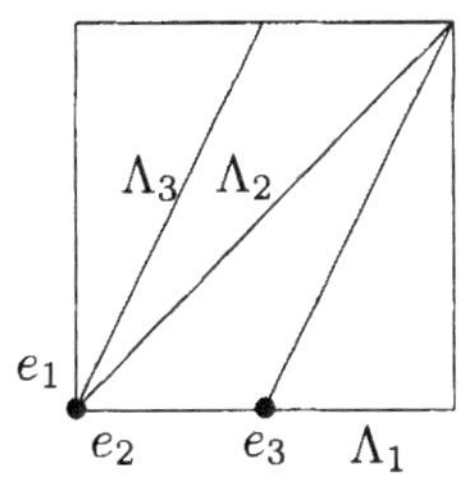

To compute the composition in the Fukaya category, we lift our diagram to $\mathbb{R}^2$ in all possible ways and count the number of triangles with appropriate weights determined by the Kähler form. These will give the structure coefficients.

Of course, triangles which are translates of another triangle are equivalent. Therefore, we need to pin down one of the vertices that we will take to be the origin, i.e. e_1. We need to compute the coefficients of e_2 and e_3. The coefficient of e_2 comes from counting triangles with vertices on integer translates of the origin. Since we have pinned down the first vertex we are looking at triangles in $\mathbb{R}^2$ given by $(0,0), (n,0), (2n,2n)$. Each such triangle has area τn^2. Summing over them we get

$$\sum_{n=-\infty}^{\infty} \exp(2\pi\imath\tau n^2) = \theta[0,0](2\tau,0)$$

The coefficient of e_3 comes from triangles with vertices at

$$(0,0),(0,n+1/2),(2n+1,2n+1).$$

We get

$$\sum_{n=-\infty}^{\infty} \exp(2\pi\imath\tau(n+1/2)^2) = \theta[1/2,0](2\tau,0).$$

Thus, we have

$$m_2(e_1,e_1) = \theta[0,0](2\tau,0)e_2 + \theta[1/2,0](2\tau,0)e_3.$$

Let us now investigate the other side. We are looking at the square of the classical theta function. Since Φ is a functor, we need to show that

$$\theta(\tau,z)^2 = \theta[0,0](2\tau,0)\theta[0,0](2\tau,2z) + \theta[1/2,0](2\tau,0)\theta[1/2,0](2\tau,2z)$$

This is simply an application of the addition formula for theta functions [41]. This is a basic example but it nicely illustrates the correspondence.

Using the equivalence between $\mathrm{Coh}(E_\tau)$ and the subcategory of $H^0(\mathrm{Fuk}(E^\tau))$ with objects whose grading is in the interval $(-1/2,1/2)$, one can then extend the functor to $\Phi : D^b(\mathrm{Coh}(E_\tau)) \to H^0(\mathrm{Fuk}(E^\tau))$ uniquely by requiring that it be compatible with Serre and Poincaré duality. Thus, we get

Theorem 4.2 *$D^b(\mathrm{Coh}(E_\tau))$ and $H^0(\mathrm{Fuk}(E^\tau))$ are equivalent as graded categories.*

The next step is to determine if the triangulated structures coincide. The triangulated structure descends from a cone construction at the chain level. Thus, the natural way to proceed is to show that as A_∞-categories $D^b(E_\tau)$ and $H^0(\mathrm{Fuk}(E^\tau))$ are quasi-isomorphic. But recall that the definition of the Fukaya category is missing $\mathrm{Hom}(\mathcal{L},\mathcal{L})$. We must restrict our class of composable morphisms to Hom-spaces between different objects. With this limitation in mind, one can then attack the problem.

In [45], the possible minimal A_∞-structures on $D^b(\mathrm{Coh}(E_\tau))$ compatible with Serre duality (i.e. cyclic) and the tensor structure and with a suitably large class of allowed compositions are classified up to homotopy. The problem then reduces to computing a single triple product as the test of homotopy equivalence. The triple products on $D^b(\mathrm{Coh}(E_\tau))$ and $H^0(\mathrm{Fuk}(E^\tau))$ coincide. Thus, we get the following result.

Theorem 4.3 *Consider $D^b(\mathrm{Coh}(E_\tau))$ with the admissible class of morphisms determined by the equivalence with $H^0(\mathrm{Fuk}(E^\tau))$. The A_∞-structures induced on this category by the dg-category of line bundles and the Fukaya category coincide.*

In fact, the result is a little stronger. If one can extend the definition of the Fukaya category to include endomorphism sets while preserving the A_∞-structure already present on composable morphisms, then one knows that the resulting A_∞-category will be homotopy equivalent to $D^b(E_\tau)$ with the minimal A_∞-structure determined by the dg-category of line bundles. Of course, one can extend the A_∞-structure in such a way since the A_∞-structure inherited from the dg-category of line bundles allows all possible compositions. So such an extension certainly exists; it is just not of a symplectic nature yet. Modulo this small detail we have homological mirror symmetry for elliptic curves.

5 Further results

In this section, we give a rapid review of the established cases of homological mirror symmetry beyond dimension one. By no means is this meant to be a comprehensive treatment. The interested reader is encouraged to consult the references.

5.1 A-branes on (near) Fano manifolds versus B-branes on LG-models.

5.1.1 *Fano surfaces.* In the previous examples, we were fortunate to be able to compute compositions in Fukaya categories. The Landau-Ginzburg mirror to $\mathbb{P}^1_k$ was an extremely simple situation since the fibers of the potential were zero-dimensional manifolds as were the Lagrangians involved. For $\mathbb{P}^1_{\mathbb{C}}$, any choice of almost complex structure was automatically integrable and regular. Thus, we could apply the Riemann mapping theorem and reduce our counts to purely topological considerations.

If we consider A-branes on the mirror of $\mathbb{P}^2$, the fibers are no longer zero-dimensional but are now non-compact Riemann surfaces. Thus, the counting can again be reduced to topological considerations. This at least gives one the hope of applying previous methods to this problem. Moreover, if we replace $\mathbb{P}^2$ by any other two-dimensional smooth variety (or stack), the Landau-Ginzburg mirror (if it exists) should be approachable via the same methods.

In fact, this is done in a pair of papers [4, 5], see also [58]. In [4], Auroux, Katzarkov, and Orlov tackle weighted projective surfaces as stacks and demonstrate the following.

Theorem 5.1 *(Auroux-Katzarkov-Orlov) Let a, b, c be mutually prime positive integers and consider the stack $\mathbb{CP}(a, b, c)$. Then $D^b(\mathrm{Coh}(\mathbb{CP}(a, b, c))$ is triangle equivalent to $D^\pi(FS(W))$ where $W : \{x^a y^b z^c = 1 | x, y, z \in \mathbb{C}\} \to \mathbb{C}$ is given by $W(x, y, z) = x + y + z$.*

Moreover, they also address how homological mirror symmetry behaves with respect to non-commutative deformations of the homogeneous coordinate ring of the weighted projective plane. In addition, using the results for weighted projective planes, they prove a similar result for Hirzebruch surfaces.

In [5], Auroux, Katzarkov, and Orlov continue the study of this side of mirror symmetry for surfaces. They investigate the effect of blowing up $\mathbb{CP}^2$ at ≤ 9 points. To properly address this, they must compactify the mirror to $\mathbb{CP}^2$ and deform it.

Theorem 5.2 *(Auroux-Katzarkov-Orlov) Let K be a set of k points in $\mathbb{CP}^2$. Consider the del Pezzo surface X_K given by blowing up $\mathbb{CP}^2$ at those points. Then $D^b(\mathrm{Coh}(X_K))$ is triangle equivalent to $D^\pi(FS(W_K))$ where $W_K : M_K \to \mathbb{C}$ is an elliptic fibration obtained by deforming a compactified mirror of $\mathbb{CP}^2$.*

For more details, see [4, 5]. In [58], Ueda studies toric blow-ups of $\mathbb{CP}^2$.

5.1.2 *Projective, smooth toric varieties.* In [1], Abouzaid takes a slightly different approach to the study of homological mirror symmetry for smooth projective toric varieties. Motivated by the results on Fano surfaces and the underlying physics, one can speculate that the mirror of a such a toric variety X_Δ determined by the polytope Δ is the Landau-Ginzburg model $W_\Delta : (\mathbb{C}^\times)^{\dim X_\Delta} \to \mathbb{C}$ where W is the Laurent polynomial whose Newton polytope is Δ. In fact, symplectically, one has a lot of freedom in choosing the coefficients of W_Δ. This allows one to take a limit that corresponds to a tropical degeneration. In this limit, one can identify

A and B branes on either side with certain Morse chain complexes on the moment polytope of X_Δ. Morally, one reduces mirror symmetry in this setting to data on the shared moment polytope. Of course, a bit of work goes into relating these "tropical Lagrangians" back to the usual Lagrangians. In the end, one gets the following result. Let $\mathrm{Fuk}(W_\Delta)$ denote the A_∞-(pre)-category of admissible Lagrangians in $(\mathbb{C}^\times)^{\dim X_\Delta}$. Essentially, these are Lagrangians with boundary on a regular fiber over a point p of W_Δ which project under W_Δ, near p, to curves emanating from p.

Theorem 5.3 *(Abouzaid) There is a full and faithful triangle functor*

$$i : D^b(\mathrm{Coh}(X_\Delta)) \to D^\pi(\mathrm{Fuk}(W_\Delta)).$$

Moreover, if X_Δ is ample, the functor is essentially surjective.

For more details, see [2].

5.2 A-branes on Fano manifolds versus B-branes on LG-models. Although there are few results on the Fukaya category of a toric variety, similar to subsection 3.4, one can define a symplectic category that works. In [12], Cho and Oh define a version of Floer cohomology called adapted Floer cohomology which uses the standard complex structure on the toric variety instead of a generic almost complex structure. If the toric variety is convex, then this reduces to the usual Floer cohomology. They also restrict themselves to the Lagrangian torus fibers (with flat line bundles) of the moment map fibration.

As in subsection 3.4, there are only a finite number of torus fibers that have non-vanishing Floer cohomology. These are called balanced torus fibers. Cho and Oh show that balanced torus fibers are in one-to-one correspondence with the critical points of the mirror LG superpotential. There is no Floer cohomology between two distinct balanced torus fibers. And, given a balanced torus fiber L, the $\mathbb{Z}/2\mathbb{Z}$-graded Floer cohomology is $HF(L, L)$ is a Clifford algebra on $\dim L$ generators with non-degenerate pairing given by the holomorphic Hessian of W at the critical point p corresponding to L. As in subsection 3.4, the intrinsic formality of the Clifford algebra uniquely determines any underlying A_∞-structure on $HF(L, L)$.

For B-branes on the mirror, a simple computation shows that the skyscraper sheaves at the singular points are mutually orthogonal, and that the endomorphism algebra of any of these objects is a Clifford algebra on $\dim L$ generators with the bilinear form given by the holomorphic Hessian at the corresponding point. As before, to match up completely, we need to change variables from $\mathbb{C}^\times$ to $\mathbb{C}$ using the exponential map. Nevertheless, we get the following.

Theorem 5.4 *For a convex smooth projective toric variety X_Δ with mirror W_Δ, the smallest triangulated thick subcategory contaning the skyscraper sheaves at the singular points of W_Δ in $D_{sing}(W_\Delta)$ is triangle equivalent to the smallest triangulated thick subcategory containing the balanced torus fibers in $D^\pi(\mathrm{Fuk}(X_\Delta))$.*

5.3 A-branes versus B-branes on Calabi-Yau manifolds.

5.3.1 *Abelian varieties.* Following the results of [46], the natural inclination was to consider higher dimensional abelian varieties A using similar methods. And, indeed, in [17], this is what is done. Fukaya is able to demonstrate that the corresponding Hom-spaces match up and, in the case where all three Lagrangians are mutually transverse, the m_3-compositions match also.

Another approach was undertaken in [35]. Kontsevich and Soibelman use non-Archimedean analysis to skirt convergence issues and, similar to [1], use a degeneration to the base of a torus fibration to show that $D^b(\mathrm{Coh}(A))$ embeds as a triangulated subcategory of $D^\pi(\mathrm{Fuk}(A^\vee))$ where $A^\vee$ is the mirror abelian variety to A. Specifically, it is equivalent to the category of Lagrangians transverse to the base.

5.3.2 *Quartic surface.* In [50], Seidel implements his plan from [54] which is to use deformation theory and results about directed categories of vanishing cycles to deduce homological mirror symmetry for a Calabi-Yau variety. The idea is as follows. The computations on the algebro-geometric side are not so bad. On the symplecto-geometric side, all the relevant Lagrangians sit in an affine patch of the Calabi-Yau, in this case a quartic surface in $\mathbb{P}^3$, as vanishing cycles of a Picard-Lefschetz fibration. Here they can be viewed as matching paths corresponding to lower-dimensional Picard-Lefschetz fibrations. This viewpoint allows one to automate an inductive procedure that computes Floer homology (Fukaya categories) associated to these Lagrangians from the directed category of vanishing cycles associated to the Picard-Lefschetz fibration. Then, to pass from the Lagrangians in the affine patch to the Lagrangians in the total space one has to include pseudo-holomorphic discs that hit the divisor at infinity, i.e. the complement of the affine patch. Counting these discs can be viewed as a formal deformation of the A_∞-structure associated endomorphism algebra of the vanishing cycles in the affine patch. Then, the endomorphism algebra of the vanishing cycles in the compact Calabi-Yau is obtained by specialising this formal deformation to a non-zero value of the parameter. The key to making this result feasible is that the deformation spaces involved are small, i.e. one-dimensional, and carry a $k^\times$ action. Thus, there are really only two different deformations possible, the trivial one and the non-trivial one. The issue of triviality of the deformation can be reduced to a computation involving Hochschild cohomology of the underlying A_∞-algebra which can be checked on one side via explicit computations and on the other by the non-degeneration of a certain spectral sequence. Of course, Seidel still needs to avoid the convergence problem, so he works with Novikov rings.

Theorem 5.5 *(Seidel) Let X be a smooth quartic surface in $\mathbb{P}^3_{\mathbb{C}}$ and let $\Lambda_{\mathbb{Q}}$ be the rational Novikov field over $\mathbb{C}$. Take the quartic surface in $\mathbb{P}^3_{\Lambda_{\mathbb{Q}}}$ defined by*

$$x^4 + y^4 + z^4 + w^4 + q(xyzw) = 0,$$

quotient by the standard action of $(\mathbb{Z}/4\mathbb{Z})^2$, take the minimal crepant resolution, and denote it by Y. Then there is an equivalence of triangulated categories

$$D^\pi(\mathrm{Fuk}(X)) \cong \hat{\psi}^* D^b(\mathrm{Coh}(Y))$$

where ψ is a continuous automorphism of $\Lambda_{\mathbb{N}}$.

For more details, see [50].

6 How is homological mirror symmetry related to mirror symmetry?

Mirror symmetry is a wide and varied subject. Homological mirror symmetry is a single attempt to unify a majority of the various topics. As such, it must prove its utility in other mirror symmetric mathematical investigations. In this section, we present the relations between homological mirror symmetry and three other prominent mathematical aspects of mirror symmetry.

6.1 SYZ. Perhaps the most tantalising approach to mirror symmetry was outlined by Strominger, Yau, and Zaslow in [56]. Here, mirror symmetry of two Calabi-Yau manifolds could be realised geometrically, as opposed to homologically. Namely, given a Calabi-Yau manifold X over $\mathbb{C}$ there should exist a fibration by special Lagrangian tori. Dualising this torus fibration (and accounting for the singular fibers appropriately) should realise the mirror Calabi-Yau Y. Such an explicit geometric operation should provide an explicit functor between the categories of branes on each side of the mirror correspondence. Unfortunately, special Lagrangian torus fibrations are hard to come by on Calabi-Yau manifolds. Of course, many expect mirror symmetry to hold not necessarily throughout the whole moduli spaces of theories on each side, but in a small neighborhood of some special singularities on each side. Thus, one can try to implement SYZ in conjunction with a degeneration of the structures. Indeed, motivated by the SYZ conjecture, M. Gross and B. Siebert have begun a program [26, 24, 25] to explain mirror symmetry as a Legendre transform of dual affine structures obtained from toric degenerations. For an introduction see [23]. We have already seen such logic employed in the previous sections in investigating homological mirror symmetry both in the case of Calabi-Yau varieties, specifically abelian varieties, and also in the case of Fano varieties. Indeed, perhaps the most spectacular application of SYZ-inspired ideas to homological mirror symmetry is found in [1], where one can actually see how the branes on each side correspond by repackaging them as data on the base moment polytope. And SYZ informs our understanding of the other side of homological mirror symmetry for Fano varieties. It is under SYZ that the balanced torus fibers of [12] correspond to the critical points of the mirror superpotential. This is explored in [30].

6.2 Mirror maps. Another common area of investigation in mirror symmetry is the determination of the mirror map. The mirror map is an isomorphism between neighborhoods of two special singular points in moduli on each side of the mirror correspondence. Aside from its evident importance, mirror maps turn out to possess rich arithmetic properties. Mirror maps are interesting and elusive. If indeed homological mirror symmetry provides the fullest explanation for the mirror symmetry phenomenon, then we should somehow be able to extract the mirror map. Indeed, in some cases, one can do exactly that.

In [3], Aldi and Zaslow use the following idea of Seidel. A projective variety X is determined by its homogeneous coordinate ring. The data of the homogeneous coordinate ring sits inside $D^b(\mathrm{Coh}(X))$ and can be extracted assuming we know the very ample line bundle L corresponding to the embedding and the autoequivalence corresponding to $- \otimes L$. Then, the homogeneous coordinate ring is $\oplus_{n>0} \mathrm{Hom}_{D^b(\mathrm{Coh}(X))}(L, L^{\otimes n})$. Given a triangulated category, an object, and an autoequivalence we can always form such a ring but there is no reason to expect commutativity. But, if we know that homological mirror symmetry holds, then we know what object and functor to look at — the mirror to L and the mirror to $- \otimes L$. In general, it is expected that the mirror autoequivalence to tensoring by a line bundle L is given by the Dehn twist τ_S about a vanishing cycle S mirror to L. Assuming homological mirror symmetry, we can extract the homogeneous coordinate ring of the mirror as $\oplus_{n>0} \mathrm{Hom}_{D^\pi(\mathsf{Fuk}(Y))}(S, \tau^n(S))$. The structure of this ring depends only on the symplectic structure of Y. Thus, as we vary the symplectic structure of Y, we should see the variation of the complex structure of the mirror X manifested in the change in the homogeneous coordinate ring of X.

In this way, Aldi and Zaslow of [3] extract the mirror map in a number of examples and check that the logic is correct.

The upshot is that if one can check homological mirror symmetry in a given case by exhibiting an explicit equivalence, one should be able to compute the corresponding mirror map and extract the wealth of arithmetic information that is expected to lie within.

6.3 Instanton numbers. Mirror symmetry broke onto the mathematical scene by properly counting curves [9] on Calabi-Yau threefolds, and curve, or instanton, counts have been a central feature of mirror symmetry since. Indeed, mirror symmetry was originally viewed as the study of these counts and their relation to solutions of certain differential equations. If homological mirror symmetry provides the fullest explanation of the mirror symmetry phenomenon, then one should be able to extract these instanton counts solely from the categorical data. Physically, the instanton counts come from the closed string sector of our topological field theory, whereas what has been discussed earlier concerns the open string sector. The objects in each of our categories are appropriate boundary conditions for these open strings in different theories. One needs a mathematical construction that relates the two. In [13, 14], Costello considers the notion of a open-closed topological conformal field theory inside which sit open and closed TCFTs. He shows that an open TCFT is (morally) the same as an A_∞-category with a cyclically-invariant inner pairing and demonstrates a means to construct, from an open TCFT, an open-closed TCFT whose space of closed states is the Hochschild chain complex $C_*(A, A)$ of the A_∞-category associated to the initial open TCFT. He also demonstrates how to build a Gromov-Witten potential from a TCFT and thus gives instanton counts associated to an A_∞-category with a cyclically-invariant inner pairing. In the case of the Fukaya category, the passage to Hochschild homology has a more natural explanation. For an A_∞-category with a cyclically invariant inner product, one has a quasi-isomorphism between $C_*(A, A)$ and $C^*(A, A)$, the Hochschild chain complex. The Hochschild cohomology of an A_∞-category A is the same as the space of endomorphisms of A in the derived category of A-bimodules; here a bimodule over a category is simply an autofunctor of the category. Using the tensor product, we see that A corresponds to the identity functor. For the Fukaya category of a symplectic manifold (M, ω), we can make autofunctors from Lagrangians in $(M \times M, -\omega \oplus \omega)$. Morphisms between these functors correspond to Floer cohomology between the associated Lagrangians. The identity corresponds simply to the diagonal and it is now well known that $HF(L, L)$ is isomorphic to the quantum cohomology of L. For more on Lagrangian correspondences, see [59].

References

[1] Mohammed Abouzaid, *Morse Homology, Tropical Geometry, and Homological Mirror Symmetry for Toric Varieties*, arXiv:math.SG/0610004.

[2] Mohammed Abouzaid, *Homogeneous coordinate rings and mirror symmetry for toric varieties*, Geom. Topol. **10** (2006), 1097–1157 (electronic).

[3] Marco Aldi and Eric Zaslow, *Seidel's mirror map for abelian varieties*, Adv. Theor. Math. Phys. **10** (2006), no. 4, 591–602.

[4] Denis Auroux, Ludmil Katzarkov, and Dmitri Orlov, *Mirror symmetry for weighted projective planes and their noncommutative deformations*, arXiv:math.AG/0404281.

[5] Denis Auroux, Ludmil Katzarkov, and Dmitri Orlov, *Mirror symmetry for del Pezzo surfaces: vanishing cycles and coherent sheaves*, Invent. Math. **166** (2006), no. 3, 537–582.

[6] Hyman Bass and Amit Roy, *Lectures on topics in algebraic K-theory*, Notes by Amit Roy. Tata Institute of Fundamental Research Lectures on Mathematics, No. 41, Tata Institute of Fundamental Research, Bombay, 1967.

[7] Alexei Bondal and Mikhail Kapranov, *Representable functors, Serre functors, and reconstructions*, Izv. Akad. Nauk SSSR Ser. Mat. **53** (1989), no. 6, 1183–1205, 1337.

[8] Alexei Bondal and Dmitri Orlov, *Reconstruction of a variety from the derived category and groups of autoequivalences*, Compositio Math. **125** (2001), no. 3, 327–344.

[9] Philip Candelas, Xenia de la Ossa, Paul Green, and Linda Parkes, *A pair of Calabi-Yau manifolds as an exactly soluble superconformal theory*, Nuclear Phys. B **359** (1991), no. 1, 21–74.

[10] Cheol-Hyun Cho, *Holomorphic discs, spin structures, and Floer cohomology of the Clifford torus*, Int. Math. Res. Not. (2004), no. 35, 1803–1843.

[11] ———, *Products of Floer cohomology of torus fibers in toric Fano manifolds*, Comm. Math. Phys. **260** (2005), no. 3, 613–640.

[12] Cheol-Hyun Cho and Yong-Geun Oh, *Floer cohomology and disc instantons of Lagrangian torus fibers in Fano toric manifolds*, Asian J. Math. **10** (2006), no. 4, 773–814.

[13] Kevin Costello, *The Gromov-Witten potential associated to a TCFT*, arXiv:math.QA/0509264.

[14] Kevin Costello, *Topological conformal field theories and Calabi-Yau categories*, Adv. Math. **210** (2007), no. 1, 165–214.

[15] Vin de Silva, *The index of a family of pseudo-holomorphic disks*, Preprint, 2000.

[16] David Eisenbud, *Homological algebra on a complete intersection, with an application to group representations*, Trans. Amer. Math. Soc. **260** (1980), no. 1, 35–64.

[17] Kenji Fukaya, *Mirror symmetry of abelian varieties and multi-theta functions*, J. Algebraic Geom. **11** (2002), no. 3, 393–512, Revised version available at the author's homepage.

[18] Kenji Fukaya, Yong-Geun Oh, Hiroshi Ohta, and Kaoru Ono, *Lagrangian intersection Floer theory: anomaly and obstruction*, Preprint available at an author's homepage, 2000.

[19] Kenji Fukaya and Kaoru Ono, *Arnold conjecture and Gromov-Witten invariant for general symplectic manifolds*, The Arnoldfest (Toronto, ON, 1997), Fields Inst. Commun., vol. 24, Amer. Math. Soc., Providence, RI, 1999, pp. 173–190.

[20] Sergei Gelfand and Yuri Manin, *Methods of homological algebra*, Springer Monographs in Mathematics, Springer-Verlag, Berlin, 2003.

[21] Ezra Getzler and John D.S. Jones, *A_∞-algebras and the cyclic bar complex.*, Ill. J. Math. **34** (1990), no. 2, 256–283 (English).

[22] Mikhail Gromov, *Pseudoholomorphic curves in symplectic manifolds*, Invent. Math. **82** (1985), no. 2, 307–347.

[23] Mark Gross, *The Strominger-Yau-Zaslow conjecture: From torus fibrations to degenerations*, arXiv:0802.3407.

[24] Mark Gross and Bernd Siebert, *From real affine geometry to complex geometry*, arXiv:math/0703822.

[25] ———, *Mirror symmetry via logarithmic degeneration data. II*, arXiv:0709.2290.

[26] ———, *Mirror symmetry via logarithmic degeneration data. I*, J. Differential Geom. **72** (2006), no. 2, 169–338.

[27] Robin Hartshorne, *Residues and duality*, Lecture notes of a seminar on the work of A. Grothendieck, given at Harvard 1963/64. With an appendix by P. Deligne. Lecture Notes in Mathematics, No. 20, Springer-Verlag, Berlin, 1966.

[28] ———, *Algebraic geometry*, Springer-Verlag, New York, 1977, Graduate Texts in Mathematics, No. 52.

[29] Helmut Hofer and Dietmar A. Salamon, *Floer homology and Novikov rings*, The Floer memorial volume, Progr. Math., vol. 133, Birkhäuser, Basel, 1995, pp. 483–524.

[30] Kentaro Hori, *Mirror symmetry and quantum geometry*, Proceedings of the International Congress of Mathematicians, Vol. III (Beijing, 2002) (Beijing), Higher Ed. Press, 2002, pp. 431–443.

[31] Tornike Kadeišvili, *On the theory of homology of fiber spaces*, Uspekhi Mat. Nauk **35** (1980), no. 3(213), 183–188, International Topology Conference (Moscow State Univ., Moscow, 1979).

[32] Bernhard Keller, *Introduction to A-infinity algebras and modules.*, Homology Homotopy Appl. **3** (2001), no. 1, 1–35 (English).

[33] ———, *On differential graded categories*, International Congress of Mathematicians. Vol. II, Eur. Math. Soc., Zürich, 2006, pp. 151–190.

[34] Maxim Kontsevich, *Homological algebra of mirror symmetry*, Proceedings of the International Congress of Mathematicians, Vol. 1, 2 (Zürich, 1994) (Basel), Birkhäuser, 1995, pp. 120–139.

[35] Maxim Kontsevich and Yan Soibelman, *Homological mirror symmetry and torus fibrations*, Symplectic geometry and mirror symmetry (Seoul, 2000), World Sci. Publ., River Edge, NJ, 2001, pp. 203–263.

[36] Gang Liu and Gang Tian, *Floer homology and Arnold conjecture*, J. Differential Geom. **49** (1998), no. 1, 1–74.

[37] Martin Markl, *Transferring A_∞ (strongly homotopy associative) structures*, Rend. Circ. Mat. Palermo (2) Suppl. (2006), no. 79, 139–151.

[38] Dusa McDuff and Dietmar Salamon, *Introduction to symplectic topology*, second ed., Oxford Mathematical Monographs, The Clarendon Press Oxford University Press, New York, 1998.

[39] ———, *J-holomorphic curves and symplectic topology*, American Mathematical Society Colloquium Publications, vol. 52, American Mathematical Society, Providence, RI, 2004.

[40] Sergei Merkulov, *Strong homotopy algebras of a Kähler manifold*, Internat. Math. Res. Notices (1999), no. 3, 153–164.

[41] David Mumford, *Tata lectures on theta. II*, Progress in Mathematics, vol. 43, Birkhäuser Boston Inc., Boston, MA, 1984, Jacobian theta functions and differential equations, With the collaboration of C. Musili, M. Nori, E. Previato, M. Stillman and H. Umemura.

[42] August Newlander and Louis Nirenberg, *Complex analytic coordinates in almost complex manifolds*, Ann. of Math. (2) **65** (1957), 391–404.

[43] Yong-Geun Oh, *Second variation and stabilities of minimal Lagrangian submanifolds in Kähler manifolds*, Invent. Math. **101** (1990), no. 2, 501–519.

[44] Dmitri Orlov, *Triangulated categories of singularities and D-branes in Landau-Ginzburg models*, Tr. Mat. Inst. Steklova **246** (2004), no. Algebr. Geom. Metody, Svyazi i Prilozh., 240–262.

[45] Alexander Polishchuk, *A_∞-structures on an elliptic curve*, Comm. Math. Phys. **247** (2004), no. 3, 527–551.

[46] Alexander Polishchuk and Eric Zaslow, *Categorical mirror symmetry: the elliptic curve*, Adv. Theor. Math. Phys. **2** (1998), no. 2, 443–470.

[47] Raphaël Rouquier, *Dimensions of triangulated categories*, arXiv:math.CT/0310134.

[48] Yongbin Ruan, *Virtual neighborhoods and the monopole equations*, Topics in symplectic 4-manifolds (Irvine, CA, 1996), First Int. Press Lect. Ser., I, Int. Press, Cambridge, MA, 1998, pp. 101–116.

[49] Paul Seidel, *Fukaya categories and Picard-Lefschetz theory*, To appear in ETH Lecture Notes series of the European Math. Soc.

[50] Paul Seidel, *Homological mirror symmetry for the quartic surface*, arXiv:math.SG/0310414.

[51] Paul Seidel, *Graded Lagrangian submanifolds*, Bull. Soc. Math. France **128** (2000), no. 1, 103–149.

[52] ———, *More about vanishing cycles and mutation*, Symplectic geometry and mirror symmetry (Seoul, 2000), World Sci. Publ., River Edge, NJ, 2001, pp. 429–465.

[53] ———, *Vanishing cycles and mutation*, European Congress of Mathematics, Vol. II (Barcelona, 2000), Progr. Math., vol. 202, Birkhäuser, Basel, 2001, pp. 65–85.

[54] ———, *Fukaya categories and deformations*, Proceedings of the International Congress of Mathematicians, Vol. II (Beijing, 2002) (Beijing), Higher Ed. Press, 2002, pp. 351–360.

[55] Bernd Siebert, *Gromov-Witten invariants of general symplectic manifolds*, arXiv:arXiv:dg-ga/9608005.

[56] Andrew Strominger, Shing-Tung Yau, and Eric Zaslow, *Mirror symmetry is T-duality*, Winter School on Mirror Symmetry, Vector Bundles and Lagrangian Submanifolds (Cambridge, MA, 1999) (Providence, RI), AMS/IP Stud. Adv. Math., vol. 23, Amer. Math. Soc., 2001, pp. 333–347.

[57] Richard Thomas and Shing-Tung Yau, *Special Lagrangians, stable bundles and mean curvature flow*, Comm. Anal. Geom. **10** (2002), no. 5, 1075–1113.

[58] Kazushi Ueda, *Homological mirror symmetry for toric del Pezzo surfaces*, Comm. Math. Phys. **264** (2006), no. 1, 71–85.

[59] Katrin Wehrheim and Chris Woodward, *Functoriality for Floer theory in Lagrangian correspondences*, Preprint, 2007.

[60] Charles A. Weibel, *An introduction to homological algebra*, Cambridge Studies in Advanced Mathematics, vol. 38, Cambridge University Press, Cambridge, 1994.

Fields Institute Communications
Volume 54, 2008

Orbifold Gromov-Witten Invariants and Topological Strings

Vincent Bouchard
Harvard University
Jefferson Physical Laboratory
17 Oxford St.
Cambridge, MA 02138, USA
bouchard@physics.harvard.edu

Abstract. In this contribution I propose a (hopefully) pedagogical approach to the computation of orbifold Gromov-Witten invariants using mirror symmetry and topological string theory, focusing on the orbifold $\mathbb{C}^3/\mathbb{Z}_3$. Recent B-model developments on the mirror side, which led to predictions for "open orbifold Gromov-Witten invariants" of $\mathbb{C}^3/\mathbb{Z}_3$, are also addressed. This contribution is based on the results of [1] and [9].

1 Introduction

1.1 General idea. Historically, mirror symmetry has proved to be very successful in addressing problems of enumerative geometry, starting with the seminal work of Candelas et al [13] on the number of rational curves in the quintic threefold. The leitmotiv of the physical approach to Gromov-Witten theory is to map the problem to the mirror side, where computationally efficient techniques such as special geometry and the holomorphic anomaly equations of [7] are available.

Our goal is to pursue this line of thought in the context of enumerative geometry of orbifolds. Since direct evaluation of the Hodge integrals entering into the definition of orbifold Gromov-Witten invariants is rather complex (see for instance [8]), we aim at using mirror symmetry to map the problem of computing orbifold Gromov-Witten invariants to the mirror side.

1.2 Geometry. More precisely, we consider orbifolds of the form $\overline{X} = \mathbb{C}^3/G$, where G is a finite abelian group, $G \subseteq SU(3)$; such orbifolds admit a toric Calabi-Yau crepant resolution X. The prototypical example that we follow throughout this paper is $\overline{X} = \mathbb{C}^3/\mathbb{Z}_3$, which has a unique crepant resolution given by the total space of the canonical bundle over $\mathbb{P}^2$. Another example that could be studied

2000 *Mathematics Subject Classification.* Primary 14N35; Secondary 14J32, 81T30, 81T45, 11F37.

Key words and phrases. Gromov-Witten invariant, orbifold, topological string theory, mirror symmetry, orbifold point, open Gromov-Witten invariant.

along similar lines is $\mathbb{C}^3/\mathbb{Z}_4$, which has a toric crepant resolution given by the total space of the canonical bundle over F_2 — the second Hirzebruch surface.

1.3 Strategy. The usual mirror symmetric approach consisted in mapping the local problem of computing A-model topological string amplitudes near the large radius point in the stringy Kähler moduli space $\mathcal{KM}(X)$ of X — that is, generating functions of Gromov-Witten invariants of X — to the often simpler problem of computing B-model topological string amplitudes near the mirror point in the (suitably compactified) complex structure moduli space $\mathcal{M}(Y)$ of the mirror threefold Y. The main ingredients were the mirror map near the large radius point, and a formalism to compute the B-model amplitudes.

When X is the crepant resolution of an orbifold $\overline{X}$, the stringy Kähler moduli space $\mathcal{KM}(X)$ contains both a large radius point, where the A-model amplitudes generate Gromov-Witten invariants of X, and an orbifold point, where the A-model amplitudes now generate orbifold Gromov-Witten invariants of $\overline{X}$. Since mirror symmetry provides a global isomorphism between $\mathcal{KM}(X)$ and $\mathcal{M}(Y)$, we should be able to parallel the large radius procedure near the orbifold point to compute orbifold Gromov-Witten invariants of $\overline{X}$ from mirror symmetry. In order to make such an attempt successful, we need to understand the mirror map near the orbifold point, and we must provide a formalism to compute the B-model amplitudes at the mirror point. This is precisely the aim of this paper.

1.4 Opening up the strategy. In fact, we go even further. From a physics perspective, closed topological string theory can hardly live without its best friend open topological string theory. The latter is related to enumerative problems involving open Gromov-Witten invariants, which are concerned with maps from Riemann surfaces with boundaries. Again, one can use (an open version of) mirror symmetry at large radius combined with the open B-model to compute these invariants.

Using the new B-model formalism developed in [9, 33], as in the closed case we are able to extend this procedure to the orbifold point: we obtain predictions for generating functions of "open orbifold Gromov-Witten invariants". While open Gromov-Witten invariants are well-defined for smooth target spaces (see [24] for a computational definition using localization), it is not clear whether and how they can be generalized to the orbifold setting. However, the physics calculation provides an incentive for looking for a mathematical definition of such invariants.

1.5 Outline. In this contribution we propose a (hopefully) pedagogical approach to these ideas, which were mostly developed in [1] with Mina Aganagic and Albrecht Klemm and in [9] with Albrecht Klemm, Marcos Mariño and Sara Pasquetti. In section 2, we review the usual mirror symmetry approach at large radius. In section 3 we propose a parallel procedure to map orbifold Gromov-Witten theory to the mirror side, and use it to compute (high genus) orbifold Gromov-Witten invariants of $\mathbb{C}^3/\mathbb{Z}_3$. These two sections follow relatively closely the physics presentation in [8]. Section 4 is then devoted to the study of open amplitudes, using the B-model formalism of [9, 33]. As an application, we compute the "disk amplitude" of the orbifold $\mathbb{C}^3/\mathbb{Z}_3$.

Acknowledgments. I would like to thank heartily M. Aganagic, R. Cavalieri, A. Klemm, M. Mariño and S. Pasquetti for very enjoyable collaborations leading to the results presented in this contribution. I would also like to thank J. Bryan and T. Coates for interesting discussions. Finally, thanks are due to the organizers of the Banff workshop on *String dualities and modular forms* for the stimulating research environment.

2 Mirror symmetry at large radius

Let us start by reviewing general features of mirror symmetry at large radius, and how it can be used to compute Gromov-Witten invariants of smooth Calabi-Yau threefolds. Good references on mirror symmetry include the two books [19, 28], while topological string theory is discussed in the book [34].

2.1 General statement. The main characters are:

- (X, Y): a mirror pair of smooth Calabi-Yau threefolds, where X is a (non-compact) toric Calabi-Yau threefold which is the crepant resolution of an orbifold $\overline{X}$ of the form $\mathbb{C}^3/G$, with $G \subseteq SU(3)$ a finite abelian group;
- $\mathcal{M}(Y)$: a suitable compactification of the complex structure moduli space of Y;
- $\mathcal{KM}(X)$: a suitable compactification of the complexified Kähler moduli space of X (the so-called *enlarged* or *stringy* Kähler moduli space).

Mirror symmetry provides a local isomorphism, called the *mirror map*, between $\mathcal{KM}(X)$ and $\mathcal{M}(Y)$, which maps a neighborhood of a maximally unipotent boundary point $q_0 \in \mathcal{M}(Y)$ to a neighborhood of a corresponding large radius point $p_0 \in \mathcal{KM}(X)$. Moreover, mirror symmetry tells us that the mirror map lifts to an isomorphism between the *A-model* amplitudes at $p_0 \in \mathcal{KM}(X)$, and the *B-model* amplitudes at $q_0 \in \mathcal{M}(Y)$. Of course, we have not yet defined what the A- and the B-model amplitudes are; we will come back to that in a minute. Let us start by expanding a little more on how the mirror map is defined.

2.2 The mirror map. The isomorphism in the neighborhood of a large radius point $p_0 \in \mathcal{KM}(X)$ can be described as follows. $H^2(X, \mathbb{C})$ is spanned by

$$t_1 T_1 + \ldots + t_r T_r, \tag{2.1}$$

where $T_1, \ldots, T_r \in H^2(X, \mathbb{C})$ is a basis of generators for the cone σ containing the large radius point $p_0 \in \mathcal{KM}(X)$ corresponding to X. The complexified Kähler parameters t_i parameterize $\mathcal{KM}(X)$ near p_0.

On the mirror side, when Y is compact, $\mathcal{M}(Y)$ is projective special Kähler,[1] hence can be parameterized by the periods of the holomorphic volume-form Ω on Y. More precisely, choose a symplectic basis of three-cycles $A^I, B_J \in H_3(Y, \mathbb{Z})$, $I, J = 0, \ldots, r$, and define the periods

$$\omega^I = \oint_{A^I} \Omega, \qquad \frac{\partial \mathcal{F}}{\partial \omega^I} = \oint_{B_I} \Omega, \tag{2.2}$$

where $\mathcal{F}$ is the prepotential. The periods are solutions of the Picard-Fuchs equations, with the following properties. In terms of coordinates q_i, $i = 1, \ldots, r$, centered at the maximally unipotent boundary point $q_0 \in \mathcal{M}(Y)$, there is a unique period which is holomorphic, say ω^0, and r periods have logarithmic behavior,

$$\omega^i = \frac{\omega^0}{2\pi i} \log(q_i) + \mathcal{O}(q_i), \qquad i = 1, \ldots, r. \tag{2.3}$$

There are r other periods which are quadratic in the logarithm, and one is cubic.

[1]See [23] for a mathematical description of special geometry.

The mirror map then consists in choosing appropriate combinations of these periods which are mapped to the complexified Kähler parameters t_i. It turns out that the mirror map reads

$$(t_1, \dots, t_r) \mapsto \frac{1}{\omega^0}(\omega^1, \dots, \omega^r), \tag{2.4}$$

which are sometimes called the *flat coordinates* on the projective special Kähler manifold $\mathcal{M}(Y)$.

When X and Y are noncompact, as in the case of a toric threefold X, $\mathcal{M}(Y)$ is not projective special Kähler anymore. However, various properties of special geometry still hold, such as the existence of a prepotential $\mathcal{F}$, and the parameterization of $\mathcal{M}(Y)$ by "periods" — once those are properly defined in the noncompact case — of the holomorphic volume-form. This can be understood either by seeing Y as the limit of a compact threefold [16], or intrinsically for the noncompact geometry (see for instance [30] for the latter point of view). Furthermore, in the noncompact case the mirror map is simplified by the fact that $\omega^0 = 1$, hence the t^i are directly identified with the logarithmic periods ω^i.

2.3 A-model on X. We now describe the A-model amplitudes on X. Recall that X is a toric Calabi-Yau threefold, which can be described as a symplectic quotient

$$X = \{X_1, \dots, X_{3+k} \in (\mathbb{C}^{3+k} - Z) | \sum_{i=1}^{3+k} L_i^a |X_i|^2 = r^a \quad, a = 1, \dots, k\}/(S^1)^k, \tag{2.5}$$

where the k S^1's act as

$$X_i \mapsto e^{iL_i^a \theta_a} X_i. \tag{2.6}$$

The charge matrix L gives the toric data of X, and for X to be Calabi-Yau the charges must satisfy

$$\sum_{i=1}^{3+k} L_i^a = 0, a = 1, \dots, k. \tag{2.7}$$

Z is the union of the singular sets Z_a implied by the relations $\sum_{i=1}^{3+k} L_i^a |X_i|^2 = r^a$, and the $r^a \in \mathbb{R}^+$ are the Kähler parameters of X.

To describe the A-model amplitudes, start with a theory — a nonlinear sigma model — of maps $f : \Sigma \to M$ from Riemann surfaces Σ to a target Calabi-Yau threefold M. There are two ways of twisting this sigma model to obtain topological theories, namely the A- and the B-models. The A-model does not depend on complex moduli, while the B-model is independent of Kähler moduli.

The A-model on X becomes a theory of holomorphic maps $f : \Sigma \to X$, which can be reformulated in terms of Gromov-Witten invariants of the target space X. In the neighborhood of $p_0 \in \mathcal{KM}(X)$, the A-model genus g amplitudes F_g become generating functions for the unmarked genus g Gromov-Witten invariants $N_{g,\beta}$ of X, that is,

$$F_g = \sum_{\beta \in H_2(X)} N_{g,\beta} Q^\beta, \tag{2.8}$$

where

$$Q^\beta = e^{2\pi i \int_\beta \omega}, \tag{2.9}$$

and ω is a complexified Kähler class of X. In terms of the basis of $H^2(X,\mathbb{C})$ introduced above, we can write

$$Q^\beta = \prod_{i=1}^{r} Q_i^{\int_\beta T_i}, \tag{2.10}$$

with the exponentiated complexified Kähler parameters

$$Q_i = \mathrm{e}^{2\pi i t_i}. \tag{2.11}$$

2.4 B-model on Y. Following the general procedure of Hori-Vafa [29], the mirror Y of a toric Calabi-Yau threefold X has the form

$$Y = \{ww' = G(x, y; q_1, \ldots, q_r)\}, \tag{2.12}$$

where $w, w' \in \mathbb{C}$, $x, y \in \mathbb{C}^*$, and $q_1, \ldots, q_r$ are coordinates on $\mathcal{M}(Y)$ centered around the corresponding maximally unipotent boundary point. $G(x, y; q_1, \ldots, q_r)$ is a polynomial in x and y which describes a family of punctured Riemann surfaces embedded in $\mathbb{C}^* \times \mathbb{C}^*$, parameterized by the q_i's. In other words, Y is given by a conic fibration over $\mathbb{C}^* \times \mathbb{C}^*$, where the fiber degenerates to two lines over the family of Riemann surfaces

$$\Sigma(q_1, \ldots, q_r) = \{G(x, y; q_1, \ldots, q_r) = 0\}. \tag{2.13}$$

In the following, for brevity we will often omit the dependence on the q_i's, and call Σ a Riemann surface, understanding implictly that it is in fact a family of Riemann surfaces parameterized by the q_i's. Σ is generally called the *mirror curve* of X.

The B-model on Y localizes on constant maps, and becomes a theory of variations of complex structures of the target space Y. In particular, the genus 0 amplitude F_0 is simply given by the prepotential $\mathcal{F}$ of special geometry introduced above. Let us be a little more specific on special geometry for the noncompact Y, following [2]. When Y has the form (2.12), we consider only complex structure deformations that only involve varying the mirror curve Σ. The periods of the holomorphic volume-form

$$\Omega = \frac{\mathrm{d}w \mathrm{d}x \mathrm{d}y}{wxy} \tag{2.14}$$

over three-cycles reduce to integrals

$$\int_D \frac{\mathrm{d}x}{x} \wedge \frac{\mathrm{d}y}{y}, \tag{2.15}$$

where $D \subset \mathbb{C}^* \times \mathbb{C}^*$ is a real two-dimensional domain such that $\partial D \subset \Sigma$. The above integral reduces, in a given coordinate patch, to

$$\int_\gamma \log y \frac{\mathrm{d}x}{x}, \tag{2.16}$$

where $\gamma \in \Sigma$ is a one-cycle. Thus, the periods become integrals of the one-form

$$\lambda = \log y \frac{\mathrm{d}x}{x} \tag{2.17}$$

over one-cycles on the Riemann surface Σ. These periods govern the complex structure deformations, and, as in the compact case, are annihilated by a system of Picard-Fuchs equations which can be determined following the work of Chiang, Klemm, Yau and Zaslow [16]. Moreover, they satisfy special geometric relations, which define a prepotential $\mathcal{F}$ as before.

The higher genus B-model amplitudes are harder to describe mathematically. The genus 1 amplitude F_1 can be defined in terms of Ray-Singer torsion of Y. For $g \geq 2$, one can use the holomorphic anomaly equations of [7] — which may be understood as some sort of higher genus generalization of special geometry — to reconstruct the amplitudes recursively in the neighborhood of $q_0 \in \mathcal{M}(Y)$, up to an unknown holomorphic function at each genus depending on a finite number of constants. External data, such as boundary conditions, must be used to fix these functions.

2.5 Computing Gromov-Witten invariants of X. The mirror symmetric isomorphism between the A-model amplitudes at a large radius point and the B-model amplitudes near a maximally unipotent boundary point provides a concrete way of computing Gromov-Witten invariants of X, which has proved very succesful historically. This was the approach first used by Candelas et al [13] to compute the number of rational curves in the quintic threefold, which was then extended to higher genus in [7]. The two main ingredients entering in the calculation are:

- the mirror map near the large radius point;
- a framework to compute the B-model amplitudes near q_0, such as special geometry and the holomorphic anomaly equations.

Recall however that the holomorphic anomaly equations, in themselves, do not provide a complete framework to compute the higher genus amplitudes F_g, due to the holomorphic ambiguity persisting at each genus. It must be supplemented by additional data. However, recently boundary conditions for the amplitudes have been found which fully fix the ambiguities in some local geometries (such as local $\mathbb{P}^2$), and allow computations of high (but finite) genus amplitudes for compact threefolds (for instance, $g = 51$ for the quintic) [25, 26, 27].

2.6 Example: local $\mathbb{P}^2$. The main example that we will study in this paper is the orbifold $\overline{X} = \mathbb{C}^3/\mathbb{Z}_3$. Its unique crepant resolution is the toric threefold $X = \mathcal{O}(-3) \to \mathbb{P}^2$, which is the total space of the canonical bundle over $\mathbb{P}^2$. It is usually called local $\mathbb{P}^2$ in the physics literature. Let us describe mirror symmetry at the large radius point in the Kähler cone of local $\mathbb{P}^2$.

X is toric; its fan is generated by the one-dimensional cones:

$$\{(0,0,1),(0,-1,1),(-1,0,1),(1,1,1)\}. \tag{2.18}$$

These satisfy the linear relation

$$-3(0,0,1)+(0,-1,1)+(-1,0,1)+(1,1,1)=(0,0,0), \tag{2.19}$$

hence the charge matrix is simply $L = (-3,1,1,1)$.

The mirror threefold Y is given by

$$Y = \{ww' = y^2 + y(1+x) + qx^3\}, \tag{2.20}$$

where $w, w' \in \mathbb{C}$, $x, y \in \mathbb{C}^*$ and q is a coordinate on $\mathcal{M}(Y)$ centered at the maximally unipotent boundary point $q_0 := \{q = 0\} \in \mathcal{M}(Y)$. That is, the mirror curve reads

$$\Sigma = \{y^2 + y(1+x) + qx^3 = 0\}, \tag{2.21}$$

which has genus 1 and three punctures. It can be seen pictorially by fattening the toric diagram of local $\mathbb{P}^2$, as in figure 1.

Following [16], the Picard-Fuchs system in the coordinate q reads

$$\mathcal{D}_q = \Theta_q^3 + 3q(3\Theta_q + 2)(3\Theta_q + 1)\Theta_q, \tag{2.22}$$

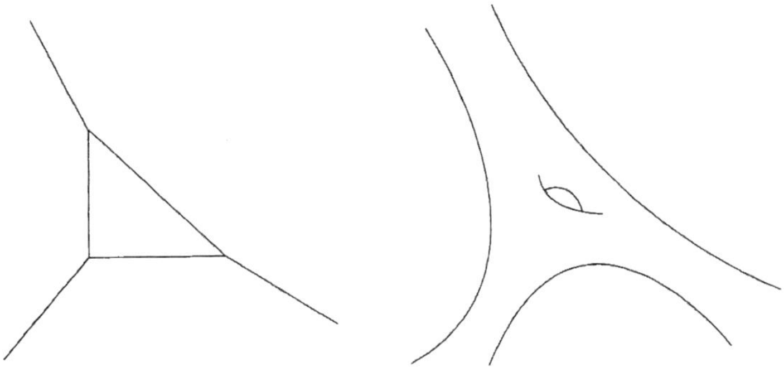

$\mathcal{O}(-3) \to \mathbb{P}^2$ $\Sigma : \{y^2 + y(1+x) + qx^3 = 0\}$

Figure 1 The toric diagram of local $\mathbb{P}^2$ and the mirror curve $\Sigma \subset \mathbb{C}^* \times \mathbb{C}^*$.

where we introduced the logarithmic derivative $\Theta_q = q\partial_q$. Solving $\mathcal{D}_1\Pi = 0$, we get the solution vector

$$\Pi = (1, \omega(q), \pi(q)), \tag{2.23}$$

where 1 is just the constant solution, $\omega(q)$ is a logarithmic solution and $\pi(q)$ is doubly logarithmic. The mirror map is given by

$$t = \omega(q) = \frac{1}{2\pi i}\left(\log q + \sum_{n=1}^{\infty} \frac{(-1)^n}{n}\frac{(3n)!}{(n!)^3}q^n\right), \tag{2.24}$$

where t is the complexified Kähler parameter of local $\mathbb{P}^2$. The period t_D dual to t is given by the combination

$$t_D = \frac{1}{2}\pi(q) - \frac{1}{2}\omega(q) - \frac{1}{4} = -3\frac{\partial\mathcal{F}}{\partial t}, \tag{2.25}$$

which defines the prepotential $\mathcal{F}$.[2]

The prepotential $\mathcal{F}$ computes the genus 0 amplitude F_0 at the large radius point of local $\mathbb{P}^2$. To compute the higher genus amplitudes, one can solve the holomorphic anomaly equations and supplement them with boundary conditions to fix the holomorphic ambiguity at each genus [32].

3 Mirror symmetry at the orbifold point

So far we only gave a local description of mirror symmetry, near a large radius point of $\mathcal{KM}(X)$. However, from a physics point of view, mirror symmetry should be global, in the sense that $\mathcal{KM}(X)$ should be globally isomorphic to $\mathcal{M}(Y)$, and similarly for the A- and the B-model amplitudes.

3.1 A global formulation of mirror symmetry. To study this global description we need to understand better the enlarged Kähler moduli space $\mathcal{KM}(X)$. Generically, the compactification $\mathcal{KM}(X)$ of the Kähler moduli space has a rather complicated structure which goes beyond the Kähler cone of X. Roughly speaking, $\mathcal{KM}(X)$ is obtained by gluing along common walls the Kähler cones of threefolds

[2]The unusual factor of -3 here comes from the fact that since Y is noncompact, it is not possible to find a symplectic basis of three-cycles; instead, the A- and the B-cycles have intersection number -3.

birationally equivalent to X. Some of these cones correspond to smooth threefolds related to X by flops; each such cone then contains a large radius point, which is mapped to a corresponding maximally unipotent boundary point in $\mathcal{M}(Y)$ on the mirror side. However, some other patches correspond to "non-geometric phases", by which we mean that they are obtained from X by contracting some cycles.

When X is toric, it turns out that $\mathcal{KM}(X)$ is also toric and is easily described by the *secondary fan* associated to X (see for instance [19], section 3.4 and chapter 6, for a more precise discussion). In toric geometry, the secondary fan of a toric manifold X is the fan generated by the one-dimensional cones encoding the linear relations between the one-dimensional cones in the fan of X. In other words, the one-dimensional cones in the secondary fan of X are given by the columns of the charge matrix L of X.

In this paper, we are interested in toric Calabi-Yau threefolds X which are the crepant resolutions of orbifolds $\overline{X}$ of the form $\mathbb{C}^3/G$. In these cases, $\mathcal{KM}(X)$ comprises a patch which contains an *orbifold point* $p_{orb} \in \mathcal{KM}(X)$, where the cycles of X are contracted to yield the orbifold $\overline{X}$. On the mirror side, the orbifold point is mapped to the point $q_{orb} \in \mathcal{M}(Y)$ at the intersection of the *orbifold divisors* in $\mathcal{M}(Y)$, which are the divisors around which the periods have finite monodromy. We will call q_{orb} the point of *finite monodromy*. When $\mathcal{M}(Y)$ is one-dimensional, as for local $\mathbb{P}^2$, the point of finite monodromy is simply the orbifold divisor itself, around which the periods have finite mondoromy.

We studied mirror symmetry near the large radius point p_0 corresponding to X; we now want to understand mirror symmetry near the orbifold point p_{orb} corresponding to $\overline{X}$. Our aim is to use mirror symmetry in this neighborhood to compute orbifold Gromov-Witten invariants of $\overline{X}$. In order to do so, we first need to describe the mirror map near the orbifold point, and then propose a formalism to compute the B-model amplitudes near this point.

3.2 The orbifold mirror map. Recall that the mirror map at large radius was defined by providing a map between the parameters spanning $H^2(X, \mathbb{C})$ in a basis canonically defined at p_0 to the logarithmic solutions of the Picard-Fuchs equations near q_0. In fact, the "full" mirror map may be understood as mapping the parameters spanning the even cohomology groups $H^{2n}(X, \mathbb{C})$, $n = 0, 1, 2, 3$, to corresponding solutions of the Picard-Fuchs equations.

The correct notion of cohomology of an orbifold $\mathbb{C}^3/G$ consists in the Chen-Ruan orbifold cohomology ring [14, 15]. With respect to ordinary cohomology, orbifold cohomology contains extra classes incorporating combinations of geometric and representation theoretic data associated to the action of G on $\mathbb{C}^3$. In particular, although $\mathbb{C}^3/G$ has no compact cycles, its orbifold cohomology ring is not trivial. The extra classes correspond to *twisted sectors* in physics language, and are in correspondence with the even cohomology classes at the large radius point.

The mirror map at the orbifold point can then be defined by mapping the parameters spanning the orbifold cohomology ring of $\mathbb{C}^3/G$ in a basis canonically defined at p_{orb} to corresponding solutions of the Picard-Fuchs solutions near the point of finite monodromy q_{orb}. The burden of the work consists in finding a canonical basis for the orbifold cohomology ring, and a good basis of solutions for the Picard-Fuchs equations. We claim that for simple orbifolds such as $\mathbb{C}^3/\mathbb{Z}_3$ we can determine the orbifold mirror map uniquely, up to a scale factor, simply by requiring that the B-model genus 0 amplitude be invariant under the $\mathbb{Z}_3$-monodromy around

q_{orb}. An equivalent prescription will be to match the representation theoretic data in the orbifold cohomology ring, keeping track of the action of $\mathbb{Z}_3$, to the action of the $\mathbb{Z}_3$-monodromy on the periods.

3.3 B-model at the point of finite monodromy. The next step is to provide a framework to compute the B-model amplitudes at the point of finite monodromy $q_{orb} \in \mathcal{M}(Y)$. In section 4, we will propose a new formalism to compute unambiguously all open and closed B-model amplitudes at any point in the moduli space, which can then be used to calculate the closed B-model amplitudes at q_{orb}. However, for the purpose of computing only the closed B-model amplitudes, we do not need such a formalism, the holomorphic anomaly equations being sufficient. Let us now explain why.

3.3.1 *The holomorphic anomaly equations and modularity.* For simplicity, in the following we focus on the case where $\mathcal{M}(Y)$ is one-dimensional; the results are easily generalized to the multi-dimensional case. Parameterize $\mathcal{M}(Y)$ by a period t, for instance the logarithmic period canonically chosen by the mirror map near a maximally unipotent boundary point. It turns out that the B-model amplitudes $\widehat{F}_g$,[3] when expanded in terms of the period t, are *not* holomorphic. Their anti-holomorphic dependence is in fact encoded in the holomorphic anomaly equation, which reads, in the noncompact case,

$$\partial_{\bar{t}} \widehat{F}_g = \frac{1}{2} C_{\bar{t}}^{(0)tt} \left(D_t \partial_t \widehat{F}_{g-1} + \sum_{n=1}^{g-1} \partial_t \widehat{F}_n \partial_t \widehat{F}_{g-n} \right), \tag{3.1}$$

where both $C_{\bar{t}}^{(0)tt}$, which is related to the Yukawa coupling, and the covariant derivative D_t can be defined in terms of special geometric data. We refer the reader to [7, 25] for precise definitions of these objects.

The crucial point for us is that (3.1) can be integrated, either using an iterative Feynman graph procedure as in [7, 1], or by direct integration of (3.1) using modular properties of the amplitudes [25]. For $g > 1$, the result is [1, 25]

$$\widehat{F}_g(\tau, \bar{\tau}) = (\partial_t \tau)^{2g-2} \sum_{k=0}^{3g-3} \widehat{E}_2^k(\tau, \bar{\tau}) c_k^{(g)}(\tau), \tag{3.2}$$

where we now expressed the amplitudes in terms of the period matrix

$$\tau = -\frac{1}{4\pi} \frac{\partial t_D}{\partial t} = -\frac{1}{4\pi} \frac{\partial^2 \mathcal{F}}{\partial t^2}, \tag{3.3}$$

with t_D the period dual to t and $\mathcal{F}$ the prepotential, to make the modular properties of the amplitudes manifest. In (3.2), we introduced the standard non-holomorphic extension of the second Eisenstein series $E_2(\tau)$, defined by

$$\widehat{E}_2(\tau, \bar{\tau}) = E_2(\tau) - \frac{3}{\pi \mathrm{Im}\,(\tau)}. \tag{3.4}$$

The $c_k^{(g)}(\tau)$ are holomorphic modular forms of weight $6(g-1) - 2k$ with respect to the monodromy group of the periods — that is, the monodromy group of the Picard-Fuchs equations. It can be shown that $\partial_t \tau$ is a holomorphic modular form

---

[3]The reason for the hat here is that these amplitudes are the *physical* B-model topological string amplitudes, which are defined globally all over the moduli space $\mathcal{M}(Y)$. To make contact with the objects considered so far we will need to take their limits at special points in $\mathcal{M}(Y)$, such as maximally unipotent boundary points and points of finite monodromy.

of weight -3, while $\widehat{E}_2(\tau,\bar{\tau})$ is an almost holomorphic modular form of weight 2. As a result, we get that the $\widehat{F}_g$'s are almost holomorphic modular forms of weight 0. Their only anti-holomorphic dependence appears through $\widehat{E}_2(\tau,\bar{\tau})$.

Moreover, the $c_k^{(g)}(\tau)$, for $k \geq 1$ and $g > 1$, are entirely fixed by the holomorphic anomaly equation. On the one hand, by solving the equation iteratively using a Feynman procedure, as in [7, 1], one can show that the $c_k^{(g)}(\tau)$ are determined in terms of derivatives of lower genus amplitudes; see [1] for the explicit formulae. On the other hand, since we know that the $c_k^{(g)}(\tau)$ are modular forms of weight $6(g-1)-2k$, we can express them as polynomials in the generators of the ring of modular forms of the given weight, and find the coefficients by directly integrating the holomorphic anomaly equation. The second procedure, which was studied in [25], is a much more efficient way to fix the forms $c_k^{(g)}(\tau)$ for $k \geq 1, g > 1$.

However, at each genus the form $c_0^{(g)}(\tau)$, which is a holomorphic modular form of weight $6(g-1)$, is undetermined by the holomorphic anomaly equation. It corresponds to the so-called holomorphic ambiguity at each genus, which comes from the fact that the equation (3.1) only fixes the anti-holomorphic derivative of the $\widehat{F}_g$'s. This is the main hindrance when using the holomorphic anomaly equations to compute the amplitudes F_g; additional data must be used to calculate the $c_0^{(g)}(\tau)$'s.

3.3.2 *Going to special points in $\mathcal{M}(Y)$.* As mentioned above, so far we discussed the *physical* amplitudes $\widehat{F}_g$, which are defined globally all over the moduli space. To make contact with the rest of the paper, we must consider the limits of these amplitudes near special points in $\mathcal{M}(Y)$.

Recall that the holomorphic anomaly equation (3.1) was written in terms of a chosen period t. Suppose that t is the logarithmic period near a maximally unipotent boundary point, canonically chosen by the mirror map. Then, it was explained in [7] that the limit of the amplitudes at this point, which are mirror dual to the Gromov-Witten generating functions at large radius, is obtained by sending $\bar{t} \to \infty$. In terms of the period matrix τ, this corresponds to

$$F_g = \lim_{\mathrm{Im}(\tau)\to\infty} \widehat{F}_g. \tag{3.5}$$

In mathematical language, since $\widehat{F}_g$ is an almost holomorphic modular form, it can be written as a finite power series in $\mathrm{Im}(\tau)^{-1}$, starting with a constant term. This limit is given by keeping only the constant term; this is the usual isomorphism between the ring of almost holomorphic modular forms and the ring of quasi-modular forms; see [31]. The F_g's are then quasi-modular forms of weight 0 under the monodromy group.

Now suppose we consider another period, call it σ, instead of t, and rewrite the holomorphic anomaly equations (3.1) in terms of σ. For instance, let σ be the period canonically chosen by the mirror map at the point of finite monodromy. Again, the limit of the amplitudes at the orbifold point will be given by sending $\mathrm{Im}(\tau_\sigma) \to \infty$, where τ_σ is now defined by (3.3) in terms of the dual periods σ and σ_D. Taking this limit should give the B-model amplitudes F_g^{orb} at the point of finite monodromy, which should be mirror dual to the generating functionals of orbifold Gromov-Witten invariants.

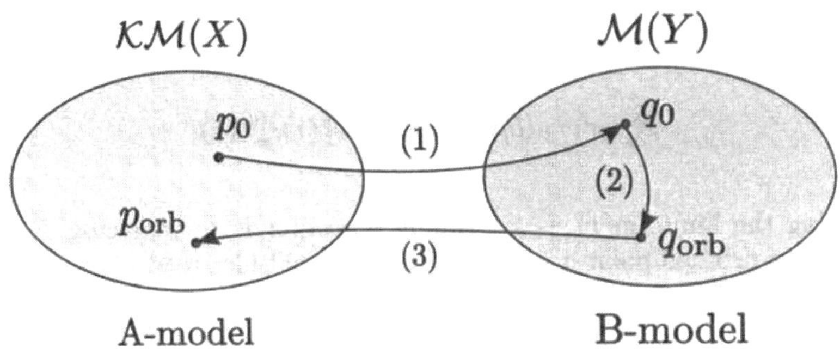

Figure 2 A schematic illustration of our strategy to compute orbifold Gromov-Witten invariants.

Hence, the holomorphic anomaly equation can be used to determine the F_g's at any point in the moduli space, including the point of finite monodromy, up to the holomorphic ambiguities $c_0^{(g)}$ at each genus.

3.4 Computing orbifold Gromov-Witten invariants of $\overline{X}$. Now, what kind of additional data can we use to fix the ambiguities at the point of finite monodromy? Well, the simple realization of [1] is that we in fact do not need any new data! Indeed, a crucial point is that the ambiguities at each genus, which correspond to the combinations

$$h_g = (\partial_t \tau)^{2g-2}\, c_0^{(g)}, \tag{3.6}$$

are globally defined holomorphic functions on the moduli space $\mathcal{M}(Y)$. Hence, if we know the amplitudes at a large radius point $q_0 \in \mathcal{M}(Y)$, we can fix the h_g, and use them, in conjunction with (3.2), to compute the amplitudes at the point of finite monodromy $q_{orb} \in \mathcal{M}(Y)$.

Our strategy to compute orbifold Gromov-Witten invariants should now be clear. Consider a toric Calabi-Yau threefold X which is the crepant resolution of an orbifold $\overline{X} = \mathbb{C}^3/G$; the enlarged Kähler moduli space $\mathcal{KM}(X)$ contains an orbifold point $p_{orb} \in \mathcal{KM}(X)$ corresponding to $\overline{X}$. We first determine the mirror maps near the large radius point $p_0 \in \mathcal{KM}(X)$ and the orbifold point $p_{orb} \in \mathcal{KM}(X)$. The calculation then proceeds in three steps, which are illustrated in figure 2.

1. We compute the generating functionals of Gromov-Witten invariants of X, using for instance the topological vertex [3], or localization of Hodge integrals. These are mapped by mirror symmetry at large radius to the B-model amplitudes near $q_0 \in \mathcal{M}(Y)$.
2. From these amplitudes we fix the holomorphic functions h_g, which are valid all over the moduli space and can be used to compute the B-model amplitudes at q_{orb} through (3.2).
3. Finally, we use the orbifold mirror map to extract the orbifold Gromov-Witten invariants of $\overline{X}$ from the B-model amplitudes at q_{orb}.

3.4.1 *An alternative strategy.* Note that there is a second method that can be used to compute the B-model amplitudes at q_{orb}, which was emphasized in [1]. The amplitudes at the maximally unipotent boundary point q_0, which we denote by

$F_g^\infty(\tau)$, can be written as

$$F_g^\infty(\tau) = (\partial_t \tau)^{2g-2} \sum_{k=0}^{3g-3} E_2^k(\tau) c_k^{(g)}(\tau), \tag{3.7}$$

since taking the limit $\mathrm{Im}(\tau) \to \infty$ simply corresponds to replacing $\widehat{E}_2(\tau,\bar{\tau})$ by $E_2(\tau)$. At the orbifold point q_{orb}, similarly the amplitudes read

$$F_g^{orb}(\tau_\sigma) = (\partial_\sigma \tau_\sigma)^{2g-2} \sum_{k=0}^{3g-3} E_2^k(\tau_\sigma) \tilde{c}_k^{(g)}(\tau_\sigma), \tag{3.8}$$

where now τ_σ is defined from the periods σ, σ_D canonically chosen at the point q_{orb}. By definition of the period matrix, τ and τ_σ must be related by a symplectic transformation, that is,

$$\tau_\sigma = \frac{a\tau + b}{c\tau + d} \quad \text{for some} \quad \begin{pmatrix} a & b \\ c & d \end{pmatrix} \in SL(2,\mathbb{C}). \tag{3.9}$$

By using modular properties of the objects entering into (3.7) one can implement the symplectic transformation to calculate the amplitudes (3.8). That method relates directly, on the mirror side, the orbifold amplitudes to the amplitudes of its crepant resolution, in the spirit of the crepant resolution conjecture [11, 18, 35].

Note however that this procedure may seem a little bit strange from a modular forms point of view, since the transformation is generically in $SL(2,\mathbb{C})$ rather than in $SL(2,\mathbb{Z})$. But the transformation of the amplitudes from the point q_0 to the point q_{orb} can be made precise without direct reference to modular forms, as in [1], using wavefunction properties of the amplitudes. There it was also shown that this alternative approach is indeed equivalent to the first method proposed above.

3.5 Example: $\mathbb{C}^3/\mathbb{Z}_3$. Let us come back to our main example. Recall that $X = \mathcal{O}(-3) \to \mathbb{P}^2$ is the unique crepant resolution of the orbifold $\overline{X} = \mathbb{C}^3/\mathbb{Z}_3$. The enlarged Kähler moduli space $\mathcal{KM}(X)$ is described by the secondary fan associated to X. The secondary fan is generated by the two one-dimensional cones,

$$\{(-3),(1)\}. \tag{3.10}$$

Hence $\mathcal{KM}(X)$ is one-dimensional and has two patches, one of which contains the large radius point p_0 associated to X, and the other of which includes the orbifold point p_{orb} associated to $\overline{X}$.

The mirror threefold Y was described in (2.20), in terms of the coordinate q centered at the maximally unipotent boundary point $q_0 = \{q = 0\} \in \mathcal{M}(Y)$. The point of finite monodromy $q_{orb} \in \mathcal{M}(Y)$ is located at $q \to \infty$. A natural coordinate centered at q_{orb} is

$$\psi = q^{-1/3}, \tag{3.11}$$

as can be read off from the secondary fan. Note that under $\mathbb{Z}_3$-monodromy around $q_{orb} = \{\psi = 0\}$, ψ undergoes

$$\psi \mapsto \mathrm{e}^{2\pi i/3} \psi. \tag{3.12}$$

3.5.1 *The orbifold mirror map.* We first need to fix the mirror map near the orbifold point $p_{orb} \in \mathcal{KM}(X)$. Writing the Picard-Fuchs equations (2.22) in terms of the coordinate ψ, we get

$$\mathcal{D}_\psi = \psi^3\Theta_\psi^3 + 27(\Theta_\psi - 2)(\Theta_\psi - 1)\Theta_\psi, \tag{3.13}$$

with $\Theta_\psi = \psi\partial_\psi$. A solution vector to $\mathcal{D}_\psi\Pi^{orb} = 0$ is given by

$$\Pi^{orb} = (1, B_1(\psi), B_2(\psi)),$$

with

$$B_k(\psi) = \frac{(-1)^{k+1}\psi^k}{k}\, {}_3F_2\left(\frac{k}{3}, \frac{k}{3}, \frac{k}{3}; \frac{2k}{3}, 1 + \frac{k}{3}; \left(-\frac{\psi}{3}\right)^3\right). \tag{3.14}$$

Using the explicit expansion of the hypergeometric system we get

$$B_k(\psi) = \sum_{n\geq 0} \frac{(-1)^{3n+k+1}\psi^{3n+k}}{(3n+k)!}\left(\frac{\Gamma\left(n+\frac{k}{3}\right)}{\Gamma\left(\frac{k}{3}\right)}\right)^3. \tag{3.15}$$

As described in section 3.2, to get the orbifold mirror map we need to find linear combinations of the solutions above that are mapped to a basis for the orbifold cohomology of $\mathbb{C}^3/\mathbb{Z}_3$. The orbifold cohomology $H^*_{\mathrm{orb}}(\mathbb{C}^3/\mathbb{Z}_3)$ has basis $\mathbf{1}_1$, $\mathbf{1}_\omega$ and $\mathbf{1}_{\omega^2}$, which are indexed by the elements $1, \omega, \omega^2 \in \mathbb{Z}_3$, with $\omega = \mathrm{e}^{2\pi i/3}$. The basis elements have degrees

$$\deg(\mathbf{1}_1) = 0, \qquad \deg(\mathbf{1}_\omega) = 2, \qquad \deg(\mathbf{1}_{\omega^2}) = 4. \tag{3.16}$$

Hence $H^*_{\mathrm{orb}}(\mathbb{C}^3/\mathbb{Z}_3)$ is spanned by

$$\sigma_0\mathbf{1}_1 + \sigma_1\mathbf{1}_\omega + \sigma_2\mathbf{1}_{\omega^2}. \tag{3.17}$$

The orbifold mirror map will be given by mapping σ_1 to an appropriate combination of 1, $B_1(\psi)$ and $B_2(\psi)$.

Recall that monodromy around q_{orb} is given by $\psi \mapsto \mathrm{e}^{2\pi i/3}\psi$, which implies

$$(1, B_1(\psi), B_2(\psi)) \mapsto (1, \mathrm{e}^{2\pi i/3}B_1(\psi), \mathrm{e}^{4\pi i/3}B_2(\psi)). \tag{3.18}$$

But $\mathbf{1}_\omega$ corresponds to the element $\omega \in \mathbb{Z}_3$; thus, it is clear that σ_1 should be mapped to $B_1(\psi)$ directly, up to an overall scale factor. More precisely, we claim that the mirror map is given by

$$(\sigma_1, \sigma_2) = (B_1(\psi), B_2(\psi)). \tag{3.19}$$

Another way of arguing for this mirror map is by computing the genus 0 amplitude, as we do next. Up to scale, the above mirror map is the only map that yields a genus 0 amplitude which is invariant under orbifold monodromy. Note that this is also the mirror map that was proved in [17].

3.5.2 *Genus* 0 *amplitude.* Before computing the genus 0 amplitude, let us clarify the relation between the A-model amplitudes and Gromov-Witten theory at the orbifold point. At large radius, the genus g A-model amplitudes became generating functions for unmarked genus g Gromov-Witten invariants $N_{g,\beta}$ in homology classes $\beta \in H_2(X, \mathbb{Z})$. At the orbifold point, $\mathbb{C}^3/\mathbb{Z}_3$ contains no compact curve, hence the only invariants correspond to constant maps $\beta = 0$. However, the A-model amplitudes now become generating functions for marked orbifold Gromov-Witten invariants; more precisely

$$F_g^{orb} = \sum_{n=0}^{\infty} \frac{1}{(3n)!} N_{g,n}\sigma_1^{3n}, \tag{3.20}$$

where the invariants $N_{g,n}$ are given by $3n$ insertions of the orbifold cohomology class $\mathbf{1}_\omega$. Note that the unmarked ($n = 0$) invariants are only well-defined for $g \geq 2$. Moreover, we incorporated implicitly the fact that only invariants with $3n$ insertions are non-zero, which ensures that the amplitudes are invariant under orbifold monodromy.

The genus 0 amplitude is as usual given by the prepotential of special geometry, in the basis canonically chosen at q_{orb}. That is,

$$\sigma_2 = -3\frac{\partial \mathcal{F}^{orb}}{\partial \sigma_1} = -3\frac{\partial F_0^{orb}}{\partial \sigma_1}. \tag{3.21}$$

Integrating σ_2, we get:

$$N_{0,1} = \frac{1}{3}, \quad N_{0,2} = -\frac{1}{3^3}, \quad N_{0,3} = \frac{1}{3^2}, \quad N_{0,4} = -\frac{1093}{3^6}, \dots \tag{3.22}$$

This potential has now been computed mathematically in three independent ways by Coates, Corti, Iritani and Tseng in [17], Bayer and Cadman in [6] and by Cadman and Cavalieri [12]. Agreement with these results for the genus 0 amplitude fixes the normalization of the mirror map (3.19).

3.5.3 *Higher genus amplitudes.* To extract the higher genus amplitudes of $\mathbb{C}^3/\mathbb{Z}_3$, we need to compute the holomorphic functions h_g at each genus g. This can be done easily at large radius, by first computing the A-model amplitudes through the topological vertex, and then mapping them to the B-model side using the usual mirror map at large radius. We obtain, for the marked invariants:

$$F_g^{orb} = \sum_{n=1}^{\infty} \frac{1}{(3n)!} N_{g,n} \sigma_1^{3n}, \tag{3.23}$$

with the numbers:[4]

g	$n=1$	2	3	4
0	$\frac{1}{3}$	$-\frac{1}{3^3}$	$\frac{1}{3^2}$	$-\frac{1093}{3^6}$
1	0	$\frac{1}{3^5}$	$-\frac{14}{3^5}$	$\frac{13007}{3^8}$
2	$\frac{1}{2^4 \cdot 3^4 \cdot 5}$	$-\frac{13}{2^4 \cdot 3^6}$	$\frac{20693}{2^4 \cdot 3^8 \cdot 5}$	$-\frac{12803923}{2^4 \cdot 3^{10} \cdot 5}$
3	$-\frac{31}{2^5 3^5 5 \cdot 7}$	$\frac{11569}{2^5 3^9 5 \cdot 7}$	$-\frac{2429003}{2^5 3^{10} 5 \cdot 7}$	$\frac{871749323}{2^4 3^{11} 5 \cdot 7}$
4	$\frac{313}{2^7 3^9 5^2}$	$-\frac{1889}{2^7 3^9}$	$\frac{115647179}{2^6 3^{13} 5^2}$	$-\frac{29321809247}{2^8 3^{12} 5^2}$
5	$-\frac{519961}{2^9 3^{11} 5^2 7 \cdot 11}$	$\frac{196898123}{2^9 3^{12} 5^2 7 \cdot 11}$	$-\frac{339157983781}{2^9 3^{14} 5^2 7 \cdot 11}$	$\frac{78658947782147}{2^9 3^{16} 5 \cdot 7}$
6	$\frac{14609730607}{2^{12} 3^{13} 5^3 7^2 11}$	$-\frac{258703053013}{2^{10} 3^{15} 5^1 7^2 11}$	$\frac{2453678654644313}{2^{12} 3^{14} 5^3 7^2 11}$	$-\frac{4001577419396960 1803}{2^{11} 3^{18} 5^3 7^2 11}$

The unmarked invariants ($n = 0$) for $g \geq 2$ (these are not well-defined for $g = 0, 1$) can also be calculated, and read

$$N_{2,0} = \frac{-1}{2160} + \frac{\chi(X)}{5760}, \quad N_{3,0} = \frac{1}{544320} - \frac{\chi(X)}{1451520}, \tag{3.24}$$
$$N_{4,0} = -\frac{7}{41990400} + \frac{\chi(X)}{87091200}, N_{5,0} = \frac{3161}{77598259200} - \frac{\chi(X)}{2554675200}, \dots$$

[4]We would like to thank A. Klemm for the computation of the invariants with $g \geq 3$.

where $\chi(X)$ is the "Euler number" of $X = \mathcal{O}(-3) \to \mathbb{P}^2$. Although $\chi(X)$ is not really well defined mathematically since X is noncompact, its natural value can be found by noting that any vector bundle retracts to its zero section; hence, since $X = \mathcal{O}(-3) \to \mathbb{P}^2$, we obtain that $\chi(X) = \chi(\mathbb{P}^2) = 3$. Note that the numbers $N_{2,0}$ and $N_{3,0}$ have now been computed independently by myself and R. Cavalieri in [8], by evaluating directly the Hodge integrals entering in the definition of orbifold Gromov-Witten invariants.

3.5.4 *Symplectic transformation.* If we wanted to implement the second strategy to compute the invariants, as in [1], we would need the symplectic transformation between the basis of periods $(t_D, t, 1)$ at q_0 and the basis of periods $(\sigma_2, \sigma_1, 1)$ at q_{orb} canonically chosen by the mirror maps. The transformation is easily found by analytic continuation of the periods; see for example [20]. Define

$$c_1 = -\frac{1}{2\pi i}\frac{\Gamma(1/3)}{\Gamma(2/3)^2}, \qquad c_2 = \frac{1}{2\pi i}\frac{\Gamma(2/3)}{\Gamma(1/3)^2}, \qquad \beta = \frac{1}{(2\pi i)^3}, \qquad \omega = e^{2\pi i/3}. \tag{3.25}$$

We get the transformation

$$\begin{pmatrix} t_D \\ t \\ 1 \end{pmatrix} = \begin{pmatrix} \frac{\beta\omega^2}{c_1} & \frac{\beta\omega}{c_2} & \frac{1}{3} \\ -c_2 & c_1 & 0 \\ 0 & 0 & 1 \end{pmatrix} \begin{pmatrix} \sigma_2 \\ \sigma_1 \\ 1 \end{pmatrix}. \tag{3.26}$$

Note that this transformation is not quite symplectic, since its determinant is $-\beta$; that is, it changes the scale of the symplectic form. However, this can be taken into account by renormalizing the string coupling constant; see [1] for the details of this procedure.

4 Open mirror symmetry

So far we considered A- and B-model closed topological string theory, and used mirror symmetry and the holomorphic anomaly equations to compute orbifold Gromov-Witten invariants. In this section we would like to extend this line of thought to open topological string theory, following the recent work of [9].

The general idea is the same: we want to use mirror symmetry and B-model topological string theory to compute the A-model amplitudes at the orbifold point, which are related to Gromov-Witten theory of the orbifold $\overline{X}$. However, we now consider open topological string theory, which is a theory of maps with boundaries ending on submanifolds (usually called *branes*) of the target space. At large radius, A-model open topological strings compute the so-called open Gromov-Witten invariants of X [24]. On the contrary, to the best of our knowledge it is not clear what A-model open topological string theory computes in orbifold enumerative geometry; it should be some sort of generalization of open Gromov-Witten invariants to the orbifold setting. Keeping this in mind, we use the mirror symmetric physical approach to compute some generating functions for such as yet undefined "open orbifold Gromov-Witten invariants".

The main differences in considering open topological string theory are:

- the mirror map now comprises an open sector, which maps the moduli associated to the branes on the A- and the B-model sides;
- the holomorphic anomaly equations are no longer sufficient to compute the open B-model amplitudes. Hence, we need a new formalism to compute the amplitudes at the orbifold point.

We (partially) overcame these two obstacles in [9] for simple orbifolds such as $\mathbb{C}^3/\mathbb{Z}_3$. Let us first review general features of topological open string theory and the new B-model formalism of [9], which is based on the ideas of [33]. We then study how the open/closed mirror map can be fixed at the orbifold point, and compute some "open orbifold Gromov-Witten invariants" of $\mathbb{C}^3/\mathbb{Z}_3$.

4.1 Open A-model on X. A-model topological open string theory is a theory of holomorphic maps $f : \Sigma_{g,h} \to X$, where $\Sigma_{g,h}$ is a genus g Riemann surface with h holes, and X is a Calabi-Yau threefold, such that f maps the boundaries of $\Sigma_{g,h}$ to a special Lagrangian submanifold $L \subset X$, which is called a *brane*. Near a large radius point $p_0 \in \mathcal{KM}(X)$, the A-model produces the generating functionals

$$A_h^{(g)}(X_1, \dots, X_h) = \sum_{w_i \in \mathbb{Z}} F_{g,w} X_1^{w_1} \cdots X_h^{w_h}, \tag{4.1}$$

where the X_i are the open moduli associated to the brane. The $F_{g,w}$ here are generating functions for open Gromov-Witten invariants of (L, X), counting maps at genus g, indexed by their relative homology class $\beta \in H_2(X, L)$, and with winding numbers w_i, $i = 1, \dots, h$, specifying how many times the i-th boundary wraps around the one-cycle in L:

$$F_{g,w} = \sum_{\beta \in H_2(X,L)} N_{g,w,\beta}\, Q^\beta. \tag{4.2}$$

To define the open A-model amplitudes we needed to fix the brane L. When X is a toric threefold, the usual branes considered are noncompact and have topology $\mathbb{C} \times S^1$; we will call these branes *toric branes*.[5]

4.2 Open B-model on Y. In the B-model, the Lagrangian branes are mapped to holomorphic submanifolds of Y. More precisely, the toric brane L introduced above maps to a one-complex dimensional holomorphic submanifold of Y, given by

$$G(x, y; q_1, \dots, q_r) = 0 = w'. \tag{4.3}$$

Hence, it is parameterized by w, and its moduli space corresponds to the mirror curve

$$\Sigma = \{G(x, y; q_1, \dots, q_r) = 0\}. \tag{4.4}$$

The open string moduli X_i thus become variables on the mirror curve Σ.

We proposed in [9], following the ideas of [33], a complete recursive formalism to generate the B-model open and closed amplitudes. The formalism is based on the recursion relations of Eynard and Orantin which solve the loop equations of matrix models [22]. We refer the reader to [9] for the details of the formalism. To be concise, let us simply say that the open amplitudes are encoded by some meromorphic differentials living on the mirror curve Σ, while the closed amplitudes are functions on Σ. The recursive process can be understood as some sort of gluing process for the open amplitudes, whereby the gluing procedure consists in taking residues of some meromorphic differentials at the ramification points of the projection map $\Sigma \to \mathbb{C}^*$. It is important to note that, in contrast to the holomorphic anomaly equations for closed amplitudes (or their generalization to open amplitudes [36]), the recursive formalism proposed in [9] is complete, in the sense that there is

[5]See [9] for an extensive discussion of these branes and the different phases in the open moduli space.

no ambiguity at each genus, and no additional data is needed to fix the topological string amplitudes.

The input of the recursive process is rather simple. Using the notation above, to generate all the open and closed amplitudes near a given point in $\mathcal{M}(Y)$, say a maximally unipotent boundary point $q_0 \in \mathcal{M}(Y)$, one only needs to know the disk amplitude $A_1^{(0)}$ and the annulus amplitude $A_2^{(0)}$ near q_0.

4.2.1 *Disk amplitude.* Following the work of [4, 5], we know that the disk amplitude is simply given by the Abel-Jacobi map of the curve $\Sigma \subset \mathbb{C}^* \times \mathbb{C}^*$:

$$A_1^{(0)}(x) = \int \lambda = \int \log y(x) \frac{\mathrm{d}x}{x}, \tag{4.5}$$

where $y(x)$ is obtained by solving the defining equation of Σ for y; x here is the B-model open string modulus. Note that $A_1^{(0)}$ is globally defined all over the moduli space $\mathcal{M}(Y)$, and can be expanded both at maximally unipotent boundary points and the point of finite monodromy.

4.2.2 *Annulus amplitude.* The annulus amplitude is slightly more complicated. Generically, it is given by

$$A_2^{(0)}(x_1, x_2) = \int \left(B(x_1, x_2) - \frac{\mathrm{d}x_1 \mathrm{d}x_2}{(x_1 - x_2)^2} \right), \tag{4.6}$$

where $B(x_1, x_2)$ is the Bergman kernel of the Riemann surface Σ. This is defined to be the unique meromorphic differential on Σ with a double pole at $x_1 = x_2$ with no residue, and no other pole, and normalized such that

$$\oint_{A_I} B(x_1, x_2) = 0, \tag{4.7}$$

where (A_I, B^I) is a canonical basis of one-cycles on Σ. Note that by definition the Bergman kernel depends on a choice of canonical basis of cycles, or periods, on Σ.

In fact, we can be more precise on how the Bergman kernel depends on the choice of cycles. Under an $\mathrm{Sp}(2g, \mathbb{Z})$ transformation of the periods, the Bergman kernel transforms with a shift, as follows:

$$B(x_1, x_2) \mapsto B(x_1, x_2) - 2\pi i \omega(x_1)(C\tau + D)^{-1} C \omega(x_2), \tag{4.8}$$

with

$$\begin{pmatrix} A & B \\ C & D \end{pmatrix} \in \mathrm{Sp}(2g, \mathbb{Z}), \tag{4.9}$$

and τ is the period matrix. Here, $\omega(p)$ is the holomorphic differentials put in vector form. Hence, $B(x_1, x_2)$ provides an "open analog" – since it is a differential on Σ – to the second Eisenstein series $E_2(\tau)$, which also transforms with a shift. For more on the modular properties of the Bergman kernel, see [21, 22].

Therefore, once one knows the disk and the annulus amplitude at a given point in $\mathcal{M}(Y)$, one can generate unambiguously all other open and closed amplitudes at this point, using the recursive process of [9]. However, to map the results to the more interesting A-model where the amplitudes are related to Gromov-Witten theory, one must also know the open/closed mirror map near this point, which we now turn to.

4.3 The open/closed mirror map. In the context of open topological string theory, the usual mirror map must be supplemented with an open sector, which relates the A-model open modulus X to the B-model open modulus x. Consider first a large radius point $p_0 \in \mathcal{KM}(X)$ and its corresponding maximally unipotent boundary point $q_0 \in \mathcal{M}(Y)$. The open mirror map is defined such that, in conjunction with the closed mirror map, it lifts to an isomorphism of the open and closed A- and B-model amplitudes near p_0 and q_0.

Since x and X are $\mathbb{C}^*$ coordinates, as is customary in the physics literature we write $x = e^u$, $X = e^U$. While the closed mirror map is given by the integrals of the one-form λ over one-cycles on the mirror curve Σ (the "periods" of λ), the open mirror map, as argued in [4], is given by a chain integral of the one-form λ:

$$U = \int_{\alpha_u} \lambda, \tag{4.10}$$

where α_u is a given chain on Σ. In analogy with the flat coordinates on $\mathcal{M}(Y)$ — which we now call "closed flat coordinates" to avoid confusion — we will call this integral the *open flat coordinate* on the open/closed moduli space. As explained in [9], once evaluated, the chain integral always takes the form

$$U = u + 2\pi i \sum_{i=1}^{r} r_i^u \left(\frac{1}{2\pi i} \log q_i - \omega^i \right), \tag{4.11}$$

where the q_i and ω^i were introduced in section 2.2; the ω^i are the logarithmic periods (the closed flat coordinates) at $q_0 \in \mathcal{M}(Y)$, while the q_i are the usual coordinates on $\mathcal{M}(Y)$ centered at q_0. The coefficients r_i^u are rational numbers. In physics language, what this statement means is that the open string coordinates only receive corrections coming from closed string instantons.

What we are now interested in is determining the open mirror map at the point of finite monodromy $q_{orb} \in \mathcal{M}(Y)$ mirror dual to the orbifold point $p_{orb} \in \mathcal{KM}(X)$. By definition, we know that the open flat coordinate at q_{orb}, which we denote by U_{orb}, will always be given by a linear combination of the period integrals of λ and the chain integral (4.10). We claim that, as in the closed case, for simple orbifolds such as $\mathbb{C}^3/\mathbb{Z}_3$, the good linear combination can be uniquely determined, up to scale, by requiring that the disk amplitude be invariant under finite monodromy around q_{orb}.

Let us now put these ideas into practice for our orbifold friend $\mathbb{C}^3/\mathbb{Z}_3$.

4.4 Example: $\mathbb{C}^3/\mathbb{Z}_3$. Consider again the crepant resolution $X = \mathcal{O}(-3) \to \mathbb{P}^2$ of $\mathbb{C}^3/\mathbb{Z}_3$. We now fix a toric brane, that is, a special Lagrangian submanifold of X with topology $\mathbb{C} \times S^1$; for example, we consider an "outer brane" with zero framing, in the terminology of [9]. The mirror curve reads

$$\Sigma = \{y^2 + y(1+x) + qx^3 = 0\}. \tag{4.12}$$

The maximally unipotent boundary point is at $q = 0$. Recall that the only logarithmic period is given by

$$\omega(q) = \frac{1}{2\pi i} \left(\log q + \sum_{n=1}^{\infty} \frac{(-1)^n}{n} \frac{(3n)!}{(n!)^3} q^n \right). \tag{4.13}$$

The chain integral (4.10) giving the open mirror map at large radius was performed in [4], and the result is

$$U = u + \frac{2\pi i}{3}\left(\frac{1}{2\pi i}\log q - \omega(q)\right) = u - \frac{1}{3}\sum_{n=1}^{\infty}\frac{(-1)^n}{n}\frac{(3n)!}{(n!)^3}q^n. \tag{4.14}$$

To compute the disk amplitude we need to integrate the one-form λ. Solving (4.12) for y, we get

$$y(x) = -\frac{1+x}{2} - \frac{1}{2}\sqrt{(1+x)^2 - qx^3}, \tag{4.15}$$

where we kept the branch of the square root which is relevant for the disk amplitude. The B-model disk amplitude at q_0 is then given by expanding

$$A_1^{(0)} = \int \log y(x)\frac{\mathrm{d}x}{x} \tag{4.16}$$

around $q = 0$. To get the A-model amplitude at large radius, we simply plug in the open and closed mirror maps.

4.4.1 *Orbifold mirror map.* We now want to compute the amplitude at the orbifold point $p_{orb} \in \mathcal{KM}(X)$. We first need to compute the B-model amplitude at the point of finite monodromy $q_{orb} \in \mathcal{M}(Y)$. Recall that q_{orb} is located at $\psi = 0$, with

$$\psi = q^{-1/3}. \tag{4.17}$$

The disk amplitude at q_{orb} is simply given by writing (4.15) in terms of ψ, and then expanding (4.16) around $\psi = 0$. However, to obtain the A-model amplitude at the orbifold point we need to plug in the open and closed mirror maps near the orbifold point. We already found the closed mirror map in (3.19), which is given by

$$(\sigma_1, \sigma_2) = (B_1(\psi), B_2(\psi)). \tag{4.18}$$

We now need to find the open mirror map.

Well, the open flat parameter must be given by a linear combination of the periods of λ and the chain integral (4.10). It turns out that the only combination, up to scale, that gives a disk amplitude invariant under the $\mathbb{Z}_3$-monodromy is

$$U_{orb} = U + \frac{2\pi i}{3}\omega(q) = u + \frac{1}{3}\log q. \tag{4.19}$$

In terms of exponentiated parameters, $X_{orb} = \mathrm{e}^{U_{orb}}$ and $x = \mathrm{e}^u$, we get

$$X_{orb} = xq^{1/3} = \frac{x}{\psi}. \tag{4.20}$$

This gives the open mirror map at the orbifold point, where X_{orb} is the A-model open modulus and x is the B-model open modulus.

4.4.2 *Orbifold disk amplitude.* Now, expanding the disk amplitude around $\psi = 0$ and plugging in the open and closed orbifold mirror maps (4.18) and (4.20) we obtain

$$A_1^{(0)} = \left(\sigma_1 + \frac{\sigma_1^4}{648} - \frac{29\,\sigma_1^7}{3674160} + \frac{6607}{71425670400}\sigma_1^{10} + \dots\right) X_{orb}$$
$$+ \left(-\frac{\sigma_1^2}{4} - \frac{\sigma_1^5}{1296} + \frac{197\,\sigma_1^8}{58786560} - \frac{5737}{142851340800}\sigma_1^{11} + \dots\right) X_{orb}^2$$
$$+ \left(-\frac{1}{3} + \frac{\sigma_1^3}{9} + \frac{\sigma_1^6}{1944} - \frac{\sigma_1^9}{544320} + \dots\right) X_{orb}^3 + \mathcal{O}(X_{orb}^4). \tag{4.21}$$

Up to the scale of X_{orb}, which we could not fix in the mirror map (4.20), we predict that this amplitude should generate "open orbifold Gromov-Witten invariants" of $\mathbb{C}^3/\mathbb{Z}_3$ at genus 0 with one hole; it would be fascinating to understand what these invariants really are. Finally, note that (4.21) is indeed invariant under the $\mathbb{Z}_3$-monodromy given by $\psi \mapsto e^{2\pi i/3}\psi$, as it should be.

A little note of care may be added here. In the B-model, the open amplitudes explicitly depend on a choice of parameterization for the mirror curve Σ. At large radius, this corresponds on the mirror A-model side to a choice of framing and phase for the Lagrangian brane [9]. However, for the A-model amplitudes at the orbifold point it is unclear what this freedom means. In particular, we should be more precise and say that, in the terminology of [9], the orbifold disk amplitude presented above should be for an "outer brane with zero framing" in the orbifold $\mathbb{C}^3/\mathbb{Z}_3$, even though we do not really understand what this means yet.

4.4.3 *Orbifold annulus amplitude.* To compute the remaining amplitudes at the orbifold point, we first need to obtain the B-model annulus amplitude at the point of finite monodromy. In turn, this involves computing the Bergman kernel at this point. The strategy goes as follows; we will use our knowledge of the Bergman kernel at q_0 to compute it at q_{orb}.

In terms of closed periods, moving from q_0 to q_{orb} involves a symplectic transformation of the periods; that is, the periods $(t_D, t, 1)$ at q_0 and the periods $(\sigma_2, \sigma_1, 1)$ at q_{orb} are related by the transformation (3.26). However, we have seen that the Bergman kernel is *not* modular under symplectic transformations of the periods; instead, it transforms with a shift, see (4.8). Hence, we can use this transformation formula, in conjunction with (3.26), to extract the Bergman kernel at q_{orb} from the Bergman kernel at q_0. This also involves analytically continuing the large radius Bergman kernel to the point $q_{orb} : \{\psi = 0\}$, which can be done with standard methods. In fact, a consistency check is that the analytic continuation of the large radius annulus amplitude, when expanded in terms of σ_1 after plugging in the open/closed orbifold mirror map, does not have rational coefficients; however, after shifting it using (4.8), the result should have a rational expansion.

We are presently in the process of computing the annulus amplitude at q_{orb} [10]. When this is done, we can generate all open and closed amplitudes at the point of finite monodromy q_{orb} unambiguously, and then plug in the open and closed orbifold mirror maps found above to compute the open and closed A-model amplitudes at the orbifold point. We hope to report on that in the near future.

References

[1] M. Aganagic, V. Bouchard and A. Klemm, "Topological Strings and (Almost) Modular Forms," Commun. Math. Phys. **277**, 771 (2008) [arXiv:hep-th/0607100].

[2] M. Aganagic, R. Dijkgraaf, A. Klemm, M. Mariño and C. Vafa, "Topological strings and integrable hierarchies," Commun. Math. Phys. **261**, 451 (2006) [arXiv:hep-th/0312085].

[3] M. Aganagic, A. Klemm, M. Mariño and C. Vafa, "The topological vertex," Commun. Math. Phys. **254**, 425 (2005) [arXiv:hep-th/0305132].

[4] M. Aganagic, A. Klemm and C. Vafa, "Disk instantons, mirror symmetry and the duality web," Z. Naturforsch. A **57**, 1 (2002) [arXiv:hep-th/0105045].

[5] M. Aganagic and C. Vafa, "Mirror symmetry, D-branes and counting holomorphic discs," arXiv:hep-th/0012041.

[6] A. Bayer and C. Cadman, "Quantum cohomology of $[\mathbb{C}^n/\mu_r]$," arXiv:0705.2160.

[7] M. Bershadsky, S. Cecotti, H. Ooguri and C. Vafa, "Kodaira-Spencer theory of gravity and exact results for quantum string amplitudes," Commun. Math. Phys. **165**, 311 (1994) [arXiv:hep-th/9309140].

[8] V. Bouchard and R. Cavalieri, "On the mathematics and physics of high genus invariants of $[\mathbb{C}^3/\mathbb{Z}_3]$," arXiv:0709.3805 [math.AG].

[9] V. Bouchard, A. Klemm, M. Mariño and S. Pasquetti, "Remodeling the B-model," arXiv:0709.1453 [hep-th].

[10] V. Bouchard, A. Klemm, M. Mariño and S. Pasquetti, work in progress.

[11] J. Bryan and T. Graber, "The crepant resolution conjecture," arXiv:math.AG/0610129.

[12] C. Cadman and R. Cavalieri, "Gerby localization, $\mathbb{Z}_3$-Hodge integrals and the GW theory of $\mathbb{C}^3/\mathbb{Z}_3$," arXiv:0705.2158.

[13] P. Candelas, X. C. De La Ossa, P. S. Green and L. Parkes, "A pair of Calabi-Yau manifolds as an exactly soluble superconformal theory," Nucl. Phys. B **359**, 21 (1991).

[14] W. Chen and Y. Ruan, "Orbifold Gromov-Witten Theory," in *Orbifolds in mathematics and physics (Madison, WI, 2001)*, volume 310 of *Contemp. Math.*, pages 25–85. Amer. Math. Soc., Providence, RI, 2002 [arXiv:math.AG/0103156].

[15] W. Chen and Y. Ruan, "A new cohomology theory of orbifolds," *Comm. Math. Phys.*, 248(1):1–31, 2004.

[16] T. M. Chiang, A. Klemm, S. T. Yau and E. Zaslow, "Local mirror symmetry: Calculations and interpretations," Adv. Theor. Math. Phys. **3**, 495 (1999) [arXiv:hep-th/9903053].

[17] T. Coates, A. Corti, H. Iritani and H.-H. Tseng, "Computing Genus-Zero Twisted Gromov-Witten Invariants," arXiv:math.AG/0702234.

[18] T. Coates, A. Corti, H. Iritani and H.-H. Tseng, "The crepant resolution conjecture for type A surface singularities," arXiv:0704.2034 [math.AG].

[19] D. A. Cox and S. Katz, *Mirror symmetry and algebraic geometry*, Providence, USA, AMS (2000) 469 p.

[20] D. E. Diaconescu and J. Gomis, "Fractional branes and boundary states in orbifold theories," JHEP **0010**, 001 (2000) [arXiv:hep-th/9906242].

[21] B. Eynard, M. Mariño and N. Orantin, "Holomorphic anomaly and matrix models," JHEP **0706**, 058 (2007) [arXiv:hep-th/0702110].

[22] B. Eynard and N. Orantin, "Invariants of algebraic curves and topological expansion," Comm. Numb. Theor. Phys., **1**, 347 (2007) [arXiv:math-ph/0702045].

[23] D. S. Freed, "Special Kähler manifolds," Commun. Math. Phys. **203**, 31 (1999) [arXiv:hep-th/9712042].

[24] T. Graber and E. Zaslow, "Open string Gromov-Witten invariants: Calculations and a mirror 'theorem'," arXiv:hep-th/0109075.

[25] T. W. Grimm, A. Klemm, M. Mariño and M. Weiss, "Direct integration of the topological string," arXiv:hep-th/0702187.

[26] M. X. Huang and A. Klemm, "Holomorphic anomaly in gauge theories and matrix models," arXiv:hep-th/0605195.

[27] M. X. Huang, A. Klemm and S. Quackenbush, "Topological string theory on compact Calabi-Yau: modularity and boundary conditions," arXiv:hep-th/0612125.

[28] K. Hori *et al.*, *Mirror symmetry*, Providence, USA, AMS (2003) 929 p.

[29] K. Hori and C. Vafa, "Mirror symmetry," arXiv:hep-th/0002222.

[30] S. Hosono, "Central charges, symplectic forms, and hypergeometric series in local mirror symmetry," arXiv:hep-th/0404043.

[31] M. Kaneko and D. Zagier, "A generalized Jacobi theta function and quasimodular forms," in *The moduli space of curves*, Progr. Math., 129, Birkhauser, Boston, MA, 1995, pp. 165–172.

[32] A. Klemm and E. Zaslow, "Local mirror symmetry at higher genus," arXiv:hep-th/9906046.

[33] M. Mariño, "Open string amplitudes and large order behavior in topological string theory," arXiv:hep-th/0612127.

[34] M. Mariño, *Chern-Simons theory, matrix models, and topological strings*, Oxford, UK, Clarendon (2005) 197 p.
[35] Y. Ruan, "Cohomology ring of crepant resolutions of orbifolds," arXiv:math.AG/0108195.
[36] J. Walcher, "Extended Holomorphic Anomaly and Loop Amplitudes in Open Topological String," arXiv:0705.4098 [hep-th].

Fields Institute Communications
Volume **54**, 2008

Conformal Field Theory and Mapping Class Groups

Terry Gannon
Department of Mathematics
University of Alberta
Edmonton, AB Canada T6G 2G1
tgannon@math.ualberta.ca

Abstract. Rational conformal field theories produce a tower of finite-dimensional representations of surface mapping class groups, acting on the conformal blocks of the theory. We review this formalism. We show that many recent mathematical developments can be fitted into the first 2 floors of this tower. We also review what is known in higher genus.

1 Introduction

Conformal field theory (CFT) is an exceptionally symmetric quantum field theory. It takes classical mathematical structures, and 'loops' or 'complexifies' them, to produce infinite-dimensional structures such as affine algebras or vertex operator algebras. Its importance to math – which is considerable – is that the resulting structures tend to straddle several areas, such as geometry, algebra, number theory, functional analysis, ...

One of its most beautiful but least appreciated accomplishments is the reorganisation of several recent mathematical developments, such as Monstrous Moonshine, Jones' knot invariant, the modularity of affine Kac-Moody characters, and braid group monodromy of the KZ equation, into the first two floors of an infinite tower. This paper describes the resulting picture. For more details see the book [18] and references therein.

2 Mapping class groups

Up to homeomorphism, a (connected oriented real) surface is completely characterised by its genus g and number n of boundary components (punctures). In fact, there's only one way (up to equivalence) to give it a real-differential structure. But a real surface can also be a complex curve – it can usually be given a complex-differential (equivalently, a conformal) structure in infinitely many different ways.

For example, the torus $\mathbb{R}^2/\mathbb{Z}^2$ can be given a complex structure by replacing $\mathbb{R}^2$ with $\mathbb{C}$ and $\mathbb{Z}^2$ by $\mathbb{Z}+\tau\mathbb{Z}$ for any $\tau\in\mathbb{C}$ with nonzero imaginary part. In fact,

2000 *Mathematics Subject Classification.* Primary 81R10, 11F11, 57B60; Secondary 57M27, 81T40, 32G15.

any torus is conformally equivalent to one of the form $T_\tau := \mathbb{C}/(\mathbb{Z}+\tau\mathbb{Z})$, where τ lies in the upper half-plane $\mathbb{H} := \{x+\mathrm{i}y \in \mathbb{C} \,|\, y>0\}$. Moreover, the tori T_τ and $T_{\tau'}$ are themselves conformally equivalent iff $\tau' = \frac{a\tau+b}{c\tau+d}$ for a matrix $\left(\begin{smallmatrix} a & b \\ c & d \end{smallmatrix}\right) \in \mathrm{SL}_2(\mathbb{Z})$.

The set of possible complex structures on the torus forms the *moduli space* $\mathfrak{M}_{1,0}$, so labelled because the torus has genus 1 and 0 punctures. This moduli space can be identified with the orbifold $\mathbb{H}/\mathrm{SL}_2(\mathbb{Z})$: we call $\mathrm{SL}_2(\mathbb{Z}) =: \Gamma_{1,0}$ its *mapping class group*, and $\mathbb{H} =: \mathfrak{T}_{1,0}$ its *Teichmüller space*. Similarly, $\Gamma_{g,n}$, $\mathfrak{T}_{g,n}$, and $\mathfrak{M}_{g,n} = \mathfrak{T}_{g,n}/\Gamma_{g,n}$ can be defined for any other $g,n \geq 0$. In particular, the Teichmüller space $\mathfrak{T}_{g,n}$ (a simply connected complex manifold) accounts for 'continuous' conformal equivalences, while the mapping class group $\Gamma_{g,n}$ (almost always an infinite discrete group) contains the left-over 'discontinuous' ones. The moduli spaces usually have conical singularities, corresponding to surfaces with extra symmetries; taking into account these stabilisers, $\Gamma_{g,n}$ will be the (orbifold) fundamental group of $\mathfrak{M}_{g,n}$.

For example, $\mathfrak{M}_{1,0}$ is a sphere with a puncture (corresponding to the cusp $\mathbb{Q}\cup \mathrm{i}\infty$), and conical singularities at $\tau = \mathrm{i}$ and $e^{2\pi\mathrm{i}/3}$. Because T_τ can also be interpreted as a torus with a special point, namely the additive identity 0, we also have $\mathfrak{T}_{1,1} = \mathbb{H}$ and $\Gamma_{1,1} = \mathrm{SL}_2(\mathbb{Z})$.

The surfaces relevant to our story possess additional structure. Let Σ be a compact genus-g surface with n marked points $p_i \in \Sigma$. About each point p_i choose a local coordinate z_i, vanishing at p_i – this identifies a neighbourhood of p_i with a neighbourhood of $0 \in \mathbb{C}$. We call $(\Sigma, \{p_i\}, \{z_i\})$ an *enhanced surface* of type (g,n). The resulting moduli space $\widehat{\mathfrak{M}}_{g,n}$ is infinite-dimensional, but its mapping class group $\widehat{\Gamma}_{g,n}$ is an extension of $\Gamma_{g,n}$ by $\mathbb{Z}^n$. For example, $\widehat{\Gamma}_{1,1}$ is the braid group $\mathcal{B}_3$.

As we will see below, a rational conformal field theory gives finite-dimensional representations of each $\widehat{\Gamma}_{g,n}$ – merely projective for $n=0$, but it seems a true one for $n \geq 1$ (though a proof of trueness is to my knowledge only available for $g \leq 1$). Enhanced surfaces are important because they have canonical sewings. Nevertheless it is common to restrict instead to the projective representations of $\Gamma_{g,n}$, and pay at most lip service to the coordinates z_i.

3 Conformal field theory

This section introduces the correlation functions and chiral blocks of conformal field theory.

A conformal field theory (CFT) is a quantum field theory, usually on a two-dimensional space-time Σ, whose symmetries include the conformal transformations (so conformally equivalent space-times are identified). We restrict to compact orientable Σ. The same CFT lives simultaneously on all such Σ. See e.g. [9,16,19,27], and Chapter 4 of [18] for reviews.

Two dimensions are special for CFT because the local conformal maps, which form the Lie algebra $\mathfrak{so}_{n+1,1}(\mathbb{R})$ in $\mathbb{R}^n$ for $n>2$, becomes infinite-dimensional in $\mathbb{R}^2$ (thanks to their identification with (anti-)holomorphic maps). The conformal algebra in two dimensions consists of two commuting copies of the Witt algebra $\mathfrak{Witt}$ (one for the holomorphic maps, and the other for anti-holomorphic ones). $\mathfrak{Witt}$ is the infinite-dimensional Lie algebra of vector fields on S^1, and has a basis ℓ_n, $n \in \mathbb{Z}$, satisfying

$$[\ell_m, \ell_n] = (m-n)\ell_{m+n} \,. \tag{3.1}$$

Its unique nontrivial central extension is the Virasoro algebra $\mathfrak{Vir}$, with basis L_n, C satisfying $[L_n, C] = 0$ and

$$[L_m, L_n] = (m-n)L_{m+n} + \delta_{n,-m}\frac{m\,(m^2-1)}{12}\,C\,. \tag{3.2}$$

Basic data in the CFT are the quantum fields $\varphi(z)$, called *vertex operators*, living on space-time $\Sigma = S^2 = \mathbb{C}\cup\{\infty\}$ and centred at $z = 0$. Being quantum fields, these φ are 'operator-valued distributions' on Σ, acting on the space $\mathcal{H}$ of states for Σ. The most important vertex operators are the *stress-energy tensors* $T(z), \overline{T}(z)$, which are the conserved currents of the conformal symmetry, as promised by Noether's Theorem; the corresponding conserved charges are operators $L_n, \overline{L}_m$ defining a $\mathfrak{Vir}$-representation, with central term C given by scalars $cI, \overline{c}I$ called the *central charges*.

In a typical quantum field theory, a theoretical physicist makes contact with experiment by computing transition amplitudes $\langle \text{out}|\text{in}\rangle$ between incoming and outgoing states, given mathematically by a Hermitian product $(|\text{out}\rangle, S|\text{in}\rangle)$ in the Hilbert space $\mathcal{H}$ of states, for some operator S called the scattering matrix. In practise these can only be calculated in infinite time ($t \to \pm\infty$) limits. The typical way ('LSZ reduction formulae') to express these asymptotic amplitudes is via artifacts sometimes called *correlation functions*. The theory is regarded as solved if all correlation functions can be computed. We are interested in the correlation functions

$$\langle \varphi_1(z_1)\,\varphi_2(z_2)\cdots\varphi_n(z_n)\rangle_{\Sigma;p_1,\dots,p_n} \tag{3.3}$$

of CFT, for any vertex operators φ_i and any enhanced surface $(\Sigma, \{p_i\}, z_i)$ (so $\varphi_i(z_i)$ is 'centred' at $p_i \in \Sigma$). The remainder of this section explains how physicists think of these CFT correlation functions. Their intuition is provided by string theory.

In a typical quantum theory, correlation functions are calculated perturbatively by Taylor-expanding in some coupling constant. For this purpose, Feynman's path integral formulation – the quantisation of Hamilton's action principle in classical mechanics – is convenient. Each term in this perturbation series is computed separately using Feynman diagrams and rules. The Feynman diagrams of quantum field theory are graphs, with a different kind of edge for each species of particle, and a different kind of vertex for every term in the interaction part of the Lagrangian; Feynman's rules describe how to go from these diagrams to certain integral expressions and hence to the individual terms in the Taylor series expansion of the given correlation function. Feynman diagrams are combinatorial artifacts describing 'virtual' (non-real) processes; topologically equivalent ones are identified, and in practise only the simplest are ever considered.

Applying this perturbation formalism to string theory recovers CFT. Consider for convenience closed strings. Then CFT lives on the world-sheet Σ (string theory's Feynman diagrams) traced by the strings as they virtually evolve, colliding and separating, through time: string amplitudes (in e.g. 26-dimensional space-time) can be expressed as correlation functions of a (point-particle) CFT in two dimensions. The boundaries of these world-sheets are the incoming and outgoing strings; the world-sheets for asymptotic amplitudes have semi-infinite end-tubes and can be conformally mapped to compact surfaces with punctures p_i (one for every external string). The corresponding Feynman integral is over moduli space $\widehat{\mathfrak{M}}_{g.n}$. The data of those external strings are stored in the appropriate vertex operator attached to that point p_i. The Witt algebra arises here as infinitesimal reparametrisations of the string (a circle).

Everything in CFT comes in a combination of strictly holomorphic, and strictly anti-holomorphic, quantities. Here, 'holomorphic' is in terms of space-time Σ (which locally looks like $\mathbb{C}$), or the appropriate moduli space (which usually locally looks like $\mathbb{C}^\infty$). These holomorphic and anti-holomorphic building blocks are called *chiral.* A CFT is studied by first analysing its chiral parts, and then determining explicitly how they piece together to form the physical(=bi-chiral) quantities. Almost all attention by mathematicians has focused on the chiral (as opposed to physical) data. In string theory, this holomorphic/anti-holomorphic alternative corresponds to classical ripples travelling clockwise/anti-clockwise around the string.

A typical vertex operator $\varphi(z)$ depends neither holomorphically nor anti-holomorphically on z. Let $\mathcal{V}$ consist of all the holomorphic, and $\overline{\mathcal{V}}$ the anti-holomorphic, ones. For example, the stress-energy tensor $T(z)$ and all of its derivatives lie in $\mathcal{V}$. In the very simplest CFTs, called the minimal models, $\mathcal{V}$ consists only of $T(z)$ and its derivatives.

These *chiral algebras* $\mathcal{V}, \overline{\mathcal{V}}$ have a rich mathematical structure, with a 'multiplication' coming from the so-called operator product expansion, and are examples of *vertex operator algebras* (see [28], or Chapter 5 of [18]). $\mathcal{V}$ and $\overline{\mathcal{V}}$ mutually commute and the full symmetry 'algebra' of the CFT can be identified with $\mathcal{V} \oplus \overline{\mathcal{V}}$. Since quantum fields act on state-space $\mathcal{H}$, it carries a representation of $\mathcal{V} \oplus \overline{\mathcal{V}}$ and decomposes into a direct integral of irreducible $\mathcal{V} \oplus \overline{\mathcal{V}}$-modules. A *rational conformal field theory* (RCFT) is one whose state-space $\mathcal{H}$ decomposes in fact into a *finite direct sum*

$$\mathcal{H} = \oplus_{M\in\Phi,\overline{N}\in\overline{\Phi}}\, \mathcal{Z}_{M,\overline{N}}\, M \otimes \overline{N}\ , \tag{3.4}$$

where Φ and $\overline{\Phi}$ denote the (finite) sets of irreducible $\mathcal{V}$- and $\overline{\mathcal{V}}$-modules, and the $\mathcal{Z}_{M,\overline{N}} \geq 0$ are multiplicities. The RCFT are especially symmetric and well-defined quantum field theories and are the CFTs we're interested in. The name 'rational' arises because their central charges $c, \overline{c}$ lie in $\mathbb{Q}$.

The correlation functions (3.3) can be expressed in terms of purely chiral quantities called *conformal* or *chiral blocks*, denoted

$$\langle \mathcal{I}_1(v_1, z_1)\, \mathcal{I}_2(v_2, z_2) \cdots \mathcal{I}_n(v_n, z_n)\rangle_{(\Sigma;p_1,\dots,p_n;M^1,\dots,M^n)}\ . \tag{3.5}$$

As usual, $(\Sigma, \{p_i\}, \{z_i\})$ is an enhanced surface, and to each p_i we assign a $\mathcal{V}$-module $M^i \in \Phi$ and a state $v_i \in M^i$. The holomorphic field $\mathcal{I}_i(v_i, z_i)$ centred at p_i is an operator-valued distribution called an *intertwiner* sending ('intertwining') one $\mathcal{V}$-module (say $M \in \Phi$) with another (say $N \in \Phi$). For fixed M, N, M^i, the dimension of the space of intertwiners is called the *fusion coefficient* $\mathcal{N}_{M^i,M}^N$, and is given by Verlinde's formula

$$\mathcal{N}_{M^i,M}^N = \sum_{P\in\Phi} \frac{S_{M^iP}\, S_{MP}\, \overline{S_{NP}}}{S_{\mathcal{V}P}}\ , \tag{3.6}$$

where the matrix S (no relation to the scattering matrix) is defined in (5.1) below. For example, $\mathcal{V}$ is always a module for itself and $\mathcal{N}_{\mathcal{V},\mathcal{V}}^{\mathcal{V}} = 1$; the unique (up to scaling) intertwiner $\mathcal{I}(v, z)$ bijectively associates states v with vertex operators $\varphi(z) = \mathcal{I}(v, z)$ (the so-called 'state-field correspondence'). Thus intertwiners generalise vertex operators $\varphi \in \mathcal{V}$.

To solve a given RCFT, the strategy then is to:

(a) construct all possible chiral blocks (3.5); and
(b) construct the correlation functions (3.3) from those chiral blocks.

4 The chiral blocks of RCFT

To a fixed $(g,n;M^1,\ldots,M^n)$, an RCFT assigns a finite-dimensional space $\mathfrak{F}^{(g,n)}_{(M^i)}$ of chiral blocks (3.5). Chiral blocks are important for RCFT because finite combinations of them are the correlation functions, and knowing the latter is equivalent to solving the theory.

Each chiral block depends multilinearly on the states $v_i\in M^i$, and holomorphically on the z_i, provided branchcuts in Σ between p_i are made; locally, it can be regarded as a holomorphic function on $\widehat{\mathfrak{M}}_{g,n}$. The dimension of this space is given by Verlinde's formula

$$\dim\mathfrak{F}^{(g,n)}_{(M^i)}=\sum_{P\in\Phi}\frac{S_{M^1P}}{S_{\mathcal{V}P}}\cdots\frac{S_{M^nP}}{S_{\mathcal{V}P}}S_{\mathcal{V}P}^{2(1-g)}\ ,\tag{4.1}$$

a generalisation of (3.6).

Moore and Seiberg [30] – see also [2] – isolated the data (linear transformations between finite-dimensional vector spaces) defining each chiral half of RCFT, and provided a complete set of relations they satisfy. Huang is pursuing the explicit construction for all sufficiently nice chiral algebras $\mathcal{V}$ (see e.g. [21]).

A basis for $\mathfrak{F}^{(g,n)}_{(M^i)}$ is found by performing the following Feynman rules (called 'conformal bootstrap'). Fix a surface Σ of type (g,n). The space $\mathfrak{F}^{(0,3)}_{(M,N,P)}$ consists of intertwiners; arbitrarily fix bases for all those spaces. Now, dissect Σ into pairs-of-pants, as in Figure 1; assign a dummy label $N_j\in\Phi$ to each internal cut in the dissection; for each vertex in your dissection, choose an intertwining operator from the basis of the appropriate space $\mathfrak{F}^{(0,3)}$; 'evaluate' the corresponding chiral block – e.g. for each cut, a trace is taken of the product of intertwiners. Repeating, by running through all possible values of the dummy labels, the result is a basis of chiral blocks.

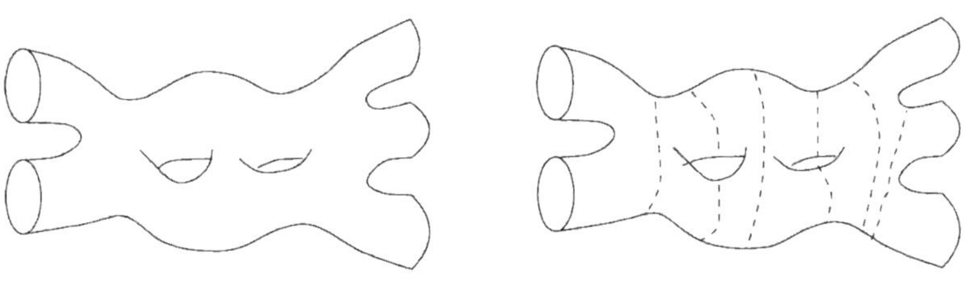

Figure 1 Dissecting a surface into pairs-of-pants

For an important example, let Σ be the torus $\mathbb{C}/(\mathbb{Z}+\tau\mathbb{Z})$ with one puncture p (say at 0), assigned the module $M^1=\mathcal{V}$ and state $v_1=|0\rangle$ (the vacuum, the state of lowest energy). One cut suffices to unfold it into a sphere with 3 punctures, assigned $\mathcal{V}$-modules $\mathcal{V},M,M\in\Phi$ (M is the dummy label). The fusion coefficient $\mathcal{N}_{\mathcal{V},M}^{M}$ always equals 1, and so for each $M\in\Phi$ there is a unique intertwiner, say $\mathcal{I}_1^{(M)}$. Hence $\dim\mathfrak{F}_{\mathcal{V}}^{(1,1)}=\|\Phi\|$. These Feynman rules yield the chiral block

$$\chi_M(\tau):=\mathrm{tr}_M e^{2\pi i\tau\,(L_0-c/24)}\ ,\tag{4.2}$$

where c is the central charge and the Virasoro element L_0 corresponds to energy (the trace comes from the dissection). These span $\mathfrak{F}^{(1,0)}$.

One of the simplest RCFT is the Ising model, a minimal model. It has central charge $c = \bar{c} = \frac{1}{2}$, and its chiral algebra has 3 irreducible modules, which we'll label $\Phi = \overline{\Phi} = \{\mathcal{V}, \epsilon, \sigma\}$. Its toroidal chiral blocks (4.2) are

$$\begin{aligned}\chi_{\mathcal{V}}(\tau) &= q^{-1/48}\,(1 + q^2 + q^3 + 2q^4 + 2q^5 + 3q^6 + 3q^7 + \cdots)\,, \\ \chi_{\epsilon}(\tau) &= q^{23/48}\,(1 + q + q^2 + q^3 + 2q^4 + 2q^5 + 3q^6 + 3q^7 + \cdots)\,, \\ \chi_{\sigma}(\tau) &= q^{1/24}\,(1 + q + q^2 + 2q^3 + 2q^4 + 3q^5 + 4q^6 + 5q^7 + \cdots)\,, \end{aligned} \tag{4.3}$$

where $q = e^{2\pi i\tau}$.

Other important RCFT are the *Wess–Zumino–Witten* (WZW) models. These correspond to strings living on a compact Lie group G. The chiral algebra $\mathcal{V}$ is closely related to the affine Kac–Moody algebra $\mathfrak{g}^{(1)}$ (see [23]), where $\mathfrak{g}$ is the Lie algebra of G ($\mathfrak{g}^{(1)}$ is the nontrivial central extension of the loop algebra $\mathfrak{g} \otimes \mathbb{C}[z^{\pm 1}]$). Its modules $M \in \Phi$ can be identified with the integrable highest-weight modules $L(\lambda)$ at a level k determined by the central charge c. The chiral blocks $\chi_M(\tau)$ for the WZW models are a specialisation of the corresponding affine algebra $\mathfrak{g}^{(1)}$-character $\chi_\lambda(h)$, and for this reason $\chi_M(\tau)$ in any RCFT is called the character of the $\mathcal{V}$-module M. We'll return to the WZW and Ising models shortly.

Each dissection produces a basis for the space $\mathfrak{F}^{(g,n)}_{(M^i)}$. However, any Σ can be dissected in different ways. The over-used term *duality* refers here to the invertible matrices relating the chiral blocks of different dissections. For example, the left dissection in Figure 2 of the $(g,n) = (0,4)$ surface corresponds to a matrix $F\left[\begin{smallmatrix} M & N \\ L & P\end{smallmatrix}\right]$ of size $n \times n$ for $n = \dim \mathfrak{F}^{(0,4)}_{(L,M,N,P)}$ (for an appropriate orientation of surfaces and punctures – a minor technicality we've been ignoring), called the fusing matrix. Likewise, the right dissection defines the braiding matrix B.

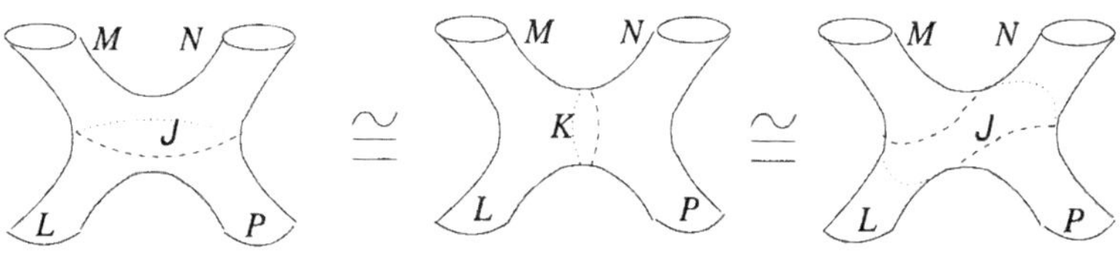

Figure 2 The fusing and braiding matrices F, B

All duality transformations are built up from a few elementary ones, like B and F. By decomposing surfaces in different ways, we get relations between these elementary dualities. For example, the B-matrices obey an equation of the form (symbolically) $BBB = BBB$ called the Yang–Baxter equation, and this is the source of its name 'braiding'.

Consider four marked points w_i on the sphere $\mathbb{C} \cup \{\infty\}$. Using the Möbius (=conformal) symmetry of the sphere, move w_i to $0, w, 1, \infty$, respectively, where w is the cross-ratio $\frac{(w_1-w_2)(w_3-w_4)}{(w_1-w_3)(w_2-w_4)}$. If we label all four marked points with the Ising module $\sigma \in \Phi$, then the space $\mathfrak{F}^{(0,4)}_{(\sigma\sigma\sigma\sigma)}$ of chiral blocks is two-dimensional; choosing the smallest energy state in σ, the chiral blocks are spanned by

$$\mathcal{F}_1(w) = \frac{\sqrt{1+\sqrt{1-w}}}{(w(1-w))^{1/8}},$$
$$\mathcal{F}_2(w) = \frac{\sqrt{1-\sqrt{1-w}}}{(w(1-w))^{1/8}}. \tag{4.4}$$

Fusing interchanges $w_1 = 0$ and $w_3 = 1$, hence involves the Möbius transformation $w \mapsto (1-w)/(1-0) = 1-w$. Likewise, braiding interchanges w_2 with w_3, and so involves $w \mapsto (0-1)/(0-w) = 1/w$. The braiding and fusing matrices here become

$$B\begin{bmatrix} \sigma & \sigma \\ \sigma & \sigma \end{bmatrix} = \frac{e^{-\pi i/8}}{\sqrt{2}}\begin{pmatrix} 1 & i \\ i & 1 \end{pmatrix},$$
$$F\begin{bmatrix} \sigma & \sigma \\ \sigma & \sigma \end{bmatrix} = \frac{1}{\sqrt{2}}\begin{pmatrix} 1 & 1 \\ 1 & -1 \end{pmatrix}. \tag{4.5}$$

5 Monodromy in RCFT

The fractional powers in the Ising blocks (4.4) tell us they have branch-point singularities – we must make cuts in the w-plane to get holomorphic functions there. If instead we analytically continue these functions along a closed curve, the value of the block need not return to the same value. For example, consider the circle $w(t) = r\,e^{2\pi i t}$ for r small: the value of $\mathcal{F}_i(w)$ at $t = 1$ is $e^{-\pi i/4}$ times its value at $t = 0$. This factor $e^{-\pi i/4}I$ is the monodromy about $w = 0$. Likewise, their monodromy about $w = 1$ is

$$\begin{pmatrix} \mathcal{F}_1(w) \\ \mathcal{F}_2(w) \end{pmatrix} \mapsto \begin{pmatrix} 0 & e^{-2\pi i/8} \\ e^{-2\pi i/8} & 0 \end{pmatrix}\begin{pmatrix} \mathcal{F}_1(w) \\ \mathcal{F}_2(w) \end{pmatrix}.$$

Reintroducing the four coordinates w_i, the chiral blocks $\mathcal{F}_i$ will be holomorphic on the universal cover of the configuration space $\mathfrak{C}_4(S^2) = \{(w_1, w_2, w_3, w_4) \in (S^2)^4 \,|\, w_i \neq w_j\}$ of the Riemann sphere. Analytically continuing along any closed path γ in $\mathfrak{C}_4(S^2)$ defines an action of the fundamental group $\pi_1(\mathfrak{C}_4(S^2))$ – the pure braid group of the sphere with four strands – on the space $\mathfrak{F}^{(0,4)}_{(\sigma\sigma\sigma\sigma)}$ of chiral blocks. For example, the monodromy about $w = 1$ found above corresponds to the pure braid σ^2_{1w}, where σ_{1w} is the twist of the 1-strand with the w-strand (each strand corresponds to one of the points $0, 1, w, \infty$). Actually, the full spherical braid group $\mathfrak{B}_4(S^2)$ acts: $\beta \in \mathfrak{B}_4(S^2)$ maps the space $\mathfrak{F}^{(0,4)}_{(M^1,M^2,M^3,M^4)}$ to $\mathfrak{F}^{(0,4)}_{(M^{\beta 1},M^{\beta 2},M^{\beta 3},M^{\beta 4})}$, where βi is the associated permutation. For example, the twist $\sigma_{1w} \in \mathfrak{B}_4(S^2)$ is the braiding matrix $B\left[\begin{smallmatrix} \sigma & \sigma \\ \sigma & \sigma \end{smallmatrix}\right]$.

Equivalently, as a 'function' on the configuration space, the chiral blocks form holomorphic horizontal sections of a projectively flat vector bundle. What this means is that each chiral block satisfies a system of partial differential equations (the *Knizhnik–Zamolodchikov* or KZ equations) describing how to parallel-transport it around configuration space, and flatness says it *locally* depends only on the moduli space parameters (and not on the path chosen). *Globally*, however, there will be monodromy.

More generally, a chiral block $\mathcal{F}$ for an enhanced surface Σ is a multi-valued function on the appropriate moduli space. To make it well-defined, $\mathcal{F}$ can be lifted to the corresponding Teichmüller space. There will be an action of the corresponding mapping class group $\widehat{\Gamma}_{g,n}$, coming from monodromy. In other words, the space $\mathfrak{F}^{(g,n)}_{(M^i)}$ of chiral blocks carries a representation $\rho^{(g,n)}_{(M^i)}$ of $\widehat{\Gamma}_{g,n}$. This $\widehat{\Gamma}_{g,n}$-representation is built up from the duality matrices, such as the braiding and fusing matrices. As we shall see below, this picture unifies the Jones knot polynomial, the modularity of Monstrous Moonshine, and many other phenomena.

For example, we can dissect the torus using a single vertical cross-sectional cut, or using a horizontal equatorial cut; one basis is $\chi_M(\tau)$ and the other is $\chi_M(-1/\tau)$. Duality says that they both span the same space:

$$\chi_M(-1/\tau) = \sum_{N\in\Phi} S_{MN}\,\chi_N(\tau)\,. \tag{5.1}$$

Likewise, performing a Dehn twist about the vertical cut, we obtain

$$\chi_M(\tau+1) = \sum_{N\in\Phi} T_{MN}\,\chi_N(\tau)\,. \tag{5.2}$$

Here, S,T are complex matrices. Together, $\tau\mapsto -1/\tau$ and $\tau\mapsto\tau+1$ generate the modular group $\mathrm{PSL}_2(\mathbb{Z})$, and S,T generate a true representation $\rho^{(1,0)}$ of the central extension $\mathrm{SL}_2(\mathbb{Z}) = \Gamma_{1,0}$. Hence RCFT characters χ_M form a weight-0 vector-valued modular form for $\mathrm{SL}_2(\mathbb{Z})$, with multiplier $\rho^{(1,0)}$.

For example, the matrix T for WZW models involves the quadratic Casimir of G, while the matrix S involves characters of G evaluated at elements of finite order. For the Ising model, these matrices are

$$S = \frac{1}{2}\begin{pmatrix} 1 & 1 & \sqrt{2} \\ 1 & 1 & -\sqrt{2} \\ \sqrt{2} & -\sqrt{2} & 0 \end{pmatrix}, \quad T = \begin{pmatrix} e^{-\pi i/24} & 0 & 0 \\ 0 & -e^{-\pi i/24} & 0 \\ 0 & 0 & e^{\pi i/12} \end{pmatrix}. \tag{5.3}$$

Perhaps the most elegant treatment of the finite-dimensional representations of a compact Lie group G is Borel–Weil theory, which constructs them via the G-action on line bundles over the flag manifold $G_{\mathbb{C}}/B$. Something similar happens to the Virasoro algebra $\mathfrak{Vir}$, with now the moduli spaces of curves playing the role of the flag manifold and mapping class groups taking the place of the Weyl group. A copy of $\mathfrak{Witt} = \mathrm{Vect}(S^1)$ attached to the ith puncture on an enhanced surface of type (g,n) acts naturally on the moduli space $\widehat{\mathfrak{M}}_{g,n}$: the vector field $z_i^\ell\partial/\partial z_i$ for $\ell\geq 1$ changes the local coordinate z_i; $\partial/\partial z_i$ moves the puncture; and $z_i^\ell\partial/\partial z_i$ for $\ell\leq -1$ can change the conformal structure of the surface. This infinitesimal action fills out the tangent space to any point on $\widehat{\mathfrak{M}}_{g,n}$. In this picture, the central extension of $\mathfrak{Witt}$ to $\mathfrak{Vir}$ arises geometrically as a curvature effect. The KZ equations say roughly that the desired sections of $\widehat{\mathfrak{M}}_{g,n}$-vector bundles should respect this $\mathfrak{Vir}$ action.

6 Correlation functions

Our primary interest is the $\Gamma_{g,n}$-action $\rho^{(g,n)}_{(M^i)}$ on the spaces $\mathfrak{F}^{(g,n)}_{(M^i)}$. Correlation functions – the quantities of physical interest – are sesquilinear combinations of chiral blocks which have trivial monodromy. In a typical quantum field theory the correlation functions are computed perturbatively, but in RCFT they can be found exactly.

For example, the toroidal correlation function is

$$\mathcal{Z}(\tau) := \sum_{M\in\Phi,\overline{N}\in\overline{\Phi}} \mathcal{Z}_{M,\overline{N}}\,\chi_M(\tau)\,\chi_{\overline{N}}(\overline{\tau}) \tag{6.1}$$

(recall (3.4)). It is required to be invariant under the $\mathrm{SL}_2(\mathbb{Z})$ action (5.1) and (5.2) on the chiral blocks $\chi_M(\tau)$ of the torus. For the special case of the Ising model (recall (5.2)), the unique solution to the various constraints is

$$\mathcal{Z}(\tau) = \chi_{\mathcal{V}}(\tau)\,\chi_{\mathcal{V}}(\overline{\tau}) + \chi_\epsilon(\tau)\,\chi_\epsilon(\overline{\tau}) + \chi_\sigma(\tau)\,\chi_\sigma(\overline{\tau})\ . \tag{6.2}$$

Its $\mathrm{SL}_2(\mathbb{Z})$-invariance follows from the unitarity of the matrices (5.3). The classification of possible toroidal correlation functions for WZW models involves quite interesting Lie theory and number theory – see e.g. [17] for a short proof of Cappelli–Itzykson–Zuber's A-D-E classification of SU_2.

The most elegant and general construction of correlation functions from chiral blocks uses topological field theories and the language of category theory [14].

7 Genus 0: braids and knots

In the 1980s, Jones studied the combinatorial characterisation of embedding one von Neumann algebra (a factor) in another, and as an unexpected by-product obtained new representations of braid groups $\mathcal{B}_n$. We can obtain a knotted link from a braid by gluing the n top endpoints of the braid to the corresponding bottom ones. It is possible to characterise completely (using the 'Markov moves') the different braids which yield the same knot, and remarkably Jones' $\mathcal{B}_n$-representations respect this redundancy, in the sense that Jones could obtain from his representations (using the trace in the underlying von Neumann algebra), a polynomial knot invariant [22].

Witten [32] reinterpreted Jones' braid group representations as due to the (projective) representation of the genus-0 mapping class groups $\Gamma_{0,n}$, coming from the SU_2 WZW model (as we shall see, $\Gamma_{0,n}$ is essentially a braid group). Witten showed how the $\Gamma_{0,n}$-representation of any other RCFT similarly gives rise to other knot invariants, thus generalising Jones' invariant considerably by embedding it naturally in a much broader context. von Neumann algebras arise in quantum field theory (hence RCFT) through the assignment to each region of space-time of the observables measurable in that region; when one region is a subset of another, then its algebra of observables is embedded in the other.

$\Gamma_{0,n}$ has presentation

$$\begin{aligned}\Gamma_{0,n} \;&=\; \langle \sigma_1,\ldots,\sigma_{n-1}\,|\sigma_i\sigma_j = \sigma_j\sigma_i\ (|i-j|>1),\\ &\quad\ \sigma_i\sigma_{i+1}\sigma_i = \sigma_{i+1}\sigma_i\sigma_{i+1},\\ &\quad\ \sigma_1\cdots\sigma_{n-1}\sigma_{n-1}\cdots\sigma_1 = 1 = (\sigma_1\cdots\sigma_{n-1})^n\rangle\\ &\cong \mathcal{B}_n(S^2)/\mathbb{Z}_2,\end{aligned} \tag{7.1}$$

the quotient of the spherical braid group by its centre. RCFT obtains the $\Gamma_{0,n}$-representation $\rho^{(0,n)}_{(M^i)}$ by assigning each generator σ_i to a matrix in block-diagonal form, whose blocks are braiding matrices. We get a different representation, of dimension given by (4.1), for every choice $M^1,\ldots,M^n$ of $\mathcal{V}$-modules. An element $\beta\in\Gamma_{0,n}$ sends the space $\mathfrak{F}^{(0,n)}_{(M^i)}$ to $\mathfrak{F}^{(0,n)}_{(M^{\beta i})}$, so to get a representation of the full $\Gamma_{0,n}$ we should sum $\mathfrak{F}^{(0,n)}_{(M^i)}$ over all reorderings of (M^i). It is common to lift this $\Gamma_{0,n}$-action to the braid group $\mathcal{B}_n$ in the manner clear from (7.1). In all known examples, it seems, these representations are always defined over some cyclotomic field.

The groups $\Gamma_{0,0}\cong\Gamma_{0,1}\cong 1$, $\Gamma_{0,2}\cong\mathbb{Z}_2$, $\Gamma_{0,3}\cong S_3$ are all finite and so aren't very interesting. This is because the Möbius symmetry on S^2 is triply transitive.

However, the image of $\Gamma_{0,n}$ for $n \geq 4$ will usually be infinite – e.g. in the special case of two dimensions, the question of $\Gamma_{0,4}$ having finite image reduces to Schwarz' classical analysis of the finite monodromy of the hypergeometric equation, and as such is very rare.

$\Gamma_{0,4}$ is an extension of $\mathrm{PSL}_2(\mathbb{Z})$ by $\mathbb{Z}_2 \times \mathbb{Z}_2$, and the part of $\Gamma_{0,4}$ corresponding to a trivial permutation $\beta i = i$ is isomorphic to the principal congruence subgroup $\Gamma(2)/\{\pm 1\}$. Using the $\Gamma(2)$-Hauptmodul $\theta_2(\tau)^4/\theta_3(\tau)^4$, we can lift the chiral blocks $\mathcal{F}(w)$ to the upper half-plane $\mathbb{H}$, and in this way interpret these chiral blocks as vector-valued modular forms for $\mathrm{SL}_2(\mathbb{Z})$. For example, the Ising blocks (4.4) would now become

$$\begin{aligned} \mathcal{F}_1(\tau) &= q^{-\frac{1}{16}}(1 + q^{\frac{1}{2}} + 3q + 4q^{\frac{3}{2}} + 5q^2 + 8q^{\frac{5}{2}} + 11q^3 + \cdots)\,, \\ \mathcal{F}_2(\tau) &= q^{\frac{3}{16}}(2 + 2q^{\frac{1}{2}} + 2q + 4q^{\frac{3}{2}} + 8q^2 + 10q^{\frac{5}{2}} + 12q^3 + \cdots)\,. \end{aligned} \tag{7.2}$$

These lifts $\mathcal{F}(\tau)$ will always be holomorphic in $\mathbb{H}$, but can have poles at the cusps. The weight can be rational because the $\Gamma(2)$-representation will typically be projective; in fact arbitrary rational weight is possible. The theory of these vector-valued modular forms of arbitrary rational weight for $\mathrm{PSL}_2(\mathbb{Z})$, and no restriction on the kernel of the multiplier, has been developed recently [4], [26] and is quite rich. It is interesting that RCFT produces plenty of examples of these.

The $\Gamma_{0,n}$-action coming from WZW models is especially interesting. Consider SU_2 for concreteness. Choose n distinct points $z_1, \ldots, z_n \in \mathbb{C}$ and n $\mathfrak{sl}_2(\mathbb{C})$-modules V^i; write $\widehat{V}^i$ for the corresponding $\mathfrak{sl}_2^{(1)}$-modules. Then the conformal blocks $\mathcal{F} \in \mathfrak{F}^{(0,n)}_{(\hat{V}^i)}$ are precisely the functions $\mathcal{F} : \mathfrak{C}_n(S^2) \to V_1 \otimes \cdots \otimes V_n$ satisfying the *KZ equations* [27]

$$\partial \frac{\mathcal{F}}{\partial z_i} = \frac{1}{k+2} \sum_{j \neq i} \frac{\Omega_{ij}}{z_i - z_j} \mathcal{F}\,, \tag{7.3}$$

where $\Omega_{ij}/(z_i - z_j)$ is the classical Yang–Baxter r-matrix for SU_2. As mentioned earlier, any solution to (7.3) can be parallel transported through $\mathfrak{C}_n(S^2)$; projective flatness means that this parallel transport along a closed loop depends (up to a projective factor) only on the homotopy class of the loop. In other words, the space $\mathfrak{F}^{(0,n)}_{(\hat{V}^i)}$ of solutions to (7.3) carries a projective reprsentation of the pure spherical braid group $\pi_1(\mathfrak{C}_n(S^2))$. The Drinfel'd–Kohno monodromy theorem expresses this monodromy in terms of the $6j$-symbols of the quantum group $U_q(\mathfrak{sl}_2(\mathbb{C}))$, for $q = e^{\pi i/(k+2)}$, which are straightforward to compute [25]. Something similar happens for any G.

The infinitely many irreducible finite-dimensional modules of a simple Lie algebra $\mathfrak{g}$ naturally span a symmetric monoidal category (see [31] for definitions); its representation ring is isomorphic to a polynomial ring in r variables, where r is the rank of the algebra. On the other hand, the finitely many level k irreducible integrable modules of the affine algebra $\mathfrak{g}^{(1)}$ span (among other things) a braided monoidal category; the corresponding representation ring is called a *fusion ring* and has structure constants equal to the fusion coefficients (3.6). The key ingredient in this category – the braiding – comes from the braid group monodromy of (7.3). Something similar happens for any RCFT.

8 Genus 1: modularity

The characters χ_λ of the affine algebra $\mathfrak{g}^{(1)}$ are defined exactly as for semi-simple $\mathfrak{g}$, as a sum of exponentials of the Cartan subalgebra, though the sum will now be infinite. In fact a miracle happens: the character χ_λ will be a modular function for some subgroup of $\mathrm{SL}_2(\mathbb{Z})$! One of the coordinates of the Cartan subalgebra of $\mathfrak{g}^{(1)}$ plays the role of $\tau \in \mathbb{H}$, and the others come along for the ride. The algebraic proof of this modularity makes it look accidental: the character χ_λ is expressed as a fraction; the denominator is automatically a modular form for $\mathrm{SL}_2(\mathbb{Z})$, by the simple combinatorics of affine algebras; the numerator is a modular form (in fact a lattice theta function) for some congruence subgroup of $\mathrm{SL}_2(\mathbb{Z})$, because the Weyl group of $\mathfrak{g}^{(1)}$ contains translations in a lattice; their quotient yields a modular function.

RCFT provides a much more satisfying explanation for this unexpected modularity. The characters $\chi_\lambda(\tau)$ of these affine algebra modules equal the chiral blocks (4.2) of the corresponding WZW model, and the action (5.1) and (5.2) of $\Gamma_{1,0} = \mathrm{SL}_2(\mathbb{Z})$ coming from RCFT explains their unexpected modularity. This relation with RCFT also tells us the $\mathfrak{g}^{(1)}$-modules are simultaneously $\mathfrak{Vir}$-modules. All of this was known to algebraists before the relation of affine algebras to RCFT was developed, but this relation emphasises that these properties of affine algebra modules are not accidental but naturally fit into a much broader perspective.

There is more to being a modular form or function than transforming nicely with respect to $\mathrm{SL}_2(\mathbb{Z})$. Good behaviour at the cusps of $\mathbb{H}$ is also crucial, as they compactify the domain. These cusps correspond to a pinched torus; their analogues for the other moduli spaces are surfaces with nodes (this is the Deligne–Mumford compactification). RCFT requires nice behaviour ('factorisation') of chiral blocks as we move in moduli space toward these degenerate surfaces. This connects the moduli spaces of different topologies, and tells us RCFT is naturally defined on a 'universal tower' of moduli spaces.

Interesting modularity certainly isn't restricted to affine algebras. Indeed, the Monstrous Moonshine conjectures (see e.g. [18]) relate character values of the monster finite simple group to various Hauptmoduls. For example, the first nontrivial coefficient (196884) in the j-function nearly equals the dimension 196883 of the first nontrivial representation of the monster. Conjecturally, Hauptmoduls are associated to pairs of commuting elements in the Monster – e.g. the j-function is assigned to (e, e). The starting point to our (still incomplete) understanding of these conjectures is the construction [13] of an RCFT (with central charge $c = 24$, $\|\Phi\| = 1$, and anti-holomorphic chiral algebra $\overline{\mathcal{V}} = \mathbb{C}$) whose symmetry group equals the Monster and whose single character (4.2) equals the j-function.

To some crude extent, Moonshine can then be interpreted as the conjunction of two different pictures of quantum field theory, applied to that very special RCFT: the Hamiltonian picture, which provides us a Hilbert space (state-space) carrying an action of the Monster, and an energy operator L_0 such that (4.2) is defined; and the Feynman picture, which lives in moduli space and which makes modularity manifest. As explained at the end of Section 5, the Virasoro algebra, through its action on the moduli spaces $\widehat{\mathfrak{M}}_{g,n}$, lies at the heart of Moonshine.

Can we see more directly why the RCFT characters $\chi_M(\tau)$ of (4.2) should have anything to do with modularity? The chiral blocks on the torus can be obtained from those of the plane $\mathbb{C}$, by first considering the map $z \mapsto w := e^{2\pi i z}$. Though holomorphic, it changes the global topology, sending the plane $\mathbb{C}$ to the annulus $\mathbb{C}\backslash\{0\}$, and this topology change is responsible for the $-c/24$ in (4.2). To obtain our torus, we now identify w and qw, where as always $q = e^{2\pi i\tau}$. This is equivalent to taking the finite annulus $\{w \in \mathbb{C} \mid |q| < |w| < 1\}$ and sewing together its two boundary circles appropriately. The resulting torus is conformally equivalent to $\mathbb{C}/(\mathbb{Z}+\tau\mathbb{Z})$. Applying this construction to chiral blocks, we find that those for the torus are indeed given by (4.2) (e.g. the trace comes from sewing). The proof [33] of modularity of vertex operator algebra characters follows this outline.

The $\mathrm{SL}_2(\mathbb{Z})$-representation (5.1) and (5.2) is defined over a cyclotomic field for any RCFT [8], and its kernel contains a congruence subgroup [3]. Perhaps the latter isn't so surprising, considering that $\mathfrak{F}^{(1,0)}$ has a basis (4.2) with integer q-expansions. Intimately connected with this congruence subgroup property, the matrices S, T have nice properties with respect to the cyclotomic Galois group [8,3].

$\Gamma_{1,1}$ is also $\mathrm{SL}_2(\mathbb{Z})$, and so the chiral blocks in $\mathfrak{F}_M^{(1,1)}$ also form vector-valued modular forms for $\mathrm{SL}_2(\mathbb{Z})$. The weights though can be arbitrary rational numbers and the kernel need not be of finite index. So together with the $\mathcal{F} \in \mathfrak{F}^{(0,4)}_{(M^i)}$ (as explained in the last section), RCFT is a rich source of vector-valued modular forms. These $\Gamma_{1,1}$-representations are explicitly known in terms of the duality matrices (see e.g. [30]), and so the machinery of [4] allows these chiral blocks to be explicitly found [5].

But specialising to the WZW models, we should expect a nice Lie theoretic answer, and indeed the complete answer is known for SU_2 [24,11]. The matrix representing $\tau \mapsto \tau+1$ will again be given by quadratic Casimirs, but the matrix representing $\tau \mapsto -1/\tau$ involves the continuous q-ultraspherical polynomials, also known as the Macdonald polynomials for $\mathfrak{sl}_2(\mathbb{C})$ (and as such are a natural generalisation of SU_2-characters, which describe $\tau \mapsto -1/\tau$ for $\rho^{(1,0)}$). Similarly, the chiral blocks in $\mathfrak{F}_\lambda^{(1,1)}$ can be expressed using Macdonald 'polynomials' for $\mathfrak{sl}_2^{(1)}$ (a natural generalisation of the $\mathfrak{sl}_2^{(1)}$-characters which are the chiral blocks in $\mathfrak{F}^{(1,0)}$). Something similar can be expected for the other WZW models on the punctured torus. Likewise, evaluating these chiral blocks for the Moonshine RCFT should extend Moonshine to noncommuting pairs g, h in the monster [5].

The space $\mathfrak{F}_{\mathcal{V}}^{(1,1)}$, with the $\mathcal{V}$-module $\mathcal{V}$, can be identified with $\mathfrak{F}^{(1,0)}$ except that in the former we have the freedom to evaluate the block at any state $v \in \mathcal{V}$. In particular, taking v to have the minimum energy – the vacuum – recovers the characters (4.2), but taking other v will give a vector-valued modular form of even weight, with the same $\mathrm{SL}_2(\mathbb{Z})$-multiplier $\rho^{(1,0)}$ as in (5.1) and (5.2) . [10] evaluated these for the Moonshine RCFT and found that all modular forms (of the right shape) arise. What is interesting is that the coefficients of these modular forms will have Moonshine-like interpretations involving characters of the stabiliser of the state v.

9 Higher genus

For a fixed RCFT, chiral blocks $\mathcal{F} \in \mathfrak{F}^{(g,n)}_{(M^1,\ldots,M^n)}$ yield vector-valued automorphic functions for the infinite discrete groups $\Gamma_{g,n}$, each realising a finite-dimensional representation $\rho^{(g,n)}_{(M^i)}$ of $\Gamma_{g,n}$. This tower of automorphic functions is coherent in the sense that it respects basic operations like sewing or pinching the surfaces. As mentioned earlier, $(g,n) = (1,0), (1,1), (0,4)$ all give a vector-valued modular form for $\mathrm{SL}_2(\mathbb{Z})$; for $(1,0)$ this is a classical object, being weight-0 and invariant under some congruence subgroup, but for $(1,1)$ and $(0,4)$ the weight is rational and the image of the multiplier will usually be infinite. Relatively little is known in higher genus g, and surely it is a direction for important future research. The main open challenge is to identify the special features and structures occurring here. In this sense most of the work done has been negative. In this section we sample a few of the highlights.

Most of the work has focused on the kernel and image of these representations $\rho^{(g,n)}_{(M^i)}$. The groups $\Gamma_{0,n}$, $n \leq 3$, are finite; all other $\ker \rho^{(g,n)}_{(M^i)}$ will be infinite, since $\Gamma_{g,n}$ is generated by infinite-order Dehn twists but $\rho^{(g,n)}_{(M^i)}$ maps each of these to a finite-order matrix. However, for fixed n, the intersection over all k of the kernel of $\rho^{(g,0)}$ for the SU_n WZW model at level k is trivial in any genus $g > 2$ [1].

The image of $\rho^{(1,0)}$ is always finite [3], but this is atypical: it is expected that a generic RCFT will have all other images infinite. For example, Funar [15] found that all $\mathrm{im}\,\rho^{(g,0)}$ will be infinite for SU_2 WZW models at all levels $k > 8$, and for all genera $g > 1$. Moreover, Masbaum [29] found an infinite-order matrix in $\mathrm{im}\,\rho^{(0,4)}$ for those RCFT.

In the RCFT associated to even self-dual lattices L (where the strings live on the torus $\mathbb{R}^n/L$ for $n = \dim L$), the conformal blocks in $\mathfrak{F}^{(g,0)}$ can be expressed in terms of Siegel theta functions, and the Torelli subgroup of $\Gamma_{g,0}$ is in the kernel of $\rho^{(g,0)}$. This is very atypical for RCFT, e.g. it is known to fail for SU_2 WZW at most levels.

On the other hand, these representations for all known RCFT seem to be always definable over a cyclotomic field. A notion of integrality for these representations is being developed [20].

A class of RCFT very conducive to study are the so-called holomorphic orbifolds, associated to the Drinfel'd double of a finite group G. In this case, the chiral blocks in $\mathfrak{F}^{(g,0)}$ are parametrised by $\mathrm{Hom}(\pi_1(\Sigma_g), G)/G$, i.e. group homomorphisms φ from π_1 of a genus-g surface Σ_g into G, where we identify $\varphi(\sigma)$ and $g^{-1}\varphi(\sigma)g$. $\Gamma_{g,0}$ acts naturally on $\pi_1(\Sigma_g)$ and hence $\rho^{(g,0)}$ here becomes a permutation representation. This means $\mathrm{im}\,\rho^{(g,0)}$ here is manifestly finite. However, we can see from this explicitly that $\ker \rho^{(g,0)}$ won't contain the Torelli generators listed by [6], at least for generic groups G, and so even in this extremely well-behaved theory, $\rho^{(g,0)}$ doesn't factor through to a representation of Siegel's modular group $\mathrm{Sp}_{2g}(\mathbb{Z})$. For instance [12] (though this was known earlier) show that $\mathrm{im}\,\rho^{(0,n)}$ will always be finite here, and it is tempting to guess that all $\mathrm{im}\,\rho^{(g,n)}$ are finite here.

Acknowledgements. I've benefitted greatly from communications with Peter Bántay, Gregor Masbaum, and Mark Walton. This research is supported in part by NSERC.

References

[1] J. E. Andersen, 'Asymptotic faithfulness of the quantum $SU(n)$ representations of the mapping class groups', *Ann. Math.* **163** (2006) 347–368.

[2] B. Bakalov and A. Kirillov, Jr., *Lectures on Tensor Categories and Modular Functors* (American Mathematical Society, Providence 2001).

[3] P. Bántay, 'The kernel of the modular representation and the Galois action in RCFT', *Commun. Math. Phys.* **233** (2003) 423–438.

[4] P. Bántay and T. Gannon, 'Vector-valued modular functions for the modular group and the hypergeometric equation', *Commun. Number Theory and Phys.* (to appear), arXiv:math/0705.2467.
[5] P. Bántay and T. Gannon, in preparation.
[6] J. S. Birman, 'On Siegel's modular group', *Math. Ann.* **191** (1971) 59–68.
[7] J. S. Birman, *Braids, Links, and Mapping Class Groups* (Princeton University Press, Princeton 1974).
[8] A. Coste and T. Gannon, 'Remarks on Galois in rational conformal field theories', *Phys. Lett.* **B323** (1994) 316–321.
[9] P. Di Francesco, P. Mathieu, and D. Sénéchal, *Conformal Field Theory* (Springer, New York 1996).
[10] C. Dong and G. Mason, 'Monstrous moonshine at higher weight', *Acta Math.* **185** (2000) 101–121.
[11] P. I. Etingof and A. A. Kirillov Jr, 'On the affine analogue of Jack and Macdonald polynomials', *Duke Math. J.* **78** (1995) 229–256.
[12] P. Etingof, E. Rowell and S. Witherspoon, 'Braid group reprsentations from twisted quantum doubles of finite groups', math/070327.
[13] I. Frenkel, J. Lepowsky, and A. Meurman, *Vertex Operator Algebras and the Monster* (Academic Press, San Diego 1988).
[14] J. Fuchs, I. Runkel, and C. Schweigert, 'Boundaries, defects and Frobenius algebras', *Fortsch. Phys.* **51** (2003) 850–855.
[15] L. Funar, 'On the TQFT representations of the mapping class groups', *Pacif. J. Math.* **188** (1999) 251–274.
[16] M. R. Gaberdiel, 'Introduction to conformal field theory', *Rep. Prog. Phys.* **63** (2000) 607–667.
[17] T. Gannon, 'The Cappelli–Itzykson–Zuber A-D-E classification', *Rev. Math. Phys.* **12** (2000) 739–748.
[18] T. Gannon, *Moonshine Beyond the Monster* (Cambridge University Press, 2006).
[19] K. Gawędzki, 'Conformal field theory', Séminaire Bourbaki, *Astérisque* **177-178** (1989) 95–126.
[20] P. M. Gilmer and G. Masbaum, 'Integral lattices in TQFT', arXiv: math.QA/0411029.
[21] Y.-Z. Huang, *Two-Dimensional Conformal Geometry and Vertex Operator Algebras* (Birkhäuser, Boston 1997).
[22] V. F. R. Jones, 'Hecke algebra representations of braid groups and link polynomials', *Ann. of Math.* **126** (1987), 335–388.
[23] V. G. Kac, *Infinite Dimensional Lie Algebras*, 3rd edn (Cambridge University Press, Cambridge 1990).
[24] A. A. Kirillov Jr, 'On an inner product in modular tensor categories', *J. Amer. Math. Soc.* **9** (1996) 1135–1169.
[25] A. N. Kirillov and N. Yu. Reshetikhin, 'Representations of the algebra $U_q(sl(2))$, q-orthogonal polynomials and invariants of links', in: *Infinite-dimensional Lie algebras and groups (Luminy-Marseille, 1988)* (World Scientific, Teaneck NJ 1989) 285–339.
[26] M. Knopp and G. Mason, 'Vector-valued modular forms and Poincaré series', *Illinois J. Math.* **48** (2004) 1345-1366.
[27] T. Kohno, *Conformal Field Theory and Topology* (American Math. Soc., 2002).
[28] J. Lepowsky and H. Li, *Introduction to Vertex Operator Algebras and Their Representations* (Birkhäuser, Boston 2004).
[29] G. Masbaum, 'An element of infinite order in TQFT-representations of mapping class groups', *Contemp. Math.* **233** (1999) 137–139.
[30] G. Moore and N. Seiberg, 'Classical and quantum conformal field theory', *Commun. Math. Phys.* **123** (1989) 177–254.
[31] V. G. Turaev, *Quantum Invariants of Knots and 3-manifolds* (de Gruy-ter, Berlin 1994).
[32] E. Witten, 'Quantum field theory and the Jones polynomial', *Commun. Math. Phys.* **121** (1989) 351–399.
[33] Y. Zhu, 'Modular invariance of characters of vertex operator algebras', *J. Amer. Math. Soc.* **9** (1996) 237–302.

Fields Institute Communications
Volume **54**, 2008

$SL(2,\mathbb{C})$ Chern-Simons Theory and the Asymptotic Behavior of the Colored Jones Polynomial

Sergei Gukov
Department of Physics and Mathematics
University of California
Santa Barbara, CA 93106 USA
gukov@math.ucsb.edu

Hitoshi Murakami
Department of Mathematics
Tokyo Institute of Technology
Oh-okayama, Meguro, Tokyo 152-8551, Japan
starshea@tky3.3web.ne.jp

Abstract. It has been proposed that the asymptotic behavior of the colored Jones polynomial is equal to the perturbative expansion of the Chern-Simons gauge theory with complex gauge group $SL(2,\mathbb{C})$ on the hyperbolic knot complement. In this note we make the first step toward verifying this relation beyond the semi-classical approximation. This requires a careful understanding of some delicate issues, such as normalization of the colored Jones polynomial and the choice of polarization in Chern-Simons theory. Addressing these issues allows us to go beyond the volume conjecture and to verify some predictions for the behavior of the subleading terms in the asymptotic expansion of the colored Jones polynomial.

1 Introduction

The original volume conjecture [11, 17] is a remarkable relation between the limit of the colored Jones polynomial, $J_N(K;q)$, of a knot K and the volume of the knot complement $S^3 \setminus K$:

Conjecture 1.1 (Volume Conjecture) *For a knot K,*

$$\lim_{N\to\infty} \frac{\log \left| J_N\left(K; \exp\left(2\pi\sqrt{-1}/N\right)\right) \right|}{N} = \frac{1}{2\pi}\mathrm{Vol}(S^3 \setminus K), \tag{1.1}$$

2000 *Mathematics Subject Classification.* Primary 57M27, 58J52; Secondary 57M25, 57M50, 58J28.

Key words and phrases. colored Jones polynomial, volume conjecture, A-polynomial, Chern-Simons theory.

where $\mathrm{Vol}(S^3 \setminus K)$ *is the simplicial volume of the knot complement. In particular, if* K *is hyperbolic, then* $\mathrm{Vol}(S^3 \setminus K)$ *is the hyperbolic volume.*

The physical interpretation of this relation was proposed in [9], where it was conjectured that the asymptotic expansion of the colored Jones polynomial $J_N(K;q)$ in the limit $N \to \infty$, $q \to 1$ should be equal to the partition function of the $SL(2,\mathbb{C})$ Chern-Simons theory on the knot complement $S^3 \setminus K$. In turn, this implies the following conjecture (see eqn. (6.27) in [9]):

Conjecture 1.2 *As* $N \to \infty$ *and* $k \to \infty$ *with* $u := 2\pi\sqrt{-1}(\frac{N}{k} - 1)$ *kept fixed, the colored Jones polynomial has the asymptotic expansion*

$$\log J_N(K; \exp(2\pi\sqrt{-1}/k))$$

$$\underset{N,k\to\infty}{\sim} \frac{k}{\sqrt{-1}} S(u) + \frac{1}{2}\delta_K(u)\log k + \frac{1}{2}\log\left(\frac{T_K(u)}{2\pi^2}\right) + \sum_{n=1}^{\infty}\left(\frac{2\pi}{k}\right)^n S_{n+1}(u) \quad (1.2)$$

where the function $S(u)$ *in the first term is the classical action of the Chern-Simons theory;* $T_K(u)$ *is the Ray-Singer torsion of the knot complement twisted by the flat connection corresponding to the representation* $\rho : \pi_1(S^3 \setminus K) \to SL(2;\mathbb{C})$ *determined by* u *(see [21, 22] and §3 below); the number* $\delta_K(u) \in \mathbb{Z}$ *is determined by the topology of the knot complement and the representation* ρ*; finally, the function* $S_n(u)$ *denotes the* n*-loop contribution [26, 1, 2].*

One of the most interesting special cases of this conjecture is for the hyperbolic flat connection corresponding to the discrete faithful representation. We wish to stress, however, that the physical arguments in [9] which lead to eqn. (6.27) and also to eqn. (1.2) here are more general and not limited to this special case.

In particular, this physical interpretation of the volume conjecture opens an avenue for several generalizations. First, it suggests that, for a knot K, there exists a 1-parameter family of relations like (1.1) — sometimes called the "generalized" or "parameterized" volume conjecture — which relate a family of limits of the colored Jones polynomial to the volume function $\mathrm{Vol}(K;u)$ on the character variety of the knot complement; see Conjecture 1.3 below. Moreover, the interpretation via $SL(2,\mathbb{C})$ Chern-Simons theory predicts the structure of the subleading terms in the asymptotic expansion of the colored Jones polynomial, where each term in (1.2) has an *a priori* definition and can be computed independently, before comparing to the colored Jones polynomial. In what follows, first we shall discuss the leading term in the expansion (1.2) and then return to the subleading terms in §3.

Notice that the leading term in the expansion (1.2) gives precisely the generalization of the volume conjecture, which can be stated as follows:

Conjecture 1.3 *There exists an open subset* $\mathcal{O}_K$ *of* $\mathbb{C}$ *such that for any* $u \in \mathcal{O}_K$ *the following limit exists:*

$$\lim_{N\to\infty} \frac{\log J_N\left(K; \exp\left(\left(u + 2\pi\sqrt{-1}\right)/N\right)\right)}{N} = \frac{2\pi}{\left(u + 2\pi\sqrt{-1}\right)} S(u). \quad (1.3)$$

The precise formulation of the Conjectures 1.2 and 1.3 requires fixing normalizations of the colored Jones polynomial and the partition function of the Chern-Simons theory. Moreover, in the geometric quantization approach to Chern-Simons gauge theory, the partition function is determined by quantizing the moduli space of flat connections, which involves an extra ambiguity, the choice of *polarization*.[1] In principle, all these choices should be fixed by *a priori* arguments, such as a non-perturbative formulation of the $SL(2,\mathbb{C})$ Chern-Simons theory. However, with the lack of such arguments one might use the fact that all these choices do not depend on the knot K and, therefore, can be fixed once and for all by considering, say, the leading term $S(u)$ for a particular knot. Thus, the simplest choice of polarization (that is, the choice of the symplectic potential $S(u)$) consistent with the asymptotic behavior of the colored Jones polynomial for the figure-eight knot at $\mathrm{Re}(u) = 0$ gives $\mathrm{Im}\, S(u) = \frac{1}{2\pi}\mathrm{Vol}(K;u)$, where $\mathrm{Vol}(K;u)$ is the volume function [19, 5],

$$d\,\mathrm{Vol}(K;u) = -\frac{1}{2}\left(\mathrm{Re}(u)d\,\mathrm{Im}(v) - \mathrm{Re}(v)d\,\mathrm{Im}(u)\right). \tag{1.4}$$

Here, $v(u)$ is defined implicitly as a solution to the equation $A(-e^{-v/2}, e^{u/2}) = 0$ where $A(l,m)$ is the A-polynomial of [5]. We return to the definition of these quantities in §2 below. In particular, one of the goals of §2 is to clarify the relation between $S(u)$ and the volume function $\mathrm{Vol}(K;u)$, and to explain the geometric interpretation of the zero locus of the A-polynomial and the role of the variables u and v. For now, we simply note that eqn. (1.4) determines $\mathrm{Im}\, S(u)$ only up to terms proportional to $\mathrm{Re}(u)$. One way to fix this ambiguity could be to study the asymptotic behavior of the colored Jones polynomial for $\mathrm{Re}(u) \neq 0$. However, until recently there was no single example of a hyperbolic knot for which this was done. Without concrete examples or further insights, one might simply assume that there are no additional terms proportional to $\mathrm{Re}(u)$, so that $\mathrm{Im}\, S(u) = \frac{1}{2\pi}\mathrm{Vol}(K;u)$. This gives the generalized volume conjecture originally proposed in [9].

Recently, it was realized [14] that the generalized volume conjecture proposed in [9] requires a modification, precisely by the terms proportional to $\mathrm{Re}(u)$. Indeed, the explicit study of the limit (1.3) for the figure-eight knot [18] and torus knots at $\mathrm{Re}(u) \neq 0$ leads to the version of the Conjecture 1.3 with the function $S'(u)$, where [14]:

$$\mathrm{Im}\, S'(u) = \frac{1}{2\pi}\mathrm{Vol}(K;u) + \frac{1}{2}\mathrm{Re}(u) + \frac{1}{4\pi}\mathrm{Re}(u)\,\mathrm{Im}(v). \tag{1.5}$$

where, again, u and v obey $A(-e^{-v/2}, e^{u/2}) = 0$, see §2 below. We believe it is this generalization of the volume conjecture — called the parameterized volume conjecture in [14] — which has a chance of being true. We note (and explain in more detail in §2) that in the geometric quantization approach to the $SL(2,\mathbb{C})$ Chern-Simons theory, both $S(u)$ and $S'(u)$ can be interpreted as the semi-classical action of the $SL(2,\mathbb{C})$ Chern-Simons theory, obtained in a different polarization. For this reason, the physical considerations in [9] cannot tell us whether it is $S(u)$ or $S'(u)$ which should appear in the right-hand side of (1.2) and (1.3). However, given the evidence in [14, 18], we believe it is the function $S'(u)$ which is the correct semi-classical action of the $SL(2,\mathbb{C})$ Chern-Simons theory, and both Conjectures 1.2 and 1.3 should be considered with $S'(u)$ given by (1.5). We present further evidence for this in §3.

One of the main goals of the present paper is to show that the Conjectures 1.2 and 1.3 with the function $S'(u)$ are consistent with the proposal that the asymptotic expansion of the colored Jones polynomial is equal to the loop expansion of the partition function of the $SL(2,\mathbb{C})$ Chern-Simons theory. In particular, in §2 we

[1]For a general introduction to geometric quantization see [27].

show that the function $S'(u)$ can be interpreted as the semi-classical action in the $SL(2,\mathbb{C})$ Chern-Simons theory, and differs from $S(u)$ by a choice of polarization. Note, that $S(u) = S'(u)$ for $\mathrm{Re}(u) = 0$, which for hyperbolic knots is expected to correspond to the case of cone-manifolds. Once we identify the correct choice of polarization, in §3 we present further evidence for the Conjecture 1.2 by studying the subleading terms in the asymptotic expansion (1.2) of the colored Jones polynomial and compare with the expected behavior of the partition function of the $SL(2,\mathbb{C})$ Chern-Simons theory.

Before we proceed, let us describe the representation $\rho\colon \pi_1(S^3 \setminus K) \to SL(2,\mathbb{C})$ determined by u. Following [14], we define

$$v_K(u) := 4\pi \frac{d\,S'(u)}{d\,u} - 2\pi\sqrt{-1}, \tag{1.6}$$

where $S'(u)$ is defined by (1.3). Then, ρ can be defined as a representation from $\pi_1(S^3 \setminus K)$ to $SL(2,\mathbb{C})$ sending the longitude and the meridian to the elements whose eigenvalues are $(l, m) = (-\exp(-v_K(u)/2), \exp(u/2))$. This particular relation between (l, m) and the parameter u that determines a way in which we take the limit of the colored Jones polynomial in (1.2) follows from the interpretation of the $SL(2,\mathbb{C})$ Chern-Simons theory as a three-dimensional theory of gravity (for more details see section 5.2 of [9]). The representation ρ defines a flat $SL(2,\mathbb{C})$ bundle over $S^3 \setminus K$, which we denote by E_ρ. This bundle will play an important role in §3.

Finally, we want to make a disclaimer that, aiming at the audience of physicists as well as mathematicians, sometimes we had to make painful choices between rigorous mathematical treatment and a more informal style of the presentation used in the physics literature. We hope we managed to strike a balance.

Acknowledgments The authors would like to thank Jérôme Dubois, Stavros Garoufalidis, and Toshiaki Hattori for helpful conversations. It is also a pleasure to thank the organizers of the conference "Around the Volume Conjecture" at Columbia University in March 2006 and the conference "Modular Forms and String Duality" at Banff in June 2006, which stimulated much of this work. This work was supported in part by the DOE under grant number DE-FG03-92-ER40701, in part by RFBR grant 04-02-16880, and in part by the grant for support of scientific schools NSh-8004.2006.2 (S.G.), and in part by Grant-in-Aid for Scientific Research (B) (15340019) (H.M.).

2 Choice of polarization

In this section we show that the version of the generalized volume conjecture 1.3 proposed in [14] with the function $S'(u)$ is consistent with the interpretation in terms of $SL(2;\mathbb{C})$ Chern-Simons theory suggested in [9]. In particular, we show that $S'(u)$ can be interpreted as the semi-classical action in the $SL(2,\mathbb{C})$ Chern-Simons theory, and differs from $S(u)$ by a choice of polarization. To explain this in detail, let us start by fixing notations. We use $\tilde{u}$ and $\tilde{v}$ instead of u and v used in [9] respectively. The relation between $(\tilde{u}, \tilde{v})$ and (u, v) will be explained later.

Let us recall the argument in [9]. Fix an oriented (not necessarily hyperbolic) knot K in S^3. Let M be its complement $S^3 \setminus \operatorname{int} N(K)$, where $N(K)$ is the tubular neighborhood of K and $\operatorname{int} N(K)$ is its interior. The boundary of M is denoted by Σ. Note that $\pi_1(\Sigma) \cong \mathbb{Z} \times \mathbb{Z}$ is generated by the meridian μ and the longitude λ, where μ bounds a disk in $N(K)$ and is oriented so that the linking number of K

and μ is -1, and λ is null homologous in M and parallel to K in $N(K)$. (So our orientation of the meridian here is *different* from the usual one in knot theory.)

We consider a representation ρ of $\pi_1(M)$ to $SL(2;\mathbb{C})$ and denote by $m = \exp\tilde{v}$ and $l = \exp\tilde{u}$ the eigenvalues of its images of μ and λ respectively. Then the pair (m, l) is a zero of the A-polynomial [5]. As explained in [9, §3.1], the zero locus of the A-polynomial defines a Lagrangian subvariety L (possibly singular) of $\mathcal{P}$ with respect to the 2-form ω, where $\mathcal{P} = \mathbb{C}^* \times \mathbb{C}^*$ is the representation space of $\pi_1(\Sigma)$ and ω is defined as follows:

$$\omega := -\frac{1}{\pi} d\tilde{u} \wedge d\tilde{v}$$

so that $\omega\,\big|_L = 0$. (Note that here we put $\sigma = k$ in [9, (3.7)], and rescaled ω by a factor of k which now explicitly appears as the coefficient of the classical action in (1.2).) If we put

$$\theta := \frac{1}{2\pi}\left(\tilde{v}\,d\tilde{u} - \tilde{u}\,d\tilde{v} + d\left(\tilde{u}\overline{\tilde{v}}\right)\right),$$

we have $d\theta = \omega$ [9, (3.26)]. Let S be the classical Chern–Simons action corresponding to ρ. Then S can be obtained by integrating θ over a path on L, that is, we have

$$S = \int \theta.$$

Since the Lagrangian subvariety L is quantizable the integral above is well-difined [9, §3]. Note that S depends on the choice of θ satisfying $d\theta = \omega$. So we can define S only up to a total derivative on $\mathcal{P}$ (the choice of polarization).

One possible choice of θ, consistent with $d\theta = \omega$, gives

$$S = \frac{\sqrt{-1}}{2\pi}\left(\mathrm{Vol}(l,m) + 2\pi^2\sqrt{-1}\,\mathrm{CS}(l,m)\right), \tag{2.1}$$

where $\mathrm{Vol}(l,m)$ and $\mathrm{CS}(l,m)$ are the volume and the Chern–Simons invariant of the representation ρ. According to (1.2), the leading term of the log of the N-colored Jones polynomial evaluated at $e^{2\pi\sqrt{-1}/k}$ should be $-\sqrt{-1}kS$ when $N \to \infty$ and $k \to \infty$. This leads to a generalization of the volume conjecture [9, (5.12)]

$$\lim_{N\to\infty, k\to\infty} \frac{\log J_N\left(K; \exp\left(2\pi\sqrt{-1}/k\right)\right)}{k} = \frac{1}{2\pi}\left(\mathrm{Vol}(l,m) + 2\pi^2\sqrt{-1}\,\mathrm{CS}(l,m)\right). \tag{2.2}$$

Now, as we pointed out earlier, the dependence on the choice of polarization is related to the choice of the 1-form θ such that $d\theta = \omega$. In particular, we can consider the following 1-form:

$$\theta' := \frac{1}{2\pi}\left(-2\tilde{u}\,d\tilde{v} + 2\pi\sqrt{-1}d\tilde{v}\right).$$

Note that $d\theta = d\theta' = \omega$. Let S' be the classical Chern–Simons action obtained from θ' ($S' := \int\theta'$). Then

$$\begin{aligned} dS - dS' &= \theta - \theta' \\ &= \frac{1}{2\pi} d\left(\tilde{u}\tilde{v} + \tilde{u}\overline{\tilde{v}} - 2\pi\sqrt{-1}\tilde{v}\right) \\ &= \frac{1}{\pi} d\left(\tilde{u}\,\mathrm{Re}(\tilde{v}) - \pi\sqrt{-1}\tilde{v}\right). \end{aligned}$$

Therefore from (2.1) we have

$$\begin{aligned}\mathrm{Vol}(l,m) &= 2\pi\,\mathrm{Im}(S) \\ &= 2\pi\,\mathrm{Im}(S') + 2\,\mathrm{Im}(\tilde{u})\,\mathrm{Re}(\tilde{v}) - 2\pi\,\mathrm{Re}(\tilde{v}).\end{aligned} \tag{2.3}$$

Now we consider the pair (u,v) used in the previous section. As described in [18], u and v are related to $\tilde{u}$ and $\tilde{v}$ as follows:

$$\begin{cases}\tilde{v} &= \dfrac{u}{2}, \\ \tilde{u} &= -\dfrac{v}{2}.\end{cases}$$

So from (2.3) we have

$$\mathrm{Vol}(l,m) = 2\pi\,\mathrm{Im}(S') - \pi\,\mathrm{Re}(u) - \frac{1}{2}\,\mathrm{Re}(u)\,\mathrm{Im}(v).$$

Comparing with (1.5) and (1.2), $-\sqrt{-1}kS'$ gives the leading term of the log of the N-colored Jones polynomial if we use θ' to define the classical action S',

$$\lim_{N\to\infty,k\to\infty}\frac{\log\left|J_N\left(K;\exp\left(2\pi\sqrt{-1}/k\right)\right)\right|}{k} = \frac{1}{2\pi}\,\mathrm{Vol}(K;u)+\frac{1}{2}\,\mathrm{Re}(u)+\frac{1}{4\pi}\,\mathrm{Re}(u)\,\mathrm{Im}(v)$$

This is precisely the version of the generalized volume conjecture proposed in [14].

3 Beyond the leading order

Now let us discuss the subleading terms in the asymptotic expansion (1.2) of the colored Jones polynomial.

The simplest knot to consider is the unknot U. Since $J_N(U;q) = [N] := \left(q^{N/2}-q^{-N/2}\right)/\left(q^{1/2}-q^{-1/2}\right)$, we have

$$\begin{aligned}&\log J_N(U;\exp(2\pi\sqrt{-1}/k)) \\ &= \log\sin(N\pi/k) - \log\sin(\pi/k) \\ &\sim \log k - \log\pi + \log\sin\bigl(u/(2\sqrt{-1})\bigr) + \text{terms of the order } k^{-1} \text{ or lower}\end{aligned} \tag{3.1}$$

for large N and k with fixed $N/k = u/(2\pi\sqrt{-1}) + 1$ and $u \neq 0$.

Since the colored Jones polynomial vanishes at $u = 0$, in this case one should use the *reduced* colored Jones polynomial $V_N(K;q) := J_N(K;q)/J_N(U;q)$ to study the asymptotic expansion (1.2). We should note, however, that it is $J_N(K;q)$ which naturally appears in Chern-Simons theory.

3.1 The logarithmic term. In this subsection we study the logarithmic term in the asymptotic expansion (1.2).

For a knot K in S^3, let M_K be the complement of the interior of the regular neighborhood of K. Denote by Σ_K the boundary torus of M_K. Let $H^i\left(M_K;E_\rho\right) \cong H^i\left(S^3\setminus K;E_\rho\right)$ be the i-th cohomology group of M_K with coefficients in the flat $SL(2,\mathbb{C})$ bundle E_ρ which was described at the end of §1. It is well known that $H^i\left(M_K;E_\rho\right)$ is isomorphic to the cohomology $H^i\left(M_K;sl_2(\mathbb{C})\right)$ with coefficients the Lie algebra $sl_2(\mathbb{C})$ twisted by the adjoint action of ρ. We will mainly use $H^i\left(M_K;sl_2(\mathbb{C})\right)$ for calculation.

Define

$$\delta_K^{\mathrm{rep}}(\rho) := 3 + h^1(M_K;\rho) - h^0(M_K;\rho), \tag{3.2}$$

where $h^0(M_K;\rho) := \dim H^0(M_K;\rho)$ and

$$h^1(M_K;\rho) := \dim \left(\mathrm{Ker}[H^1(M_K;\rho) \to H^1(\Sigma_K;\rho')]\right),$$

where the map is induced by the inclusion $\Sigma_K \to M_K$ and ρ' is the restriction of ρ to $\pi_1(\Sigma_K)$. From [10, Page 42] and [20, Démonstration de Proposition 3.7 (Page 72)], we have

$$\begin{aligned} h^1(M_K;\rho) &= \dim H^1(M_K;\rho) - \frac{1}{2}\dim H^1(\Sigma_K;\rho') \\ &= \begin{cases} \dim H^1(M_K;\rho) - 3 & \text{if } \rho' \text{ is trivial,} \\ \dim H^1(M_K;\rho) - 1 & \text{if } \rho' \text{ is non-trivial,} \end{cases} \end{aligned} \tag{3.3}$$

where ρ' is trivial if it sends every element of $\pi_1(\Sigma_K)$ to $\pm I$ with I the identity matrix.

Among other things, the identification of the asymptotic expansion (1.2) with the partition function in the Chern-Simons theory implies (see *e.g.* [7, 9, 26])

$$\delta_K(u) = \delta_K^{\mathrm{rep}}(\rho), \tag{3.4}$$

where ρ is determined by u as described in §1. The first term (equal to 3) in (3.2) comes from the normalization by the partition function of S^3,

$$\begin{aligned} \log Z(S^3) &= \log \sqrt{2/k}\,\sin(\pi/k) \\ &\sim -\frac{3}{2}\log k + \cdots . \end{aligned}$$

To evaluate (3.2), it is useful to note that

$$h^0(K;\rho) = \dim(H_\rho), \tag{3.5}$$

where H_ρ is the isotropy group of E_ρ, the subgroup of $SL(2,\mathbb{C})$ that commutes with the holonomies of flat connections on E_ρ. In other words,

$$H_\rho = \{g \in SL(2;\mathbb{C}) \mid g\rho(\gamma) = \rho(\gamma)g \text{ for any } \gamma \in \pi_1(M_K)\}.$$

Another useful fact is that $h^1(M_K;\rho)$ counts the infinitesimal deformations of E_ρ for a fixed representation of $\pi_1(\Sigma_K)$ [23, 24, 25].

We can show that $h^1(M_K;\rho) = 0$ if K is a hyperbolic knot or a torus knot for a typical representation ρ.

Definition 3.1 [20, Definition 3.21] Let γ be a simple closed curve on Σ_K for a knot K. Then a representation ρ of $\pi_1(M_K)$ in $SL(2;\mathbb{C})$ is called γ-regular if

1. the map $H^1(M_K;\rho) \to H^1(\gamma;\rho_0)$ induced by the inclusion $\gamma \to M_K$ is injective, and
2. if $\mathrm{Tr}(\rho(\pi_1(\Sigma_K))) = \pm 1$, then $\rho(\gamma) \neq \pm I$.

The following representations are known to be γ-regular [20, Page 83].

1. The holonomy representation corresponding to the complete hyperbolic structure of a hyperbolic knot is μ-regular, where μ is the meridian.
2. For a hyperbolic knot, if the Dehn surgery along a simple closed curve γ is hyperbolic then the holonomy representation induced by γ is γ-regular.
3. For a torus knot, any irreducible representation is both μ-regular and λ-regular, where λ is the longitude [6, Example 1].

We can show the following proposition. (Although this proposition and the next lemma follow from [20, §3.1.4 and §3.2], here we write proofs for completeness.)

Proposition 3.2 *Let ρ be a γ-regular representation of $\pi_1(M_K)$ for a knot K for some simple closed curve γ on Σ_K. Then $\delta_K^{rep}(\rho) = 3 - h^0(K;\rho)$.*
Moreover if ρ is non-Abelian, then $\delta_K^{rep}(\rho) = 3$.

Proof Let ρ' be the restriction of ρ to $\pi_1(\Sigma_K)$, where Σ_K is the boundary torus of the regular neighborhood of K.

Since ρ is γ-regular, the composition

$$H^1(M_K;\rho) \to H^1(\Sigma_K;\rho') \to H^1(\gamma;\rho_0)$$

is injective and so is the map $H^1(M_K;\rho) \to H^1(\Sigma_K;\rho')$. Therefore we have $h^1(M_K;\rho) = \dim \operatorname{Ker}[H^1(M_K;\rho) \to H^1(\Sigma_K;\rho')] = 0$ and the formula follows.

If ρ is non-Abelian, we have $h^0(K;\rho) = 0$ (see for example [20, Lemme 0.7 (ii)]). □

As a corollary we have

Corollary 3.3 *If K is a hyperbolic knot or a torus knot and ρ is a non-Abelian representation of $\pi_1(S^3 \setminus K)$ in $SL(2;\mathbb{C})$, then $\delta_K^{rep} = 3$.*

3.1.1 *Abelian representations.* We will study the case where ρ is Abelian.

Lemma 3.4 *For any knot K, there exist open sets $U_1 \ni I$ and $U_2 \ni -I$ of $SL(2;\mathbb{C})$ such that if ρ is an Abelian representation sending the meridian into U_1 or U_2 then $\delta_K^{rep}(\rho) = 2$.*

Proof First note that $H_1(M_K;\mathbb{Z}) \cong \mathbb{Z}$ is generated by the meridian. We choose an element $\mu \in \pi_1(M_K)$ that is mapped to the meridian by the Abelianization. If ρ is Abelian, it is determined by the image $A := \rho(\mu) \in SL(2;\mathbb{C})$, since it factors through $H_1(M_K;\mathbb{Z})$.

We can calculate $H^i(M_K;\rho)$ by using the infinite cyclic covering space $\widetilde{M_K}$ of M_K. We have the following chain complex of $\widetilde{M_K}$ as $\mathbb{C}[t,t^{-1}]$-modules by using the Fox differential calculus as described in the proof of [3, Theorem 6.1]:

$$C_2 \xrightarrow{d_2} C_1 \xrightarrow{d_1} C_0,$$

where C_2 is generated by $\{r_1, r_2, \dots, r_{n-1}\}$, C_1 is generated by $\{s_1, s_2, \dots, s_n\}$, C_0 is generated by $\{p\}$, and d_2 and d_1 are given as follows. We define d_1 by $d_1(s_i) := (t-1)p$ for $1 \le i \le n$. Let $F(t)$ be the $n \times (n-1)$ matrix with entries in $Z[t,t^{-1}]$ given by the Fox free differential calculus. Then $d_2(r_i) = \sum_{j=1}^{n} F_{ji}(t)s_j$, where $F_{ji}(t)$ is the (j,i) entry of $F(t)$. Note that the sum of each column of $F(t)$ is zero (and so $d_1 \circ d_2 = 0$), and that the determinant of any $(n-1)\times(n-1)$ matrix obtained from $F(t)$ by deleting any row gives the Alexander polynomial $\Delta(K;t)$ of K. (See for example [12, Chapter 11].)

Then the twisted cohomology $H^i(M_K; sl_2(\mathbb{C}))$ is calculated from the following cochain complex:

$$\{0\} \to \operatorname{Hom}_{\mathbb{C}[t,t^{-1}]}(C_0; sl_2(\mathbb{C})) \xrightarrow{d_1^*} \operatorname{Hom}_{\mathbb{C}[t,t^{-1}]}(C_1; sl_2(\mathbb{C})) \xrightarrow{d_2^*} \operatorname{Hom}_{\mathbb{C}[t,t^{-1}]}(C_2; sl_2(\mathbb{C})). \quad (3.6)$$

Here an element $\varphi \in \operatorname{Hom}_{\mathbb{C}[t,t^{-1}]}(C_0; sl_2(\mathbb{C}))$ is given by $\varphi(p) \in sl_2(\mathbb{C})$ in such a way that

$$\varphi(g(t)p) := g(\operatorname{Ad}(A))\varphi(p),$$

where $g(t) \in \mathbb{C}[t,t^{-1}]$ is a Laurent polynomial and $\mathrm{Ad}(A)$ is the adjoint representation. Since

$$d_1^*(\varphi)(s_i) := \varphi(d_1(s_i)) = \varphi((t-1)p) = (\mathrm{Ad}(A)-I)\varphi(p) = A\varphi(p)A^{-1} - \varphi(p)$$

for any i, we have $\dim \mathrm{Ker}\, d_1^* = 1$ if $A \neq \pm I$. Thus we have $h^0(M_K;\rho) = 1$. (Another way to show this is to use (3.5).)

Similarly, if $\psi \in \mathrm{Ker}\, d_2^*$, then

$$d_2^*(\psi)(r_j) := \psi(d_2(r_j)) = \psi\left(\sum_{k=1}^{n} F_{kj}(t)s_k\right) = \sum_{k=1}^{n} F_{kj}(\mathrm{Ad}(A))\psi(s_k) = O$$

for $j = 1,2,\ldots,n-1$, where O is the 2×2 zero matrix. Since $\sum_{k=1}^{n} F_{kj} = 0$ for any j, we have

$$\sum_{k=1}^{n-1} F_{kj}(\mathrm{Ad}(A))\{\psi(s_k) - \psi(s_n)\} = O. \tag{3.7}$$

By conjugation we may assume that A is of the form $\begin{pmatrix} a & b \\ 0 & a^{-1}\end{pmatrix}$. Note that for any $X := \begin{pmatrix} x & y \\ z & -x\end{pmatrix} \in sl_2(\mathbb{C})$ the $(2,1)$-entry of $A^m X A^{-m}$ is equal to $a^{-2m}z$. So comparing the $(2,1)$-entries of (3.7) we have

$$\sum_{k=1}^{n-1} F_{kj}\left(a^{-2}\right)\{\psi_{2,1}(s_k) - \psi_{2,1}(s_n)\} = 0$$

for $j = 1,2,\ldots,n-1$, where $\psi_{2,1}(s_k)$ is the $(2,1)$ entry of $\psi(s_k)$. Now we know the determinant of the $(n-1)\times(n-1)$ matrix $(F_{kj}(t))_{1\le k,j\le n-1}$ is $\Delta(K;t)$. Since $\Delta(K;\pm 1)$ is odd for any knot K (see for example [12, Chapter 6]), there exist open sets $U_1 \ni I$ and $U_2 \ni -I$ in $SL(2;\mathbb{C})$ such that $\Delta(K;a^{-2}) \neq 0$ if a is an eigenvalue of a matrix in $U_1 \cup U_2$. Thus if $A \in U_1 \cup U_2$, the matrix $(F_{kj}(t))_{1\le k,j\le n-1}$ is non-singular and we have $\psi_{2,1}(s_k) = \psi_{2,1}(s_n)$ for any $1 \le k \le n-1$ from (3.7). Similarly, we have $\psi_{1,1}(s_k) = \psi_{1,1}(s_n)$ and $\psi_{1,2}(s_k) = \psi_{1,2}(s_n)$, which implies $\psi(s_k) = \psi(s_n)$. This means that the 1-cocycle group Z^1 of (3.6) is given by

$$Z^1 = \left\{\psi \in \mathrm{Hom}_{\mathbb{C}[t,t^{-1}]}(C_1; sl_2(\mathbb{C})) \mid \psi(s_1) = \psi(s_2) = \cdots = \psi(s_n)\right\}$$

and so $\dim Z^1 = 3$. Since $\dim \mathrm{Ker}\, d_1^* = 1$, the dimension of the 1-coboundary group B^1 is 2. We finally have $h^1(M_K;\rho) = \dim Z^1 - \dim B^1 = 1$.

From (3.3) we have $\delta_K^{\mathrm{rep}}(\rho) = 2$. □

In [8], S. Garoufalidis and T. Le prove that for any knot K the limit

$$\lim_{N\to\infty} V_N\left(K; \exp((u + 2\pi\sqrt{-1})/N)\right)$$

exists if u is sufficiently close to $-2\pi\sqrt{-1}$, which was first proved for the figure-eight knot by the second author in [15]. This means that $\delta_K(u) = 2$ from (3.1). We expect that such u determines an Abelian representation ρ.

3.1.2 *Connected-sums.* We will discuss the behavior of $\delta_K(u)$ and $\delta_K^{\mathrm{rep}}(\rho)$ under connected-sum and satellite. Let us denote by $K_1\sharp K_2$ the connected-sum of two knots K_1 and K_2. Then we have $J_N(K_1\sharp K_2;q) = J_N(K_1;q)J_N(K_2;q)/J_N(U;q)$. Therefore we have

$$\delta_{K_1\sharp K_2}(u) = \delta_{K_1}(u) + \delta_{K_2}(u) - 2 \tag{3.8}$$

if $u \neq 0$. Correspondingly, for $u = 0$ we have $V_N(K_1\sharp K_2;q) = V_N(K_1;q)V_N(K_2;q)$ which implies

$$\delta_{K_1\sharp K_2}(0) = \delta_{K_1}(0) + \delta_{K_2}(0) \tag{3.9}$$

Now, let us compare this with $\delta^{\mathrm{rep}}_{K_1\sharp K_2}(\rho)$. Note that the complement $M_{K_1\sharp K_2}$ is obtained from M_{K_1} and M_{K_2} by glueing along an annulus. More precisely, $M_{K_1\sharp K_2}$ is obtained from M_{K_1} and M_{K_2} by identifying annuli $A_1 \in T_{K_1}$ and $A_2 \in T_{K_2}$, where A_1 (resp. A_2) is the regular neighborhood of the meridian μ_1 (resp. μ_2) of K_1 (resp. K_2).

We first calculate $H^0(S^1;\rho)$.

Lemma 3.5 *If a representation* $\rho\colon \pi_1(S^1) \to SL(2;\mathbb{C})$ *is not* $\pm I$ *then we have*

$$H^0(S^1;\rho) = \mathbb{C},$$

Otherwise, we have

$$H^0(S^1;\rho) = \mathbb{C}^3,$$

Proof We use the interpretation of h^0 described in (3.5). If ρ is not $\pm I$ its isotropy group is one-dimensional and so $h^0(S^1;\rho) = 1$. In the case where $\rho = \pm I$ the equality follows since the cohomology group is the usual one. □

Next we calculate $H^i(M_{K_1\sharp K_2};\rho)$ for $i = 0$ and 1 under some assumptions.

Lemma 3.6 *Let* $\rho\colon \pi_1(M_{K_1\sharp K_2}) \to SL(2;\mathbb{C})$ *be a representation, and* ρ_1, ρ_2, *and* ρ_0 *the restrictions of* ρ *to* $\pi_1(M_{K_1})$, $\pi_1(M_{K_2})$, *and* $\pi_1(A)$ *respectively, where* $A := M_{K_1} \cap M_{K_2} (\cong A_1 \cong A_2)$. *Suppose that* ρ_1 *and* ρ_2 *are* μ_1*- and* μ_2*-regular respectively.*

Then

$$\delta^{rep}_{K_1\sharp K_2}(\rho) = \delta^{rep}_{K_1}(\rho_1) + \delta^{rep}_{K_2}(\rho_2) - 2 \tag{3.10}$$

if $\rho_0 \neq \pm I$, *and*

$$\delta^{rep}_{K_1\sharp K_2}(\rho) = \delta^{rep}_{K_1}(\rho_1) + \delta^{rep}_{K_2}(\rho_2) \tag{3.11}$$

if $\rho_0 = \pm I$.

Proof The Mayer–Vietoris exact sequence for $M_{K_1\sharp K_2} = M_{K_1} \cup M_{K_2}$, M_{K_1}, M_{K_2} and A gives the following exact sequence:

$$\begin{aligned}\{0\} &\to H^0(M_{K_1\sharp K_2};\rho) \to H^0(M_{K_1};\rho_1) \oplus H^0(M_{K_2};\rho_2) \to H^0(A;\rho_0)\\ &\to H^1(M_{K_1\sharp M_2};\rho) \to H^1(M_{K_1};\rho_1) \oplus H^1(M_{K_2};\rho_2) \xrightarrow{j_1^*-j_2^*} H^1(A;\rho_0),\end{aligned}$$

where j_1^* and j_2^* are induced by the inclusions $j_1\colon A \to M_{K_1}$ and $j_2\colon A \to M_{K_2}$ respectively. Therefore we have

$$\begin{aligned}&\dim H^0(A;\rho_0)\\ &= \left\{\dim H^0(M_{K_1};\rho_1) + \dim H^0(M_{K_2};\rho_2) - \dim H^0(M_{K_1\sharp K_2};\rho)\right\}\\ &+ \left\{\dim H^1(M_{K_1\sharp K_2};\rho) - \dim \operatorname{Ker}(j_1^* - j_2^*)\right\}.\end{aligned}$$

If $\rho_0 \neq \pm I$, then from Lemma 3.5 and (3.3) we have

$$\begin{aligned}\delta^{\mathrm{rep}}_{K_1\sharp K_2} &= 3 + \dim \operatorname{Ker}\left[H^1(M_{K_1\sharp K_2};\rho) \to H^1(T_{K_1\sharp K_2};\rho')\right] - \dim H^0(M_{K_1\sharp K_2};\rho)\\ &= 3 + \dim H^1(M_{K_1\sharp K_2};\rho) - 1 - \dim H^0(M_{K_1\sharp K_2};\rho)\\ &= 2 + \dim H^0(A;\rho_0) + \dim \operatorname{Ker}(j_1^* - j_2^*) - h^0(M_{K_1};\rho_1) - h^0(M_{K_2};\rho_2)\\ &= 3 + \dim \operatorname{Ker}(j_1^* - j_2^*)\end{aligned}$$

Now since j_1^* and j_2^* are injective and $\dim H^1(A;\rho_0) = 1$, we conclude that $\dim \operatorname{Ker}(j_1^* - j_2^*) = \dim(\operatorname{Im} j_1^* \cap \operatorname{Im} j_2^*) = 1$. Therefore

$$\delta^{\mathrm{rep}}_{K_1\sharp K_2} = 4 = \delta^{\mathrm{rep}}_{K_1}(\rho_1) + \delta^{\mathrm{rep}}_{K_2}(\rho_2) - 2.$$

If $\rho = \pm I$, then since $\dim H^0(A;\rho_0) = 3$ we have

$$\begin{aligned}\delta^{\mathrm{rep}}_{K_1\sharp K_2} &= \dim H^1(M_{K_1\sharp K_2};\rho) - \dim H^0(M_{K_1\sharp K_2};\rho)\\ &= 3 + \dim \operatorname{Ker}(j_1^* - j_2^*) - h^0(M_{K_1};\rho_1) - h^0(M_{K_2};\rho_2).\end{aligned}$$

But in this case we have

$$\begin{aligned}&\dim \operatorname{Ker}(j_1^* - j_2^*)\\ =&3 + \dim \operatorname{Ker}[H^1(M_{K_1};\rho_1) \to H_1(T_{K_1})] + \dim \operatorname{Ker}[H^1(M_{K_2};\rho_2) \to H_1(T_{K_2})]\end{aligned}$$

and so

$$\begin{aligned}\delta^{\mathrm{rep}}_{K_1\sharp K_2} &= 6 + h^1(M_{K_1},\rho_1) + h^1(M_{K_2};\rho_2) - h^0(M_{K_1},\rho_1) + h^0(M_{K_2};\rho_2)\\ &= \delta^{\mathrm{rep}}_{K_1}(\rho_1) + \delta^{\mathrm{rep}}_{K_2}(\rho_2).\end{aligned}$$

□

The above lemma also follows from Proposition 0.11 of [20] for the Mayer-Vietoris exact sequence.

Observation 3.7 *Equation* (3.8) *should be compared with its "physics" counterpart, Equation* (3.10). *On the other hand, Equations* (3.9) *and* (3.11), *which appear to have the same form, actually have very different origins.*

Remark 3.8 The assumption of Lemma 3.6 is true for hyperbolic knots and for torus knots. In particular, it is true for hyperbolic knots with representations close to the holonomy representations corresponding to their complete hyperbolic structures.

3.1.3 *Satellite knots.* We consider satellite knots.

Let K be a knot in a solid torus D. If $e\colon D \to S^3$ is an embedding, then the image $e(K)$ forms a knot. We call $e(K)$ a satellite of the knot C with companion K, where C is the image of the core of D. Then the complement $M_{e(K)}$ of the interior of the regular neighborhood of K is obtained from $S^3 \setminus \operatorname{Int} D$ and $D_K := D \setminus \operatorname{Int} N(K)$ by pasting along their boundaries ∂M_C and ∂D, where Int denotes the interior, $N(K)$ is the regular neighborhood of K in D, and $M_C := S^3 \setminus \operatorname{Int} N(C)$. Note that ∂D_K consists of two tori; ∂D and $\partial M_{e(K)}$.

We can compute $\delta^{\mathrm{rep}}_{e(K)}(\rho)$ under some assumptions.

Lemma 3.9 *Let $e(K)$ be a satellite of a knot C with companion K and $\rho\colon \pi_1(S^3 \setminus e(K)) \to SL(2;\mathbb{C})$ a non-trivial representation. We assume the following four conditions:*

$$h^0(M_C;\rho_1) = 0, \tag{i}$$

$$h^1(M_C;\rho_1) = 0, \tag{ii}$$

$$h^0(D_K;\rho_2) = 0, \tag{iii}$$

$$\operatorname{Ker}\left[H^1(D_K;\rho_2) \to H^1(\partial D_K;{\rho'}_2)\right] = 0, \tag{iv}$$

where ρ_1, ρ_2, *and* ${\rho'}_2$ *are the restrictions of* ρ *to* $\pi_1(S^3 \setminus C)$, $\pi_1(D \setminus K)$ *and* ∂D_K *respectively, and the map in* (iv) *is induced by the includion. We also assume all the induced representations* ρ_0, ρ_1, *and* ρ_2 *are non-trivial.*

Then we have $\delta_K^{rep}(\rho) = 4$.

Proof From the Mayer–Vietoris exact sequence for $M_{e(K)} = M_C \cup D_K$, M_C, D_K and $T = M_C \cap D_K$, we have

$$\{0\} \to H^0(M_{e(K)};\rho) \to H^0(M_C;\rho_1) \oplus H^0(D_K;\rho_2) \to H^0(T;\rho_0)$$
$$\to H^1(M_{e(K)};\rho) \to H^1(M_C;\rho_1) \oplus H^1(D_K;\rho_2) \xrightarrow{j_1^* - j_2^*} H^1(T;\rho_0),$$

where ρ_0 is the restriction of ρ to $\pi_1(T)$, and j_1 and j_2 are inclusions. From the assumptions (i) and (iii), we have $h^0(M_{e(K)};\rho) = \dim H^0(M_{e(K)};\rho) = 0$.

We note that $h^1(M_C;\rho) = \dim \operatorname{Ker} j_1^* = 0$. We also note that

$$\begin{aligned}\dim \operatorname{Ker} j_2^* &= \dim \operatorname{Ker}\left[H^1(D_K;\rho_2) \to H^1(\partial D_K;{\rho'}_2) \to H^1(T;\rho_0)\right]\\ &= \dim \operatorname{Ker}\left[H^1(\partial D_K;{\rho'}_2) \to H^1(T;\rho_0)\right]\\ &= 1\end{aligned}$$

since $\dim H^1(T;\rho_0) = 1$ from [20, Démonstration de Proposition 3.7 (Page 72)]. Therefore the kernel of the map $j_1^* - j_2^*$ is one-dimensional, and we have

$$H^1(M_{e(K)};\rho) = H^0(T;\rho_0) \oplus \mathbb{C}.$$

Since $\pi_1(T)$ is abelian and $\rho_0 \neq \pm I$, we have $H^0(T;\rho_0) = \mathbb{C}$ from (3.5). From (3.3), we finally get $h^1(M_{e(K)};\rho) = 1$.

Thercfore $\delta_{e(K)}(u) = 4$ as stated. □

Observation 3.10 *This can be compared with a result of Zheng* [28, Theorem 1.4], *where he studies Whitehead doubles of non-trivial torus knots and proves that* $\delta_K(0) = 4$ *for such knots.*

In this case D_K *is homeomorphic to the complement of (the interior of the regular neighborhood of) the Whitehead link, which is hyperbolic. So from* [10, Corollary 1.2], *the assumption* (iv) *is satisfied if* ρ *is a small deformation of the holonomy representation. The other conditions also hold.*

Remark 3.11 The assumption $h^1(S^3 \setminus C;\rho_1) = h^0(S^3 \setminus C;\rho_1) = 0$ of Lemma 3.9 is true if C is a hyperbolic knot and ρ_1 is close to the holonomy representation corresponding to the complete hyperbolic structure. The assumption $h^1(D \setminus K;\rho_2) = h^0(D \setminus K;\rho_2) = 0$ may be true if $D \setminus K$ possesses a complete hyperbolic structure and ρ_2 is close to the holonomy representation.

Remark 3.12 It would be interesting to study the asymptotic behavior of the colored Jones polynomial of the Hopf link H. Since $J_N(H;q) = [N^2]$ (see for example [12, Lemma 14.2]), we have

$$\begin{aligned}\log J_N(H;\exp 2\pi\sqrt{-1}/k) &= \log\sin\left(\pi\left(\frac{N}{k}\right)^2 k\right) - \log\sin(\pi/k)\\ &\sim 2\log k + 2\log(N/k)\end{aligned}$$

if N/k is *very small.* This formula would mean that $\delta_H(u) = 4$, which is consistant with the fact that $3 + h^1(T \times I) - h^0(T \times I) = 4$ since the complement of the Hopf link is $T \times I$.

3.2 Ray–Singer torsion. The next interesting term in (1.2) is the term containing the Ray–Singer torsion $T_K(u) := T(S^3 \setminus K, \rho)$. Following the conventions used in the literature on Chern-Simons theory, we define the Ray–Singer torsion of a 3-manifold M with respect to a flat bundle E_ρ corresponding to ρ by

$$T(M, \rho) = \exp\left(-\frac{1}{2}\sum_{n=0}^{3} n(-1)^n \log \det{}' \Delta_n^{E_\rho}\right), \tag{3.12}$$

where $\Delta_n^{E_\rho}$ is the Laplacian on n-forms with coefficients in E_ρ, and $\det{}'\Delta_n^{E_\rho}$ is the regularized determinant of the restriction of the orthocomplement of its kernel. Using Poincaré duality, one finds

$$T(M, \rho) = \frac{\left(\det{}'\Delta_0^{E_\rho}\right)^{3/2}}{\left(\det{}'\Delta_1^{E_\rho}\right)^{1/2}}. \tag{3.13}$$

We recall that the relation between u and the corresponding representation ρ was discussed at the end of §1. Note that the definition of the Ray–Singer torsion is particularly simple when $h^0(S^3 \setminus K; \rho) = h^1(S^3 \setminus K; \rho) = 0$; in this case the Laplacians $\Delta_n^{E_\rho}$ have empty kernels.

Very much like the leading term in the expansion (1.2), in general, $T_K(u)$ is a non-trivial function on the character variety. In view of the Cheeger-Müller theorem [4, 13], it would be interesting to compare the Ray–Singer torsion as a function on the character variety to the Reidemeister torsion studied by Porti [20]. For example, for the figure-eight knot E and $\mathrm{Re}(u) = 0$ one has

$$T_E(u) = \frac{1}{\sqrt{(3/2 - \cos\alpha)(1/2 + \cos\alpha)}} \tag{3.14}$$

if $u \neq 0$, and

$$T_E(0) = \frac{\pi^2}{\sqrt{(3/2 - \cos\alpha)(1/2 + \cos\alpha)}} \tag{3.15}$$

where $\alpha = |\sqrt{-1}u|$ is the singular (cone) angle of the cone manifold $M = S^3 \setminus K$. Note that in [20, §5.3, Exemple 1] the torsion is given as $\pm 1/T_E(u)$ for $0 < \sqrt{-1}u < 2\pi/3$. Note also that we use the *reduced* colored Jones polynomial V_N when $u = 0$. The difference between (3.14) and (3.15) comes from $-\log\pi$ in (3.1).

Let us consider the following function of N and $r := N/k = 1 + u/(2\pi\sqrt{-1})$.

$$\mathrm{Re}\left\{\log J_N\left(E; \exp(2\pi r\sqrt{-1}/N)\right)\right\} - \mathrm{Re}\left\{\frac{N}{r\sqrt{-1}}S(r) + \frac{3}{2}\log\left(\frac{N}{r}\right) + \frac{1}{2}\log\left(\frac{T_E(u)}{2\pi^2}\right)\right\},$$

for the figure-eight knot E. Here we use the result of [16]. We expect that it vanishes when $N \to \infty$. In Figure 1–6 we use MATHEMATICA to plot the graphs of this function for $N = 100$, 200, 300, 400, 500 and 1000.

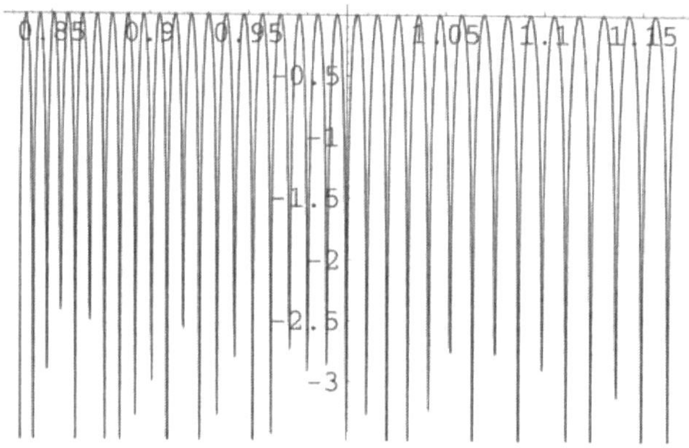

Figure 1 $N = 100$

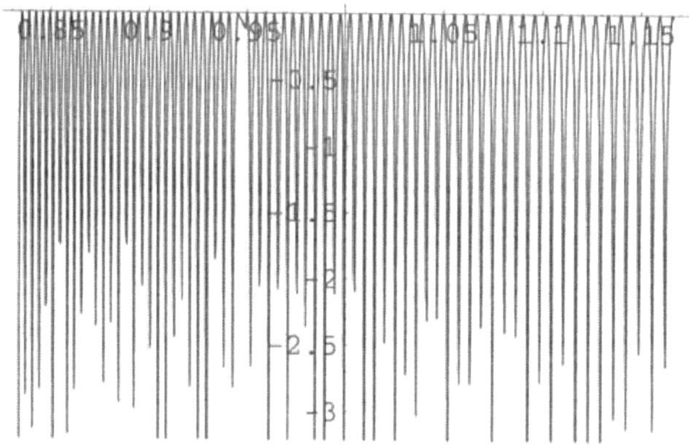

Figure 2 $N = 200$

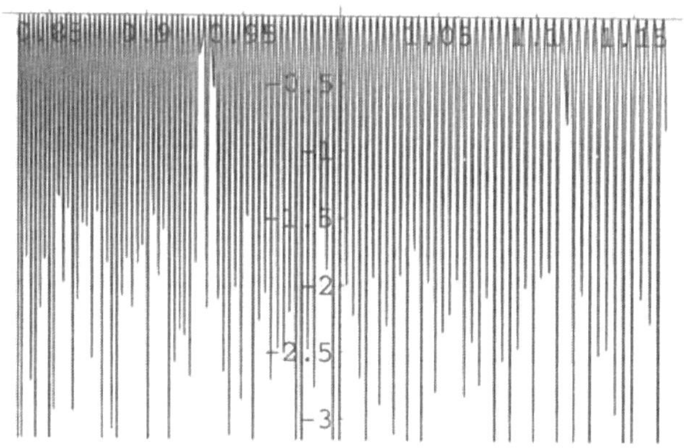

Figure 3 $N = 300$

We also draw the graph of

$$\operatorname{Re}\left\{\log J_N\bigl(E;\exp(2\pi\sqrt{-1}/N)\bigr)\right\} - \operatorname{Re}\left\{\frac{N}{\sqrt{-1}}S(0) + \frac{3}{2}\log N + \frac{1}{2}\log\frac{T_E(0)}{2\pi^2}\right\}$$

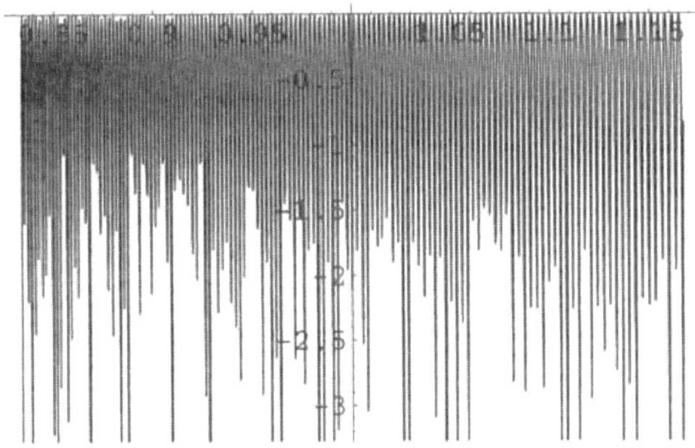

Figure 4 $N = 400$

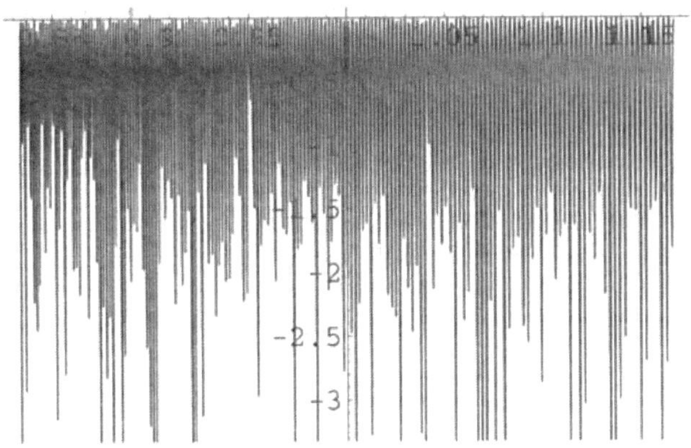

Figure 5 $N = 500$

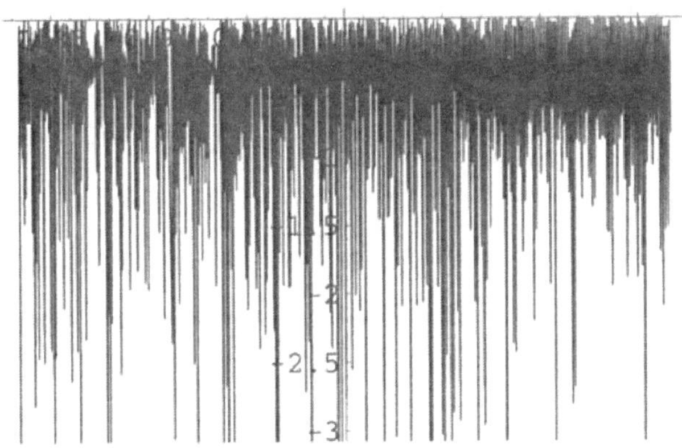

Figure 6 $N = 1000$

for $N = 100 \times n$ with $1 \leq n \leq 100$.

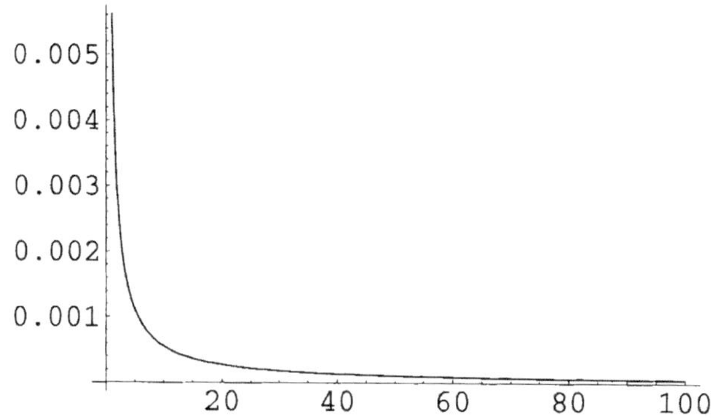

Figure 7 The horizontal axis corresponds to $N/100$.

References

[1] S. Axelrod and I. M. Singer, *Chern-Simons perturbation theory*, Proceedings of the XXth International Conference on Differential Geometric Methods in Theoretical Physics, Vol. 1, 2 (New York, 1991) (River Edge, NJ), World Sci. Publishing, 1992, pp. 3–45.

[2] D. Bar-Natan, *Perturbative Aspects of the Chern-Simons Topological Quantum Field Theory*, Ph.D. thesis, Princeton Univeristy, June 1991.

[3] H. U. Boden, C. M. Herald, P. A. Kirk, and E. P. Klassen, *Gauge theoretic invariants of Dehn surgeries on knots*, Geom. Topol. **5** (2001), 143–226 (electronic).

[4] J. Cheeger, *Analytic torsion and the heat equation*, Ann. of Math. (2) **109** (1979), no. 2, 259–322.

[5] D. Cooper, M. Culler, H. Gillet, D. D. Long, and P. B. Shalen, *Plane curves associated to character varieties of 3-manifolds*, Invent. Math. **118** (1994), no. 1, 47–84.

[6] J. Dubois, *Non abelian twisted Reidemeister torsion for fibered knots*, Canad. Math. Bull. **49** (2006), no. 1, 55–71.

[7] D. S. Freed and R. E. Gompf, *Computer calculation of Witten's 3-manifold invariant*, Comm. Math. Phys. **141** (1991), no. 1, 79–117.

[8] S. Garoufalidis and T. T. Q. Le, *An analytic version of the Melvin-Morton-Rozansky Conjecture*, arXiv:math.GT/0503641.

[9] S. Gukov, *Three-dimensional quantum gravity, Chern-Simons theory, and the A-polynomial*, Comm. Math. Phys. **255** (2005), no. 3, 577–627.

[10] C. D. Hodgson and S. P. Kerckhoff, *Rigidity of hyperbolic cone-manifolds and hyperbolic Dehn surgery*, J. Differential Geom. **48** (1998), no. 1, 1-59.

[11] R. M. Kashaev, *The hyperbolic volume of knots from the quantum dilogarithm*, Lett. Math. Phys. **39** (1997), no. 3, 269–275.

[12] W. B. R. Lickorish, *An introduction to knot theory*, Graduate Texts in Mathematics, vol. 175, Springer-Verlag, New York, 1997.

[13] W. Müller, *Analytic torsion and R-torsion of Riemannian manifolds*, Adv. in Math. **28** (1978), no. 3, 233–305.

[14] H. Murakami, *A version of the volume conjecture*, Adv. Math. **211** (2007), no. 2, 678–683.

[15] ———, *The colored Jones polynomials and the Alexander polynomial of the figure-eight knot*, J.P.J. Geom. Topol. **7** (2007), no. 2, 219–269.

[16] ———, *Some limits of the colored Jones polynomials of the figure-eight knot*, Kyungpook Math. J. **44** (2004), no. 3, 369–383.

[17] H. Murakami and J. Murakami, *The colored Jones polynomials and the simplicial volume of a knot*, Acta Math. **186** (2001), no. 1, 85-104.

[18] H. Murakami and Y. Yokota, *The colored Jones polynomials of the figure-eight knot and its Dehn surgery spaces*, to appear in J. Reine Angew. Math., arXiv:math.GT/0401084.

[19] W. D. Neumann and D. Zagier, *Volumes of hyperbolic three-manifolds*, Topology **24** (1985), no. 3, 307–332.

[20] J. Porti, *Torsion de Reidemeister pour les variétés hyperboliques*, Mem. Amer. Math. Soc. **128** (1997), no. 612, x+139.

[21] D. B. Ray and I. M. Singer, *R-torsion and the Laplacian on Riemannian manifolds*, Advances in Math. **7** (1971), 145–210.

[22] ———, *Analytic torsion for complex manifolds*, Ann. of Math. (2) **98** (1973), 154–177.

[23] A. Weil, *On discrete subgroups of Lie groups*, Ann. of Math. (2) **72** (1960), 369–384.

[24] ———, *On discrete subgroups of Lie groups. II*, Ann. of Math. (2) **75** (1962), 578–602.

[25] ———, *Remarks on the cohomology of groups*, Ann. of Math. (2) **80** (1964), 149–157.

[26] E. Witten, *Quantum field theory and the Jones polynomial*, Comm. Math. Phys. **121** (1989), 351–399.

[27] N. M. J. Woodhouse, *Geometric quantization*, Oxford Mathematical Monographs, The Clarendon Press Oxford University Press, New York, 1992.

[28] H. Zheng, *Proof of the volume conjecture for Whitehead doubles of a family of torus knots*, Chin. Ann. Math. Ser. B. **28** (2007), no. 4, 375–388.

Fields Institute Communications
Volume **54**, 2008

Open Strings and Extended Mirror Symmetry

Johannes Walcher
School of Natural Sciences
Institute for Advanced Study
Princeton, NJ 08540 USA
walcher@itp.phys.ethz.ch

Abstract. The classical mirror theorems relate the Gromov-Witten theory of a Calabi-Yau manifold at genus 0 to the variation of Hodge structure of an associated mirror manifold. I review recent progress in extending these closed string results to the open string sector. Specifically, the open Gromov-Witten theory of a particular Lagrangian submanifold of the quintic hypersurface is related to the Abel-Jacobi map for a particular object in the derived category of coherent sheaves of the mirror quintic. I explain the relevance for the homological mirror symmetry program.

CONTENTS

1 Introduction

Mirror symmetry is one of the best-developed examples of the renewed interaction between theoretical physics and certain areas of pure mathematics that has taken place over the past several decades. From the string theory point of view, it is natural to expect that there exists a far reaching generalization of classical concepts of geometry that will incorporate the idea that the fundamental building blocks are extended objects with some degree of non-locality. There is at present no precise idea about the form such a theory should ultimately take. One has to be content that most results such as those derived from mirror symmetry take the

2000 *Mathematics Subject Classification.* Primary 14N35; Secondary 14J32, 14O07, 14K30, 81T35, 81T45.

form of unusual looking statements about classical geometric objects. Mirror symmetry, in its simplest form, relates certain properties of two Calabi-Yau manifolds[1] X and Y of different topology. The basic necessary condition for X and Y to form a *mirror pair* is that the Hodge numbers satisfy

$$h^{p,q}(Y) = h^{n-p,q}(X) \qquad \text{for } 0 \leq p,q \leq n \tag{1.1}$$

but this relation has to be enhanced with a lot more life before we call it mirror symmetry.

The tightest possible connection between mirror manifolds X and Y is the one offered by string theory: The physical theory associated with the string background defined by X is isomorphic to the physical theory associated with the string background defined by Y. The problem with this statement is that at present there is no complete definition of string theory, so this is best viewed either as a fond wish or as a constraint on a future definition. The statement of perturbative string theory, which is well-understood, is that the two-dimensional so-called sigma-models with X and Y as target spaces are isomorphic as $\mathcal{N} = 2$ superconformal field theories (SCFTs). This can also be expressed as saying that X and Y are identical when viewed in "string geometry". Although there are precise mathematical definitions of such superconformal field theories, the process of attaching an SCFT to a Calabi-Yau manifold is fairly complicated and mathematically out of reach at the moment.

Mirror symmetry celebrated its first success after a computation by Candelas, de la Ossa, Green and Parkes [1]. These authors used the above conjectural equivalence of $\mathcal{N} = 2$ SCFT (as well as some other information about the string theory interpretation of the Calabi-Yau background) to make a prediction about the number of rational curves on a generic quintic threefold. That prediction was expressed in terms of the periods of an assumed mirror manifold (called the mirror quintic) previously constructed by Greene and Plesser [2]. Thus mirror symmetry entered the world of mathematics as the challenge to understand and verify the predictions of Candelas et al.

The theory relevant to the computations on the mirror manifold ("B-model") was understood to be related to classical Hodge theory [3]. Based on the development of Gromov-Witten theory, the enumerative predictions ("A-model") of Candelas et al. were verified in subsequent years, culminating in the proof of the now classical "mirror theorems" [4, 5, 6], which amount to

$$\text{Gromov-Witten theory of } X \xleftrightarrow{\text{solved by}} \text{Hodge theory of } Y \tag{1.2}$$

We refer to [7, 8] for textbook treatments of these developments. What (1.2) does not provide, however, is an intrinsic "explanation" of why these theorems would hold, in other words, answer the question "What is mirror symmetry?"

There are two proposals that are in a sense midway between the classical mirror theorems and the full equivalence of the $\mathcal{N} = 2$ SCFTs. The first of these is known as the "homological mirror symmetry" program (or conjecture) of Kontsevich [9]. The basic idea is a kind of "categorification" of the enumerative predictions. To any Calabi-Yau manifold, one can attach, if not a full SCFT, at least two categories. The B-model category, $D^b(Y)$ is the bounded derived category of coherent sheaves,

[1] For us here, a Calabi-Yau manifold is a Kähler manifold with trivial canonical bundle and whose Hodge numbers satisfy $h^{k,0} = 0$ for $0 < k < n$, where n denotes the complex dimension of the manifold. We are interested in the case $n = 3$.

while the A-model category, Fuk(X), is a certain derived version of a symplectic category constructed by Fukaya (et al.) using (Lagrangian intersection) Floer theory. Homological mirror symmetry amounts to the statement that if X and Y are a mirror pair, the associated categories are equivalent up to interchange of A- and B-models,

$$\mathrm{Fuk}(X) \cong D^b(Y) \tag{1.3}$$

(More precisely, the full symmetry includes the statement $D^b(X) \cong \mathrm{Fuk}(Y)$. One is normally interested only in (1.3), especially when studying manifolds that are not Calabi-Yau.)

The second approach to understanding mirror symmetry from the mathematical point of view was proposed by Strominger, Yau and Zaslow (SYZ) [10]. Roughly speaking, X and Y are a mirror pair if they can both be written as fibrations by (generically) special Lagrangian tori over a common base, in such a way that the tori for X are (generically) dual to the tori for Y.

$$\begin{array}{ccccc} T^n & \longrightarrow & X & \searrow & \\ & & & & B \\ (T^n)^\vee & \longrightarrow & Y & \nearrow & \end{array} \tag{1.4}$$

This SYZ conjecture thus gives a purely geometric way of understanding the relation between the two manifolds in a mirror pair.

The present mathematical challenges of mirror symmetry include making the statements (1.3), (1.4) more precise, proving them, and finding relations between either of them and (1.2). All these conjectures are strongly backed by physical ideas from string theory, centrally involving D-branes. D-branes are boundary conditions where open strings can end and are required when the $\mathcal{N}=2$ SCFTs are considered on two-dimensional spaces with boundary. The objects in the A- and B-model categories appearing in (1.3) are models of (topological) D-branes, and the equivalence of $\mathcal{N}=2$ SCFTs would imply the equivalence of D-brane categories. In a similar vein, the special Lagrangian tori appearing in (1.4) are a special type of D-branes, referred to as "BPS", and the original argument of [10] deduced the geometric picture from the equivalence of physical theories and the properties of BPS branes therein.

The purpose of the present contribution is to review recent progress in extending classical mirror symmetry (1.2) to the open string sector, in a way that takes it closer to the homological mirror symmetry conjecture (1.3). The underlying ideas have been worked out for the particular case of the real quintic and its mirror (definitions below) in the papers [11, 12, 13], and efforts are under way to adapt the technology to other situations.[2]

Remark. We concentrate in this review on the progress made on the compact Calabi-Yau quintic threefold. Tremendous progress has been made on open Gromov-Witten theory, and mirror symmetry, for non-compact (local) Calabi-Yau manifolds. The starting point of those developments is [14]; see [15] for a review. Relations to the present work are suggestive, but have not been worked out in detail yet.

[2]During the Banff meeting, it was pointed out by G. Almkvist that results similar to those of [11] should hold for many other one-parameter Calabi-Yau manifolds. An extension of the corresponding Picard-Fuchs equations by an inhomogeneous term $\sim \sqrt{z}$ leads to an integral open string expansion similar to the quintic. It would be interesting to identify the correct normalization of these computations.

Acknowledgments. I would like to thank Noriko Yui, Helena Verrill, and Chuck Doran for organizing the productive week in Banff, and for the invitation to speak at the meeting, as well as the other participants for many useful comments. This work was supported in part by the Roger Dashen Membership at the IAS and by the NSF under grant number PHY-0503584.

2 A-model

Let X be a generic quintic Calabi-Yau threefold

$$X = \{V(x_1, \dots, x_5) = 0\} \subset \mathbb{P}^4 \tag{2.1}$$

where V is a homogeneous polynomial of degree 5 in 5 variables. In the A-model, one is interested in X as a symplectic manifold, so the precise choice of V, which corresponds to the choice of complex structure, does not matter. Namely, the *Fukaya category* $\mathrm{Fuk}(X)$ does not depend on V (at least as long as it is non-singular). It is nevertheless useful to remember V, for the following reason. A systematic way to define Lagrangian submanifolds of Calabi-Yau manifolds such as X is as the fixed point locus of an anti-holomorphic involution with respect to some complex structure. The generic quintic X in (2.1) admits an anti-holomorphic involution, given simply by complex conjugation on the homogeneous coordinates in $\mathbb{P}^4$, if V is real, and we can then consider the real locus

$$L = \{x_i = \bar{x}_i\} \subset X \tag{2.2}$$

This definition has the problem that while the A-model should not depend on V, the real locus L does. Both the topological type of L as well as even its homology class in $H_3(X;\mathbb{Z})$ change along certain singular loci in the moduli space of quintics. This problem does not affect the considerations in the present section, but for the application to mirror symmetry, it is necessary to fix a particular choice of V.

We will use the term "the real quintic" for the Lagrangian submanifold $L \subset X$ defined as the real locus of X in the complex structure at the Fermat point,

$$V_{\text{Fermat}} = x_1^5 + x_2^5 + x_3^5 + x_4^5 + x_5^5 \tag{2.3}$$

Topologically, the real quintic is isomorphic to $\mathbb{RP}^3$. This follows from the fact that the equation $V = 0$ over the reals has a unique solution for x_5 (say) in terms of $x_1, \dots, x_4$.

In this section, we review the open Gromov-Witten theory of the pair (X, L), and describe its solution in terms of a certain differential equation that is an inhomogeneous version (or extension) of the ordinary hypergeometric differential equation governing closed string mirror symmetry for the quintic.

2.1 Open Gromov-Witten invariants. Classical Gromov-Witten theory is concerned with studying intersection theory on the moduli space of stable maps

$$\overline{\mathcal{M}}_{g,n}(X, \beta) \tag{2.4}$$

Here, the data are the genus g of the domain curve (Riemann surface) with some number n of marked points, a target manifold X, and a (non-zero) homology class $\beta \in H_2(X;\mathbb{Z})$. The target X is either as taken as a (projective) algebraic variety or a symplectic manifold. Cohomology classes on $\overline{\mathcal{M}}_{g,n}(X, \beta)$ are obtained by pulling back classes from X via evaluation at the marked points, or from the moduli space of stable curves via the projection $\overline{\mathcal{M}}_{g,n}(X, \beta) \to \overline{\mathcal{M}}_{g,n}$ (being propitious with that last statement).

One of the main motivations to develop Gromov-Witten theory was to verify the physicists' enumerative predictions about rational curves on the quintic. Calabi-Yau threefolds are special from the point of view of Gromov-Witten theory, as the (virtual) dimension of $\overline{\mathcal{M}}_{g,0}(X,\beta)$ is zero for any g and β. Since $H_2(X,\mathbb{Z}) \cong \mathbb{Z}$, we can encode the target class β in an integer degree d. The *Gromov-Witten invariants* are defined by the integral

$$\tilde{N}_d^{(g)} = \int_{[\overline{\mathcal{M}}_{g,0}(X,d)]^{\mathrm{vir}}} 1 \tag{2.5}$$

that formally computes the degree of the "virtual fundamental class". A nice feature of $g = 0$, any d, is that the Gromov-Witten theory of the quintic can be obtained from the Gromov-Witten theory of $\mathbb{P}^4$ by the "Euler class formula",

$$\tilde{N}_d \equiv \tilde{N}_d^{(0)} = \int_{\overline{\mathcal{M}}_{0,0}(\mathbb{P}^4,d)} \mathbf{e}(\mathcal{E}_d) \tag{2.6}$$

where

$$\mathcal{E}_d = \pi_* f^* \mathcal{O}_{\mathbb{P}^4}(5) \tag{2.7}$$

is the obstruction bundle whose fiber at $f \in \overline{\mathcal{M}}_{0,0}(\mathbb{P}^4,d)$ is the space of global sections $H^0(\mathbb{P}^1, f^*\mathcal{O}(5))$ of the pull-back of the bundle of quintics on $\mathbb{P}^4$. The $\tilde{N}_d$ are in general rational numbers, but the Aspinwall-Morrison multi-cover formula,

$$\tilde{N}_d = \sum_{k|d} \frac{1}{k^3} N_{d/k} \tag{2.8}$$

allows the definition of "enumerative" invariants N_d, whose integrality was the main early reason for confidence in the Candelas et al. computation, but has been established rigorously only recently [16, 17].

It is suggested by the physics of D-branes or by applications to homological mirror symmetry that one should try to extend Gromov-Witten theory to the situation when the domain curve, Σ, is a Riemann surface with non-empty boundary, $\partial\Sigma \neq \emptyset$. The prototypical example is the unit disk, or the upper half-plane in $\mathbb{C}$. One should impose Dirichlet boundary conditions that require the boundary of Σ to map to a Lagrangian submanifold L of the target space X (as a symplectic manifold). One then speaks of the Gromov-Witten theory of the pair (X, L).

Remark. The physics of orientifolds further suggests the definition of Gromov-Witten theory when Σ is not orientable (the prototypical example being the real projective space $\mathbb{RP}^2$). One then requires the data of an anti-symplectic involution of the target space and the maps are equivariant maps from the orientation double cover of Σ to X. This has been studied in non-compact situations, *e.g.*, in [18, 19, 20].

The problems with defining Gromov-Witten theory when the domain curve has non-empty boundary or is non-orientable is that moduli spaces tend to be real manifolds. This means that one has to keep the symplectic side in the definitions, which is technically slightly more demanding. But more fundamentally, it results in the appearance of special (singular) loci in the moduli spaces in *real codimension one*. One needs to find suitable boundary conditions at these boundaries in moduli space in order to define an intersection theory. A general attitude towards this problem is that the physicists' concept of a D-brane includes not only the boundary conditions on the maps, but also boundary conditions at these boundaries in moduli space. However, precise prescriptions have been given only in a limited number of cases so far, such as [21, 22].

Solomon [22] has worked out a definition of open Gromov-Witten invariants when the worldsheet is the disk, and the target space data is a pair (X, L), where L is a Lagrangian submanifold that is the fixed point set of an anti-symplectic involution.[3] The essential idea is to eliminate (some of) the boundaries of moduli space by replacing any bubbling disk by its image under the anti-symplectic involution.

It is shown in [12] that for the real quintic pair (X, L) described above, and *odd degree*, $d \in 2\mathbb{Z}+1$,[4] the open Gromov-Witten invariants of [22] can moreover be computed from the open Gromov-Witten theory of the pair $(\mathbb{CP}^4, \mathbb{RP}^4)$, in a way similar to (2.6). The involution (complex conjugation on $\mathbb{CP}^4$) induces an involution on $\overline{\mathcal{M}}_{0,0}(\mathbb{P}^4, d)$. Over the real locus, $\overline{\mathcal{M}}^{\mathbb{R}}_{0,0}(\mathbb{P}^4, d)$, the involution induces a real structure on $\mathcal{E}_d$, and one has the Euler class formula,

$$\tilde{n}_d = \int_{\overline{\mathcal{M}}^{\mathbb{R}}_{0,0}(\mathbb{P}^4)} \mathbf{e}(\mathcal{E}_d^{\mathbb{R}}) . \tag{2.9}$$

The difficult parts in the proof and evaluation of this formula arise from subtle orientation issues; see [12] for details. Finally, we note that the analogue of (2.8) is

$$\tilde{n}_d = \sum_{k|d} \frac{1}{k^2} n_{d/k} \tag{2.10}$$

which is essentially the multi-cover formula for the disk first predicted by Ooguri and Vafa [23]. Given the results for the $\tilde{n}_d$ below, and the proof of integrality of the closed string invariants N_d [17], it should be possible to prove integrality of the n_d as well.

2.2 Schubert computation in degree 1. As an elementary application of the definition (2.9), let us compute the simplest case, $n_1 = \tilde{n}_1$. First we review the complex case, which can be done by Schubert calculus [24].

The space of lines in $\mathbb{P}^4$ is isomorphic to the space of 2-dimensional linear subspaces of $\mathbb{C}^5$, *i.e.*, the complex Grassmannian $G(2,5)$. This Grassmannian comes equipped with the tautological rank two bundle U, fitting into the sequence

$$U \to \mathbb{C}^5 \to V \tag{2.11}$$

where $\mathbb{C}^5$ stands for the trivial rank 5 bundle, and V is the universal quotient bundle. The cohomology ring of $G(2,5)$ is generated in degree 2 and 4 by the Chern classes of U, $c(U) = (1+a)(1+b) = 1 + u_1 + u_2$, where $a+b = u_1$, $a \cdot b = u_2$. The relations follow from (2.11), $c(U) \cdot c(V) = 1$. As is well-known, these relations can be integrated to a potential,

$$V(u_1, u_2) = \frac{a^6}{6} + \frac{b^6}{6} = \frac{u_1^6}{6} - u_2 u_1^4 + \frac{3}{2} u_2^2 u_1^2 - \frac{1}{3} u_2^3 \tag{2.12}$$

Namely, the intersection ring of $G(2,5)$ is $\mathbb{C}[u_1, u_2]/(\partial_1 V, \partial_2 V)$. A quintic on $\mathbb{P}^4$ induces a section of the bundle $\mathcal{E}_1 = \mathrm{Symm}^5(U))$ over $G(2,5)$. We have $c(\mathcal{E}_1) =$

[3]This involution might appear similar to, but should not be confused with, the involution appearing in the above remark on orientifolds.

[4]This degree is defined as the relative cohomology class in $H_2(X, L; \mathbb{Z}) \cong \mathbb{Z}$.

$\prod_{n=0}^{5}(1+na+(5-n)b)$, from which

$$c_6(\mathcal{E}_1) = 5a(4a+b)(3a+2b)(2a+3b)(a+4b)5b$$
$$= 25u_2(9u_2^3 + 24u_2u_1^4 + 58u_2^2u_1^2) \quad (2.13)$$

Thus,

$$N_1 = \int_{G(2,5)} c_6(\mathrm{Symm}^5 U^*) = 2875 \quad (2.14)$$

where we used the relations from (2.12), as well as $\int_{G(2,5)} u_2^3 = 1$. This last relation can be derived by expressing the Euler class of $G(2,5)$ in terms of u_1, u_2, and using $\chi(G(2,5)) = 10$.

To repeat the exercise over the reals, we use the Grassmannian of (oriented) real lines, $G^{\mathbb{R}}(2,5)$, which comes with its own "tautological sequence",

$$U_{\mathbb{R}} \to \mathbb{R}^5 \to V_{\mathbb{R}} \quad (2.15)$$

Actually, the sequence (2.15) gives easy access only to cohomology with $\mathbb{Z}_2$ coefficients, see [25]. For our purposes, it is more useful to think of $G^{\mathbb{R}}(2,5)$ as the Hermitian symmetric space

$$G^{\mathbb{R}}(2,5) \cong \frac{SO(5)}{SO(2)\times SO(3)} \quad (2.16)$$

which can also be represented as the quadric

$$G^{\mathbb{R}}(2,5) \cong \{z_1^2+z_2^2+z_3^2+z_4^2+z_5^2=0\} \subset \mathbb{CP}^4 \quad (2.17)$$

The cohomology ring of $G^{\mathbb{R}}(2,5)$ has one generator x in degree 2 and one generator y in degree 3, and relations that can be integrated to the potential

$$V_{\mathbb{R}}(x,y) = \frac{1}{4}x^4 + xy^2 \quad (2.18)$$

From the presentation (2.17), we find $\chi(G^{\mathbb{R}}(2,5)) = 4 = \int 2x^3$. We also have $\mathbf{e}(U) = x$, and by setting $b=-a$ in (2.2), and taking a square root, we find

$$\mathbf{e}(\mathrm{Symm}^5(U_{\mathbb{R}})) = 15x^3 \quad (2.19)$$

which integrates to

$$n_1 = \int_{G^{\mathbb{R}}(2,5)} \mathbf{e}(\mathrm{Symm}^5(U)) = 30 \quad (2.20)$$

This is the result first obtained by Solomon [22]. The complete formula for all open Gromov-Witten invariants of the disk for the real quintic has been predicted in [11], and was proved in [12].

2.3 Result. Recall that classical mirror symmetry for the quintic is governed by the differential operator

$$\mathcal{L} = \theta^4 - 5z(5\theta+1)(5\theta+2)(5\theta+3)(5\theta+4)\,, \quad (2.21)$$

where $\theta = zd/dz$. Four linearly independent solutions, $\varpi_i(z)$, $i=0,1,2,3$, of (2.21) can be obtained from the hypergeometric series

$$\mathfrak{F}(z,H) = \sum_{n=0}^{\infty} z^{H+n}\frac{\prod_{r=1}^{5n}(5H+r)}{\prod_{r=1}^{n}(H+r)^5} \quad (2.22)$$

as the coefficients of the Taylor series around $H = 0$

$$\mathfrak{F}(z,H) = \sum_{i=0}^{3} \varpi_i(z) H^i \bmod H^4 \tag{2.23}$$

The prediction of Candelas et al. [1] is that the generating function of genus 0 Gromov-Witten invariants on the quintic,

$$F(t) = \frac{5}{6}t^3 + \sum_{d=1}^{\infty} \tilde{N}_d q^d \tag{2.24}$$

is given by a certain "magical" combination of solutions of the differential equation

$$F(t(z)) = \frac{5}{2}\left(\frac{\varpi_1(z)}{\varpi_0(z)}\frac{\varpi_2(z)}{\varpi_0(z)} - \frac{\varpi_3(z)}{\varpi_0(z)}\right) \tag{2.25}$$

where the variables $q \equiv e^t$ are related to the variable z of (2.21) via the "mirror map",

$$q(z) = \exp\bigl(\varpi_1(z)/\varpi_0(z)\bigr)\,. \tag{2.26}$$

Turning to the real quintic, we form the generating function of open Gromov-Witten invariants,

$$T(t) = \sum_{\substack{d=1 \\ d \text{ odd}}}^{\infty} \tilde{n}_d q^{d/2} \tag{2.27}$$

and introduce a "mirror" function by

$$\tau(z) = 2\sum_{d \text{ odd}} \frac{(5d)!!}{(d!!)^5} z^{d/2} \tag{2.28}$$

This function is a solution of the inhomogeneous Picard-Fuchs equation

$$\mathcal{L}\tau(z) = \frac{15}{8}\sqrt{z} \tag{2.29}$$

and is also related to the hypergeometric series (2.22) as

$$\tau(z) = 30\,\mathfrak{F}(z,1/2) \tag{2.30}$$

Theorem A. *Let $T(t)$ be the generating function of open Gromov-Witten invariants of the pair (X,L), $q(z)$ be the mirror map, and $\tau(z)$ the solution of the inhomogeneous Picard-Fuchs equation* (2.29), *all as defined above. We have*

$$T(t(z)) = \frac{\tau(z)}{\varpi_0(z)} \tag{2.31}$$

2.4 Proof by localization. Kontsevich has given a formula that computes the Gromov-Witten invariants $\tilde{N}_d$ from (2.6) via localization on $\overline{\mathcal{M}}_{0,0}(\mathbb{P}^4,d)$ with respect to the action of the five-dimensional torus

$$\mathbb{T}^5 \subset GL(5,\mathbb{C}) : \mathbb{CP}^4 \to \mathbb{CP}^4 \tag{2.32}$$

The fixed loci of the torus action are encoded combinatorially in certain decorated graphs Γ and can be described geometrically as moduli spaces of curves with marked points (and trivial target space). The integrals over the fixed loci are known explicitly, and allow the evaluation of the Atiyah-Bott equivariant localization formula

$$\tilde{N}_d = \sum_{\Gamma} \frac{1}{|\mathrm{Aut}\,\Gamma|} \int_{\overline{\mathcal{M}}_\Gamma} \frac{\mathbf{e}(\mathcal{E}_d)}{\mathbf{e}(\mathcal{N}_\Gamma)} \tag{2.33}$$

The essential idea to prove the analogous result for the open Gromov-Witten invariants is to use the presentation of $\tilde{n}_d$ as an integral over the real moduli space; see (2.9). Inspection shows that the involution σ (complex conjugation) commutes with a two-dimensional subtorus of (2.32)

$$\mathbb{T}^2 \subset GL(5,\mathbb{R}) : \mathbb{RP}^4 \to \mathbb{RP}^4 \tag{2.34}$$

The fixed loci of this $\mathbb{T}^2$ action on the real moduli space are identical to the fixed loci of the $\mathbb{T}^5$ action which are invariant under the involution. Thus, we need to evaluate a sum over the subset of graphs in (2.33) satisfying $\sigma(\Gamma) = \Gamma$. Inspection shows that these graphs all have a distinguished "central" edge that is invariant under the involution and is decorated by an odd integer degree, call it e. We cut the graphs through this distinguished edge into two pieces, γ and $\sigma(\gamma)$, each of which represents a disk of degree d. Finally, we recall that if $E = E_{\mathbb{R}} \otimes \mathbb{C}$ is a complex vector bundle with a real structure, we have $\mathbf{e}(E_{\mathbb{R}}) = \pm\sqrt{\mathbf{e}(E)}$, up to orientation. This allows the easy evaluation of the (equivariant) Euler class of the real bundles in (2.9). Since the moduli spaces of pointed curves all originate away from that distinguished edge, the integral over the fixed loci can be evaluated as before. We thus obtain the formula,

$$\tilde{n}_d = 2\sum_{\gamma} \frac{1}{|\mathrm{Aut}\,\gamma|} \int_{\overline{\mathcal{M}}_\gamma} \frac{\mathbf{e}(\mathcal{E}_d^{\mathbb{R}})}{\mathbf{e}(\mathcal{N}_\gamma^{\mathbb{R}})} = \sum_{\Gamma=\sigma(\Gamma)} \pm\sqrt{\frac{2}{e|\mathrm{Aut}\,\Gamma|}}\sqrt{\int_{\overline{\mathcal{M}}_\Gamma} \frac{\mathbf{e}(\mathcal{E}_d)}{\mathbf{e}(\mathcal{N}_\Gamma)}} \tag{2.35}$$

where the sign of the square root is fixed by careful consideration of the orientation of the various spaces involved.

To handle the combinatorics of (2.35), we view the graphs γ as graphs representing spheres of degree $(d-e)/2$ with one marked point at which is attached an "intersection disk". This observation allows us to reduce the sum in (2.35) to the evaluation of a certain equivariant correlator in closed Gromov-Witten theory. Formulas for such correlators were proved in [5, 26], and the rest is straightforward; see [12]. □

3 B-model

Let W denote the one-parameter family of degree five polynomials in five variables

$$W = \frac{1}{5}\left(x_1^5 + x_2^5 + x_3^5 + x_4^5 + x_5^5\right) - \psi x_1 x_2 x_3 x_4 x_5 \tag{3.1}$$

Let μ_5 be the multiplicative group of fifth roots of unity. The family (3.1) is invariant under the action of

$$\Gamma = \mathrm{Ker}\big((\mu_5)^5 \xrightarrow{p} \mu_5\big)/\mu_5 \tag{3.2}$$

where p is the product of the five fifth roots and the identification is by the diagonal subgroup. The mirror quintic, Y, is constructed by blowing up the singular loci in the quotient of the one-parameter family of quintics by the group Γ

$$Y = \widetilde{\{W = 0\}/\Gamma} \tag{3.3}$$

We will usually confuse the manifold Y for a fixed value of ψ with the family of manifolds $Y \xrightarrow{\pi} M$ varying over the base curve $M \ni \psi$.

In this section, we are concerned with the "B-model" on Y. This means that we are interested in Y as an algebraic variety and in the variation of various data attached to Y as the parameter ψ is varied. The precise way in which we do the resolution (3.3) is part of the Kähler data. It affects the A-model on Y, but not any of the considerations here.

3.1 Variation of Hodge structure. The third cohomology group of Y is $H^3(Y;\mathbb{C}) \cong \mathbb{C}^4$, and has the *Hodge decomposition*

$$H^3(Y;\mathbb{C}) = H^{3,0}(Y) \oplus H^{2,1}(Y) \oplus H^{1,2}(Y) \oplus H^{0,3}(Y) \tag{3.4}$$

As Y varies with ψ, the decomposition of $H^3(Y;\mathbb{C}) \cong H^3(Y;\mathbb{Z})\otimes\mathbb{C}$ (as a topological invariant) into the Dolbeault cohomology groups $H^{p,q}(Y) \cong H^q_{\bar\partial}(\Omega^p_Y)$ changes. This gives rise to a so-called *variation of Hodge structure*; see *e.g.*, [27]. The first key fact in that theory is that the Hodge decomposition (3.4) does not vary holomorphically over $M \ni \psi$. However, the *Hodge filtration*

$$F^pH^3(Y) = \bigoplus_{p'\geq p} H^{p',3-p'}(Y) \tag{3.5}$$

does vary holomorphically. The next key element in the theory of variation of Hodge structure is the flat ("Gauss-Manin") connection ∇ which originates in the local triviality of $H^3(Y;\mathbb{Z})$ over M, and satisfies Griffiths transversality

$$\nabla F^pH^3(Y) \subset F^{p-1}H^3(Y) \otimes \Omega_M \tag{3.6}$$

In this context, the Picard-Fuchs equation $\mathcal{L}\varpi(z) = 0$ that we have met (in the A-model) in the previous section can be understood as follows. Since Y is Calabi-Yau, there exists a unique-up-to-scale section of the canonical bundle, in other words, a holomorphic $(3,0)$-form, Ω. As ψ varies, we are dealing with a section, also denoted by Ω, of the holomorphic line bundle $F^3H^3(Y)$ over M.

Now given any topological three-cycle $\Gamma \in H_3(Y;\mathbb{Z})$, we can consider the *period integral*

$$\varpi = \int_\Gamma \Omega \tag{3.7}$$

The Picard-Fuchs equation is a differential equation satisfied by any period integral, as a function of the complex structure parameters. For the one-parameter family of quintics (3.1), and a particular section, $\hat\Omega$, of $F^3H^3(Y)$, the Picard-Fuchs equation takes the form $\mathcal{L}\varpi(z) = 0$, with

$$\mathcal{L} = \theta^4 - 5z(5\theta+1)(5\theta+2)(5\theta+3)(5\theta+4)\,, \tag{3.8}$$

as in (2.21). Here, $z = (5\psi)^{-5}$ and $\theta = zd/dz$. This equation is the same before or after quotienting by the group $\Gamma = (\mu_5)^3$.

The structure above is not special to any particular dimension; we have just written it out in dimension 3 for convenience. The central character in closed string mirror symmetry whose properties are special to dimension 3 is the so-called *Yukawa coupling*. It arises from the third iterate of the differential period mapping, $H^1(TY) \to \bigoplus \mathrm{Hom}\big(H^{p,q}(Y), H^{p-1,q+1}(Y)\big)$ by contraction with your choice of section of F^3H^3, and is usually written as

$$\kappa = \int_Y \Omega\wedge\nabla^3\Omega \tag{3.9}$$

Classical closed string mirror symmetry provides an identification between the generating function of genus 0 Gromov-Witten invariants of some Calabi-Yau X and the Yukawa coupling of the mirror manifold Y. There are two key steps. The first is to provide the *mirror map* between the "formal" (Kähler) parameters of X and the complex structure parameters of Y. The second is a normalization of the Yukawa coupling, *i.e.*, the identification of a particular distinguished choice of holomorphic three-form. Both steps can be formalized by studying the degeneration of the variation of Hodge structure at certain special points in moduli space known under the name "maximal unipotent monodromy". This is explained for the quintic in [3]; for full details; see *e.g.*, [7].

We now turn to the part of Hodge theory on Y that provides the B-model counterpart to the generating function for the open Gromov-Witten invariants of the real quintic studied in the previous section.

3.2 Normal functions. Let $\big(H^3(Y;\mathbb{Z}), F^*H^3(Y)\big)$ be an integral variation of Hodge structure of weight 3 as we had in the previous subsection. The *Griffiths intermediate Jacobian fibration* is the fibration $J^3(Y) \to M$ of complex tori

$$J^3(Y) = \frac{H^3(Y)}{F^2H^3(Y) \oplus H^3(Y;\mathbb{Z})} \tag{3.10}$$

An equivalent definition of J^3 is obtained by noting that by Poincaré duality we have an isomorphism $H^3(Y)/F^2H^3(Y) \cong (F^2H^3(Y))^*$, under which $H^3(Y;\mathbb{Z}) \cong H_3(Y;\mathbb{Z})$. Thus,

$$J^3(Y) = (F^2H^3(Y))^*/H_3(Y;\mathbb{Z}) \tag{3.11}$$

Next, a *Poincaré normal function* of the variation of Hodge structure is a holomorphic section ν of $J^3(Y)$ satisfying *Griffiths transversality for normal functions,*

$$\nabla\tilde{\nu} \in F^1H^3(Y) \otimes \Omega_M \tag{3.12}$$

where $\tilde{\nu}$ is an arbitrary lift of ν from $J^3(Y)$ to $H^3(Y)$ (the condition (3.12) does not depend on the lift).

An important source of normal functions are *homologically trivial algebraic cycles,* where in our situation, we are interested in cycles of codimension 2. Let C be such a homologically trivial algebraic cycle (formal integral combination of holomorphic curves). We are here assuming that C varies in a nice family with Y over M, again confusing some notation. C being homologically trivial means that there exists a three-chain Γ, varying over M and unique modulo $H_3(Y;\mathbb{Z})$ such that for all $m \in M$, $\partial\Gamma_m = C_m$.

To define a normal function ν_C attached to C, we use the definition (3.11). Any class $[\omega] \in F^2H^3(Y)$ can be represented locally on M by a three-form ω pointing in the fiber direction with at least two holomorphic indices. One can show that the definition

$$\nu_C([\omega])_m := \int_{\Gamma_m} \omega_m \tag{3.13}$$

satisfies all conditions from the above definition of a normal function (holomorphicity and transversality). This association defines a map from algebraic cycles modulo rational equivalence to the intermediate Jacobian $J^3(Y)$, known as the *Abel-Jacobi map.*

3.3 Extended Picard-Fuchs equation. After these general definitions, we consider again the one-parameter family of quintics (3.1). It contains for any ψ a twin family of algebraic curves of degree 2 given by

$$C_\pm = \{x_1 + x_2 = 0, x_3 + x_4 = 0, x_5^2 \pm \sqrt{5\psi} x_1 x_3 = 0\} \subset \{W = 0\} \tag{3.14}$$

Since the second Betti number of the quintic is equal to 1, those two curves must lie in the same homology class, and we can consider the normal function attached to $C = C_+ - C_-$. More precisely, we are here interested in the mirror quintic, so we should really be discussing this after taking the quotient by μ_5^3. However, this requires some explicit choices for the resolution of the singularities, which is somewhat cumbersome. We refer to [13] for details.

Given the normal function attached to $C = C_+ - C_-$, we can consider its "truncation", namely, the restriction of the integral to $[\omega] \in F^3 H^3$,[5]

$$\mathcal{T}_B = \mathcal{T}_B(z) = \int_\Gamma \hat{\Omega} \tag{3.15}$$

where $\hat{\Omega}$ is a particular choice of holomorphic three-form on Y, defined as Poincaré residue by the formula

$$\hat{\Omega} = \Big(\frac{5}{2\pi i}\Big)^3 \psi \operatorname{Res}_{W=0} \frac{\alpha}{W} \tag{3.16}$$

where α is the four-form on projective space

$$\alpha = \sum_i (-1)^{i-1} x_i dx_1 \wedge \cdots \wedge \widehat{dx_i} \wedge \cdots \wedge dx_5 \tag{3.17}$$

The choice of holomorphic three-form in (3.16) is precisely the one for which the Picard-Fuchs equation of the mirror quintic takes the form (3.8).

Theorem B. *Let $\mathcal{T}_B$ be the truncated normal function of the algebraic cycle $C = C_+ - C_-$, for the choice of holomorphic three-form given in* (3.16). *Let $\mathcal{L}$ be the Picard-Fuchs operator in* (3.8). *Then*

$$\mathcal{L}\mathcal{T}_B(z) = \frac{15}{16\pi^2}\sqrt{z} \tag{3.18}$$

The proof [13] is an explicit computation that uses the algorithm of Griffiths-Dwork to derive the Picard-Fuchs equation of hypersurfaces, and keeps careful track of all boundary terms originating from the fact that $\partial\Gamma = C$. One begins with the definition of $\hat{\Omega}$ as a residue to write the integral $\int_\Gamma \hat{\Omega}$ as an integral over a four-chain that is a little tube over Γ around the hypersurface $W = 0$. Application of $\mathcal{L}$ reduces this to an integral of a certain meromorphic three-form over a tube around the algebraic cycle C. The latter can be evaluated based on the following fact. The curves C_+, C_- lie entirely in the plane $P = \{x_1 + x_2 = 0, x_3 + x_4 = 0\}$. If one could restrict the tube to lie entirely in P as well, the integral of any meromorphic three-form over it would vanish. The reason one cannot restrict the computation to P is that the intersection of P with the hypersurface is the reducible degree five curve $x_5^5 - 5\psi x_5 x_1^2 x_3^2$. Any tube in P around say C_+ will intersect one of the other components, and therefore not encircle $W = 0$ as it should. For the same reason, one can reduce the computation to the intersection points of the components of $P \cap \{W = 0\}$, which is a straightforward computation in an affine patch. □

[5] Recall that $z = (5\psi)^{-5}$.

4 Mirror symmetry

Let's summarize what we have learned so far. By comparing (2.29) with (3.18) we learn that the generating function for open Gromov-Witten invariants of the real quintic satisfies, up to an overall normalization, the same inhomogeneous Picard-Fuchs equation as the truncated normal function associated with the algebraic cycle $C = C_+ - C_-$ on the mirror quintic. In this section, we put this result in the context of mirror symmetry.

We begin by presenting explicitly two objects in the category of B-branes on Y that we conjecture are equivalent under mirror symmetry to the two objects in the Fukaya category of X that can be defined from the real quintic. We will then explain the relation between those objects and the normal function ν_C. Finally, we will complete the generating function of open Gromov-Witten invariants to a full identification of the normal function in the A-model.

We also include two somewhat speculative considerations. The first is a formal statement relating real sections of the quintic with objects in the category of B-branes for the mirror. The second is another application of this proposed correspondence and concerns the A_∞ structure or Floer homology of the real quintic. A better understanding of these issues could be a possible starting point for establishing homological mirror symmetry for the quintic (and other hypersurfaces).

4.1 Matrix factorizations. Let $V \in \mathbb{C}[x_1, x_2, \ldots, x_5]$ be a polynomial. A *matrix factorization* of V is a $\mathbb{Z}_2$-graded free $\mathbb{C}[x_1, \ldots, x_5]$-module M equipped with an odd endomorphism $Q : M \to M$ of square V,

$$Q^2 = V \cdot \mathrm{id}_M \tag{4.1}$$

The category $\mathrm{MF}(V)$ is the triangulated category of matrix factorizations with morphisms given by Q-closed morphisms of free modules, modulo Q-exact morphisms. Matrix factorizations have been well-known objects since the mid '80's, see in particular [28], and it was proposed by Kontsevich that $\mathrm{MF}(V)$ should be a good description of B-type D-branes in a Landau-Ginzburg model based on the worldsheet superpotential V [29, 30, 31]. To apply this to the case of interest, we need a little bit of extra structure.

When V is of degree 5, the so-called homological Calabi-Yau/Landau-Ginzburg correspondence [32, 33, 34, 35] states that the derived category of coherent sheaves of the projective hypersurface $X = \{V = 0\} \subset \mathbb{P}^4$ is equivalent to the graded, equivariant category of matrix factorizations of the corresponding Landau-Ginzburg superpotential,

$$D^b(X) \cong \mathrm{MF}(V/\mu_5) \tag{4.2}$$

where μ_5 is the group of fifth roots of unity acting diagonally on $x_1, \ldots, x_5$. The analogous statement for the mirror quintic (specialize V to W from (3.1)) is

$$D^b(Y) \cong \mathrm{MF}(W/(\mu_5)^4) \tag{4.3}$$

where $(\mu_5)^4$ is the subgroup of phase symmetries of W whose product is equal to 1 (*i.e.*, $\mathrm{Ker}(p)$ in the notation of (3.2)).

To describe an object mirror to the real quintic, we begin by finding a matrix factorization of the one-parameter family of polynomials (3.1). If $S \cong \mathbb{C}^5$ is a

5-dimensional vector space, we can associate to its exterior algebra a $\mathbb{C}[x_1,\ldots,x_5]$-module $M = \bigwedge^* S \otimes \mathbb{C}[x_1,\ldots,x_5]$. It naturally comes with the decomposition

$$M = M_0 + M_1 + M_2 + M_3 + M_4 + M_5\,, \quad \text{where } M_s = \bigwedge^s S \otimes \mathbb{C}[x_1,\ldots,x_5], \tag{4.4}$$

and the $\mathbb{Z}_2$-grading $(-1)^i$. Let η_i be a basis of S and $\bar\eta_i$ the dual basis of S^*, both embedded in $\mathrm{End}(M)$. We then define two families of matrix factorizations $(M, Q_\pm)$ of W by

$$Q_\pm = \frac{1}{\sqrt5}\sum_{i=1}^{5}(x_i^2\eta_i + x_i^3\bar\eta_i) \pm \sqrt\psi \prod_{i=1}^{5}(\eta_i - x_i\bar\eta_i) \tag{4.5}$$

To check that $Q_\pm^2 = W\cdot \mathrm{id}_M$, one uses that η_i, $\bar\eta_i$ satisfy the Clifford algebra

$$\{\eta_i, \bar\eta_j\} = \delta_{ij} \tag{4.6}$$

as well as the ensuing relations

$$\{(x_i^2\eta_i + x_i^3\bar\eta_i), (\eta_i - x_i\bar\eta_i)\} = 0 \qquad \text{and} \qquad (\eta_i - x_i\bar\eta_i)^2 = -x_i \tag{4.7}$$

The matrix factorization (4.5) is quasi-homogeneous ($\mathbb{C}^*$-gradable), but we will not need this data explicitly.

Now to specify objects in $\mathrm{MF}(W/\Gamma)$, where $\Gamma = \mu_5$ or $(\mu_5)^4$ for the quintic and mirror quintic, respectively, we have to equip M with a representation of Γ such that $Q_\pm$ is equivariant with respect to the action of Γ on the x_i. Since $Q_\pm$ is irreducible, this representation of Γ on M is determined up to a character of Γ by a representation on S, *i.e.*, an action on the η_i. For $\gamma\in\Gamma$, we have $\gamma(x_i) = \gamma_i x_i$ for some fifth root of unity γ_i. We then set $\gamma(\eta_i) = \gamma_i^{-2}\eta_i$, making $Q_\pm$ equivariant. As noted, this representation is unique up to an action on M_0, *i.e.*, a character of Γ.

For the mirror quintic, $\Gamma = \mathrm{Ker}((\mu_5)^5 \to \mu_5)$, so $\Gamma^* = (\mu_5)^5/\mu_5$, and we label the characters of Γ as $[\chi]$. The corresponding objects of $\mathrm{MF}(W/\Gamma)$ constructed out of $Q_\pm$ (4.5) are classified as $Q_\pm^{[\chi]} = (M, Q_\pm, \rho_{[\chi]})$, where $\rho_{[\chi]}$ is the representation on M we just described.

4.2 A homological mirror symmetry conjecture. In section 2, we have introduced the real quintic L as the Lagrangian defined as the real points of the Fermat quintic. To get closer to homological mirror symmetry, we need to make two small amendments to this definition.

First, to define an object in the Fukaya category, $\mathrm{Fuk}(X)$, we need to specify not only a Lagrangian submanifold, but also a flat $U(1)$ local system on L. Since $H_1(L\cong \mathbb{RP}^3;\mathbb{Z}) = \mathbb{Z}_2$, there are two possible choices, leading to two objects in $\mathrm{Fuk}(X)$ which we denote by $L_\pm$.

Second, we note that the Fermat quintic is invariant under more than one anti-holomorphic involution. If as before μ_5 denotes the multiplicative group of fifth roots of unity, we define for $\chi = (\chi_1,\ldots,\chi_5)\in(\mu_5)^5$ an anti-holomorphic involution σ_χ of $\mathbb{P}^4$ by its action on homogeneous coordinates

$$\sigma_\chi : x_i \to \chi_i\bar x_i \tag{4.8}$$

The Fermat quintic is invariant under any σ_χ. The involution and the fixed point locus only depend on the class of χ in $(\mu_5)^5/\mu_5 \cong (\mu_5)^4$, and we obtain in this way $5^4 = 625$ pairs of objects $L_\pm^{[\chi]}$ in $\mathrm{Fuk}(X)$.

We emphasize again that although we have defined the Lagrangians $L_{\pm}^{[\chi]}$ as fixed point sets of anti-holomorphic involutions of the Fermat quintic, we can think of the corresponding objects of Fuk(X) without reference to the complex structure.

Conjecture 1. *There is an equivalence of categories* $\mathrm{Fuk}(X) \cong \mathrm{MF}(W/(\mu_5)^4)$ *which identifies the 625 pairs of objects* $L_{\pm}^{[\chi]}$ *with the 625 pairs of equivariant matrix factorizations* $Q_{\pm}^{[\chi]}$.

Remark. One can formulate a similar conjecture for any hypersurface in weighted projective space which has a Fermat point in its complex structure moduli space.

There are at present two pieces of evidence for the above conjecture. The first is a matching of the intersection indices, defined in the A-model by the geometric (transversal) intersection number of Lagrangian submanifolds, and in the B-model as the Euler characteristic on the morphism spaces. That evidence was first noted in [36], and is reviewed in detail also in [13]. One of the central consequences of that computation is that the images of $L_{\pm}^{[\chi]}$ in homology actually generate $H_3(X;\mathbb{Z})$. This follows from the fact that the rank of the 625×625-dimensional intersection matrix is equal to 204, which is the rank of $H_3(X;\mathbb{Z})$.

The second piece of evidence comes from the matching between open Gromov-Witten theory of the real quintic and the normal function ν_C. Before filling in the connection, we note that this structure does not depend on the group theoretic data, so we can return to focus on the ordinary real quintic $L_\pm = L_{\pm}^{[\chi=1]}$ and its mirror object $Q_\pm = Q_{\pm}^{[\chi=1]}$.

4.3 Assembly. Recall that the theory of algebraic Chern classes provides a map

$$c_i^{\mathrm{alg}} : D^b(Y) \to \mathrm{CH}^i(Y) \tag{4.9}$$

from the derived category of coherent sheaves to the Chow groups of algebraic cycles modulo rational equivalence. The Chern classes factor through algebraic K-theory, and in particular, split any exact triangle in $D^b(Y)$. The image of c_i^{alg} in rational cohomology coincides with the topological Chern class $c_i \equiv c_i^{\mathrm{top}}$. Combining this with the Abel-Jacobi map described in section 3, we obtain the basic construction that associates a normal function to any object in $D^b(Y)$ of trivial topological Chern class. Since $\mathrm{MF}(W/(\mu_5)^4) \cong D^b(Y)$, we can extend this definition to the category of matrix factorizations.

The following result derives from [13].

Proposition C. *Let* $Q_\pm$ *be the pair of matrix factorizations defined in* (4.5). *Let* $C_\pm \subset Y$ *be the pair of algebraic curves studied in section 3. We have*

$$c_2^{\mathrm{alg}}(Q_+) - c_2^{\mathrm{alg}}(Q_-) = [C_+ - C_-] \in \mathrm{CH}^2(Y) \tag{4.10}$$

The essential idea that explains the coincidence of Gromov-Witten theory of the real quintic with the normal function ν_C associated with $C = C_+ - C_-$ should now be clear. Following mirror symmetry, the A-model has been equipped with a variation of Hodge structure isomorphic to that defined by the B-model on the mirror quintic [37]. For homological mirror symmetry to make sense, there ought to exist an A-model version of the Abel-Jacobi map into the intermediate Jacobian of the A-model variation of Hodge structure.

Based on considerations in string theory, it was argued in [11] that for the object $L_+ - L_-$, the truncated normal function in the A-model takes the form

$$\mathcal{T}_A(t) = \frac{t}{2} + \Big(\frac{1}{4} + \frac{1}{2\pi^2} \sum_{d \text{ odd}} \tilde{n}_d q^{d/2}\Big) \tag{4.11}$$

where $q = e^{2\pi i t}$, and $\tilde{n}_d$ are the open Gromov-Witten invariants of the real quintic, as we had it before (with slightly different conventions on t).

Recalling eq. (2.29), we see that

$$\mathcal{L}\big(\varpi_0(z)\mathcal{T}_A(t(z))\big) = \frac{15}{16\pi^2}\sqrt{z} \tag{4.12}$$

which is the same equation satisfied by $\mathcal{T}_B(z)$ from (3.18). The boundary conditions on $\mathcal{T}_B(z)$ are most conveniently fixed around $z^{-1} = 0$, namely [13]

$$\mathcal{T}_B(z) = -\frac{4}{3}\sum_{m=0}^{\infty} \frac{\Gamma(-3/2-5m)}{\Gamma(-3/2)} \frac{\Gamma(1/2)^5}{\Gamma(1/2-m)^5} z^{-(m+1/2)} \tag{4.13}$$

Note that this definition implicitly assumes a particular lift of the quotient by $H_3(Y;\mathbb{Z})$ in the definition of the intermediate Jacobian.

The following result is proved in [11].

Proposition D. *Let $\mathcal{T}_B(z)$ be the truncated normal function of the cycle $C_+ - C_-$. Let $\mathcal{T}_A(t)$ be given by the series in (4.11), extended to the z-plane by the mirror map and analytic continuation. We have*

$$\varpi_0(z)\mathcal{T}_A(t(z)) = \mathcal{T}_B(z)\,, \tag{4.14}$$

up to an integral period, $\int_{\Gamma^c}\hat{\Omega}(z)$, for some $\Gamma^c \in H_3(Y;\mathbb{Z})$

4.4 Floer homology of the real quintic. There is an interesting feature of the one-parameter family of matrix factorizations $Q_\pm$ that we have so far ignored, although it was really the initial motivation to investigate open mirror symmetry for the real quintic.

Consider the endomorphism algebra $\mathrm{Hom}^*(Q,Q)$ of the matrix factorization $Q = Q_+$, as objects in $\mathrm{MF}(W/(\mu_5)^4)$. This algebra is $\mathbb{Z}$-graded thanks to the homogeneity of W [32]. We also have $\mathrm{Hom}^0(Q,Q) \cong \mathbb{C}$ since Q is irreducible, and this implies $\mathrm{Hom}^3(Q,Q) \cong \mathbb{C}$ by Serre duality. Finally, it is shown in [38, 39] that

$$\mathrm{Hom}^1(Q,Q) = \mathrm{Hom}^2(Q,Q) = \begin{cases} 0 & \psi \neq 0 \\ \mathbb{C} & \psi = 0 \end{cases} \tag{4.15}$$

The appearance of an additional cohomology element in $\mathrm{Hom}^1(Q,Q)$ is signaled by the fact that $\mathcal{T}_B(z)$ in (4.13) vanishes at $\psi = 0$ (recall $z = (5\psi)^{-1}$).

To interpret (4.15) in the A-model, we recall that the morphism algebra of objects in the Fukaya category is defined using Lagrangian intersection Floer homology [40]. For the endomorphism algebra of a single Lagrangian, Floer homology is essentially a deformation of ordinary Morse homology by holomorphic disks.

For example, consider the real quintic $L \cong \mathbb{RP}^3$. Think of $\mathbb{RP}^3$ as $S^3/\mathbb{Z}_2$, and embed the S^3 in $\mathbb{R}^4 \ni (y_0,\ldots,y_3)$ as $y_0^2 + y_1^2 + y_2^2 + y_3^2 = 1$. A standard Morse function for $\mathbb{RP}^3$ in this presentation is given by $f = y_1^2 + 2y_2^2 + 3y_3^2$ restricted to the

S^3. This Morse function is self-indexing and has one critical point in each degree $i = 0, 1, 2, 3$. The Morse complex takes the form

$$C^0 \xrightarrow{0} C^1 \xrightarrow{\delta} C^2 \xrightarrow{0} C^3 \tag{4.16}$$

Working with integer coefficients, $C^i \cong \mathbb{Z}$ for all i, we have $\delta = 2$, and the complex (4.16) computes the well-known integral cohomology of $\mathbb{RP}^3$.

To compute Floer homology of the real quintic, we have to deform (4.16) by holomorphic disks, *i.e.*, $\delta = 2 + \mathcal{O}(e^{-t/2})$. In the standard treatments, such as [40], but also the more recent works such as [41], this requires taking coefficients from a certain formal (Novikov) ring with uncertain convergence properties. In other words, Floer homology is at present only defined in an infinitesimal neighborhood of the large volume point in moduli space (which leads to the often heard remark that $\mathrm{HF}^*(L, L)$ is isomorphic to $H^*(L)$). However, to make contact with (4.15), we have to understand $\mathrm{HF}^*(L, L)$ at the opposite end of moduli space, $z^{-1} = 0$. The results about the enumeration of disks ending on the real quintic that we have reviewed in this paper make one optimistic that this difficulty can be overcome, and support the conjecture made in [39]

Conjecture 2. *The Floer homology* $\mathrm{HF}^*(L, L)$ *of the real quintic can be defined in an open neighborhood of large volume. The complex* (4.16), *with complex coefficients, can be analytically continued over the entire* z*-plane. The differential* δ *has a single zero at the point* $z^{-1} = 0$.

References

[1] P. Candelas, X. C. De La Ossa, P. S. Green and L. Parkes, "A pair of Calabi-Yau manifolds as an exactly soluble superconformal theory," Nucl. Phys. B **359**, 21 (1991).

[2] B. R. Greene and M. R. Plesser, "Duality in Calabi-Yau moduli space," Nucl. Phys. B **338**, 15 (1990).

[3] D. R. Morrison, "Mirror symmetry and rational curves on quintic three-folds: a guide for mathematicians," J. Amer. Math. Soc. **6** (1993) 223–247, [`http://xxx.arxiv.org/abs/alg-geom/9202004` arXiv:alg-geom/9202004].

[4] M. Kontsevich, "Enumeration of rational curves via torus actions," in *The Moduli Space of Curves*, 335–368, Prog. Math. **129**, Birkhäuser, Boston 1995.

[5] A. B. Givental, "Equivariant Gromov-Witten invariants," Internat. Math. Res. Notices 1996, no. 13, 613–663. [`http://xxx.arxiv.org/abs/alg-geom/9603021` arXiv:alg-geom/9603021]

[6] B. .H. Lian, K. Liu, S.-T. Yau, "Mirror Principle I," Surv. Differ. Geom., **5**, (1999) [`http://xxx.arxiv.org/abs/alg-geom/9712011` arXiv:alg-geom/9712011]

[7] D. A. Cox, S. Katz, "Mirror symmetry and algebraic geometry," Mathematical Surveys and Monographs, 68. American Mathematical Society, Providence, RI, 1999.

[8] K. Hori et al., "Mirror Symmetry," Clay Mathematics Monographs, Vol. 1, AMS 2003.

[9] M. Kontsevich, "Homological algebra of mirror symmetry," Proceedings of I.C.M., (Zürich, 1994) 120–139, Birkhäuser, 1995. [`http://xxx.arxiv.org/abs/math.AG/9411018` arXiv:math.ag/9411018].

[10] A. Strominger, S. T. Yau and E. Zaslow, "Mirror symmetry is T-duality," Nucl. Phys. B **479**, 243 (1996) [`http://xxx.arxiv.org/abs/hep-th/9606040` arXiv:hep-th/9606040].

[11] J. Walcher, "Opening mirror symmetry on the quintic," Commun. Math. Phys., to appear, `http://xxx.arxiv.org/abs/hep-th/0605162` arXiv:hep-th/0605162.

[12] R. Pandharipande, J. Solomon, and J. Walcher, "Disk enumeration on the quintic 3-fold," `http://xxx.arxiv.org/abs/math.SG/0610901` arXiv:math.sg/0610901.

[13] D.R. Morrison and J. Walcher, "D-branes and normal functions," `http://xxx.arxiv.org/abs/hep-th/0709.4028` arXiv:hep-th/0709.4028

[14] M. Aganagic and C. Vafa, "Mirror symmetry, D-branes and counting holomorphic discs," `http://xxx.arxiv.org/abs/hep-th/0012041` arXiv:hep-th/0012041.

[15] M. Mariño, "Chern-Simons theory, matrix models and topological strings," International Series of Monographs on Physics, 131. Oxford University Press, Oxford, 2005.
[16] M. Kontsevich, A. S. Schwarz and V. Vologodsky, "Integrality of instanton numbers and p-adic B-model," Phys. Lett. B **637**, 97 (2006) [http://xxx.arxiv.org/abs/hep-th/0603106 arXiv:hep-th/0603106].
[17] V. Vologodsky, "Integrality of instanton numbers," http://xxx.arxiv.org/abs/0707.4617 arXiv:0707.4617 [math.AG]
[18] B. Acharya, M. Aganagic, K. Hori and C. Vafa, "Orientifolds, mirror symmetry and superpotentials," http://xxx.arxiv.org/abs/hep-th/0202208 arXiv:hep-th/0202208.
[19] V. Bouchard, B. Florea and M. Mariño, "Topological open string amplitudes on orientifolds," JHEP **0502**, 002 (2005) [http://xxx.arxiv.org/abs/hep-th/0411227 arXiv:hep-th/0411227].
[20] V. Bouchard, B. Florea and M. Marinõ, "Counting higher genus curves with crosscaps in Calabi-Yau orientifolds," JHEP **0412**, 035 (2004) [http://xxx.arxiv.org/abs/hep-th/0405083 arXiv:hep-th/0405083].
[21] S. Katz, C.-C. M. Liu, "Enumerative geometry of stable maps with Lagrangian boundary conditions and multiple covers of the disc," Adv. Theor. Math. Phys. **5** (2002) 1–49 http://xxx.arxiv.org/abs/math.AG/0103074 arXiv:math.ag/0103074
[22] J. Solomon, "Intersection theory on the moduli space of holomorphic curves with Lagrangian boundary conditions," http://xxx.arxiv.org/abs/math.SG/0606429 arXiv:math.sg/0606429.
[23] H. Ooguri and C. Vafa, "Knot invariants and topological strings," Nucl. Phys. B **577**, 419 (2000) [http://xxx.arxiv.org/abs/hep-th/9912123 arXiv:hep-th/9912123].
[24] H. Schubert, "Kalkül der abzählenden Geometrie," Teubner Verlag, Leipzig, 1879.
[25] J. W. Milnor, J. D. Stasheff, "Characteristic classes," Annals of Mathematics Studies, No. 76. Princeton University Press, 1974.
[26] R. Pandharipande, "Rational curves on hypersurfaces (after A. Givental)," Séminaire Bourbaki. Vol. 1997/98. Astrisque No. 252 (1998), 307–340. [http://xxx.arxiv.org/abs/math.AG/9806133 arXiv:math.ag/9806133]
[27] C. Voisin, "Hodge theory and complex algebraic geometry. I, II," Cambridge Studies in Advanced Mathematics, vol. 76,77, Cambridge University Press, 2002, 2003.
[28] R. O. Buchweitz, "Maximal Cohen-Macaulay modules and Tate cohomology over Gorenstein rings," preprint, ca. 1986
[29] A. Kapustin and Y. Li, "D-branes in Landau-Ginzburg models and algebraic geometry," JHEP **0312**, 005 (2003) [http://xxx.arxiv.org/abs/hep-th/0210296 arXiv:hep-th/0210296].
[30] D. Orlov, "Triangulated categories of singularities and D-branes in Landau-Ginzburg models," http://xxx.arxiv.org/abs/math.AG/0302304 arXiv:math.ag/0302304
[31] I. Brunner, M. Herbst, W. Lerche and B. Scheuner, "Landau-Ginzburg realization of open string TFT," JHEP **0611** (2006), 043. [http://xxx.arxiv.org/abs/hep-th/0305133 arXiv:hep-th/0305133].
[32] J. Walcher, "Stability of Landau-Ginzburg branes," J. Math. Phys. **46**, 082305 (2005) [http://xxx.arxiv.org/abs/hep-th/0412274 arXiv:hep-th/0412274].
[33] D. Orlov "Derived categories of coherent sheaves and triangulated categories of singularities" http://xxx.arxiv.org/abs/math.AG/0503632 arXiv:math.ag/0503632.
[34] P. S. Aspinwall, "The Landau-Ginzburg to Calabi-Yau dictionary for D-branes," http://xxx.arxiv.org/abs/hep-th/0610209 arXiv:hep-th/0610209
[35] M. Herbst, K. Hori, and D. Page, "Phases of $N = 2$ theories in 1+1 dimensions with boundary," (to appear), and talks at MSRI Berkeley (March 2006), LMU Munich (July 2006), http://math.ias.edu/mm/hms/K.Hori-PhasesofN2theories-slides.pdf IAS Princeton (January 2007).
[36] I. Brunner, M. R. Douglas, A. E. Lawrence and C. Romelsberger, "D-branes on the quintic," JHEP **0008**, 015 (2000) [http://xxx.arxiv.org/abs/hep-th/9906200 arXiv:hep-th/9906200].
[37] D. R. Morrison, "Compactifications of moduli spaces inspired by mirror symmetry," Journées de Géométrie Algébrique d'Orsay, Astérisque, **218**, Société Mathématique de France, 1993, pp. 243–271, [http://xxx.arxiv.org/abs/alg-geom/9304007 arXiv:alg-geom/9304007].

[38] K. Hori and J. Walcher, "F-term equations near Gepner points," HEP **0501**, 008 (2005) [http://xxx.arxiv.org/abs/hep-th/0404196 arXiv:hep-th/0404196].
[39] K. Hori and J. Walcher, "D-branes from matrix factorizations," Talk at Strings '04, June 28–July 2 2004, Paris. Comptes Rendus Physique **5**, 1061 (2004) [http://xxx.arxiv.org/abs/hep-th/0409204 arXiv:hep-th/0409204].
[40] K. Fukaya, Y.-G. Oh, H. Ohta and K. Ono, "Lagrangian intersection Floer theory—anomaly and obstruction", http://www.math.kyoto-u.ac.jp/ fukaya/fooo.dvipreprint (2000)
[41] O. Cornea, F. Lalonde, "Cluster homology: an overview of the construction and results," Electron. Res. Announc. Amer. Math. Soc. **12** (2006), 1–12 http://xxx.arxiv.org/abs/math.SG/0508345 arXiv:math.sg/0508345.